INTRODUCTION TO RANDOM PROCESSES

Other McGraw-Hill Books of Interest

Antoniou • DIGITAL FILTERS: ANALYSIS AND DESIGN
Candy • SIGNAL PROCESSING: THE MODEL-BASED APPROACH
Candy • SIGNAL PROCESSING: THE MODERN APPROACH
Bristow • ELECTRONIC SPEECH RECOGNITION
Bristow • ELECTRONIC SPEECH SYNTHESIS
Davenport • PROBABILITY AND RANDOM PROCESSES
Desrochers • PRINCIPLES OF PARALLEL AND MULTIPROCESSING
Drake • FUNDAMENTALS OF APPLIED PROBABILITY THEORY
Dupraz • PROBABILITY, SIGNALS, NOISE
Guiasu • INFORMATION THEORY WITH NEW APPLICATIONS
Kuc • INTRODUCTION TO DIGITAL SIGNAL PROCESSING
Orfanidis • OPTIMUM SIGNAL PROCESSING
Papoulis • PROBABILITY, RANDOM VARIABLES, AND RANDOM STOCHASTIC PROCESSES
Papoulis • SIGNAL ANALYSIS
Papoulis • THE FOURIER INTEGRAL AND ITS APPLICATIONS
Peebles • PROBABILITY, RANDOM VARIABLES, AND RANDOM SIGNAL PRINCIPLES
Schwartz • INFORMATION TRANSMISSION, MODULATION, AND NOISE
Schwartz & Shaw • SIGNAL PROCESSING

For more information about other McGraw-Hill materials, call 1-800-2-MCGRAW in the United States. In other countries, call your nearest McGraw-Hill office.

INTRODUCTION TO RANDOM PROCESSES

With Applications to Signals and Systems

DR. WILLIAM A. GARDNER
Professor, Electrical Engineering and Computer Science
University of California, Davis
Davis, California

President, Statistical Signal Processing, Inc.
Yountville, California

Second Edition

MCGRAW-HILL PUBLISHING COMPANY
New York   St. Louis   San Francisco   Auckland   Bogotá
Caracas   Hamburg   Lisbon   London   Madrid   Mexico
Milan   Montreal   New Delhi   Oklahoma City
Paris   San Juan   São Paulo   Singapore
Sydney   Tokyo   Toronto

Library of Congress Cataloging-in-Publication Data

Gardner, William A.
   Introduction to random processes: with applications to signals and systems / William A. Gardner. — 2nd ed.
      p.   cm.
   Includes bibliographical references.
   ISBN 0-07-022855-8
   1. Stochastic processes.   2. Stochastic systems.   3. Signal processing.   I. Title.
QA274.G375   1989
519.2—dc20   89-37273
                                                                                  CIP

Copyright © 1990, 1986 by McGraw-Hill, Inc. All rights reserved. Printed in the United States of America. Except as permitted under the United States Copyright Act of 1976, no part of this publication may be reproduced or distributed in any form or by any means, or stored in a database or retrieval system, without the prior written permission of the publisher.

1234567890   DOC/DOC   89432109

ISBN 0-07-022855-8

The editors for this book were Daniel A. Gonneau and Susan Thornton and the production supervisor was Richard A. Ausburn. It was set in Times Roman by Techna Type, Inc.

Printed and bound by R. R. Donnelley & Sons Company.

Information contained in this work has been obtained by McGraw-Hill, Inc., from sources believed to be reliable. However, neither McGraw-Hill nor its authors guarantees the accuracy or completeness of any information published herein and neither McGraw-Hill nor its authors shall be responsible for any errors, omissions, or damages arising out of use of this information. This work is published with the understanding that McGraw-Hill and its authors are supplying information but are not attempting to render engineering or other professional services. If such services are required, the assistance of an appropriate professional should be sought.

For more information about other McGraw-Hill materials, call 1-800-2-MCGRAW in the United States. In other countries, call your nearest McGraw-Hill office.

Contents

PREFACE TO THE SECOND EDITION ix

PREFACE TO THE FIRST EDITION xi

PART 1: REVIEW OF PROBABILITY, RANDOM VARIABLES, AND EXPECTATION   **1**

1   PROBABILITY AND RANDOM VARIABLES   3

1.1   The Notion of Probability   3
1.2   Sets   5
1.3   Sample Space   7
1.4   Probability Space   8
1.5   Conditional Probability   9
1.6   Independent Events   10
1.7   Random Variables   11
1.8   Probability Density   12
1.9   Functions of Random Variables   20
       Exercises   22

2   EXPECTATION   29

2.1   The Notion of Expectation   29
2.2   Expected Value   30
2.3   Moments and Correlation   34
2.4   Conditional Expectation   40
2.5   Convergence   42
       Exercises   47

FURTHER READING   55

PART 2: RANDOM PROCESSES — 56

3 INTRODUCTION TO RANDOM PROCESSES — 59

- 3.1 Introduction — 59
- 3.2 Generalized Harmonic Analysis — 61
- 3.3 Signal-Processing Applications — 69
- 3.4 Types of Random Processes — 73
- 3.5 Summary — 77
- Exercises — 79

4 MEAN AND AUTOCORRELATION — 82

- 4.1 Definitions — 82
- 4.2 Examples of Random Processes and Autocorrelations — 83
- 4.3 Summary — 89
- Exercises — 90

5 CLASSES OF RANDOM PROCESSES — 95

- 5.1 Specification of Random Processes — 95
- 5.2 Gaussian Processes — 96
- 5.3 Markov Processes — 97
- 5.4 Stationary Processes — 105
- 5.5 Summary — 112
- Exercises — 114

6 THE WIENER AND POISSON PROCESSES — 124

- 6.1 Derivation of the Wiener Process — 124
- 6.2 The Derivative of the Wiener Process — 128
- 6.3 Derivation of the Poisson Process — 129
- 6.4 The Derivative of the Poisson Counting Process — 133
- 6.5 Marked and Filtered Poisson Processes — 134
- 6.6 Summary — 137
- Exercises — 138

7 STOCHASTIC CALCULUS — 148

- 7.1 The Notion of a Calculus for Random Functions — 148
- 7.2 Mean-Square Continuity — 150
- 7.3 Mean-Square Differentiability — 152
- 7.4 Mean-Square Integrability — 154

7.5	Summary	156
	Exercises	157
8	**ERGODICITY AND DUALITY**	**163**
8.1	The Notion of Ergodicity	163
8.2	Discrete and Continuous Time Averages	166
8.3	Mean-Square Ergodicity of the Mean	168
8.4	Mean-Square Ergodicity of the Autocorrelation	170
8.5	Regular Processes	174
8.6	Duality and the Role of Ergodicity	179
8.7	Summary	181
	Exercises	182
9	**LINEAR TRANSFORMATIONS, FILTERS, AND DYNAMICAL SYSTEMS**	**189**
9.1	Linear Transformation of an N-tuple of Random Variables	189
9.2	Linear Discrete-Time Filtering	191
9.3	Linear Continuous-Time Filtering	195
9.4	Dynamical Systems	200
9.5	Summary	211
	Exercises	212
10	**SPECTRAL DENSITY**	**219**
10.1	Input-Output Relations	219
10.2	Expected Spectral Density	225
10.3	Coherence	228
10.4	Time-Average Power Spectral Density and Duality	229
10.5	Spectral Density for Ergodic and Nonergodic Regular Stationary Processes	230
10.6	Spectral Density for Regular Nonstationary Processes	231
10.7	White Noise	234
10.8	Bandwidths	243
10.9	Spectral Lines	244
10.10	Summary	245
	Exercises	247
11	**SPECIAL TOPICS AND APPLICATIONS**	**260**
11.1	Sampling and Pulse Modulation	260
11.2	Bandpass Processes	266
11.3	Frequency Modulation and Demodulation	273

11.4	PSD Measurement Analysis	282
11.5	Noise Modeling for Receiving Systems	286
11.6	Matched Filtering and Signal Detection	291
11.7	Wiener Filtering and Signal Extraction	294
11.8	Random-Signal Detection	296
11.9	Autoregressive Models and Linear Prediction	300
11.10	Summary	309
	Exercises	310
12	**CYCLOSTATIONARY PROCESSES**	**323**
12.1	Introduction	323
12.2	Cyclic Autocorrelation and Cyclic Spectrum	325
12.3	Stationary and Cyclostationary Components	332
12.4	Linear Periodically Time-Variant Transformations	334
12.5	Examples of Cyclic Spectra for Modulated Signals	346
12.6	Stationary Representations	363
12.7	Cycloergodicity and Duality	367
12.8	Applications	371
12.9	Summary	402
	Exercises	404
13	**MINIMUM-MEAN-SQUARED-ERROR ESTIMATION**	**416**
13.1	The Notion of Minimum-Mean-Squared-Error Estimation	416
13.2	Geometric Foundations	418
13.3	Minimum-Mean-Squared-Error Estimation	427
13.4	Noncausal Wiener Filtering	434
13.5	Causal Wiener Filtering	442
13.6	Kalman Filtering	451
13.7	Optimum Periodically Time-Variant Filtering	461
13.8	Summary	462
	Exercises	463
	SOLUTIONS	474
	REFERENCES	524
	AUTHOR INDEX	533
	SUBJECT INDEX	535

Preface to the Second Edition

In this second edition, the pedagogical value of the exercises at the ends of the chapters has been enhanced by the addition of nearly 100 new or revised exercises for a total of over 350 exercises. This provides a broad selection ranging from drill problems and applications to verifications of theoretical results as well as extensions and generalizations of the theory. To further enhance the value of the exercises as a study aid, detailed exemplary solutions to over 50 selected exercises (designated by ★) have been included at the back of the book. In addition, a supplement containing detailed solutions to all exercises in the book can be purchased from the publisher. This manual, entitled *The Random Processes Tutor: A Comprehensive Solutions Manual for Independent Study,* is essential for full command of the subject and is highly recommended for all students and professionals.

To reflect recent advances in the application of random process theory to problems in the area of statistical signal processing, a substantial expansion of the section in Chapter 12 on applications of the theory of cyclostationary processes has been incorporated in this edition. This includes new solutions to the problems of detection and classification of weak modulated signals, selective location of multiple sources of interfering modulated signals, and distortion reduction and interference suppression for modulated signals using adaptive frequency-shift filters and blind-adaptive antenna arrays.

Other changes in this edition include a few corrections and numerous minor revisions to enhance clarity.

Acknowledgments for the Second Edition

I am grateful to Dr. Chih-Kang Chen, for his contribution to the preparation of the solutions manual from which the selection of solutions at the back of the book was obtained.

Preface to the First Edition

For the Instructor

This book is intended to serve primarily as a first course on random processes for graduate-level engineering and science students, especially those with an interest in the analysis and design of signals and systems. It is also intended to serve as a technical reference for practicing engineers and scientists.

Like any theory in engineering and science, the theory of random processes is a *tool* for solving practical problems. However, this theory has been developed into such a highly technical abstraction that expert utilization of this tool requires specialization. Moreover, even within an area of specialization, successful application of the abstract theory to practical problems in engineering and science presents a challenge that can be met only with substantial effort to appropriately conceptualize the representation of the empirical components of a problem by the abstract mathematical models with which the theory deals. A simplistic view of how a theory is used to solve a practical problem is illustrated, diagrammatically, in the figure on page xii. The point to be made is that, in practice, one does not make only one mental excursion around this circuit. For any nontrivial problem, the first excursion yields an inadequate solution. However, learning that occurs during the first excursion enables one to improve and refine the problem formulation and resultant mathematical model. Typically, it is only after many excursions of the circuit that an adequate model and a practical solution are obtained.

Although this process of conceptualization in problem solving is familiar to instructors of engineering and science, it too often receives only cursory

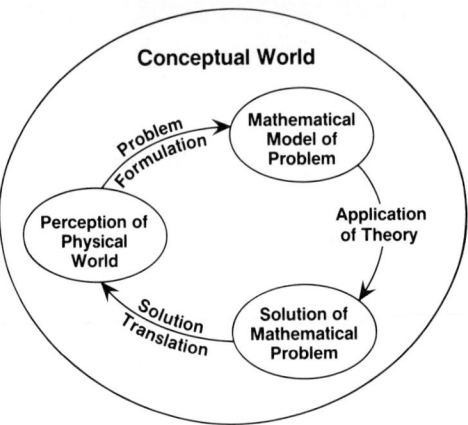

recognition in textbooks on theoretical subjects. Although a theoretical textbook that attends to this crucial problem of conceptualization and modeling can become cluttered, a textbook that all but ignores it, while clean and self-consistent, presents the student with a very limited picture of the subject as it relates to engineering and science. Thus, it is my intention, in this textbook on random processes, to strike a balance between a clean treatment of theory, on one hand, and discussion of its relation to the empirical world of signals and systems, on the other hand. The stage is set for this balance in the expository discussions throughout the book. The balance is played out by the particular selection and emphasis of topics in each chapter. For example, the concept of ergodicity is not played down—considerable attention is given to relationships between probabilistic (ensemble) averages and nonprobabilistic or deterministic (time) averages. For, as any engineer or scientist with practical signal-processing experience knows, the actual performance of signal-processing systems is often determined by time-averaging instantaneous performance measurements, whereas theoretical performance is measured by ensemble averages or *expected values*. Just as important as, but distinct from, the concept of ergodicity is the duality (described mathematically by Herman O. A. Wold's isomorphism) between those theories built on time averages and those built on ensemble averages. For example, the original theory of optimum filtering, developed by Norbert Wiener, is based on time averages, whereas the currently more popular theory of optimum filtering originated by Andrei Nikolaevich Kolmogorov, is based on ensemble averages. Similarly, the theory of spectral analysis of random processes, developed by Wiener (generalized harmonic analysis), is based on time averages, whereas the currently popular theory of spectral analysis—originated by Kolmogorov, Harald Cramér, Michel Moise Loève, and Joseph Leo Doob—is based on ensemble averages. As another example, Rudolf Emil Kalman's theory of optimum recursive filtering, which is based on ensemble averages, has as its time-average counterpart the earlier theory of recursive least-squares estimation, initiated by Carl Friedrich Gauss. Although the probabilistic theories have enjoyed more popularity in the last few decades, the earlier deterministic theories are currently gaining in popularity. This can be seen, for example, in the field of *adaptive* signal processing. In summary, a theme of this book is the exploitation of the duality that exists between probabilistic and deterministic theories, for the purpose of narrowing the gap between theory and practice. In historical perspective, this book can be viewed as a blend of the earliest and latest treatments of the subject.

Part 1 is a tutorial review which has two purposes: (1) to introduce notation and definitions for probability and random variables and (2) to refresh the reader's memory of this prerequisite material. The body of the book consists of Part 2, which is a self-contained introduction to the subject of random processes, with emphasis on the so-called *second-order theory* that deals with autocorrelation and spectral density.

With Part 1 treated as review, the entire book, with possibly some topics from the last three chapters omitted, can be covered in a 15-week course (preferably 4 hours of lecture per week). If the last two chapters are omitted, and a few selected topics from the remainder of Part 2 are only briefly treated, then a 10-week course is adequate.

For the Student

There are many physical phenomena in which the observed quantities are uniquely related, by simple laws, to a number of initial conditions and the forces causing the phenomena. Examples are the response voltage or current in an electrical circuit due to a periodic or transient excitation voltage, and the motion of a component in a mechanical device under the action of a force.

On the other hand, there is an important class of phenomena in which the observed quantities cannot be related, by simple laws, to the initial conditions and the actuating forces. Consider, for example, thermal noise in a circuit. If we attempted to relate the instantaneous values of the fluctuating noise voltage to the motions of the individual electrons, it would immediately become clear that we are faced with an impossible task. Even if we could obtain a solution to the problem by this method, its complexity would be an obstacle in the analysis of systems where thermal noise is present. For handling noise effectively, new concepts and methods are necessary.

In a dice game the prediction of the outcome of a throw, if it were possible, would require much detailed information that we do not possess. The physical properties of the dice and of the surface to which they are thrown and the initial conditions of motion of the dice are critical factors in the prediction. These factors are never fully known. Furthermore, if we assume that we have this information and are able to apply classical mechanics to the problem, it is easy to imagine how utterly complicated the solution would be. We must also remember that since the phenomenon of chance is the essence of a dice game, we really do not seek the law for the exact outcome of the game. What we desire is a mathematical description of the game in which chance is a fundamental characteristic; that is, we desire a *probabilistic* description.

In mechanics, the analysis of the behavior of a gas in a container is a problem relevant to this discussion. If the gas is assumed to be a large number of minute particles moving under their mutual interaction, then theoretically, in accordance with the principles of dynamics, the course of any particle of the gas for the infinite future is determined once the initial positions and velocities of every particle are known. But such initial conditions are impossible to obtain. Even if they were available and the solution were possible, it is beyond imagination that this extremely complex formulation of the problem would serve a useful purpose. Meaningful results in the analysis of a gas have been obtained by exploiting probabilistic concepts in addition to the concepts of classical mechanics. This has led to the *statistical theory of mechanics*.

The formation of speech waves is another example of a highly complex process that is not governed by simple laws. In fact, since the purpose of speech is to transmit information, it must have the characteristic that its variation as time passes cannot be predicted exactly by the listener. The information which the listener is to receive is unknown to him before its transmission. It is made known to him through a sequence of words selected by the sender from a collection of all possible words which are known to both sender and receiver. The message, in the form of speech, is clearly under the command of the speaker, and the listener can only make observations and cannot predict exactly, at any moment, its future course. It is true that, at times, a word or two, or even a short sequence of words, are predictable because of the rules of grammar or the occurrence of common phrases and idioms. However, precise prediction is impossible. Clearly, if speech does not possess this characteristic of unpredictability and is determinable exactly and completely, then indeed it conveys no information.

All information-bearing functions, which we shall refer to as *signals*, in the form of fluctuating quantities, such as voltages, velocities, temperatures, positions, or pressures, which are to be processed, transmitted, and utilized for the attainment of an objective, necessarily possess the characteristic that they are not subject to precise prediction. Also, in view of the foregoing discussion, thermal noise, impulse noise, errors in measurement, and other similar forms of disturbances in the transmission of a signal should be considered to have the same characteristic of unpredictability. These quantities, the signals and noises in a communication system, vary with time, either continuously or at discrete points, and are referred to as random time functions or *random processes*. Meaningful results on the analysis of random processes in communication systems have been obtained by exploiting probabilistic concepts in addition to the concepts of classical function analysis, such as harmonic analysis. This has led to the *statistical theory of communication*, and more generally to the *statistical theory of signal processing*.

The objective of this book is to present an introductory treatment of random processes, with applications to the analysis and design of signals and systems. It is assumed that the reader is familiar with Fourier analysis and the elementary theories of linear systems and of probability and random variables, at the level of typical introductory courses in a contemporary undergraduate electrical-engineering curriculum. The tutorial review of probability and random variables in Part 1 of this book is intended to serve primarily as a refresher.

Acknowledgments

It is a pleasure to express my gratitude to Professor Thomas Kailath, who read the first draft of this book and made valuable suggestions for improvement. It is also a pleasure to thank my students, who have helped to shape the style of this book. In addition, it is a pleasure to express my appreciation to Mrs. Jill M. Rojas and Mrs. Patty A. Gemulla for their excellent job of typing the entire manuscript. My deepest gratitude is expressed to my wife, Nancy, for her constant support and encouragement.

William A. Gardner

INTRODUCTION TO RANDOM PROCESSES

Part 1

REVIEW OF PROBABILITY, RANDOM VARIABLES, AND EXPECTATION

1

Probability and Random Variables

THIS FIRST CHAPTER is a brief review of the elementary theory of probability and random variables, which focuses on probabilistic concepts that are particularly relevant to the treatment of random processes in Part 2. From the point of view put forth in this book, *probability* is simply an average value, and *probability theory* is simply a calculus of averages. The ways in which such a calculus of averages can be employed for design and analysis of signal-processing systems is described in the introductory discussion in Part 2.

1.1 *The Notion of Probability*

Let us begin by considering the phenomenon of thermal noise in electrical devices. A metallic resistor contains an extremely large number of ion cores, in a nonuniform lattice, among which nearly free valence electrons move. Thermal energy agitates the entire structure, and the resultant electrical chaos creates a randomly fluctuating voltage across the terminals of the resistor, as illustrated simplistically in Figure 1.1. This *thermal noise voltage* consists of an extremely dense (at temperatures well above absolute zero) superposition of very brief positive and negative pulses, and is therefore a

4 • Review of Probability, Random Variables, and Expectation

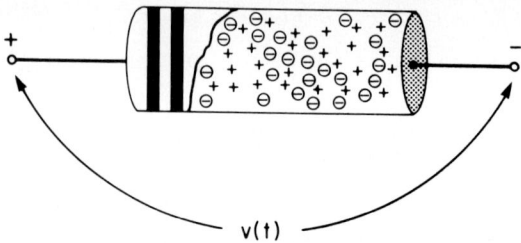

Figure 1.1 Thermal Motion of Free Electrons in a Resistor

highly erratic waveform, as illustrated in Figure 1.2, whose evolution is unpredictable.

We consider the following inquiry: What is the probability that the thermal noise voltage $V(t)$ exceeds the level 1 μV at some specific time $t = t_o$? In order to answer this question, we must develop a *probabilistic model* for this physical situation. The purpose of this chapter is to review the fundamental concept of a probabilistic model.

Let the *event* of interest to us be denoted by the symbol A:

$$A : V(t_o) > 1 \, \mu V,$$

and let an *event indicator* function I_A be defined by

$$I_A = \begin{cases} 1 & \text{if } A \text{ occurs,} \\ 0 & \text{if } A \text{ does not occur.} \end{cases} \tag{1.1}$$

Further, let us consider a very large set of (say) n resistors of identical composition, and let the voltages developed across all these resistors be denoted by the set $\{v_\sigma(t_o) : \sigma = 1, 2, 3, \ldots, n\}$. Finally, let the limiting ($n \to \infty$) average of the values $\{I_A(\sigma) : \sigma = 1, 2, 3, \ldots, n\}$ of the indicator function for all these resistors be denoted by $P(A)$:

$$P(A) = \lim_{n \to \infty} \frac{1}{n} \sum_{\sigma=1}^{n} I_A(\sigma). \tag{1.2}$$

If the number of values of $I_A(\sigma)$, among the total of n values, that equal 1

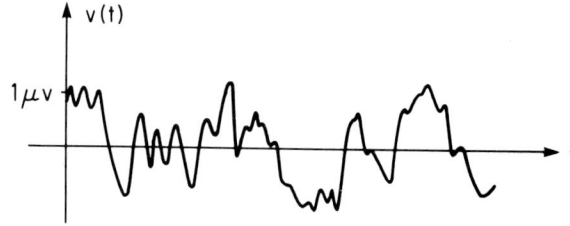

Figure 1.2 Thermal-Noise Voltage at the Terminals of a Resistor

is denoted by m, then $P(A)$ can be reexpressed as

$$P(A) = \lim_{n \to \infty} \frac{m}{n}, \tag{1.3}$$

since the remaining $n - m$ values in the sum defining $P(A)$ equal zero. We see that the average value $P(A)$ is the *relative frequency* of occurrence of the event A for a large set of identical resistors. It is said that $P(A)$ is the *probability of occurrence* of the event A, that is, the probability that $V(t_o) > 1 \, \mu V$.

In order to deal effectively with the sets underlying the concept of probability, we briefly review set notation, terminology, and operations.

1.2 Sets

A collection of entities is referred to as a *set*. The entities within the set are referred to as *elements*. If the elements of a finite set are described by $s_1, s_2, \ldots, s_n$, and the set is denoted by S, then the set is described in terms of its elements by

$$S = \{s_1, s_2, \ldots, s_n\}.$$

In practice a set S can always be viewed as a collection of some, but not all, of the elements of a larger set, say S_o. The set S can be specified by some property, Q_S, that is shared by its elements, but not by the elements in S_o that are excluded from S. This can be expressed by saying "S equals the set of all s contained in S_o such that s satisfies property Q_S." Using the symbols $\in$ for *contained in*, and : for *such that*, the specification of a set can be abbreviated to

$$S = \{s \in S_o : s \text{ satisfies } Q_S\}.$$

For example, if S is the set of all real numbers in the interval of the real line extending from 0 to 1, and S_o is the entire real line, then

$$S = \{s \in S_o : 0 \le s \le 1\}.$$

It is convenient to visualize a set as a two-dimensional planar region in planar space, and its elements as all the points within this region, as illustrated in Figure 1.3 (where only three elements are shown). Since S contains some but not all of the elements of S_o, as illustrated in Figure 1.3, it can be referred to as a *subset* of S_o, denoted by

$$S \subset S_o.$$

If there is uncertainty about S in the sense that S can possibly contain all elements in S_o (i.e., $S = S_o$), but can also possibly be a *proper subset* of S_o

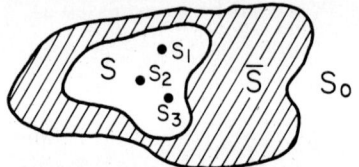

(a) Set containment and complement

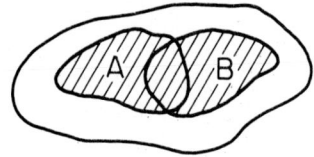

(b) Set union

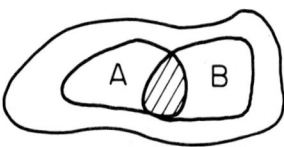

(c) Set intersection

Figure 1.3 Graphical Illustration of Set Relations

(which does not contain some of the elements in S_o), then this is denoted by

$$S \subseteq S_o.$$

The set of all the elements in S_o that are not in S is referred to as the *complement* of S (relative to S_o), and is denoted by $\overline{S}$. This is illustrated in Figure 1.3.

There are two fundamental ways that sets can be combined to form new sets, as described here. Let A and B be two subsets of S. Then their *union*, denoted by $A \cup B$, and their *intersection*, denoted by $A \cap B$, are defined by

$$A \cup B = \{s \in S : s \in A \quad \text{or} \quad s \in B\}, \tag{1.4}$$

$$A \cap B = \{s \in S : s \in A \quad \text{and} \quad s \in B\}. \tag{1.5}$$

These definitions are illustrated graphically in Figure 1.3. It is easy to verify (graphically) that these two set operations obey DeMorgan's laws

$$\overline{A \cup B} = \overline{A} \cap \overline{B}, \tag{1.6}$$

$$\overline{A \cap B} = \overline{A} \cup \overline{B}. \tag{1.7}$$

The intersection of two sets having no elements in common is the set containing no elements, which is called the *null set*, and is denoted by ∅.

1.3 Sample Space

In order to introduce additional concepts and terminology needed in the development of a probabilistic model, we return to the thermal-noise phenomenon.

1. *Experiment*: We shall think of the process of instantly observing the complete electrical state of the resistor at time $t = t_o$ as the execution of an *experiment*.
2. *Sample point*: A *sample point* is a single indecomposable outcome of an experiment, and is denoted by s. For our thermal-noise example, a sample point is the set of positions and velocities of every free electron and ion core in the resistor.
3. *Sample space*: A *sample space* is the set of all possible sample points, and is denoted by S.
4. *Event*: Loosely speaking, an *event*, say A, is anything that either does or does not happen when an experiment is performed (executed). Mathematically, an event A is the subset of S ($A \subset S$) consisting of all sample points for which a given something occurs (happens). It is said that "the event A occurs" if and only if the outcome s is in the subset A ($s \in A$). Since a single sample point can be treated as a set (containing only one element), then a sample point is also an event.

Example (*Thermal noise*): As an example of an event, we could say that if enough net negatively charged regions reside near the minus terminal of the resistor in Figure 1.1, and/or enough positively charged regions reside near the plus terminal, then the event A defined by

$$V(t_o) > 1 \, \mu V$$

will occur. Therefore, in this example, the subset A contains every set s of positions that satisfy the above loosely stated condition regarding "enough" charge separation. ∎

It is noted that in our example there are an uncountable infinity of possible sample points s, each of which is itself a very large set. This illustrates that although the *concepts* of sample point and sample space are indispensable, they are often impractical to deal with directly in applications (e.g., in computations).

A pictorial abstraction of the relationships among the entities s, A, S is shown in Figure 1.4, where s_1, s_2, s_3 and A, B are typical sample points

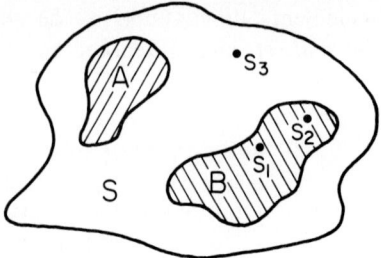

Figure 1.4 Pictorial Abstraction of Sample Space (S), Event Sets (A, B), and Sample Points (s_1, s_2, s_3)

and events. (As shown, if the experimental outcome is s_1 or s_2, then event B occurs.)

1.4 Probability Space

The sample space S is a *probability space* if and only if to every event A in the collection of *all possible events*, denoted by Σ, there is assigned a number, denoted by $P(A)$, such that the number-assigning rule [i.e., the *probability function** $P(\cdot)$] exhibits the axiomatic properties of probability, namely,

$$0 \leq P(A) \leq 1, \tag{1.8}$$
$$P(A \cup B) = P(A) + P(B) \quad \text{if and only if} \quad A \cap B = \emptyset, \tag{1.9}$$
$$P(S) = 1. \tag{1.10}$$

Strictly speaking, probability theory requires that the collection Σ be a *field*; that is, for every two sets A and B in Σ, the intersection set $A \cap B$, the union set $A \cup B$, and the complementary sets $\overline{A}$ and $\overline{B}$, must be in Σ. This requirement ensures that all reasonable operations on event sets can be accommodated by the theory; that is, all interesting event sets have a probability assigned to them.†

For a sample space that is a continuum it is convenient to visualize probability geometrically as a volume above a planar sample space. If the possibly infinitesimal probability of the event defined by an infinitesimal subset, ds, centered at the point s in a set A, is denoted by $dP(s)$, then the

---

*It should be clarified that there is ambiguity in the notation $P(A)$ when it is not clear whether A is a specific event, or simply a symbol denoting events in general, that is, a *variable*. In the former case $P(A)$ is a number, whereas in the latter case, $P(A)$ is a *function*, that is, the rule that assigns numbers to the various specific values of the variable A. When it is desired to denote the function unambiguously, the notation $P(\cdot)$ is used.

†For sample spaces containing an infinity of sample points, there is another technical requirement: Σ must contain all unions of *countable infinities* of subsets from Σ (not just finite unions), and correspondingly, the axiom (1.9) must be generalized from two (and therefore any finite number of) events to countable infinities of events.

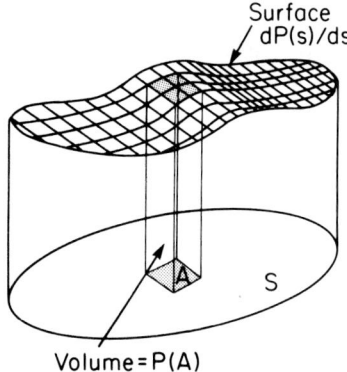

Figure 1.5 Geometrical Illustration of Probability as a Volume

probability of the event A is, from (1.9), the continuous sum (integral) of these individual probabilities over the set A,

$$P(A) = \int_{s \in A} dP(s), \qquad (1.11a)$$

since A is the union of all the disjoint subsets ds. If the ratio of $dP(s)$ to ds, $dP(s)/ds$, at the point s is visualized as the height of a surface above the sample space, then $P(A)$ is the volume defined by this surface,

$$P(A) = \int_{s \in A} \frac{dP(s)}{ds} ds, \qquad (1.11b)$$

as illustrated in Figure 1.5. In general, a planar sample space is inadequate —a larger than two-dimensional space is usually needed. For example, a six-dimensional space is needed to describe the sample space consisting of all possible positions and velocities of a *single* free electron. Thus, a *hyperplane* can be needed to describe a sample space, in which case probability can be interpreted as volume in *hyperspace*.

It is easy to verify (graphically) that for two not necessarily disjoint events A and B,

$$P(A \cup B) = P(A) + P(B) - P(A \cap B). \qquad (1.12)$$

The term subtracted corrects for the fact that the sum of the other two terms includes the volume above the intersection set $A \cap B$ twice.

1.5 Conditional Probability

In order to introduce the concept of *conditional probability*, we again return to the thermal-noise phenomenon, and we consider the following inquiry: If we are *given* the information that the charge distribution at time t_o is such that $V(t_o) > 0$, then we are given that the experimental outcome is contained in a particular subset, say B, of the sample space S; that is, we know that event B has occurred. We consider the following modification of our original inquiry in Section 1.1: Given that $V(t_o) > 0$, what is the probability that $V(t_o) > 1\ \mu V$? In other words, what is the probability of event A, given that event B has occurred? This *conditional probability*, which is denoted by $P(A|B)$, is defined by

$$P(A|B) \triangleq \frac{P(A \cap B)}{P(B)}. \qquad (1.13)$$

In this particular problem, it turns out that event A is a subset of event B ($A \subset B$), and therefore $A \cap B = A$. Hence,

$$P(A|B) = \frac{P(A)}{P(B)}$$

in this special case. Also, it is intuitively obvious for this example that $P(B) = \frac{1}{2}$, and therefore $P(A|B) = 2P(A)$.

By interchanging the roles of A and B in (1.13), the following relation, known as *Bayes' law*, can be derived:

$$P(A|B) = P(B|A)\frac{P(A)}{P(B)}. \qquad (1.14)$$

The most elegant interpretation of a conditional probability is that it is a *simple* (unconditional) probability defined on a new (conditional) probability space. Specifically, for the conditioning event B, the new sample space

$$S_B \triangleq B$$

is defined and the new probability function

$$P_B(\cdot) = P(\cdot|B) = \frac{P(\cdot \cap B)}{P(B)}$$

is defined.

1.6 Independent Events

If the occurrence of event B has no effect on the occurrence of event A, then A is said to be *independent* of B and

$$P(A|B) = P(A).$$

It follows from this and Bayes' law (1.13) that for *independent events*

$$P(A \cap B) = P(A)P(B). \qquad (1.15)$$

It also follows that
$$P(B|A) = P(B);$$
that is, if A is independent of B, then B is independent of A.

The property of independence is determined by the shape of the probability surface above the sample-space plane (Figure 1.5), but it does not have a useful geometrical interpretation. In any case, it cannot be described in terms of the sample space alone (e.g., $A \cap B = \emptyset$ does not imply A and B are independent; in fact if $P(A) > 0$ and $P(B) > 0$, then $P(A \cap B) = P(\emptyset) = 0 \neq P(A)P(B) > 0$).

1.7 Random Variables*

A real-valued function, say $X(\cdot)$, of sample points in a sample space is called a *random variable*; that is, a random variable is a rule that assigns a real number $x = X(s)$ to each sample point s. The real number $x = X(s)$ is called a *realization*, or a *statistical sample*, of $X(\cdot)$. [The abbreviated notation $X \equiv X(\cdot)$ is often used.] A pictorial abstraction of the relationship between a sample space S and a random variable X is shown in Figure 1.6.

Example (Thermal noise): As an example, the thermal-noise voltage $X = V(t_o)$ is a random variable. After the experiment has been performed, the specific value of voltage, $v(t_o)$, that occurred is a sample of the random variable $V(t_o)$ [i.e., $x = v(t_o) = V(t_o, s) = X(s)$]. ∎

More precisely, in order for a function $X(\cdot)$ to qualify as a random variable on a given probability space, it must be sufficiently well behaved that the probability of the event
$$A_x \triangleq \{s \in S : X(s) < x\}$$
is defined for all real numbers x. This is required in the mathematical development of probability theory.

As discussed in the next section, the interpretation of a random variable as a function can often be avoided, since probabilistic calculations can be carried out in terms of probability density, without regard to an underlying sample space. Nevertheless, we shall see in Chapter 13 that the interpretation of a random variable as a function is absolutely essential in order to establish the important geometrical properties possessed by random variables.

*The term *random variable* is a misnomer, because it is a *function*, not a variable, that is being referred to. *Function of a random sample point* would be more appropriate terminology, as explained in this section. The alternative term *stochastic variable* is sometimes used instead of *random variable*. The word *stochastic* derives from a Greek word meaning to aim (guess) at.

12 • Review of Probability, Random Variables, and Expectation

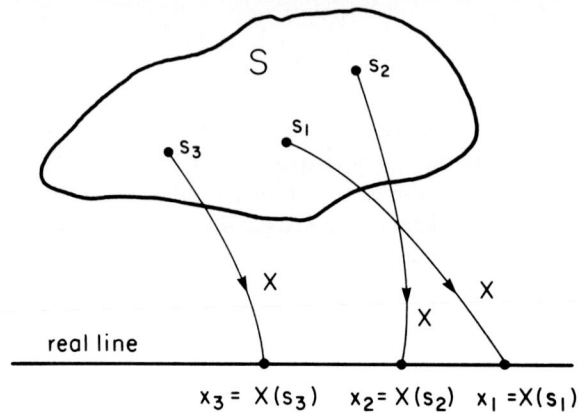

Figure 1.6 Pictorial Abstraction of a Random Variable

1.8 *Probability Density*

If our only concern with a given experiment is a particular random variable X, then the only experimental outcomes of interest to us are the samples x of X. Consequently, we can define a simplified sample space S_X for the experiment of observing X in which the sample points s are simply the sample values x. In this case, we can develop the probability-volume concept introduced in Section 1.4, using nothing more than the calculus of functions of a real variable.* However, since S_X is a one-dimensional space (the real line) rather than the two-dimensional space (plane) discussed in Section 1.4, then we shall obtain an interpretation of probability as an area rather than a volume. But for two random variables, say X and Y, the sample space S_{XY} is two-dimensional, and probability is indeed a volume, as discussed in Section 1.8.4.

1.8.1 *Distribution Function*

The probability distribution function for a random variable X is denoted by $F_X(\cdot)$, and is defined by

$$F_X(x) \triangleq \text{Prob}\{X < x\}; \tag{1.16}$$

that is, $F_X(x)$ is the probability that the random variable X will take on a value less than the number x. Using the event notation from Section 1.7,

---

*Development of the probability-volume concept in general requires the more abstract calculus of functions of abstract set points, which is based on measure theory.

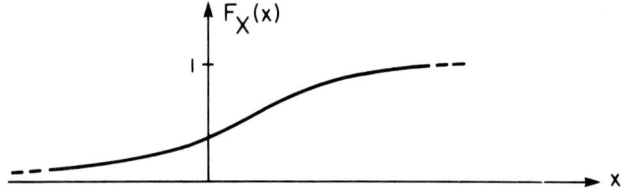

Figure 1.7 Example of a Probability Distribution Function

Equation (1.16) can be reexpressed as

$$F_X(x) \triangleq P(A_x).$$

It is easy to deduce that a distribution function is a nondecreasing function with the limiting values

$$F_X(-\infty) = 0, \qquad (1.17)$$
$$F_X(+\infty) = 1.$$

An example of a distribution function is shown in Figure 1.7.

Application (Thermal noise): As an application of the distribution function, if we let X be $V(t_o)$, with the units of volts, and we let $x = 1\ \mu V$, then A_x is the event

$$V(t_o) < 1\ \mu V = 10^{-6}\ V.$$

Thus, the complementary event $\overline{A}_x$ is

$$V(t_o) \geq 10^{-6}\ V,$$

from which it follows that

$$\text{Prob}\{V(t_o) \geq 1\ \mu V\} = P(\overline{A}_x) = 1 - F_x(10^{-6}).$$

Hence, we can obtain an answer to our original inquiry in Section 1.1, once we determine the appropriate distribution function for thermal noise voltage. We shall do this in Section 1.8.2. ∎

1.8.2 Density Function

Since the probability that a random variable X takes on a value in the interval* $[x - \varepsilon, x + \varepsilon)$ is

$$\text{Prob}\{x - \varepsilon \leq X < x + \varepsilon\},$$

then the *density of probability* at the point x, which is denoted by $f_X(x)$, is

$$f_X(x) \triangleq \lim_{\varepsilon \to 0} \frac{1}{2\varepsilon} \text{Prob}\{x - \varepsilon \leq X < x + \varepsilon\}. \qquad (1.18)$$

---
*The symbol $[\cdot, \cdot)$ denotes a *half-open interval*, for example $[a, b) = \{x : a \leq x < b\}$.

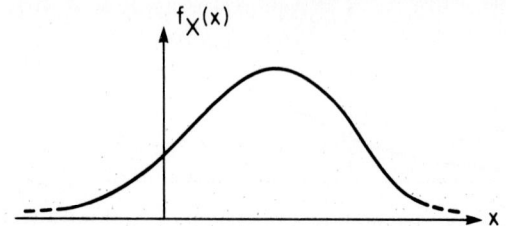

Figure 1.8 Example of a Probability Density Function

It follows from definitions (1.16) and (1.18) that the probability density and distribution functions are related by

$$f_X(x) = \frac{d}{dx} F_X(x), \tag{1.19}$$

$$F_X(x) = \int_{-\infty}^{x} f_X(y)\, dy. \tag{1.20}$$

In some cases the distribution function can contain step discontinuities, as discussed in Section 1.8.3. In this case the density function contains Dirac delta functions (impulses), and the upper limit in (1.20) must be interpreted as $x-$ rather than $x+$ in order to be consistent with the definition (1.16).

It is easy to deduce that a density function is a nonnegative function with unit area:

$$f_X(x) \geq 0, \tag{1.21}$$

$$\int_{-\infty}^{\infty} f_X(x)\, dx = 1.$$

An example of a density function is shown in Figure 1.8.

It follows from (1.16) that the probability of the event

$$A_X: x \in [x_1, x_2)$$

is

$$P(A_x) = F_X(x_2) - F_X(x_1),$$

which can be expressed as

$$P(A_X) = \int_{x_1}^{x_2} \frac{dF_X(x)}{dx}\, dx.$$

Therefore, we have

$$\text{Prob}\{x_1 \leq X < x_2\} = \int_{x_1}^{x_2} f_X(x)\, dx. \tag{1.22}$$

Hence, the probability that x is contained in some subset of real numbers, A_X, can be interpreted as the area under the probability density curve $f_X(\cdot)$ above that subset. This is a concrete example of the abstract probability-volume concept illustrated in Figure 1.5.

The Gaussian Density. An important specific example of a probability density function is the *Gaussian* (or *normal*) density function, defined by

$$f_X(x) \triangleq \frac{1}{\alpha\sqrt{2\pi}} \exp\left\{-\frac{1}{2}\frac{(x-\beta)^2}{\alpha^2}\right\} \tag{1.23}$$

for $-\infty < x < \infty$. The parameters α and β (commonly referred to as the *standard deviation* and the *mean*) are discussed in Chapter 2.

Application (Thermal Noise): Equation (1.23) has been shown to be an appropriate model for thermal-noise voltage. In this case, $\beta = 0$ and

$$\alpha^2 = 4KTBR \quad (\text{V}^2), \tag{1.24}$$

where K is Boltzmann's constant, T is the temperature (in degrees Kelvin) of the resistor, and B is the bandwidth* (in hertz) of the voltmeter used to measure the voltage, and R is the resistance in ohms. For example, at room temperature ($T = 290$ K) this reduces to

$$\alpha^2 = 1.6 \times 10^{-20} BR \text{ V}^2.$$

So, for a 100-MHz-bandwidth voltmeter, and a 100-Ω resistance, we obtain

$$\alpha^2 = 1.6 \times 10^{-10} \text{ V}^2.$$

Therefore, the answer to our original inquiry in Section 1.1 is (from Section 1.8.1)

$$\text{Prob}\{V(t_o) \geq 10^{-6} \text{V}\} = 1 - \int_{-\infty}^{10^{-6}} f_X(x)\,dx,$$

which [using (1.23)] yields

$$\text{Prob}\{V(t_o) \geq 10^{-6} \text{ V}\} = \frac{1}{2} - \text{erf}\left(\frac{10^{-6}}{\sqrt{1.6 \times 10^{-10}}}\right) \simeq 0.48,$$

where $\text{erf}(\cdot)$ denotes the *error function* which is defined by

$$\text{erf}(y) \triangleq \frac{1}{\sqrt{2\pi}} \int_0^y \exp\left\{\frac{-x^2}{2}\right\} dx.$$

---

*This formula applies for $B < 10^{12}$ Hz, as discussed in Section 10.7.

16 • Review of Probability, Random Variables, and Expectation

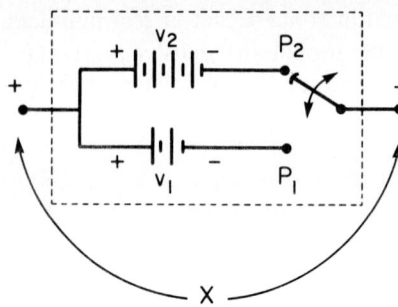

Figure 1.9 Randomly Selected Battery: An Example of a Discrete Random Variable

Also, the answer to our second inquiry (in Section 1.5) is

$$P\big[\,V(t_o) \geq 10^{-6}\,\big|\,V(t_o) > 0\,\big] = 0.96.$$

∎

1.8.3 Continuous, Discrete, and Mixed Random Variables

An important characteristic of the Gaussian distribution function is that it is a continuous function. Random variables whose distribution functions are continuous are called *continuous random variables*. At the other extreme is what is called a *discrete random variable*. Random variables of this type can take on only a countable number of distinct values.

Example (Random battery): As an example of a discrete random variable, we consider the experiment of measuring the voltage at the terminals of one of two batteries, selected at random (with a *random switch*) as depicted in Figure 1.9. This voltage, X, can take on only the values v_1 and $v_2 > v_1$, and it does this with probabilities P_1 and P_2. It follows that the distribution function for X is

$$F_X(x) = \begin{cases} 0, & -\infty < x \leq v_1, \\ P_1, & v_1 < x \leq v_2, \\ P_1 + P_2 = 1, & v_2 < x < \infty, \end{cases}$$

which is depicted in Figure 1.10.

In this case of a discrete random variable, differentiation of the piecewise constant distribution function [according to (1.19)] to obtain the density function reveals that the density function consists solely of Dirac delta functions (impulses), as shown in Figure 1.11. ∎

In between these two extreme cases of continuous and discrete random variables is what is called a *mixed random variable*. An example can be created by combining our examples of continuous and discrete random

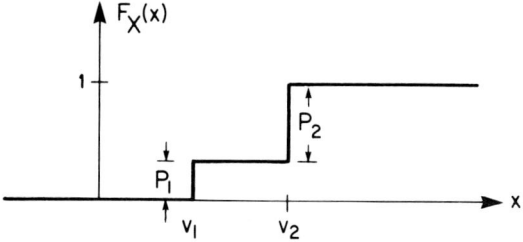

Figure 1.10 Distribution Function for the Discrete Random Variable in the Battery Example Depicted in Figure 1.9

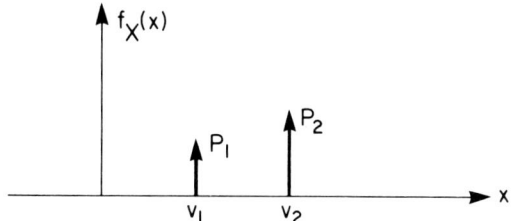

Figure 1.11 Density Function for the Discrete Random Variable in the Battery Example Depicted in Figure 1.9

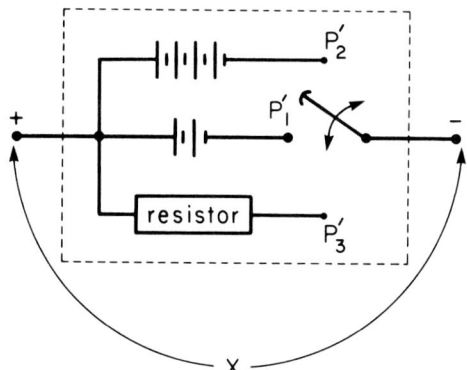

Figure 1.12 Example of a Mixed Random Variable

variables. Specifically, we consider the experiment of measuring the voltage at the terminals of one of three devices selected at random. These are, as shown in Figure 1.12, two batteries and a resistor.

It follows that the distribution function of X is an additive mixture of the distribution functions for the two-battery case and the resistor case previously considered:

$$F_X(x) = (P_1' + P_2')F_Y(x) + P_3'F_Z(x),$$

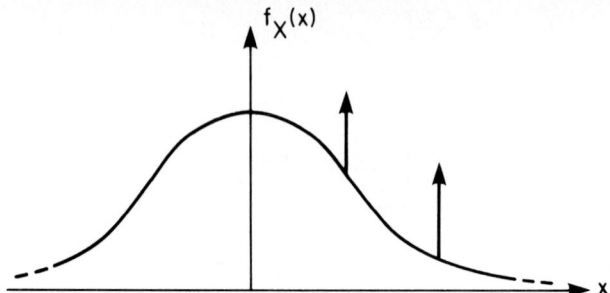

Figure 1.13 Density Function for the Mixed Random Variable in the Example Depicted in Figure 1.12

where Y is used to denote the random battery voltage, and Z is used to denote the thermal-noise voltage. The density function for the mixed random variable is also an additive mixture, and is shown in Figure 1.13.

A discrete random variable can be characterized in terms of what is called a *probability mass function*, instead of the distribution function. For the example of a random battery, the probability mass function is

$$P_X(x) = \begin{cases} P_1, & x = v_1, \\ P_2, & x = v_2, \\ 0, & \text{otherwise.} \end{cases}$$

1.8.4 Joint Distributions and Densities

The concepts and associated definitions of distribution functions and density functions extend in a natural way from a single random variable to several random variables. For example, for two random variables X and Y, the joint distribution function is defined by

$$F_{XY}(x, y) \triangleq \text{Prob}\{X < x \text{ and } Y < y\}, \tag{1.25}$$

and the joint density is the second mixed partial derivative

$$f_{XY}(x, y) = \frac{\partial^2}{\partial x \, \partial y} F_{XY}(x, y). \tag{1.26}$$

By interpreting the joint density as a surface above the (x, y) plane, we have another concrete example of the concept of abstract probability volume. The *marginal* distributions and densities can be obtained from the joint distribution and density by

$$F_X(x) = F_{XY}(x, \infty), \tag{1.27}$$

$$f_X(x) = \int_{-\infty}^{\infty} f_{XY}(x, y)\, dy, \tag{1.28}$$

and the conditional density is defined by

$$f_{X|Y}(x|Y = y) \triangleq \frac{f_{XY}(x, y)}{f_Y(y)}, \tag{1.29}$$

and similarly for the conditional distribution,

$$F_{X|Y}(x|Y < y) \triangleq \frac{F_{XY}(x, y)}{F_Y(y)}. \tag{1.30}$$

Independent Random Variables. If X and Y are statistically independent, then (and only then) (1.15) can be used to show that

$$F_{XY}(x, y) = F_X(x) F_Y(y), \tag{1.31}$$

and therefore

$$f_{XY}(x, y) = f_X(x) f_Y(y). \tag{1.32}$$

The Bivariate Gaussian Density. Two random variables, X and Y, are defined to be *jointly Gaussian* if and only if the random variable

$$Z = aX + bY$$

is Gaussian (has a Gaussian density) for every pair of real numbers, a and b. If X and Y are not *linearly dependent* (i.e., there exist no numbers c and d such that $Y = cX + d$), then it can be shown that the joint Gaussian density is given by

$$f_{XY}(x, y) = \frac{1}{2\pi \alpha \alpha' \sqrt{1 - \gamma^2}}$$

$$\times \exp\left\{-\frac{\left(\frac{x - \beta}{\alpha}\right)^2 - 2\gamma\left(\frac{x - \beta}{\alpha}\right)\left(\frac{y - \beta'}{\alpha'}\right) + \left(\frac{y - \beta'}{\alpha'}\right)^2}{2(1 - \gamma^2)}\right\}. \tag{1.33}$$

The parameters α and α' (called the *standard deviations* of X and Y), β and β' (called the *means* of X and Y), and γ (called the *correlation coefficient* of X and Y) are discussed in Chapter 2. If X and Y are linearly dependent, then $\gamma = 1$ and (1.33) does not apply. It can be shown [using

$f_{Y|X}(y|X=x) = \delta(y - cx - d)$] that in this case the joint Gaussian density is given by

$$f_{XY}(x, y) = f_X(x)\delta(y - cx - d), \quad (1.34)$$

where $f_X(x)$ is the univariate Gaussian density (1.23).

1.9 Functions of Random Variables

If a random variable Y is obtained from another random variable X by some deterministic function $g(\cdot)$,

$$Y = g(X), \quad (1.35)$$

then the probability that Y takes on some value in any set, say A, is equal to the probability that X takes on some value in the set B, where B is the set of all points x for which $g(x)$ is in the set A. Using this fact, the probability distribution for Y can be determined from the probability distribution for X. The degree of difficulty in carrying this out depends on both the sample space S_X for X and the function $g(\cdot)$. One special case of particular interest is that for which the random variable X is continuous and the inverse of $g(\cdot)$, denoted by $g^{-1}(\cdot)$, exists and is differentiable. In this case it can be shown that the probability density for Y is

$$f_Y(y) = f_X[g^{-1}(y)] \left| \frac{dg^{-1}(y)}{dy} \right|$$

$$= \frac{f_X(x)}{|dg(x)/dx|}, \quad (1.36)$$

where

$$x = g^{-1}(y). \quad (1.37)$$

More generally, if Y is a *random vector*, that is, an n-tuple of random variables,*

$$Y = [Y_1, Y_2, \ldots, Y_n]^T,$$

which is obtained from another random vector

$$X = [X_1, X_2, \ldots, X_n]^T$$

by a deterministic function $g(\cdot)$,

$$Y = g(X), \quad (1.38)$$

---

*The superscript T denotes matrix transposition; thus Y is a column vector.

and if the inverse function $g^{-1}(\cdot)$ exists and is differentiable, then the joint probability density for Y is

$$f_Y(y) = f_X[g^{-1}(y)] \left| \frac{\partial g^{-1}(y)}{\partial y} \right|, \qquad (1.39)$$

where $|\partial g^{-1}(y)/\partial y|$ is the absolute value of the determinant of the matrix of first-order partial derivatives

$$\frac{\partial g_i^{-1}(y)}{\partial y_j},$$

which is called the *Jacobian* of $g^{-1}(\cdot)$.

An important special case is that of a linear transformation

$$Y = AX + b, \qquad (1.40)$$

for which A is an $n \times n$ deterministic matrix and b is a deterministic vector. In this case, (1.39) yields

$$f_Y(y) = \frac{f_X(A^{-1}[y-b])}{|A|}, \qquad (1.41)$$

where A^{-1} is the inverse of the matrix A, and $|A|$ is the absolute value of the determinant of A.

Another important special case is that for which we are interested in only one random variable, Y, which is a linear function of an n-tuple of random variables, X. We can imbed this problem in the more general problem (1.40) considered above by defining the n random variables

$$\begin{aligned} Y_1 &= Y = a^T X + b, \\ Y_2 &= X_2, \\ Y_3 &= X_3, \\ &\vdots \\ Y_n &= X_n, \end{aligned} \qquad (1.42)$$

and then integrating the result (1.41) to obtain the marginal density for $Y = Y_1$,

$$f_Y(y) = \int_{-\infty}^{\infty} \cdots \int_{-\infty}^{\infty} f_Y(y, y_2, y_3, \ldots, y_n) \, dy_2 \, dy_3 \cdots dy_n. \qquad (1.43)$$

Example (*Sum of random variables*): Consider the sum

$$Y = X_1 + X_2. \qquad (1.44)$$

Substitution of (1.41) into (1.43) yields (exercise 14)

$$f_Y(y) = \int_{-\infty}^{\infty} f_{X_1 X_2}(y - x_2, x_2) \, dx_2. \qquad (1.45)$$

In the special case for which X_1 and X_2 are statistically independent, (1.32) and (1.45) yield

$$f_Y(y) = \int_{-\infty}^{\infty} f_{X_1}(y - x) f_{X_2}(x) \, dx \triangleq f_{X_1}(y) \otimes f_{X_2}(y), \qquad (1.46)$$

which is a *convolution* of the densities f_{X_1} and f_{X_2}. ■

Conditional Density. It can be shown that if $g(\cdot)$ in (1.38) possesses an inverse, then

$$f_{Z|Y}(z|Y = y) = f_{Z|X}(z|X = g^{-1}(y)), \qquad (1.47)$$

which reveals that conditioning on Y is equivalent to conditioning on X when Y and X are related by an invertible function.

EXERCISES

1. Reduce the following expressions for set combinations to the simplest possible forms:
   (a) $(A \cap \bar{B}) \cap (B \cap \bar{A})$
   (b) $(A \cap \bar{B}) \cup (A \cap B)$

   Use DeMorgan's laws to show that
   (c) $\overline{A \cap (B \cup C)} = (\bar{A} \cup \bar{B}) \cap (\bar{A} \cup \bar{C})$
   (d) $\overline{A \cap B \cap C} = \bar{A} \cup \bar{B} \cup \bar{C}$

2. Use a Venn diagram to prove that $P(A \cup B \cup C) = P(A) + P(B) + P(C) - P(A \cap B) - P(A \cap C) - P(B \cap C) + P(A \cap B \cap C)$.

3. (a) Among n integrated circuit chips, $m < n$ are defective. If one of these n chips is selected at random, what is the probability that it is defective? Let $n = 100$, $m = 5$ in the final result.
   (b) If the first chip selected is defective and is discarded, what is the probability that the second chip selected is defective?
   (c) If two chips are selected from the n, what are the probabilities that both are defective, one is defective, neither one is defective?

4. 100 integrated circuits of type A, 200 of type B, and 300 of type C are mixed together. Type A chips are known to be defective with probability 1/10, type B with probability 1/20, and type C with probability 1/30.

(a) If all 600 chips are mixed together and one is selected at random, what is the probability of its being defective?

(b) Given that the selected chip in part (a) is defective, what is the probability that it is of type A?

5. Between two terminals, there is a resistor connected in series with a parallel connection of two branches, one of which contains one resistor, and the other of which contains two resistors in series. If these resistors short out randomly and independently with probability p, then what is the probability of obtaining a short circuit from one terminal to the other?

6. Prove that for any n events, $P(A_1 \cup A_2 \cup \cdots \cup A_n) \leq P(A_1) + P(A_2) + \cdots + P(A_n)$. This is called the *union bound*.

7. A nuclear reactor becomes unstable if both safety mechanisms A and B fail. The probabilities of failure are $P(A) = 1/300$ and $P(B) = 1/200$. Also, if A has failed, B is then more likely to fail: $P(B|A) = 1/100$.

(a) What is the probability of the reactor going unstable?

(b) If B has failed, what is the probability of the reactor going unstable?

8. The random variable Y has the probability density

$$f_Y(y) = \begin{cases} ae^{-by}, & y \geq 0 \\ 0, & y < 0. \end{cases}$$

(a) Determine the value a in terms of b.

(b) Determine the conditional density for Y, given that $Y > c > 0$.

(c) Determine the conditional distribution for Y, given that $0 < c < Y < d$.

9. (*Binary channel*) Suppose that a single binary digit (0 or 1) is to be transmitted through a transmission channel. Because of noise in the channel, the received binary digit can be in error. A schematic of a probabilistic model for this situation is shown in Figure 1.14, for which

$$P[X = 0] = q,$$
$$P[X = 1] = p = 1 - q,$$

and

$$P[Y = 0 | X = 0] = p_0,$$
$$P[Y = 1 | X = 0] = q_0 = 1 - p_0,$$
$$P[Y = 0 | X = 1] = q_1 = 1 - p_1,$$
$$P[Y = 1 | X = 1] = p_1,$$

24 • Review of Probability, Random Variables, and Expectation

where X is the transmitted digit and Y is the received digit.
(a) Determine the probability that the digit 1 is received.

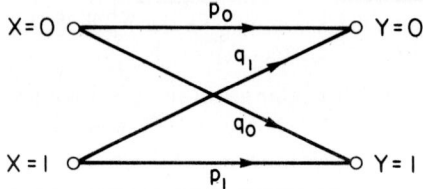

Figure 1.14 Schematic of Probabilistic Model of Binary Channel

(b) Determine the probability that an error occurs, given that the digit 1 was transmitted.
(c) Determine the unconditional probability of error.

10. Determine the probability density function for the random voltage measured across a *series* connection of the random battery in Figure 1.9 and a thermal noise resistor. Sketch a graph of this density.

11. Consider the random variable Y that is obtained from X by biasing with b and scaling with a:

$$Y = \frac{X - b}{a}.$$

Use the fact that

$$F_Y(y) = \text{Prob}\{Y < y\}$$
$$= \text{Prob}\{X < ay + b\}, \quad a > 0$$
$$= F_X(ay + b)$$

(and the analogous fact for $a < 0$) to prove that

$$f_Y(y) = |a|f_X(ay + b), \tag{1.48}$$

which indicates that the bias shifts the density, and the scalar scales the density in both height and width. Compare this result with (1.41).

★12. (a) Show that the Gaussian density (1.23) has unit area.

Hint: (i) Let $\alpha = 1$, $\beta = 0$. Then to show that

$$I \triangleq \int_{-\infty}^{\infty} f_X(x)\, dx = 1,$$

consider

$$I^2 = \int_{-\infty}^{\infty} \int_{-\infty}^{\infty} f_X(x) f_X(y) \, dx \, dy,$$

and introduce a change of variables from rectangular to polar coordinates.

$$y/x = \tan \theta,$$

$$x^2 + y^2 = r^2,$$

from which it follows that

$$dx \, dy = r \, dr \, d\theta.$$

Then use the change of variables

$$r^2/2 = t$$
$$r \, dr = dt.$$

(ii) After verifying that $I = 1$, then show that $\alpha \neq 1$, $\beta \neq 0$ does not change the result.

(b) Show that if X and Y have the bivariate density (1.33), then X has the univariate density (1.23).

Hint: Use the factorization

$$\exp\left\{ -\frac{\left(\frac{x-\beta}{\alpha}\right)^2 - 2\gamma\left(\frac{x-\beta}{\alpha}\right)\left(\frac{y-\beta'}{\alpha'}\right) + \left(\frac{y-\beta'}{\alpha'}\right)^2}{2(1-\gamma^2)} \right\}$$

$$= \exp\left\{ -\frac{1}{2}\left(\frac{x-\beta}{\alpha}\right)^2 \right\} \exp\left\{ -\frac{1}{2}w^2 \right\},$$

where

$$w \triangleq \frac{\frac{y-\beta'}{\alpha'} - \gamma\left(\frac{x-\beta}{\alpha}\right)}{\sqrt{1-\gamma^2}},$$

and make a change of variables of integration from y to w.

(c) Let X and Y have the bivariate Gaussian density as in part (b), and show that $Z = aX + bY$ has a Gaussian density.

Hint: Show that

$$f_Z(z) = \frac{1}{|a|} \int_{-\infty}^{\infty} f_{XY}\left(\frac{z - by}{a}, y\right) dy$$

and then either use

$$\int_{-\infty}^{\infty} e^{-py^2 + qy} dy = \sqrt{\frac{\pi}{p}} e^{q^2/4p}$$

to evaluate the integral, or factor the desired result—the Gaussian density with parameters $\beta'' = a\beta + b\beta'$ and $\alpha'' = [(a\alpha)^2 + (b\alpha')^2 + 2ab\gamma\alpha\alpha']^{1/2}$—out of the integral and verify that the remaining integral is unity.

(d) Let X and Y be jointly Gaussian and linearly independent, and show that $X|Y = y$ is Gaussian with parameters α'' and β'' given by

$$\beta'' = \beta + \frac{\gamma(y - \beta')\alpha}{\alpha'},$$

$$\alpha'' = \alpha\sqrt{1 - \gamma^2}.$$

Hint: Use the definition of conditional probability density and the factorization given in part (b).

13. X and Y are independent and identically distributed random variables with uniform probability densities on the interval $[0, a]$. Determine the probability density for $Z = X + Y$.

★14. Derive the formula (1.45) for the density of a sum of two random variables, by using (1.41) and (1.43).

★15. Let X_1 and X_2 be independent Gaussian random variables with zero means and unity variances. Define Y_1 and Y_2 by

$$Y_1 = [X_1^2 + X_2^2]^{1/2},$$

$$Y_2 = \tan^{-1}(X_2/X_1).$$

Then

$$X_1 = Y_1 \cos Y_2,$$

$$X_2 = Y_1 \sin Y_2.$$

Use the general result (1.39) to show that

$$f_Y(y_1, y_2) = \frac{y_1}{2\pi} \exp\left\{-\frac{y_1^2}{2}\right\}, \quad y_1 \geq 0, \quad 0 \leq y_2 < 2\pi,$$

and f_Y is zero for all other values of y_1 and y_2. Then show that

$$f_{Y_1}(y) = \begin{cases} y\exp\{-y^2/2\}, & y \geq 0, \\ 0, & y < 0, \end{cases} \qquad (1.49)$$

which is known as the *Rayleigh density*, and

$$f_{Y_2}(y) = \begin{cases} 1/2\pi, & 0 \leq y < 2\pi, \\ 0, & \text{otherwise}, \end{cases}$$

which is the *uniform density*. Finally, verify that Y_1 and Y_2 are independent.

16. Let Y_1 and Y_2 be statistically independent random variables with the Rayleigh and uniform probability densities as in exercise 15. Prove that $X_1 = Y_1 \cos Y_2$ and $X_2 = Y_1 \sin Y_2$ are statistically independent, jointly Gaussian random variables.

17. Determine the probability distribution function corresponding to the Rayleigh density (1.49).

18. Consider the following function, which represents the gain characteristic of an overdriven amplifier:

$$Y = g(X) \triangleq \begin{cases} -1, & X \leq -1 \\ X, & -1 < X < 1 \\ +1, & X \geq 1. \end{cases}$$

Determine the probability density for Y in terms of the density for X.

★19. (a) Consider the function
$$Y = g(X) \triangleq F_X(X),$$
where $F_X(\cdot)$ and its inverse $F_X^{-1}(\cdot)$ are continuous functions. Show that Y has a uniform probability density over $[0, 1]$.
(b) Consider the function
$$Z = h(Y) = F_W^{-1}(Y)$$
where F_W^{-1} is continuous and is the inverse of some distribution function F_W, which is continuous. Let Y have a uniform density over $[0, 1]$. show that Z has probability density $f_Z = f_W$.
(c) Let X have an arbitrary probability density for which the corresponding distribution and its inverse are continuous. Find a function k such that
$$Z = k(X)$$
has some other arbitrary probability density, for which the corresponding distribution and its inverse are continuous.

20. (a) A random variable X is uniformly quantized as follows:

$$Y = n, \quad n - 1/2 < X \le n + 1/2$$

for $n = 0, \pm 1, \pm 2, \ldots$. Determine the probability density for Y.
(b) It is desired to quantize X into a total of m equally probable values,

$$Y = \frac{x_n + x_{n+1}}{2}, \quad x_n < X \le x_{n+1},$$

where $x_1 < x_2 < \cdots < x_{m+1}$. Determine the quantization intervals $[x_n, x_{n+1}]$ in terms of the probability density f_X.
(c) In the general result of part (b) let f_X be the *Laplace* density

$$f_X(x) = \frac{a}{2} e^{-a|x|} \qquad (1.50)$$

and obtain an explicit solution.
(d) Use the results of (a) and part (a) of exercise 19 to show that the discrete probability mass function for the quantized random variable Y can be made uniform by preceding the quantizer with a nonlinear transformation (this is equivalent to using a nonuniform quantizer).

21. (a) Determine the joint probability density for the pair of random variables X and Y, where $Y = Z + X$, in terms of f_X and f_Z. **Hint:** Use the definition of conditional probability density.
(b) Repeat part (a) for $Y = Z/X$, where $f_X(0) = 0$.

⋆22. The output of a half-wave rectifier is given by $Y = g(X)$, where

$$g(X) \triangleq \begin{cases} X, & X \ge 0 \\ 0, & X < 0. \end{cases}$$

Determine the probability density for Y in terms of f_X.

23. For the pair of independent Gaussian random variables X_1 and X_2 with zero means and unity variances, determine the conditional joint probability density for X_1 and X_2, given that $X_1^2 + X_2^2 < a^2$.

24. Let X and Y be statistically independent random variables. Let $g(\cdot)$ and $h(\cdot)$ be any functions and show that $U \triangleq g(X)$ and $V \triangleq h(Y)$ are statistically independent. (You may assume that the inverses of $g(\cdot)$ and $h(\cdot)$ exist and are differentiable.)

25. Let $Y = X^2$. Determine $f_{X|Y}$ in terms of f_X.

2

Expectation

IN THIS CHAPTER, the fundamental notion of *expectation* is reviewed. This includes a review of *correlation*, which is the single most important notion underlying the second-order theory of random processes treated in Part 2.

2.1 The Notion of Expectation

With reference to random thermal noise as discussed in Chapter 1, we make the following inquiry: What do we *expect* the absolute value of the thermal-noise voltage $V(t_o)$ to be? In order to answer this question, we must develop a mathematical definition of the concept of *expected value*. The purpose of this chapter is to review this fundamental concept.

Let us begin by considering the simplified situation in which it is desired to determine the *expected value* of a binary random variable, such as that discussed in Section 1.8.3, for which the sample space S contains two points, s_1 and s_2:

$$X(s) = \begin{cases} x_1, & s = s_1, \\ x_2, & s = s_2. \end{cases}$$

Let us consider a very large set, called an *ensemble*, of (say) n identical experiments with outcomes denoted by $\{s(\sigma)\} = \{s(1), s(2), s(3), \ldots, s(n)\}$, and let the limiting ($n \to \infty$) average value of the random variable $X(s)$ for all these outcomes be denoted by $E\{X\}$:

$$E\{X\} = \lim_{n \to \infty} \frac{1}{n} \sum_{\sigma=1}^{n} X(s(\sigma)). \qquad (2.1)$$

If m_1 and m_2 are used to denote the numbers of random samples $s(\sigma)$, among the total of n samples, that equal s_1 and s_2, respectively, then the limit ensemble average $E\{X\}$ can be reexpressed as

$$E\{X\} = \lim_{n \to \infty} \left[\left(\frac{m_1}{n}\right) x_1 + \left(\frac{m_2}{n}\right) x_2 \right].$$

But, since

$$\lim_{n \to \infty} \frac{m_1}{n} = P(s_1),$$

and similarly for $P(s_2)$, then we see that the average value of $X(s)$ is the *probability-weighted sum of all possible values*, $X(s_1) = x_1$ and $X(s_2) = x_2$:

$$E\{X\} = X(s_1)P(s_1) + X(s_2)P(s_2).$$

It is said that $E\{X\}$ is the *expected value* of X.

2.2 Expected Value

2.2.1 Definition

The expected value of a random variable X is denoted by $E\{X\}$, and is a real (nonrandom) number defined by

$$E\{X\} \triangleq \sum_{s \in S} X(s)P(s). \qquad (2.2)$$

It is a probability-weighted average, over the entire sample space, of the sample values of the random variable. Equation (2.2) applies to discrete random variables defined on a discrete sample space. For continuous (as well as discrete and mixed) random variables defined on a continuous sample space, the discrete sum is replaced with a continuous sum (a Lebesgue integral), and the finite probability weights are replaced with infinitesimal probability weights:

$$E\{X\} \triangleq \int_{s \in S} X(s) \, dP(s). \qquad (2.3)$$

Example 1: As an example of expected value, consider the discrete sample space

$$S = \{s_1, s_2, s_3, s_4, s_5\},$$

the probability function

$$P(s_1) = \tfrac{1}{8}, \quad P(s_2) = \tfrac{1}{8}, \quad P(s_3) = \tfrac{1}{4}, \quad P(s_4) = \tfrac{3}{8}, \quad P(s_5) = \tfrac{1}{8},$$

and the random variable

$$X(s_1) = -1, \quad X(s_2) = +1, \quad X(s_3) = +1,$$
$$X(s_4) = +2, \quad X(s_5) = -1.$$

It follows from the definition (2.2) that the expected value of X is

$$E\{X\} = (-1)\tfrac{1}{8} + (+1)\tfrac{1}{8} + (+1)\tfrac{1}{4} + (+2)\tfrac{3}{8} + (-1)\tfrac{1}{8} = \tfrac{7}{8}. \quad \blacksquare$$

2.2.2 Alternative Formula

Observe that, in this example, there are only three distinct values of X, but five distinct sample points. If all nondistinct (identical) values of $X(s)$ are grouped together and their probabilities are added, an alternative formula for the expected value is obtained,

$$E\{X\} \triangleq \sum_{x \in S_X} x P_X(x). \tag{2.4}$$

In this general formula, S_X is interpreted as a sample space for the experiment of measuring X, and $P_X(\cdot)$ is the probability function for this sample space. For our example,

$$S_X = \{x_1, x_2, x_3\} = \{-1, +1, +2\},$$
$$P(x_1) = P(s_1) + P(s_5) = \tfrac{1}{4},$$
$$P(x_2) = P(s_2) + P(s_3) = \tfrac{3}{8},$$
$$P(x_3) = P(s_4) = \tfrac{3}{8},$$

and therefore,

$$E\{X\} = (-1)\tfrac{1}{4} + (+1)\tfrac{3}{8} + (+2)\tfrac{3}{8} = \tfrac{7}{8}.$$

Similarly, for the experiment of measuring the thermal noise voltage, X, the sample space is

$$S_X = \{x : -\infty < x < \infty\},$$

and the infinitesimal probability function for this infinite sample space is the differential of the distribution function

$$dP_X(x) = dF_X(x) = f_X(x)\, dx.$$

Therefore, for a continuous random variable, the alternative to (2.3) for the expected value is, analogous to (2.4),

$$E\{X\} \triangleq \int_{x \in S_X} x f_X(x)\,dx. \tag{2.5}$$

For a discrete random variable, the density function consists solely of impulses, and (2.5) reduces to (2.4). Therefore (2.5) is a generally applicable formula for the expected value. For convenience, the density function shall be defined to be zero for all values of x not contained in the sample space S_X. Equation (2.5) can then be reexpressed as

$$E\{X\} \triangleq \int_{-\infty}^{\infty} x f_X(x)\,dx. \tag{2.6}$$

Equations (2.2) and (2.3) are conceptually important in the development of the geometrical character of random variables presented in Section 13.3.3, but Equation (2.6) is often more appropriate for actual calculations, as illustrated in Part 2.

2.2.3 Properties of Expectation

Linearity. For any two random variables X and Y, and any two real numbers a and b, it can be shown that the expected value of the random variable $Z = aX + bY$ is given by

$$E\{aX + bY\} = aE\{X\} + bE\{Y\}. \tag{2.7}$$

Therefore, the transformation $E\{\cdot\}$ that maps random variables into real numbers is a *linear transformation*.

Expected Value of a Function of a Random Variable. For any function $g(\cdot)$ of a random variable X, it can be shown that the random variable $Y = g(X)$ has expected value

$$E\{g(X)\} = \int_{-\infty}^{\infty} g(x) f_X(x)\,dx. \tag{2.8}$$

This result, which is known as the *fundamental theorem of expectation*, is significantly easier to use than is the brute-force approach of determining the density function for the random variable $Y = g(X)$ and then using the definition

$$E\{g(X)\} \equiv E\{Y\} \triangleq \int_{-\infty}^{\infty} y f_Y(y)\,dy. \tag{2.9}$$

Equation (2.9) can be used to derive Equation (2.8) (see exercise 4).

Application (Thermal noise). As an application of (2.8), we let $X = V(t_o)$, the thermal-noise voltage, and we let $g(\cdot) = |\cdot|$, the absolute-value function. Then, the answer to our inquiry in Section 2.1 is

$$E\{|V(t_o)|\} = \int_{-\infty}^{\infty} |x| f_X(x) \, dx.$$

For the Gaussian density, it can be shown that this reduces to

$$E\{|V(t_o)|\} = \alpha \sqrt{\frac{2}{\pi}},$$

where, for thermal noise,

$$\alpha = 2\sqrt{KTBR}.$$

Thus, for $B = 100$ MHz, $R = 100$ Ω, $T = 290$ K (room temperature), we obtain $\alpha \simeq 1.3 \times 10^{-5}$, and therefore

$$E\{|V(t_o)|\} \simeq 10 \ \mu V. \qquad \blacksquare$$

Set Indicator Function. A conceptually important example of a random variable is the set indicator function $I_A(\cdot)$ for an arbitrary event set A,

$$I_A(s) \triangleq \begin{cases} 1 & \text{if } s \in A, \\ 0 & \text{if } s \notin A, \end{cases} \qquad (2.10)$$

which was introduced in Chapter 1. The random variable I_A takes on the value 1 whenever the event A occurs, and it takes on the value zero otherwise. By use of Equation (2.2) or (2.3) for the random variable $X = I_A$, it is easily verified* that

$$E\{I_A\} = P(A). \qquad (2.11)$$

Thus, the expected value of an event indicator is the probability of the event. Hence, *every conceivable probability can be interpreted as an expected value.*

2.2.4 Characteristic Function

A particularly useful example of the expected value of a function of a random variable is the *characteristic function* [which corresponds to the function $g(\cdot) = e^{i\omega(\cdot)}$, with parameter ω], denoted by $\Phi_X(\cdot)$, and defined by

$$\Phi_X(\omega) = E\{e^{i\omega X}\}, \qquad (2.12)$$

where $i \triangleq \sqrt{-1}$. It follows directly from (2.8) that

$$\Phi_X(\omega) = \int_{-\infty}^{\infty} f_X(x) e^{i\omega x} \, dx, \qquad (2.13)$$

---

*Recall that any set of sample points, $A = \{s\}$, is a set of mutually exclusive events; therefore, $\Sigma_{s \in A} P(s) = P(A)$.

which is the conjugate *Fourier transform* of the function $f_X(\cdot)$. A useful property of the characteristic function is that it yields the moments of the random variable; that is, the nth moment of X, $E\{X^n\}$, can be obtained by differentiation of Φ_X (exercise 10):

$$E\{X^n\} = \left(\frac{1}{i^n}\right) \left.\frac{d^n \Phi_X(\omega)}{d\omega^n}\right|_{\omega=0}. \tag{2.14}$$

2.3 Moments and Correlation

2.3.1 Mean and Variance

Since the probability density is a nonnegative function with unit area, then the expected value

$$E\{X\} = \int_{-\infty}^{\infty} x f_X(x)\, dx$$

can be interpreted as the *first moment* of the function $f_X(\cdot)$. As is well known, the first moment is a measure of the center of a function. This measure of the center of the probability density function $f_X(\cdot)$ is called the *mean* of X, and is denoted by m_X:

$$m_X \triangleq E\{X\}. \tag{2.15}$$

It is also well known that the square root of the second centralized moment of a nonnegative function is a measure of the width of the function, that is, a measure of the dispersion (spread) about the central point. This measure of width of a probability density function $f_X(\cdot)$ is called the *standard deviation* of X, and is denoted by σ_X:

$$\sigma_X \triangleq \sqrt{E\{(X - m_X)^2\}}. \tag{2.16}$$

It is easily shown (exercise 5) that the squared standard deviation, which is called the *variance*, can be expressed as

$$\sigma_X^2 \triangleq E\{(X - m_X)^2\} \equiv E\{X^2\} - m_X^2. \tag{2.17}$$

A pictorial example of the mean and standard deviation is shown in Figure 2.1.

Gaussian Density. For the Gaussian probability density, defined by

$$f_X(x) = \frac{1}{\alpha\sqrt{2\pi}} \exp\left\{-\frac{1}{2}\frac{(x-\beta)^2}{\alpha^2}\right\}, \tag{2.18}$$

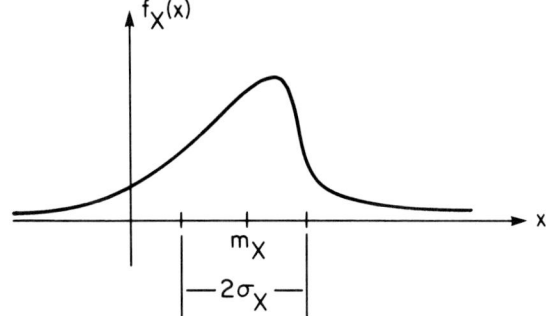

Figure 2.1 Pictorial Example of the Mean and Standard Deviation

it can be shown (exercise 6), by use of the definitions (2.15) and (2.16), that

$$m_X = \beta, \qquad \sigma_X = \alpha. \tag{2.19}$$

Application (Thermal noise). For the thermal-noise random variable,

$$m_X = 0 \quad \text{and} \quad \sigma_X = 2\sqrt{KTBR},$$

as discussed in Section 1.8.2. Therefore, the variance of the thermal-noise voltage increases linearly in temperature, resistance, and bandwidth.* ∎

2.3.2 Correlation and Covariance

The second joint moment of two random variables X and Y is called the *correlation*, and is denoted by R_{XY}:

$$R_{XY} \triangleq E\{XY\} = \int_{-\infty}^{\infty} \int_{-\infty}^{\infty} xy f_{XY}(x, y) \, dx \, dy. \tag{2.20}$$

The second joint *centralized* moment is called the *covariance*, and is denoted by K_{XY}:

$$K_{XY} \triangleq E\{(X - m_X)(Y - m_Y)\}. \tag{2.21}$$

It is easily shown (exercise 3) that

$$K_{XY} \equiv R_{XY} - m_X m_Y. \tag{2.22}$$

The correlation and covariance are each important measures of the interdependence of two random variables. If

$$K_{XY} = 0, \tag{2.23}$$

then X and Y are said to be *uncorrelated*. This terminology results from the

---

*This is true for $B < 10^{12}$ Hz, as discussed in Section 10.7.

fact that $K_{XY}/\sigma_X\sigma_Y$ is referred to as the *correlation coefficient*. If

$$R_{XY} = 0, \tag{2.24}$$

then X and Y are said to be *orthogonal*. The reason for this geometrical term is that the quantity $R_{XY} = E\{XY\}$ can be interpreted as the inner product of two vectors. This geometrical interpretation is developed in Chapter 13.

It can be shown that if two random variables are statistically independent, then they are uncorrelated, but not orthogonal unless at least one has zero mean. In contrast, if two random variables are uncorrelated, it does *not* follow that they are independent. An exception to this is the important special case for which the two random variables are jointly Gaussian, as explained in the next example.

The *correlation coefficient* for two random variables is defined by

$$\rho \triangleq \frac{K_{XY}}{\sigma_X\sigma_Y}, \tag{2.25}$$

and satisfies

$$-1 \le \rho \le +1. \tag{2.26}$$

ρ is a normalized version of the covariance. It can be shown that $|\rho| = 1$ if and only if there exist constants c and d such that $Y = cX + d$, that is, if and only if X and Y are linearly dependent.*

Example 2: As explained in Chapter 1, two random variables X and Y are defined to be *jointly Gaussian* if and only if the random variable

$$Z = aX + bY$$

is Gaussian for every pair of real numbers a and b; furthermore, it can be shown, using only this definition, that the joint probability density for jointly Gaussian random variables is given by

$$f_{XY}(x,y) = \frac{1}{2\pi\sigma_X\sigma_Y\sqrt{1-\gamma^2}}$$
$$\times \exp\left\{\frac{\left(\frac{x-m_X}{\sigma_X}\right)^2 - 2\gamma\left(\frac{x-m_X}{\sigma_X}\right)\left(\frac{y-m_Y}{\sigma_Y}\right) + \left(\frac{y-m_Y}{\sigma_Y}\right)^2}{-2(1-\gamma^2)}\right\}, \tag{2.27}$$

---

*Strictly speaking, $|\rho| = 1$ requires only that X and Y be linearly dependent in the mean-square sense, $E\{[Y - (cX + d)]^2\} = 0$.

provided that $|\gamma| \neq 1$. Moreover, it can be shown (exercise 6), by use of the definitions (2.21) and (2.25), that

$$\gamma = \rho.$$

It follows that $f_{XY}(x, y)$ factors into the product of marginals,

$$f_{XY}(x, y) = f_X(x) f_Y(y),$$

(and therefore X and Y are independent) if and only if $\rho = 0$. On the other hand, if $|\rho| = 1$, then $Y = cX + d$ for some real numbers c and d. In this case, the joint density for these jointly Gaussian random variables is given by (1.34), and X and Y are completely statistically dependent: given Y, X is known. ∎

2.3.3 Scatterplots

It is shown in Chapter 13 that the correlation coefficient ρ is in general a measure of linear dependence only. For example, if $f_X(\cdot)$ has even symmetry, and if $Y = X^2$, then $\rho = 0$, in spite of the fact that X and Y are very dependent. However, in the special case of jointly Gaussian variables, X and Y, ρ is a general measure of interdependence, not just linear dependence; that is, $\rho = 0$ if and only if the jointly Gaussian variables X and Y are statistically independent, as explained in the previous example.

A convenient method for experimental evaluation of ρ involves a graph called the *scatterplot*, which is simply a graph of each point with coordinates (x'_σ, y'_σ) given by

$$x'_\sigma = \frac{x_\sigma - m_X}{\sigma_X}, \qquad y'_\sigma = \frac{y_\sigma - m_Y}{\sigma_Y},$$

corresponding to each independent experimental (statistical) sample (x_σ, y_σ) of the pair (X, Y). Figure 2.2(a) shows a scatterplot for highly correlated variables ($\rho \simeq 1$), and Figures 2(b), (c) show scatterplots for two examples of nearly uncorrelated variables. Note that for $\rho = \pm 1$ it follows that $x'_\sigma = \pm y'_\sigma$, which yields a straight-line scatterplot with slope of ± 1.

An estimate $\hat{\rho}$ of ρ can be obtained from the statistically independent samples by using the ensemble-average estimates,

$$\hat{m}_X \triangleq \frac{1}{n} \sum_{\sigma=1}^{n} x_\sigma, \tag{2.28}$$

$$\hat{\sigma}_X^2 \triangleq \frac{1}{n} \sum_{\sigma=1}^{n} (x_\sigma - \hat{m}_X)^2, \tag{2.29}$$

$$\hat{K}_{XY} \triangleq \frac{1}{n} \sum_{\sigma=1}^{n} (x_\sigma - \hat{m}_X)(y_\sigma - \hat{m}_Y), \tag{2.30}$$

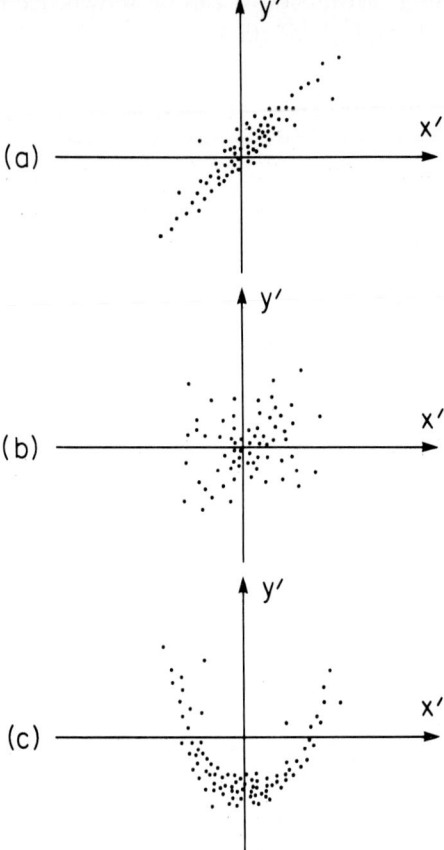

Figure 2.2 Examples of Scatterplots

in the formula

$$\hat{\rho} \triangleq \frac{\hat{K}_{XY}}{\hat{\sigma}_X \hat{\sigma}_Y}, \quad (2.31)$$

which should be compared with (2.25).

Whereas the parameters m_X, σ_X, ρ are referred to as *probabilistic parameters* (e.g., m_X is the *probabilistic mean*), the experimental estimates of these parameters, $\hat{m}_X, \hat{\sigma}_X, \hat{\rho}$, are referred to as *statistical parameters* or simply *statistics* (e.g., $\hat{m}_X$ is the *statistical mean*). Synonyms for *statistical parameter* are *empirical parameter* and *sample parameter*. There is more discussion of the relationship between probabilistic parameters and statistical parameters in Section 2.5.2.

2.3.4 Correlation Matrix

For an n-tuple of random variables

$$X = [X_1, X_2, \ldots, X_n]^T,$$

there are n^2 pairs of random variables with associated correlations

$$R_{X_i X_j} = E\{X_i X_j\}, \qquad i, j = 1, 2, \ldots, n. \tag{2.32}$$

If the (i, j)th correlation is interpreted as the (i, j)th element of an $n \times n$ matrix of correlations, denoted by R_X, then

$$R_X \triangleq E\{XX^T\}, \tag{2.33}$$

where X is a column vector, and its transpose X^T is a row vector. It follows that the matrix of covariances, with (i, j)th element

$$K_{X_i X_j} = E\{(X_i - m_{X_i})(X_j - m_{X_j})\}, \tag{2.34}$$

can be expressed as [cf. (2.22)]

$$K_X = R_X - m_X m_X^T, \tag{2.35}$$

where m_X is the n-tuple of means, with ith element $E\{X_i\}$.

Multivariate Gaussian Density. An n-tuple of random variables $\{X_i\}$ is defined to be *jointly Gaussian* if and only if every linear combination of these variables is a Gaussian random variable; that is, the random variable

$$Z \triangleq \sum_{i=1}^{n} a_i X_i \tag{2.36}$$

must be Gaussian for every n-tuple of real numbers, $\{a_i\}$. It is shown in exercise 25 using only this definition, that the joint probability density for an n-tuple of jointly Gaussian and linearly independent variables is given by

$$f_X(x) = \frac{1}{(2\pi)^{n/2} |K_X|^{1/2}} \exp\left\{-\tfrac{1}{2}(x - m_X)^T K_X^{-1}(x - m_X)\right\}. \tag{2.37}$$

The determinant $|K_X|$ is nonzero, and the matrix inverse K_X^{-1} therefore exists if and only if the n variables $\{X_i - m_{X_i}\}$ are linearly independent; that is, if and only if the only linear combination that yields $Z \equiv 0$ in (2.36) (with X_i replaced by $X_i - m_{X_i}$) is that for which $a_1 = a_2 = \cdots = a_n = 0$. Otherwise, the formula (2.37) is invalid and the density function contains

40 • Review of Probability, Random Variables, and Expectation

impulse fences (also called *line masses*), as illustrated by (1.34). Nevertheless, it is shown in exercise 25 that the density function for such jointly Gaussian linearly dependent variables can be conveniently expressed as the inverse multidimensional Fourier transform of the joint characteristic function, which, for jointly Gaussian variables, is given by

$$\Phi_X(\omega) \triangleq E\{e^{iX^T\omega}\} = \exp\{im_X^T\omega - \tfrac{1}{2}\omega^T K_X \omega\}. \tag{2.38}$$

This formula for Φ_X is valid regardless of the existence of K_X^{-1}.

Example 3: To illustrate (2.38), we consider two jointly Gaussian variables Z and Y. Their joint characteristic function is

$$\Phi_{ZY}(\omega_1, \omega_2) = E\{e^{i(Z\omega_1 + Y\omega_2)}\}$$

$$= \int_{-\infty}^{\infty}\int_{-\infty}^{\infty} e^{i(z\omega_1 + y\omega_2)} f_{ZY}(z, y)\, dz\, dy. \tag{2.39}$$

Substitution of (2.27) [with X replaced by Z] into (2.39) yields

$$\Phi_{ZY}(\omega_1, \omega_2) = \exp\{i(m_Z\omega_1 + m_Y\omega_2)$$
$$- \tfrac{1}{2}(\sigma_Z^2\omega_1^2 + \sigma_Y^2\omega_2^2 + 2\rho\sigma_Z\sigma_Y\omega_1\omega_2)\}, \tag{2.40}$$

which is of the form (2.38) with $m_X = [m_Z, m_Y]^T$ and

$$K_X = \begin{bmatrix} \sigma_Z^2 & \rho\sigma_Z\sigma_Y \\ \rho\sigma_Z\sigma_Y & \sigma_Y^2 \end{bmatrix}. \tag{2.41}$$

∎

2.4 Conditional Expectation

In order to introduce the concept of *conditional expectation*, we return, once again, to the thermal-noise phenomenon, and we consider the following inquiry: What do we expect the absolute value of the thermal noise voltage $V(t_o)$ to be, given that $V(t_o) > 0$? To answer this question, we must extend the mathematical concept of expectation to that of conditional expectation. Having already extended the concept of probability to that of conditional probability in Chapter 1, this is a minor task. In fact, since conditional probability can be reinterpreted as simple (unconditional) probability on a new (conditional) probability space, then conditional expectation can be reinterpreted as simple (unconditional) expectation for a new (conditional) random variable, defined on a new (conditional) probability space. It follows that for any conditioning event B, the general formula (2.6) for unconditional expectation is modified to

$$E\{X|B\} \triangleq \int_{-\infty}^{\infty} x f_{X|B}(x) \, dx. \tag{2.42}$$

For example, if event B is $Y = y$, for some random variable Y, then

$$E\{X|Y=y\} \triangleq \int_{-\infty}^{\infty} x f_{X|Y}(x|Y=y) \, dx. \tag{2.43}$$

Application (*Thermal noise*). For our question regarding the thermal-noise voltage, let B be the event $V(t_o) > 0$. It follows from the discussion in Section 1.5 that for $X \equiv |V(t_o)|$,

$$f_{X|B}(x) = f_X(x),$$

from which we obtain

$$E\{|V(t_o)| \,|\, V(t_o) > 0\} = E\{|V(t_o)|\}.$$

That is, knowing that $V(t_o) > 0$ does not change our expectation about $|V(t_o)|$. Would this be true if the probability density for $V(t_o)$ were not an even function? ∎

If a conditioning event is one of a set of events, say $\{B_s\}$, but the particular event from this set is not specified, then a corresponding conditional expectation has a set of possible values, one for each event B_s:

$$\{E\{X|B_s\}\}.$$

By interpreting the set of events as the sample points of a sample space, on which the probability function is defined by the probabilities $\{P(B_s)\}$ (and joint probabilities of sets of B_s), the set of conditional expectations can be interpreted as the sample values of a random variable, namely, a *random expectation*.

For example, for the set of conditioning events

$$\{Y = y : -\infty < y < \infty\},$$

we have the random expectation

$$E\{X|Y\}. \tag{2.44}$$

It is emphasized that in (2.44), the conditioning symbol is the random variable Y, not a sample y. In fact the nonrandom expectation (2.43) is a sample of the random variable (2.44).

Since $E\{X|Y\}$ is a random variable, it can have an expected value. It follows (exercise 26) from the definition of $E\{X|Y\}$ and the definition (2.6)

that

$$E\{E\{X|Y\}\} = E\{X\}. \qquad (2.45)$$

That is, the expected value of a random conditional expectation is the unconditional expectation.

Let $g(\cdot)$ be any deterministic function of a random vector Y; then it can be shown that

$$E\{Xg(Y)|Y=y\} = E\{X|Y=y\}g(y). \qquad (2.46)$$

This *factorization property* simply expresses the fact that the random vector $g(Y)$, when conditioned on $Y = y$, is nonrandom. Let $Z = g(Y)$. Then it can be shown (exercise 30) that

$$E\{E\{X|Y\}|Z\} = E\{X|Z\} \qquad (2.47)$$

and

$$E\{E\{X|Z\}|Y\} = E\{X|Z\}. \qquad (2.48)$$

An interpretation of these results is that since $g(\cdot)$ can only destroy information (it cannot create it), then conditioning on Z gives less information than conditioning on Y. In the case for which $g(\cdot)$ possesses an inverse, it follows from (1.47) that

$$E\{X|Y\} = E\{X|Z\}. \qquad (2.49)$$

2.5 Convergence

2.5.1 The Notion of Convergence

In order to introduce the concept of convergence, we return to the thermal-noise phenomenon. If we measure two time samples, $X = V(t)$ and $Y = V(t + \tau)$, of the thermal noise voltage, and if the sampling interval τ is not much greater than the mean relaxation time of electrons, then we intuitively expect these time samples to be correlated. In order to measure this correlation, we would in principle need a large number of resistors (all of the same resistance and temperature). Then we could measure, say, n statistical samples of these voltages at the two times, t and $t + \tau$, and compute the sample correlation coefficient (2.31). If we were to do this repeatedly for increasing numbers n of resistors, we would obtain a sequence, indexed by n, of correlation-coefficient measurements $\hat{\rho}_1, \hat{\rho}_2, \ldots, \hat{\rho}_n$. Intuitively we expect this sequence to *converge*, as n grows without bound, to the probabilistic correlation coefficient (2.25). This leads to the question "would this sequence converge, and if so, in what sense?" It turns out that all that can be established is that it converges in a probabilistic sense. As a matter of fact, it converges in several probabilistic senses. Before proceeding to distinguish among these, let us recall the meaning of convergence in a nonprobabilistic sense.

A sequence of real nonrandom numbers $\{x_n\} = \{x_1, x_2, x_3, \ldots\}$ converges to a number x if and only if for every positive number $\varepsilon > 0$ (no matter how small), there exists a positive integer N_ε (sufficiently large) such that for all $n > N_\varepsilon$, the difference between x_n and x is less than ε:

$$|x_n - x| < \varepsilon \qquad \text{for all} \quad n > N_\varepsilon.$$

This is abbreviated by

$$\lim_{n \to \infty} x_n = x.$$

Now let us generalize this notion of convergence to accommodate random variables.

2.5.2 Stochastic Convergence

A sequence of random variables $\{X_n\}$ is actually a family of sequences of real numbers,

$$\{\{X_n(s)\} : s \in S\},$$

together with a sequence of joint probability distributions,

$$\{F_{X_1 X_2 X_3 \ldots X_n}\}.$$

There are interesting mathematical models of sequences of random variables for which some (but not all) of the sequences in the family do not converge, but for which the entire family together exhibits some type of average convergent behavior. There are several types of average convergent behavior that are useful in probabilistic analysis. These, which are referred to as *stochastic convergence*, are described in the following paragraphs.

Convergence Almost Surely (or with Probability 1).

$$\lim_{n \to \infty} X_n(s) = X(s) \qquad \text{for all} \quad s \in \tilde{S} \subseteq S,$$

where $P(\tilde{S}) = 1$. This can be reexpressed as

$$\text{Prob}\left\{\lim_{n \to \infty} X_n = X\right\} = 1. \qquad (2.50)$$

For a sequence that converges with probability 1, there can be particular sample sequences [viz., $\{X_n(s)\}$ for $s \notin \tilde{S}$] that do not converge. However, the probability of the event that the sequence does not converge is zero:

$$P\{s \in S: s \notin \tilde{S}\} = 0.$$

Convergence in Mean Square.

$$\lim_{n \to \infty} E\{(X_n - X)^2\} = 0. \tag{2.51}$$

This "limit in mean" (square) is often abbreviated by the left-hand side of the expression

$$\operatorname*{l.i.m.}_{n \to \infty} X_n = X.$$

Alternative terminology for (2.51), comparable with "almost sure convergence" for (2.50), is *expected square convergence*.

Convergence in Probability.

$$\lim_{n \to \infty} \operatorname{Prob}\{|X_n - X| < \varepsilon\} = 1 \qquad \text{for all} \quad \varepsilon > 0. \tag{2.52}$$

Convergence in Distribution.

$$\lim_{n \to \infty} F_{X_n}(x) = F_X(x), \tag{2.53}$$

for all continuity points x of $F_X(\cdot)$.

Of the four types of convergence, (2.53) is the weakest, and (2.52) is the next weakest. For example, it is established in exercise 32 that (2.51) guarantees (2.52). However, neither (2.50) nor (2.51) is stronger than the other. There are some sequences of random variables for which there exists a limit X in the sense (2.51) but not in the sense (2.50), and there exist other sequences of random variables for which there exists a limit X in the sense (2.50) but not in the sense (2.51) (exercise 33).

Part 2 of this book is concerned with only mean-square convergence. One of the primary reasons for this is that in practice, mean squared error is the most commonly used measure of the difference between two variables associated with random signals and noises. In fact, in practice, the mean of the squared error is usually obtained by averaging over time, not over a set of statistical samples. For example, the correlation between adjacent time samples of thermal noise, $X = V(t)$ and $Y = V(t + \tau)$, is typically measured by

$$\hat{R}'_{XY} \triangleq \frac{1}{T} \int_0^T V(t + \tau + \sigma) V(t + \sigma) \, d\sigma \tag{2.54}$$

rather than by

$$\hat{R}_{XY} \triangleq \frac{1}{n} \sum_{\sigma=1}^{n} V(t+\tau,\sigma)V(t,\sigma). \qquad (2.55)$$

These two alternative approaches to measuring correlation can be seen heuristically to be closely related if the statistical samples are interpreted as time samples of a single waveform,

$$V(t,\sigma) = V(t-\sigma).$$

The relationship between time averages and ensemble averages or expected values is pursued in Part 2. The point to be made here is that in spite of the fact that it is the time averages made in practice—typically in the form of root-mean-square (rms) measurements—that motivate our preference for mean squared error as a criterion for convergence, we often study the convergence of time averages in terms of expected values.* For example, to prove that the time average (2.54) converges, and that it converges to the same quantity that the ensemble average (2.55) converges to, we can conceive of a probability space of random samples of (2.54), obtained from an ensemble of resistors, and then establish stochastic convergence in terms of the probability space. To establish the stochastic convergence of (2.55), we must conceive of random samples of (2.55), obtained from an ensemble of randomly chosen ensembles of n resistors each. This is illustrated in Section 2.5.3 for the simpler problem of measuring a mean rather than a correlation.

2.5.3 Laws of Large Numbers

Applications of these convergence concepts that enhance one's intuitive interpretation of the concept of probability are the following laws of large numbers, which strengthen the interpretation of axiomatically defined probability as a relative frequency of occurrence.

We consider the sequence of random variables

$$X_j \equiv I_A^j \triangleq \begin{cases} 1 & \text{if } A \text{ occurs,} \\ 0 & \text{otherwise,} \end{cases} \qquad (2.56)$$

where j indexes a set of n independent identical experiments, for example, measurement of a thermal-noise voltage, with $A = \{s: V(t_o) > 1 \ \mu V\}$. It is emphasized, for this example, that the time of experimentation is fixed at

---

*It should be clarified that it is not necessary to adopt a probabilistic model and study convergence in only a stochastic sense. As explained in Part 2, the duality between theories built on time averages and ensemble averages reveals that for some types of stochastic convergence based on ensemble averaging there are nonstochastic counterparts based on time averaging. The nonstochastic approach, which can be preferable for some applications, is pursued in reference [Gardner, 1987a].

$t = t_0$. Therefore, repetition of each of the n experiments requires a set of thermal-noise resistors (all of the same resistance and temperature) and voltmeters (all of the same characteristics such as bandwidth). All voltages are measured simultaneously at time t_0. Each set of n executions can be interpreted as either n statistically independent, statistically identical experiments, or one composite experiment.

The *sample mean* of the n random variables is defined by

$$\overline{X}_n = \frac{1}{n}(X_1 + X_2 + \cdots + X_n), \qquad (2.57)$$

and $\overline{X}_n$ is interpreted as a *random sample mean*; that is, it is a sum of random variables which have identical distributions but are independent; $\overline{X}_n$ is not interpreted as a sum of samples of one random variable. With $\overline{X}_n$ so defined, it can be seen that if A occurs K_n times in n trials, then

$$\overline{X}_n = K_n/n.$$

Hence $\overline{X}_n$ is the random *relative frequency of occurrence* of A. The following two laws reveal how this relative frequency is related to the *probability of occurrence*, $P(A)$.

Weak Law of Large Numbers. The sequence of random variables $\{\overline{X}_n\}$ converges in probability to the nonrandom variable $P(A)$:

$$\lim_{n \to \infty} \text{Prob}\left\{ \left| \frac{K_n}{n} - P(A) \right| < \varepsilon \right\} = 1, \qquad \varepsilon > 0. \qquad (2.58)$$

Strong Law of Large Numbers. The sequence of random variables $\{\overline{X}_n\}$ converges with probability one to the nonrandom variable $P(A)$:

$$\text{Prob}\left\{ \lim_{n \to \infty} \frac{K_n}{n} = P(A) \right\} = 1. \qquad (2.59)$$

An outline of a straightforward proof of (2.58) is given in exercise 32; however, (2.59) is considerably more difficult to prove. In the proof of (2.58), the following stronger result is obtained:

$$\underset{n \to \infty}{\text{l.i.m.}} \frac{K_n}{n} = P(A). \qquad (2.60)$$

Estimation of Mean. By way of interpretation of these laws of large numbers, it should be mentioned that since $P(A) = E\{I_A^j\}$ for each j, then these laws establish that the random sample mean (random statistical mean) of the event indicators, $\{I_A^j\}$, converges to the nonrandom mean (probabilistic mean) of I_A (which is identical for each and every j). It can be similarly shown that the random statistical mean $\hat{M}_Y$ of an arbitrary random variable Y converges to the probabilistic mean m_Y as the sample size n grows without bound (exercise 31).

2.5.4 Central Limit Theorem

A particularly important application of the concept of convergence in distribution is to the problem of determining the limiting distribution of the partial sums

$$X_n \triangleq \sum_{i=1}^{n} Z_i \qquad (2.61)$$

of a sequence $\{Z_i\}$ of random variables. Since the mean and variance of X_n can grow without bound as $n \to \infty$, the standardized variables

$$Y_n = \frac{X_n - m_n}{\sigma_n}, \qquad (2.62)$$

where m_n and σ_n^2 are the mean and variance of X_n, are considered. It can be shown that if $\{Z_i\}$ are independent and identically distributed, then Y_n converges in distribution to a Gaussian variable with zero mean and unity variance. This result is known as the *central limit theorem*. The same result also holds under various weaker hypotheses than independence and identical distributions. This theorem (in its various versions) is the primary reason that so many random phenomena are modeled in terms of Gaussian random variables. The central limit theorem comes into play when the value that a variable (such as a thermal-noise voltage) takes on is the result of a superposition of a large number of elementary effects (such as the tiny voltage impulses due to the individual ion cores and free electrons in a resistor). Another useful result regarding the convergence of partial sums (2.61), for which the $\{Z_i\}$ need not be either independent or identically distributed, is the following. It can be shown that $\{X_n\}$ converges in mean square if and only if the sequences of partial sums

$$\sum_{i=1}^{n} m_{Z_i}, \quad \sum_{i=1}^{n} \sigma_{Z_i}^2$$

both converge.

EXERCISES

1. (a) Consider a random variable X with uniform probability density over the interval $[a, b]$. Determine the mean and variance of X.

(b) Consider the random variable Y that takes on only the values $\{\Delta, 2\Delta, 3\Delta, \ldots, n\Delta\}$ with equal probability. Determine the mean and variance of Y.

(c) Let $b - a = n\Delta$ and compare (a) and (b).

2. (a) Consider the probability mass function $P(X = n) = (1 - \gamma)\gamma^n$ for $n = 0, 1, 2, 3, \ldots$, where $0 < \gamma < 1$. Determine the mean and variance of X.

(b) Consider the random variable Y with exponential density

$$f_Y(y) = \begin{cases} ae^{-ay}, & y \geq 0 \\ 0, & y < 0. \end{cases}$$

Determine the mean and variance of Y.

3. If a random variable X is scaled by a real number a, then the width, center, and height of the probability density are scaled accordingly, since for $Y = aX$,

$$f_Y(y) = \frac{1}{|a|} f_X\left(\frac{y}{a}\right).$$

Verify that $m_Y = am_X$ and $\sigma_Y = |a|\sigma_X$.

⋆4. Use the definition (2.9) to verify (2.8) for the special case in which the inverse of the function $g(\cdot)$ exists and is differentiable. Note that (2.8) is valid regardless of the existence of $g^{-1}(\cdot)$.

Hint: Use (1.36).

⋆5. (a) Verify (2.17).

(b) Use the definition (2.21) to verify the relationship (2.22).

⋆6. (a) Show that the mean and variance for the Gaussian density, (1.23), are given by $m_X = \beta$ and $\sigma_X^2 = \alpha^2$.

Hint: Use the change of variables $(x - \beta)/\alpha = y$; then for σ_X^2 use integration by parts.

(b) Show that the correlation coefficient for the bivariate Gaussian density (2.27) is given by $\rho = \gamma$.

Hint: Use the factorization in the hint for exercise 12(b) of Chapter 1 or use $R_{XY} = E\{E(X|Y)Y\}$ together with the result of exercise 12(d) of Chapter 1.

7. A continuous voltage X can take on any value between -15 V and $+15$ V with equal probability density. In order to obtain a digital representation for X, it is quantized into n equally spaced values with separation $30/(n - 1)$ V. Determine how large n must be to ensure that the mean-squared error between the quantized variable $\tilde{X}$ and the continuous variable X is no larger than 0.3 V.

8. (a) Consider the *Cauchy* probability density

$$f_X(x) = \frac{a/\pi}{x^2 + a^2}.$$

Show that the characteristic function is given by

$$\Phi_X(\omega) = e^{-a|\omega|}.$$

(b) Show that the variance of X does not exist.

9. Use the formula (2.14) to determine the first through fourth moments of a Gaussian random variable.

10. Verify (2.14).

★11. The *convolution theorem* for Fourier transforms states that the Fourier transform of the convolution of two functions is the product of their Fourier transforms. Consequently the characteristic function $\Phi_Z(\omega)$ for the sum of two independent random variables, $Z = X + Y$, is from (1.32) and (2.13)

$$\Phi_Z(\omega) = \Phi_X(\omega)\Phi_Y(\omega). \qquad (2.63)$$

Derive (2.63); that is, prove the convolution theorem.

★12. Let $\{X_1, X_2, X_3, \ldots, X_n\}$ be independent identically distributed Gaussian variables with zero means and variances σ_X^2. Determine the mean and variance of the random variable $Y = \hat{\sigma}_X^2$, where

$$\hat{\sigma}_X^2 \triangleq \frac{1}{n} \sum_{k=1}^{n} (X_k - \hat{m}_X)^2$$

and

$$\hat{m}_X \triangleq \frac{1}{n} \sum_{j=1}^{n} X_j$$

Hint: For jointly Gaussian zero-mean variables, we have

$$E\{X_iX_jX_kX_l\} = E\{X_iX_j\} E\{X_kX_l\} + E\{X_iX_k\} E\{X_jX_l\} \\ + E\{X_iX_l\} E\{X_jX_k\}.$$

13. (a) The *Poisson* probability mass function is given by

$$P(n) = \frac{\lambda^n}{n!} e^{-\lambda}, \quad n = 0, 1, 2, 3, \ldots, \qquad (2.64)$$

for some $\lambda > 0$. Consider m independent Poisson distributed random variables with parameters $\lambda_1, \lambda_2, \lambda_3, \ldots, \lambda_m$. Show that their sum is also Poisson distributed and has parameter value $\lambda = \lambda_1 + \lambda_2 + \lambda_3 + \cdots + \lambda_m$.

(b) Show that the mean and the variance of a Poisson distributed random variable both equal λ.

★**14.** One form of the *Cauchy-Schwarz* inequality is

$$\left| \iint g(x, y)h(x, y)w(x, y) \, dx \, dy \right|^2$$
$$\leq \iint [g(x, y)]^2 w(x, y) \, dx \, dy \iint [h(x, y)]^2 w(x, y) \, dx \, dy, \quad (2.65)$$

where $w(x, y)$ is a nonnegative function, and $g(x, y)$ and $h(x, y)$ are arbitrary (except that the integrals must exist). Use this inequality to prove that for any two finite-mean-square random variables X and Y,

$$|E\{XY\}|^2 \leq E\{X^2\}E\{Y^2\}.$$

Use this result to prove that the magnitude of the correlation coefficient (2.25) cannot exceed unity.

★**15.** (a) Prove that for any set of finite-variance random variables, the standard deviation of their sum can never exceed the sum of their standard deviations.

Hint: One form of the *triangle inequality* is

$$\left[\iint [g(x, y) + h(x, y)]^2 \, w(x, y) \, dx \, dy \right]^{1/2}$$
$$\leq \left[\iint [g(x, y)]^2 \, w(x, y) \, dx \, dy \right]^{1/2} \quad (2.66)$$
$$+ \left[\iint [h(x, y)]^2 \, w(x, y) \, dx \, dy \right]^{1/2},$$

where $w(x, y)$ is a nonnegative function, and $g(x, y)$ and $h(x, y)$ are arbitrary functions (except that the integrals must exist). Use this inequality to prove that for two random variables, the standard deviation of their sum cannot exceed the sum of their standard deviations. Then use this result repeatedly to obtain the desired result for any number of random variables.

(b) Prove that for statistically independent random variables, the variance of their sum equals the sum of their variances.

16. Sketch the contours of constant elevation of the bivariate Gaussian density for two random variables, X and Y, for the cases
(a) $\sigma_X = \sigma_Y = 1, \rho = 0$,
(b) $\sigma_X = \sigma_Y = 1, \rho = \frac{3}{4}$,
(c) $\sigma_X = \sigma_Y = 1, \rho = -\frac{1}{4}$,
(d) $\sigma_X = 1, \sigma_Y = 5, \rho = 0$.
What happens to the contours as $|\rho| \to 1$?

17. A pair of random variables X and Y are uniformly distributed over a rectangular region in the x-y plane. This region is centered at the origin and the longer side (length a) of the rectangle is parallel to the line $x = y$. The shorter side has length b. Solve for f_X and f_Y and the correlation coefficient ρ in terms of a and b. Under what conditions do we have $\rho = 0$ or $\rho = 1$ or $\rho = -1$?

18. (a) If X and Z are not orthogonal, then there exists a number a and a random variable N that is orthogonal to Z such that $X = aZ + N$. Prove this by minimizing with respect to a the mean-squared value of the random variable defined by $N \triangleq X - aZ$, and then showing that $E\{NZ\} = 0$.
    (b) Use the result of part (a) to show that even if X and Z are correlated and Y and Z are correlated, X and Y can be uncorrelated.

19. (a) Show that if X and Y are statistically independent, then they are uncorrelated.
    (b) To show that two variables that are uncorrelated need not be independent, show that if $f_X(\cdot)$ has even symmetry,
    $$f_X(-x) = f_X(x),$$
    then X and $Y = X^2$ are uncorrelated, but not independent.

20. Let X be a random vector with mean m_X and covariance K_X, and let c be a deterministic vector. Verify that
    $$E\left\{\left[\sum_{i=1}^{n} c_i(X_i - m_{X_i})\right]^2\right\} = c^T K_X c$$
    and
    $$E\left\{\left[\sum_{i=1}^{n} c_i X_i\right]^2\right\} = c^T K_X c + (m_X^T c)^2.$$

★21. Let $R_X^{-1/2}$ be the square root of the inverse of the correlation matrix R_X for an n-tuple X of random variables:
    $$R_X^{-1/2} R_X^{-1/2} = R_X^{-1}.$$
    Show that the random variables in the n-tuple $Y \triangleq R_X^{-1/2} X$ are mutually orthogonal and have unity mean-squared values.

22. Determine the probability $P(XY > 0)$ in terms of the means and variances for the independent Gaussian random variables X and Y.

23. (a) The *median* x_o of a random variable X is the value that is exceeded (or not exceeded) with probability of 1/2. Express x_o as the solution to an equation specified by f_X. Can you give some conditions under which the median will equal the mean? Consider the Gaussian and

Rayleigh densities. Show that $E\{|X - a|\}$ is minimum for $a = x_o$, whereas $E\{[X - a]^2\}$ is minimum for $a = m_X$.

(b) The *mode* x_* of a random variable X is the value that has the largest probability (density) of occurrence. Can you give some conditions under which the mode will equal the median and/or the mean? Consider the Gaussian and Rayleigh densities.

24. (a) Consider two resistive thermal noise sources A and B. Source A has resistance 100 Ω, temperature 290 K, and bandwidth 100 MHz. Source B has resistance 200 Ω, temperature 290 K, and bandwidth 200 MHz. Through some unusual mechanism, the two noise voltages developed by these sources have a nonzero correlation coefficient of $\rho = 1/2$. Determine the variance of the sum of these two voltages.

(b) Given that the voltage from source A is 10^{-5} V, determine the conditional mean and variance of the voltage from source B. (Assume that the voltages are jointly Gaussian.)

25. (a) Verify (2.38) for an n-tuple of jointly Gaussian random variables X.

    Hint: First use (2.13) and (2.18) to show that the characteristic function for a single Gaussian random variable Y is given by (2.38) with $m_X^T \omega$ and $\omega^T K_X \omega$ replaced by m_Y and σ_Y^2, respectively. Then let $Y \triangleq \omega^T X$. If the elements of X are jointly Gaussian, then by definition Y must be Gaussian for every n-tuple of real numbers ω. Thus, (2.38) holds with the preceding replacements (m_Y and σ_Y^2). Finally show that $m_Y = m_X^T \omega$ and $\sigma_Y^2 = \omega^T K_X \omega$ and, therefore, (2.38) holds for any n-tuple of jointly Gaussian variables X.

(b) Evaluate the n-dimensional Fourier transform of (2.38) to verify (2.37) for the case in which K_X^{-1} exists. Or, as a less technical alternative, verify the steps in Example 3, which establishes the validity of the transform pair (2.37)–(2.38) for the special case $n = 2$.

26. Use the definition (2.6) and (1.29) to verify (2.45).

27. (a) It is desired to estimate an unobservable random variable X by using a transformation of the form $\hat{X} = aY + b$ of an observable random variable Y. Find the values of a and b that minimize the mean-squared error $E\{(X - \hat{X})^2\}$.

(b) Show that if X and Y are jointly Gaussian, then the best linear estimate from part (a) is the conditional mean, $\hat{X} = E\{X|Y\}$.

(c) Express the minimized value of the mean-squared error from part (a) in terms of the correlation coefficient for X and Y.

28. Consider the random variable X that is uniformly distributed over $[0, 1)$. Determine its conditional mean value $E\{X|X > a\}$ and conditional variance $E\{(X - E\{X|X > a\})^2 | X > a\}$ for $0 < a < 1$. What happens to these as $a \to 1$?

29. Consider the joint probability density function

$$f_{XY}(x, y) = \begin{cases} ce^{-(2x+3y)}, & x \geq 0, y \geq 0 \\ 0, & \text{otherwise.} \end{cases}$$

Determine the value of c, and then determine the following:
(a) $P[X > 1/2 \text{ and } Y > 1/3]$
(b) $P[X > 1/2 | Y > 1/3]$
(c) $E\{XY\}$
(d) $E\{XY | Y > 1/3\}$
(e) $E\{XY | Y = 1/3\}$.

★30. Verify (2.47) and (2.48).

Hint: For (2.47) use the fundamental theorem of expectation and the fact that $f_{X|Y} = f_{X|Y,Z}$, since $Z = g(Y)$; for (2.48) use the fundamental theorem of expectation and the fact that $f_{Z|Y}(z|y) = \delta[z - g(y)]$ for $z = g(Y)$.

31. For a set of identically distributed independent random variables $Y_1, Y_2, \ldots, Y_n$, verify that the random sample mean

$$\hat{M}_Y \triangleq \frac{1}{n}(Y_1 + Y_2 + \cdots + Y_n)$$

has mean and variance given by

$$E\{\hat{M}_Y\} = m_Y, \qquad E\{(\hat{M}_Y - m_Y)^2\} = \sigma_Y^2/n.$$

Use this result to prove that

$$\underset{n \to \infty}{\text{l.i.m.}} \hat{M}_Y = m_Y.$$

★32. Use the result of exercise 31 together with the Bienaymé-Chebychev inequality

$$\text{Prob}\{|X - a| \geq \varepsilon\} \leq \frac{1}{\varepsilon^n} E\{|X - a|^n\} \qquad \text{(for any } a, n\text{)} \quad (2.67)$$

to prove the weak law of large numbers.

★33. Consider the sample space $\{s\} = \{1, 2, 3, \ldots\}$, with associated probability function $P(s = n) = \alpha/n^2$, and the sequence of random variables $\{X_n : n = 1, 2, 3, \ldots\}$ defined on this space by

$$X_n(s) = \begin{cases} n, & s = n, \\ 0, & s \neq n. \end{cases}$$

Prove that $\{X_n\}$ converges almost surely to $X = 0$, but $\{X_n\}$ does not converge in mean square; that is,

$$E\{(X_n - 0)^2\}$$

does not converge to zero, as $n \to \infty$.

Further Reading

Introductory

H. Cramér. *The Elements of Probability Theory and Some of Its Applications*. New York: Wiley, 1955.

W. Feller. *Probability Theory and Its Applications*. 3d Edition, Volume 1. New York: Wiley, 1968.

Intermediate

B.V. Gnedenko. *Theory of Probability*. New York: Chelsea, 1962.

E. Parzen. *Modern Probability Theory and its Applications*. New York: Wiley, 1960.

Advanced

A.N. Kolmogorov. *Foundations of the Theory of Probability* (English translation of 1933 German edition). New York: Chelsea, 1956.

M. Loéve. *Probability Theory*. 4th Edition. New York: Springer, Volume I, 1977; Volume II, 1978.

Part 2

RANDOM PROCESSES

3

Introduction to Random Processes

3.1 Introduction

In the last few decades, the probabilistic approach to the design and analysis of signals and systems has become indispensable. For example, the conceptual framework, analytical tools, and methods of design based on probability theory are now in common use for solving a wide variety of signal-processing problems involving audio signals such as speech, visual signals (images), sonar and radar signals, geophysical signals, astrophysical signals, biological signals, and so on. The probabilistic approach to the analysis and design of signals and systems gained much of its popularity through its successful application to communication systems, which led to statistical communication theory.* To illustrate how probabilistic concepts reviewed in Part 1 arise in problems of analysis and design of signals and systems, we shall consider the general communications problem.

---

*It should be mentioned that although the term *statistical* is, by convention, favored, the theory is primarily *probabilistic*. The signals and noises and quantities derived from them are the statistics, and the methods for measuring system performance in terms of these statistics are statistical, but the theory is probabilistic.

The physical phenomena to be dealt with in the typical communications problem are the fluctuating currents and voltages that appear at different stages of the communication process, that is, the inherent (background) noise that arises in the various system elements, impinging noise and interfering signals originating in the medium of propagation (the communication channel), and the information-bearing signal itself. Signal and noise phenomena of these types are characterized by an element of uncertainty, or unpredictability, or randomness; no matter how much we might know of the past history of such phenomena, we are essentially unable to predict their future behavior. (Recall the discussion in the preface for the student.) For this reason, an effective study of a communication system or device cannot generally be carried out in terms of an individual signal alone, nor can such a study safely neglect the communication-degrading effects of noise and interference. Rather one must consider the ensemble of all possible signals (for which the system is—or is to be—designed), the ensemble appropriate to the accompanying noise and interference, and the manner in which they combine in the communication process itself. Consequently, useful measures of system performance (which are essential for analysis and design) must take these ensembles into account. This is accomplished by adopting measures of performance that average over the ensembles. The averages measured in practice are *statistics*, whereas the mathematical counterparts, on which theory is built, are *expected values* (probabilistic parameters). Three of the most valuable and commonly used measures of average performance are the *signal-to-noise ratio* (SNR), which is either a ratio of squared mean to variance, or a ratio of mean squares; *the mean squared error* (MSE); and the *probability of error* (PE). (Recall from Chapter 1 that a probability is an expected value.) The SNR is a measure of the relative strengths of signal and noise. The MSE is a measure of the dissimilarity between a noisy distorted signal and its uncorrupted version. The PE is a measure of the likelihood of making an incorrect decision about which one of a finite set of possible signals (symbols) was transmitted, having received a corrupted version. Some of the ways in which SNR, MSE, and PE can be exploited in analysis and design of signals and systems are illustrated in Section 3.3.

Signal-processing systems can be broadly described in terms of the operations they perform on postulated classes of inputs—for example, linear operations such as differentiation, integration, multiplication, and convolution of input waveforms, and the counterparts of differencing, summing, etc., of discrete-time input sequences. Thus, the probabilistic study of signals and systems is, in a general sense, the study of averages of ensembles of waveforms and sequences subjected to these various operations. The underlying probability theory is, in essence, a calculus of averages.

In mathematical terminology, the probabilistic models of ensembles of waveforms and sequences are called *random processes*. The primary objective of Part 2 is to introduce theory and analytical methods for characterizing random processes and studying dynamical systems with random excitation, and the effects of linear signal-processing operations on random-process characteristics. The theory focuses on the expected values of linear and quadratic measurements on random processes, and is therefore called the *second-order theory of random processes*.

Although the averages that probability theory deals with are ensemble averages, the averages that are measured in practice are often time averages on a single member of an ensemble, that is, a single waveform. Experimental evaluation of SNR, MSE, and PE is often accomplished by time averaging. In fact, in some but not all instances, the time-average performance of a system is just as important (if not more so) as its ensemble-average performance. For example, for binary digital data-transmission systems, the PE is always measured by computing the relative frequency of received bits in error over a long stream of bits. Similarly, a channel equalizer, which removes signal distortion, is typically evaluated by measuring the time-average squared error. Similarly, the mean, variance, and mean square used in defining the SNR are typically measured by time averaging. Consequently, another objective of Part 2 is to emphasize the duality between the probabilistic theory of random processes, based on expectation or ensemble averages, and the deterministic theory, based on time averages, and also to introduce the (second-order) theory of *ergodicity*, which deals with the relationship between ensemble averages and time averages.

3.2 Generalized Harmonic Analysis

The theory of random processes based on time averages is considerably more concrete than the theory based on ensemble averages. Therefore it is helpful to gain some familiarity with the former before tackling the latter. Consequently this section provides a brief introduction to the deterministic* theory of random processes based on time averages.

---

*The term *deterministic* is used in this book as an antonym for *probabilistic*, and thus as a synonym for *nonprobabilistic*. However, the term *deterministic*, as applied to the theory described here, is somewhat a misnomer, because the theory is based on infinite limits of time averages, which are no more deterministic than are the infinite limits of ensemble averages on which the probabilistic theory is based. Some authors use the term *functional* to denote the time-average theory, and the term *stochastic* to denote the ensemble-average theory. A comprehensive development of the nonprobabilistic statistical theory of random processes, as it applies to spectral analysis, is given in [Gardner, 1987a].

With reference to the random thermal noise described in Part 1, we intuitively expect that time samples of a thermal-noise voltage waveform $V(t)$, if they are sufficiently close together in time, will be related to each other—that is, will be *correlated*. Moreover, we intuitively expect the degree of correlation to increase as the time separation τ between these time samples decreases. This correlation behavior is related to the frequencies of random collisions of ions and free electrons in the resistor, which in turn is related to the temperature of the resistor. This suggests that the frequency composition of a random waveform is related to the dependence of the correlation of time samples of the waveform on the time separation. To illustrate this relationship, we briefly compare the correlations for two different waveforms $X(t)$ and $Y(t)$. The waveforms, shown in Figure 3.1, are typical of thermal-noise voltages; however, $X(t)$ typifies thermal noise at a higher temperature than $Y(t)$. Consequently $X(t)$ contains components at higher frequencies, and is therefore more *broadband* than $Y(t)$ is. As a measure of the correlation of time samples separated by τ, we form the average of their products, denoted by

$$\hat{R}_X(\tau) \triangleq \lim_{N \to \infty} \frac{1}{2N+1} \sum_{n=-N}^{N} X(n\tau + \tau) X(n\tau). \qquad (3.1)$$

Since $X(t)$ contains more high-frequency components than $Y(t)$ does, the signs of the products $X(n\tau + \tau)X(n\tau)$ more nearly balance out than do the

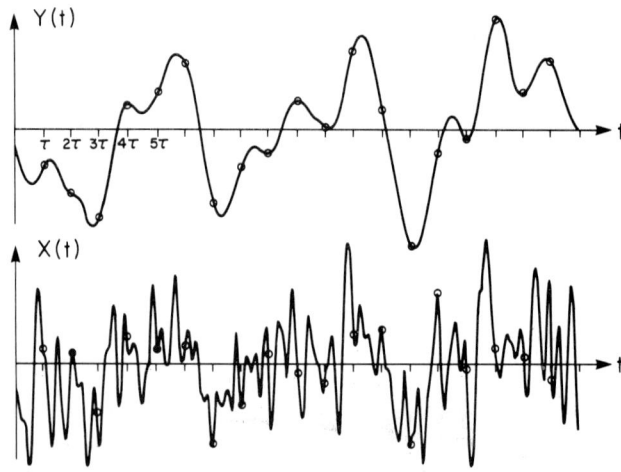

Figure 3.1 Thermal-Noise Voltage Waveforms at Low and High Temperatures: Narrowband and Broadband Frequency Compositions

signs of $Y(n\tau + \tau)Y(n\tau)$. That is, whereas the two factors $X(n\tau + \tau)$ and $X(n\tau)$ have the same sign about as often as they have opposite signs, the two factors $Y(n\tau + \tau)$ and $Y(n\tau)$ have the same sign the majority of the time. Consequently, cancellation of positive and negative terms in the sum (3.1) yields a smaller value for the correlation $\hat{R}_X(\tau)$ than for $\hat{R}_Y(\tau)$. Furthermore, as τ is decreased, the proportion of positive products from both $X(t)$ and $Y(t)$ increases, and therefore both $\hat{R}_X(\tau)$ and $\hat{R}_Y(\tau)$ increase and, in fact, approach their maximum values as τ approaches zero.

In conclusion, a comparative description of $\hat{R}_X$ and $\hat{R}_Y$ (for sufficiently small τ) would resemble the peaked graphs shown in Figure 3.2a. This suggests that, in general, broadband waveforms have narrower-peaked correlation functions than more narrowband (low-frequency) waveforms. This also suggests that the frequency composition of a random waveform can be studied in terms of its correlation, as a function of the time separation τ.

From this point forward $\hat{R}_X(\cdot)$ will be referred to as an *empirical autocorrelation function*: "empirical"* because it is a measure of the correlation between empirical variables, namely time samples (rather than probabilistic variables), and "auto-" because both variables whose correlation is being measured are from the same waveform; thus, the self-correlation of the waveform is being measured. In practice, the empirical autocorrelation can be obtained either from a discrete-time average, as in (3.1), or from a continuous-time average (centered at the time origin, $t = 0$, for convenience)

$$\hat{R}_X(\tau) \triangleq \lim_{T \to \infty} \frac{1}{T} \int_{-T/2}^{T/2} X(t + \tau) X(t) \, dt. \tag{3.2}$$

The preceding observations concerning the relationship between the frequency composition of a waveform and its empirical autocorrelation can be explained as follows. We consider the finite segment

$$X_T(t) \triangleq \begin{cases} X(t), & |t| \leq T/2, \\ 0, & |t| > T/2, \end{cases}$$

of the waveform $X(t)$, and we determine its frequency composition by *Fourier transformation*:

$$\tilde{X}_T(f) \triangleq \int_{-\infty}^{\infty} X_T(t) e^{-i2\pi ft} \, dt,$$

$$X_T(t) = \int_{-\infty}^{\infty} \tilde{X}_T(f) e^{+i2\pi ft} \, df. \tag{3.3}$$

---

*The term *empirical* as used here is somewhat a misnomer, because the limit involved in the definition of the autocorrelation is a mathematical operation, not an empirical operation. Nevertheless, the term *empirical* in contrast to the term *probabilistic* appropriately suggests a closer link with the practice of time averaging.

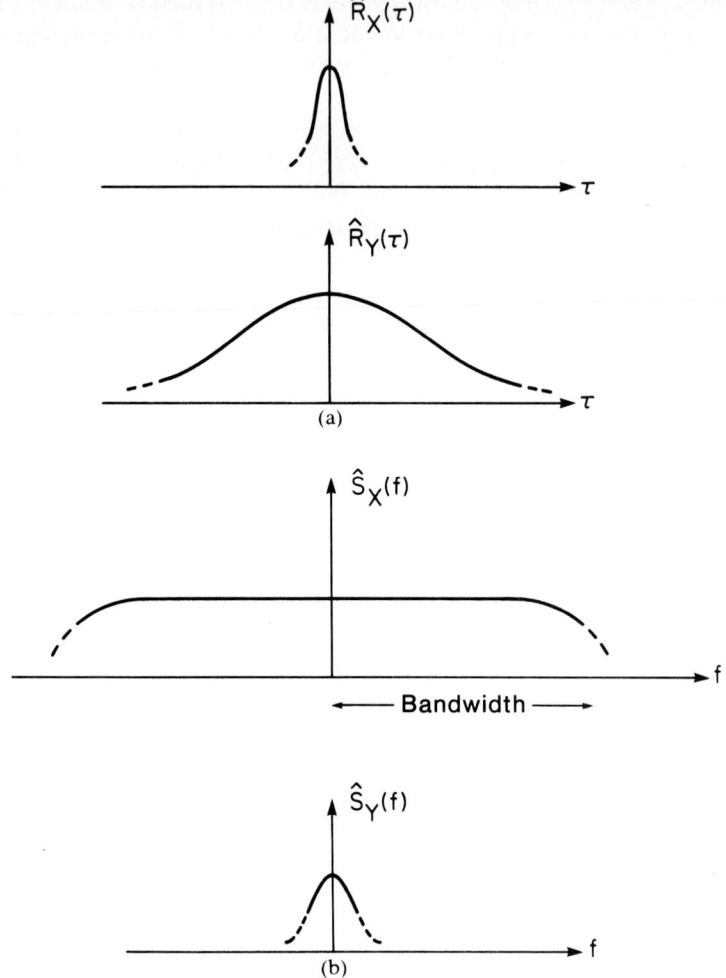

Figure 3.2 (a) Autocorrelation Functions; (b) Power Spectral Densities for Broadband and Narrowband Waveforms

It follows from the second formula (3.3) that $\tilde{X}_T(\cdot)$ is the complex density of complex sinusoids (frequency components) that compose $X_T(\cdot)$. A convenient real-valued measure of the frequency composition of X_T is the time-normalized squared magnitude, $(1/T)|\tilde{X}_T(f)|^2$, which is called the *finite-time spectrum* or *periodogram*. Application of the convolution theorem for Fourier transforms yields (exercise 3) the following characterization of this spectrum:

$$\frac{1}{T}|\tilde{X}_T(f)|^2 = \int_{-\infty}^{\infty} R_X(\tau)_T \, e^{-i2\pi f\tau} \, d\tau, \qquad (3.4)$$

where $R_X(\tau)_T$ is defined by the convolution

$$R_X(\tau)_T \triangleq \frac{1}{T}\int_{-\infty}^{\infty} X_T(t+\tau)X_T(t)\,dt = \frac{1}{T}X_T(\tau) \otimes X_T(-\tau),$$

$$= \frac{1}{T}\int_{-T/2}^{T/2-|\tau|} X(t+|\tau|)X(t)\,dt, \qquad (3.5)$$

which is called the *finite-time autocorrelation* or *correlogram*. Equation (3.4) reveals that this autocorrelation is quite obviously related to the frequency composition of the finite segment X_T. Moreover, in the limit $T \to \infty$, we obtain

$$\lim_{T\to\infty} R_X(\tau)_T = \hat{R}_X(\tau). \qquad (3.6)$$

Consequently, *the autocorrelation $\hat{R}_X$ is related to the frequency composition of X through Fourier transformation*. However, there is a technical problem at this point because for essentially all random waveforms of interest, the limit of the periodogram $(1/T)|\tilde{X}_T(f)|^2$ as $T \to \infty$ does not exist. That is, this function simply becomes more and more erratic as T is increased. Nevertheless, the limiting form of (3.4) can be obtained if the periodogram is time-averaged to remove its erratic behavior before the limit as $T \to \infty$ is taken. This can be accomplished by allowing the location of the time interval of analysis $[-T/2, T/2]$ to depend on time u to obtain $[u - T/2, u + T/2]$. The corresponding time-variant periodogram is given by $(1/T)|\tilde{X}_T(u, f)|^2$, where

$$\tilde{X}_T(u, f) \triangleq \int_{-\infty}^{\infty} X_T(u, t)e^{-i2\pi ft}\,dt = \int_{u-T/2}^{u+T/2} X(t)e^{-i2\pi ft}\,dt \qquad (3.7a)$$

and

$$X_T(u, t) \triangleq \begin{cases} X(t), & |t-u| \le T/2 \\ 0, & |t-u| > T/2. \end{cases} \qquad (3.7b)$$

Using definition (3.7) it can be shown (exercise 6) that the limit

$$\hat{S}_X(f) \triangleq \lim_{T\to\infty}\lim_{U\to\infty} \frac{1}{U}\int_{-U/2}^{U/2} \frac{1}{T}|\tilde{X}_T(u, f)|^2\,du \qquad (3.8)$$

of the time-averaged periodogram exists and is given by the Fourier transform of the limit (3.6),

$$\hat{S}_X(f) = \int_{-\infty}^{\infty} \hat{R}_X(\tau)e^{-i2\pi f\tau}\,d\tau. \qquad (3.9)$$

This function $\hat{S}_X$ is called the *power spectral density* (PSD) for the waveform X, because it can be shown (exercise 5) that it is the frequency density of the time-averaged power that the voltage $X(t)$ would dissipate in a resistance. Specifically, if the voltage waveform $X(t)$ (volts) is passed through a very narrowband filter with unity gain and center frequency f, and the filtered voltage is dropped across a resistance r (ohms), then the bandwidth-normalized time-averaged power dissipation in the resistance is

$$\frac{1}{2\Delta}\langle P\rangle \simeq \frac{\hat{S}_X(f)}{r} \quad \text{(watts)}, \tag{3.10}$$

where Δ is the very small positive-frequency bandwidth of the narrowband filter. In the limit as $\Delta \to 0$, approximation (3.10) becomes exact. Typical PSDs for the broadband and narrowband waveforms in Figure 3.1 are shown in Figure 3.2b.

Furthermore, if $X(t)$ is the input to a linear, time-invariant filter with impulse-response function $h(t)$, and $Y(t)$ is the output

$$Y(t) = \int_{-\infty}^{\infty} h(t-u)X(u)\,du = X(t) \otimes h(t), \tag{3.11}$$

then it can be shown (exercise 4) that the input and output autocorrelations are related by convolution:

$$\hat{R}_Y(\tau) = \int_{-\infty}^{\infty} \hat{R}_X(\tau-u)r_h(u)\,du = \hat{R}_X(\tau) \otimes r_h(\tau), \tag{3.12}$$

$$r_h(\tau) \triangleq \int_{-\infty}^{\infty} h(\tau+v)h(v)\,dv = h(\tau) \otimes h(-\tau). \tag{3.13}$$

It follows from (3.9), (3.12), (3.13), and the convolution theorem for Fourier transforms that the PSD satisfies the following input-output relation:

$$\hat{S}_Y(f) = \hat{S}_X(f)|H(f)|^2, \tag{3.14}$$

where

$$H(f) \triangleq \int_{-\infty}^{\infty} h(t)e^{-i2\pi ft}\,dt, \tag{3.15}$$

which is the *transfer function* for the filter. Equation (3.14) is analogous to the well-known input-output relation

$$\tilde{Y}(f) = \tilde{X}(f)H(f), \tag{3.16}$$

where $\tilde{Y}$ and $\tilde{X}$ are the Fourier transforms of transient (finite-energy) input and output waveforms X and Y. The relation (3.16) holds for persistent (infinite-energy) random waveforms in only an abstract generalized sense, as briefly discussed in Chapter 10.

Equation (3.14) can be used to derive the interpretation of (3.9) as the PSD (exercise 5), by letting H be the transfer function of a very narrowband filter and evaluating the time-averaged power

$$\langle P \rangle = \lim_{T \to \infty} \frac{1}{T} \int_{-T/2}^{T/2} P(t)\, dt, \tag{3.17}$$

where $P(t)$ is the instantaneous power of the filter response,

$$P(t) = \frac{Y^2(t)}{r}. \tag{3.18}$$

The result is (3.10).

The preceding concepts and relations, in essence, embody the subject of *generalized harmonic analysis* of random persistent waveforms, developed by Norbert Wiener [Wiener, 1930] to complement the more elementary subjects of Fourier analysis of transient waveforms and of periodic waveforms. It can be seen from the definition of autocorrelation, (3.2), and PSD, (3.8), that generalized harmonic analysis is based on time averages. However, these time averages can be reinterpreted probabilistically as expected values. To see this, we consider the following definition of *fraction-of-time amplitude distribution* for a waveform $X(t)$ (e.g., a statistical sample of a random process*):

$$\hat{F}_X(x) \triangleq \lim_{T \to \infty} \frac{1}{T} \int_{-T/2}^{T/2} u[x - X(t)]\, dt, \tag{3.19}$$

for which $u(\cdot)$ is the unit step function (here defined to be continuous from the left), and therefore $u[x - X(t)]$ is the indicator function for the event $X(t) < x$,

$$u[x - X(t)] \triangleq \begin{cases} 1, & X(t) < x, \\ 0, & X(t) \geq x, \end{cases} \tag{3.20}$$

and $\hat{F}_X(x)$ is the limit relative frequency of this event. The *fraction-of-time amplitude density* for $X(t)$ is defined by

$$\hat{f}_X(x) \triangleq \frac{d\hat{F}_X(x)}{dx}, \tag{3.21}$$

and the expected value of the amplitude of $X(t)$ is defined by

$$\hat{E}\{X(t)\} \triangleq \int_{-\infty}^{\infty} x \hat{f}_X(x)\, dx. \tag{3.22}$$

Substitution of (3.19) into (3.21) into (3.22), and use of the fact that the

---

*The notational convention of using a lowercase letter for a statistical sample is violated here.

derivative of the unit step function is the impulse function, yields

$$\hat{E}\{X(t)\} = \int_{-\infty}^{\infty} x \frac{d}{dx} \lim_{T \to \infty} \frac{1}{T} \int_{-T/2}^{T/2} u[x - X(t)] \, dt \, dx$$

$$= \lim_{T \to \infty} \frac{1}{T} \int_{-T/2}^{T/2} \int_{-\infty}^{\infty} x \delta[x - X(t)] \, dx \, dt$$

$$= \lim_{T \to \infty} \frac{1}{T} \int_{-T/2}^{T/2} X(t) \, dt. \tag{3.23}$$

Thus, the time average (3.23) can be reinterpreted as an expected value (3.22) on the basis of the fraction-of-time distribution (3.19). The same reinterpretation can be developed for other time averages, such as that used to define the autocorrelation (3.2) and PSD (3.8) (cf. Section 8.6). Hence, generalized harmonic analysis can be given probabilistic interpretations. Moreover, a more general *probabilistic harmonic analysis* that is independent of time-averages can be developed. In fact, a major objective of Part 2 is to develop probabilistic counterparts to the empirical autocorrelation function $\hat{R}_X$ in (3.2), the PSD $\hat{S}_X$ in (3.9), and the input-output relations (3.12) and (3.14) for linear time-invariant filters, and bring to light the duality that exists between the deterministic theory of generalized harmonic analysis and its probabilistic counterpart.

To see why we might be interested in a probabilistic counterpart, the reader is reminded that, as discussed in Section 3.1, we need a theory that deals with predictable signal averages rather than with unpredictable signals. As we shall see, the empirical autocorrelation (3.2) is predictable for many types of unpredictable signals, but not for all types. For example, for the noise-modulated sine wave

$$X(t) = N(t) \cos(\Omega t + \Theta),$$

for which the sine wave has been chosen from an ensemble of sine waves with different phases and frequencies, the empirical autocorrelation is

$$\hat{R}_X(\tau) = \tfrac{1}{2} \hat{R}_N(\tau) \cos \Omega \tau.$$

Thus, the unpredictable noise $N(t)$ is described (on the average) by the autocorrelation $\hat{R}_N(\tau)$, which can be shown [cf. Section 10.7] to be predictable, and the unpredictable (unknown) phase Θ of the sine wave vanishes in the autocorrelation $\hat{R}_X(\tau)$. However, since the frequency Ω of the sine wave is unpredictable, then the autocorrelation $\hat{R}_X(\tau)$ is also unpredictable. This illustrates the fact that time averaging does not always remove all unpredictability. However, by adopting a probabilistic model in which Ω is a random variable, and using ensemble instead of time averaging, the unpredictability (randomness of Ω) vanishes in the probabilistic autocorrelation. Another example of an unpredictable empirical autocorrelation function results from random filtering, rather than random modulation as in the preceding example. Specifically, if the filter impulse response h is random, then so too is its *finite autocorrelation* r_h, and therefore the autocorrelation (3.12) of the filtered waveform is random. A random filter is an appropriate model for the ensemble of channels encountered in a large communication system such as a telecommunication network.

A second important reason for interest in a probabilistic theory is that it accommodates averages that vary with time. For example, if the temperature of a resistor varies rapidly with time, then its effect on the statistical behavior of the thermal-noise voltage cannot be characterized in terms of time averages, since these remove all time-varying effects. Thus, to study the statistical behavior of time-varying phenomena, we must rely on a probabilistic model.* A third reason for interest in a probabilistic theory is that it is a more natural setting for decision problems, such as the problem of detecting the presence of a random finite-energy signal masked by noise. Examples of signal detection are described in the next section.

3.3 *Signal-Processing Applications*

To provide motivation for the theory developed in subsequent chapters, we consider here several problems that are of fundamental importance in signal processing, and for which the theory of autocorrelation and PSD provides solutions. Because the probabilistic theory is more abstract than its deterministic counterpart, and since we already have in hand the fundamental definitions on which the deterministic theory is based, these problems shall be formulated in terms of time averages, so that their solutions can be similarly described. The ensemble-average counterparts of these and many other problems are formulated and solved in Chapter 11.

3.3.1 *Interpolation of Time-Sampled Waveforms*

Since many signal-processing systems are implemented in discrete time, so that digital technology can be exploited, an important question arises as to the conditions under which an accurate approximation to a waveform can be recovered from its time samples by interpolation. Let $X(t)$ be a random (unpredictable) waveform, let $\{ X(iT) : i = 0, \pm 1, \pm 2, \ldots \}$ be its time samples, and let $p(t)$ be an interpolating pulse. Consider the following approximation to the waveform:

$$X(t) \simeq \hat{X}(t) \triangleq \sum_{i=-\infty}^{\infty} X(iT) p(t - iT). \tag{3.24}$$

---

*One important exception to this is the case of periodic time variation, as briefly explained in Chapter 12.

Let us adopt time-averaged squared error (average power) as a measure of the size of the error in this approximation:

$$\text{error power} = \left\langle [X(t) - \hat{X}(t)]^2 \right\rangle. \tag{3.25}$$

We would like to know the conditions under which this average error power is zero. We shall see in Chapter 11 that the error power is in general zero if and only if the PSD of $X(t)$ is bandlimited to less than half the sampling rate,

$$\hat{S}_X(f) = 0, \quad |f| \geq B < \frac{1}{2T}, \tag{3.26}$$

and the interpolating pulse is an appropriately designed bandlimited pulse, namely

$$p(t) = \frac{\sin(\pi t/T)}{\pi t/T}. \tag{3.27}$$

3.3.2 Noise Immunity of Frequency Modulation

It is a well established fact that frequency-modulated (FM) signals can be highly immune to noise, and are therefore of fundamental importance in the transmission of information. This was experimentally established long before the phenomenon had been accurately described analytically. Two important components of the analytical characterization of this noise immunity are the PSD of an FM signal and the gain in SNR (ratio of signal power to noise power) achieved by the process of frequency demodulation. These are briefly described here. We consider the FM signal

$$X(t) = \cos[2\pi f_o t + \Phi(t)], \tag{3.28}$$

which has *instantaneous frequency*

$$f(t) = f_o + \frac{\Psi(t)}{2\pi}, \tag{3.29a}$$

where

$$\Psi(t) \triangleq \frac{d\Phi(t)}{dt} \quad \text{and} \quad \Phi(t) = \int_{-\infty}^{t} \Psi(u)\, du. \tag{3.29b}$$

This signal is most immune to noise when the power of the random-phase process specified by $\Phi(t)$ is large:

$$\langle P \rangle = \hat{R}_\Phi(0) \gg 1. \tag{3.30}$$

In this case, the noise immunity is characterized by the PSDs of the modulated signal $X(t)$ and the modulating signal $\Psi(t)$. In particular, the ratio of output SNR to input SNR for the demodulator is shown in Chapter 11 to be closely approximated by

$$\frac{\text{SNR}_{\text{out}}}{\text{SNR}_{\text{in}}} \simeq 3\left(\frac{B_X}{B_\Psi}\right)^3, \qquad (3.31)$$

where B_X and B_Ψ are the bandwidths of the PSDs of $X(t)$ and $\Psi(t)$. Now, for a given model for a modulating signal, B_Ψ is known but B_X must be determined. It is shown in Chapter 11 that under the condition (3.30), the PSD of $X(t)$ is closely approximated by

$$\hat{S}_X(f) \simeq \tfrac{1}{4}[\hat{f}_\Psi(f - f_o) + \hat{f}_\Psi(f + f_o)], \qquad (3.32)$$

where $\hat{f}_\Psi$ is the fraction-of-time amplitude density of $\Psi(t)$. Thus, B_X is given by the width of the amplitude density $\hat{f}_\Psi$, which is a measure of the amount of frequency deviation about the carrier frequency f_o.

3.3.3 Signal Detection

The problem of detecting the possible presence of a finite-energy signal of known form buried in noise arises in many areas of signal processing, including for example, radar, sonar, communications, and telemetry. A common approach to designing signal detectors is based on maximization of the SNR. For this purpose, an appropriate measure of the SNR is the ratio of the square of the detector output Y at a given time instant, say t_o, when the signal alone is present, to the time-average power of the detector output when the noise alone is present. It is shown in Chapter 11 that the detection filter that maximizes this ratio has transfer function given by

$$H(f) = \frac{S^*(f)e^{-i2\pi f t_o}}{\hat{S}_N(f)}, \qquad (3.33)$$

where $S(f)$ is the Fourier transform of the signal to be detected, and $\hat{S}_N(f)$ is the PSD of the noise. Since this filter is matched to the signal (and matched inversely to the noise), it is called a *matched filter*. The value of the maximized SNR for this optimum detection statistic is given by

$$\text{SNR}_{\text{max}} = \int_{-\infty}^{\infty} \frac{|S(f)|^2}{\hat{S}_N(f)}\, df. \qquad (3.34)$$

This result applies when the signal is of known form. When the signal to be detected is random, this same general approach is applicable, but is more conveniently formulated probabilistically. Also, for random-signal detection, considerably better performance can be obtained using a quadratic detection device, rather than a linear device (filter). It is shown in Chapter

11 that an appropriately defined measure of the SNR is maximized by the detection statistic

$$Y = \int_{-\infty}^{\infty} \left[\frac{S_S(f)}{S_N^2(f)} \right] \left[\frac{1}{T} |\tilde{X}_T(f)|^2 \right] df, \qquad (3.35)$$

where $S_S(f)$ and $S_N(f)$ are the probabilistic PSDs of the signal and noise, and the other factor in the integrand is the measured periodogram of the noisy waveform. The value of the maximized SNR is given, to a close approximation (for long measurement time T), by

$$\text{SNR}_{\max} \simeq \left[\frac{T}{2} \int_{-\infty}^{\infty} \left(\frac{S_S(f)}{S_N(f)} \right)^2 df \right]^{1/2}. \qquad (3.36)$$

3.3.4 Signal Extraction

A problem of perhaps even greater importance in signal processing than the signal detection problem is the problem of extracting a random signal buried in noise. Let $X(t)$ be the noise-corrupted signal,

$$X(t) = S(t) + N(t), \qquad (3.37)$$

and let $\hat{S}(t)$ be a filtered version of $X(t)$,

$$\hat{S}(t) = X(t) \otimes h(t). \qquad (3.38)$$

Let us adopt the time-averaged squared error (average power) as a measure of the error in approximation between the extracted signal and the desired signal:

$$\text{error power} = \left\langle [S(t) - \hat{S}(t)]^2 \right\rangle = \text{MSE}. \qquad (3.39)$$

We would like to determine the filter transfer function $H(f)$ that minimizes this error power. We shall see in Chapter 11 that the error power is minimized by the filter

$$H(f) = \frac{\hat{S}_S(f)}{\hat{S}_S(f) + \hat{S}_N(f)}, \qquad (3.40)$$

and the minimized error power is given by

$$\text{MSE}_{\min} = \int_{-\infty}^{\infty} \frac{\hat{S}_S(f) \hat{S}_N(f)}{\hat{S}_S(f) + \hat{S}_N(f)} df. \qquad (3.41)$$

It follows from (3.40) that the filter has highest attenuation at those frequencies for which the noise power dominates the signal power, and it has lowest attenuation at those frequencies for which the opposite is true.

3.3.5 Signal Prediction

Predicting a future value of a discrete-time random process, given its past values, has a diversity of applications including forecasting in economics and meteorology, speech analysis and synthesis, and high-resolution spectral analysis and model fitting in radio astronomy, oceanography, geophysics, and many other fields. Let $\{X([k - i]T) : i = 0, 1, 2, \ldots, n - 1\}$ be the present and past $n - 1$ time samples of a random waveform. Consider the use of the linear combination of these observations

$$\hat{X}([k + p]T) = \sum_{i=0}^{n-1} h_i X([k - i]T) \qquad (3.42)$$

to predict the value $X([k + p]T)$ p units of time into the future. Let us adopt time-averaged squared error as a measure of the accuracy of the predictions for all times $\{kT : k = 0, \pm 1, \pm 2, \ldots\}$:

$$\text{error power} = \langle \{X([k + p]T) - \hat{X}([k + p]T)\}^2 \rangle = \text{MSE}, \quad (3.43)$$

where $\langle \cdot \rangle$ denotes discrete-time average as in (3.1). This is the discrete-time counterpart of the MSE measure of accuracy used in the previous example. We would like to determine the n prediction coefficients $\{h_i : i = 0, 1, 2, \ldots, n - 1\}$ that minimize this MSE. We shall see in Chapters 11 and 13 that the error power is minimized by the n prediction coefficients that satisfy the set of n simultaneous linear equations

$$\sum_{i=0}^{n-1} \hat{R}_X([j - i]T) h_i = \hat{R}_X([j + p]T), \quad j = 0, 1, 2, \ldots, n - 1, \qquad (3.44)$$

and that the minimized error power is given by

$$\text{MSE}_{\min} = \hat{R}_X(0) - \sum_{i=0}^{n-1} h_i \hat{R}_X([i + p]T), \qquad (3.45)$$

in which the coefficients $\{h_i\}$ are the solutions to (3.44).

In summary, it is clear from the preceding problem solutions that the PSD and autocorrelation play a fundamental role in theories of the analysis and design of signals and systems.

3.4 Types of Random Processes

A *random process*, denoted by $X(\cdot, \cdot)$, is a random function of time, and is therefore a function of two variables: a time variable t and a sample-point variable s. That is, a random process is an underlying experiment with an

74 • Random Processes

associated sample space in which each sample point s identifies a function of time $X(\cdot, s)$, referred to as a *sample function* or a *sample path*. Also, each time point t identifies a function of sample points, $X(t, \cdot)$, that is, a random variable. The alternative term *stochastic process* is sometimes used instead of *random process*. The complete collection of sample functions of a random process is called the *ensemble*.

Example 1: The random thermal-noise voltage $V(t) \equiv V(t, \cdot)$ (t fixed) is a time sample of the random process $X(\cdot, \cdot) = V(\cdot, \cdot)$. A statistical sample of this time sample is the number $v(t) = V(t, s)$. The random process $V(\cdot, \cdot)$ is, in essence, the collection (family) of all the random voltages $\{V(t): -\infty < t < +\infty\}$. Several typical statistical samples of this process are shown in Figure 3.3. ∎

As exemplified by the model of thermal noise, a random process has two equally valid interpretations. It is both a random function of time, and a

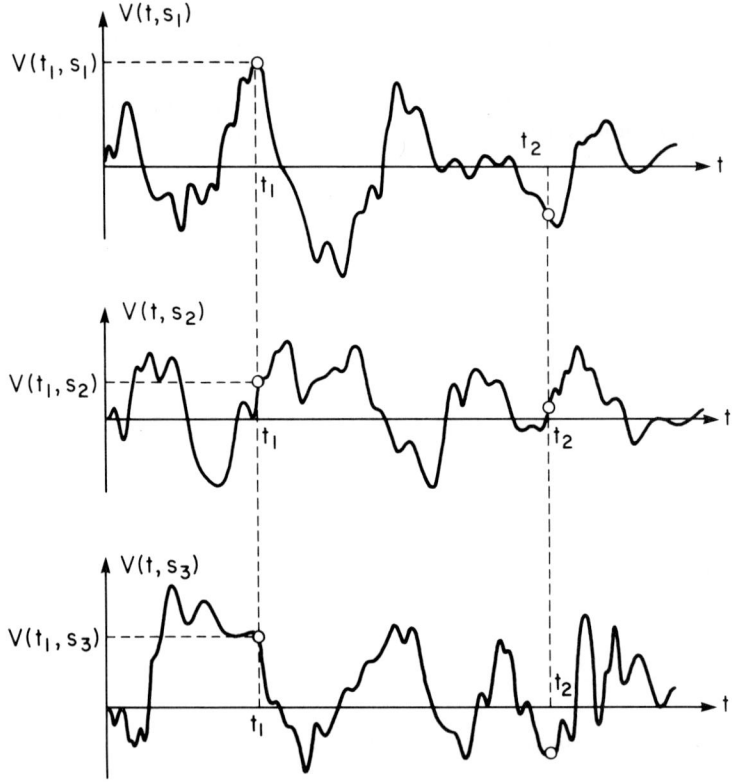

Figure 3.3 Statistical Samples of a Continuous-Time Random Process

time-indexed family of random variables, $\{X(t): -\infty < t < \infty\}$. The former interpretation is most useful in relating an evolutionary physical phenomenon to its probabilistic model as a random process. The latter interpretation is most useful for development of mathematical methods and tools of analysis for random processes.

3.4.1 Continuous- and Discrete-Time Processes

The preceding discussion is couched in terms of the concept of a *continuous-time random process*. Equally important is the concept of a *discrete-time random process*, which is a random sequence, and is denoted by $\{X_n(\cdot): n = 0, \pm1, \pm2, \ldots\}$.

Example 2:   The infinite sequence of time samples of a thermal-noise voltage, $X_n = V(n\tau)$, is a discrete-time random process. Several typical statistical samples of this process are shown in Figure 3.4. ∎

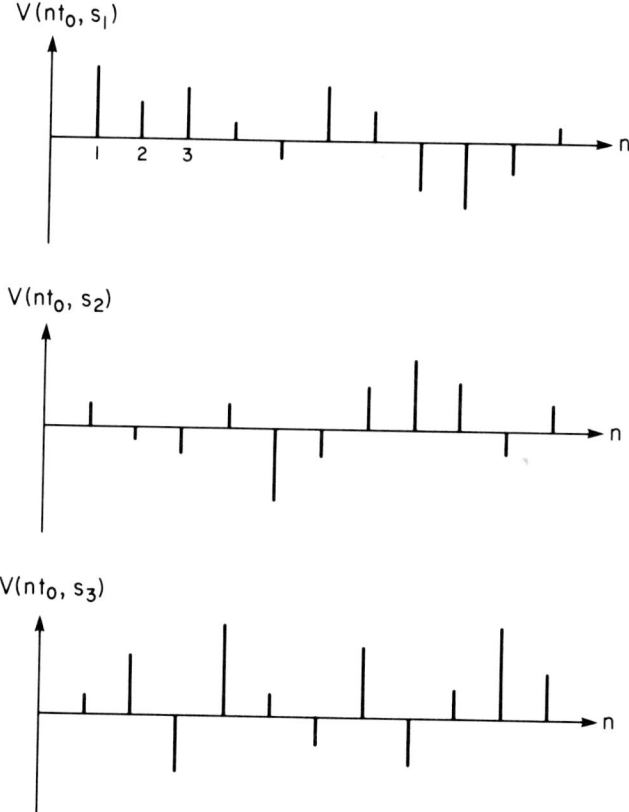

Figure 3.4   Statistical Samples of a Discrete-Time Random Process

There are several alternative notations for random processes. The alternatives to $X(\cdot, \cdot)$ are the abbreviations $X(\cdot)$, $X(t)$ (t not fixed), and X. The alternatives to $\{X_n(\cdot)\}$ are $\{X_n\}$, X_n (n not fixed), and X. These abbreviations can be used when the context prevents ambiguity.

3.4.2 Continuous- and Discrete-Amplitude Processes

In the preceding examples of random processes, the random variables of which the random process is composed are continuous random variables (cf. Section 1.8). Also of importance is the concept of a random process composed of discrete random variables.

Example 3: If a thermal-noise voltage is clipped to obtain a two-valued (± 1) waveform, denoted by $\tilde{v}(t)$,

$$\tilde{v}(t) \triangleq \begin{cases} +1, & v(t) \geq 0, \\ -1, & v(t) < 0, \end{cases}$$

then $\tilde{v}(t)$ can be modeled as a statistical sample of a discrete-amplitude continuous-time random process $\tilde{V}$. Several typical statistical samples of this process are shown in Figure 3.5. ∎

Needless to say, the concept of a *discrete-amplitude, discrete-time random process* is also of importance. An obvious example is a periodically time-sampled, clipped thermal-noise voltage, which is a random sequence of binary (two-valued) variables. Such binary random sequences arise in a number of practical situations of interest in signal processing, particularly in digital systems.

Example 4: Consider a random signal waveform, such as speech or music, which has been periodically time-sampled, amplitude-quantized, and coded into (± 1) binary words. The resultant analog-to-digital (A/D) converted signal is a binary sequence which is appropriately modeled as a statistical sample of a discrete-amplitude, discrete-time random process. Several typical statistical samples of this process are shown in Figure 3.6. ∎

The clipped and time-sampled noise process, the A/D-converted speech process, and the A/D-converted music process are all binary discrete-time random processes, and their sample paths appear to be quite similar (at least locally). Nevertheless, the statistical properties of each are unique; for example, the statistical means, variances, covariances, and so on, of the binary random variables are in general different for each process. One reason for these differences is that the frequency composition of the original

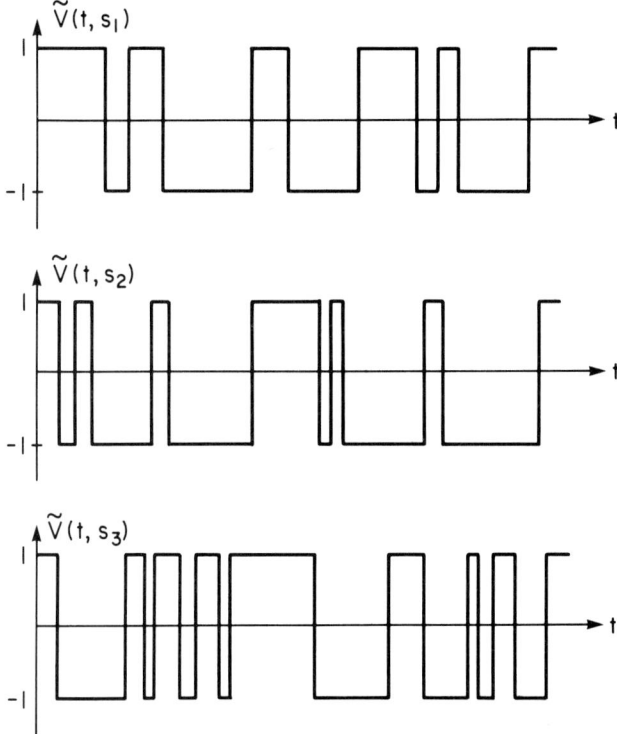

Figure 3.5 Statistical Samples of a Discrete-Valued, Continuous-Time Random Process

waveform process is quite different for each. Another reason is that the method of conversion from analog to digital is quite different for the noise process and for the speech and music processes. The next chapter focuses on the characterization of random processes in terms of probabilistic parameters such as the mean, variance, and covariance.

3.5 Summary

The role of random processes in the design and analysis of signals and systems is discussed at some length in the introductory section, and the concept of *duality* between the *probabilistic theory* of random processes, based on expectation or ensemble averages, and the *deterministic theory* of random processes, based on time averages, is introduced. The latter theory is then briefly surveyed to give some insight into the roles of the *autocorrelation function* and its Fourier transform, the *power spectral density function*. The *fraction-of-time distribution* is introduced as a means for obtaining probabilistic interpretations of the deterministic theory. Several fundamen-

78 • **Random Processes**

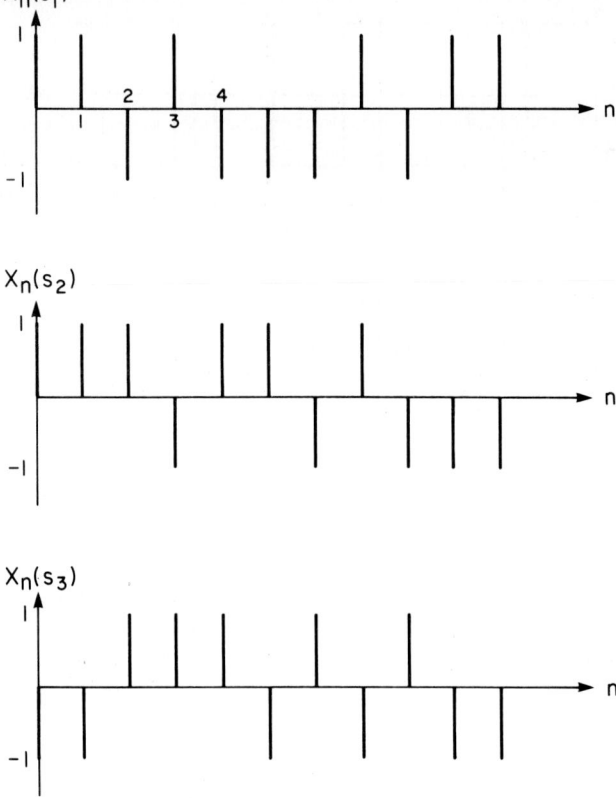

Figure 3.6 Statistical Samples of a Discrete-Valued, Discrete-Time Random Process

tal signal-processing applications are described to illustrate some ways in which the theory of autocorrelation and power spectral density can provide solutions to practical problems. These include the problems of interpolation of time-sampled waveforms, quantification of the noise-immunity property of frequency-modulated sine waves, detection of signals in noise, extraction of signals from noise, and signal prediction.

It is explained that in the probabilistic theory to be pursued in subsequent chapters, a random process is a function of two variables, a time variable and a sample space variable, and this gives rise to two interpretations of a random process. One interpretation is a time-indexed family of random variables, and the other is a sample-point-indexed family of waveforms. It is also explained that random processes of interest can be continuous or discrete in both time and amplitude.

EXERCISES

1. Consider a symmetrical square wave with period T,
$$X(t) = \begin{cases} -1, & -T/2 \le t < 0, \\ +1, & 0 \le t < T/2, \end{cases}$$
which repeats indefinitely in both directions of time. Determine its empirical autocorrelation function.

2. Consider a sine wave $X(t) = \sin(2\pi f_o t + \theta_o)$. Determine its empirical autocorrelation function $\hat{R}_X$ and power spectral density $\hat{S}_X$.

3. Use the convolution theorem to derive (3.4).

★4. Derive (3.12).

   Hint: Let $u = t - v$ in (3.11). Then replace t with $t + \tau$ and v with v_1 to obtain $Y(t + \tau)$, and also replace v with v_2 to obtain $Y(t)$. Substitute these two expressions into (3.2), with $X(t)$ in (3.2) replaced by $Y(t)$, interchange the order of the limit and the integrals, and evaluate the limit. Show that the remaining double integral is (3.12) and (3.13).

★5. Let $H(v)$ be the transfer function for a very narrowband filter with input $X(t)$ and output $Y(t)$, as depicted in Figure 3.7. Use the inverse of (3.9) and also (3.17)–(3.18) with $r = 1$ to show that $\langle P \rangle$ is the integral over all frequencies of $\hat{S}_Y(f)$, and then use this and (3.14) to derive (3.10).

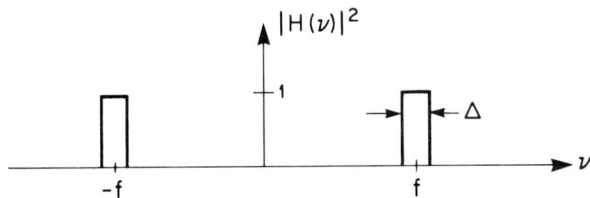

Figure 3.7 Bandpass Filter for Determination of Spectral Density

★6. Verify (3.9) as follows. Use the generalization (3.7b) of $X_T(t)$ to show that the correlogram (3.5) generalizes to the time-variant correlogram
$$R_X(u, \tau)_T = \frac{1}{T} \int_{-T/2}^{T/2 - |\tau|} X(t + u + |\tau|) X(t + u) \, dt. \qquad (3.46)$$
The generalization of (3.4), in which $\tilde{X}_T(f)$ and $R_X(\tau)_T$ are replaced with $\tilde{X}_T(u, f)$ and $R_X(u, \tau)_T$, follows from the method of exercise 3. Time average both sides of this generalized version of (3.4), interchange the order of the time-average and Fourier transformation operations

in the right-hand side, and use the fact that the time average of (3.46) is $(1 - |\tau|/T)\hat{R}_X(\tau)$. Finally, take the limit as $T \to \infty$ of both sides of the time-averaged generalized version of (3.4). This yields (3.8) and (3.9).

7. (a) A waveform $X(t)$ is called *white noise* if its PSD is flat for all frequencies, $\hat{S}_X(f) = N_o$. Determine the PSD for the output $Y(t)$ of a first-order low-pass filter with impulse-response function $h(t) = e^{-bt}$, $t \geq 0$ (where $b > 0$), and white noise input.

   (b) Determine the empirical autocorrelation function for the output $Y(t)$ in part (a). Then determine the average power dissipated in a 1-ohm resistor by this output voltage.

8. A white-noise waveform with empirical autocorrelation $\hat{R}_X(\tau) = \delta(\tau)$ is passed through a sliding-window averaging device with impulse-response function

$$h(t) = \begin{cases} 1, & 0 \leq t \leq T, \\ 0, & \text{otherwise.} \end{cases}$$

Determine the empirical autocorrelation of the waveform at the output of this device.

9. (a) Show that the average value,

$$\langle \hat{R}_X(\tau) \rangle \triangleq \lim_{Z \to \infty} \frac{1}{Z} \int_{-Z/2}^{Z/2} \hat{R}_X(\tau) \, d\tau,$$

of the autocorrelation equals the square of the mean value,

$$\hat{m}_X = \langle X(t) \rangle.$$

That is,

$$\langle R_X(\tau) \rangle = \hat{m}_X^2.$$

   (b) For any Fourier transformable function $g(t)$, with average value $\langle g \rangle \neq 0$, the Fourier transform $G(f)$ contains an additive impulse at $f = 0$ of strength $\langle g \rangle$. Use this fact and the result of (a) to show that the PSD for a process $X(t)$ has an impulse at $f = 0$ with strength equal to the squared mean of the process.

10. How much larger than the bandwidth of a modulating signal must the bandwidth of a frequency-modulated signal be if the demodulator output SNR is to exceed the input SNR by a factor of 81? What happens to the ratio of output SNR to input SNR if the bandwidth of the frequency modulated signal is doubled?

11. Consider the problem of detecting a sine wave burst of duration 1 s commencing at time zero with zero phase, amplitude 1 V, and frequency

100 Hz, in the presence of white noise with PSD $\hat{S}_N(f) = 1$ watt/Hz (on a 1-ohm basis). Determine the impulse-response function of the filter that maximizes SNR at time $t_o = 1$ s. Also determine the value of the maximized SNR. How would this SNR change if the sine wave amplitude were doubled. What if the frequency were doubled?

12. Consider the problem of extracting a random signal with empirical autocorrelation $\hat{R}_S(\tau) = (S_o/2\tau_o)e^{-|\tau|/\tau_o}$ from an additive-white-noise background with PSD $\hat{S}_N(f) = N_o$. Determine the transfer function for the filter that minimizes the time-averaged error power between its output and the uncorrupted signal, and consider what happens when $S_o/N_o \to \infty$. Also determine the value of the minimized error power. Use $S_o = 10 N_o$ in the final result.

13. Consider a random waveform with empirical autocorrelation function given by $\hat{R}_X(\tau) = \sigma^2 e^{-|\tau|/\tau_o}$. Use the general solution (3.44)–(3.45) to solve for the best second-order ($n = 2$) one-step ($p = 1$) linear predictor and its mean-squared prediction error. Let the sampling interval T be given by $T = \tau_o/2$. Would this result change if you considered a third-order ($n = 3$) predictor?

4

Mean and Autocorrelation

IN THIS CHAPTER several elementary and basic examples of random-process models are introduced, and their mean functions and autocorrelation functions are calculated.

4.1 Definitions

4.1.1 Mean

The *mean of the random process* X, at time t, is simply the mean of the random variable $X(t)$, and is denoted by

$$E\{X(t)\} \triangleq m_X(t). \qquad (4.1)$$

In contrast to the notation (from Chapter 2),

$$E\{X(t)\} = m_{X(t)}, \qquad (4.2)$$

which emphasizes the family-of-random-variables interpretation of a random process, (4.1) emphasizes the random-function interpretation; that is, the nonrandom time function $m_X(\cdot)$ is the mean of the random time function $X(\cdot)$. The function $m_X(\cdot)$ can be thought of as the mean waveform.

4.1.2 Autocorrelation

The probabilistic *autocorrelation of the random process* X, at the two times t_1 and t_2, is simply the correlation of the two random variables $X(t_1)$ and $X(t_2)$, and is denoted by

$$E\{X(t_1)X(t_2)\} \triangleq R_X(t_1, t_2). \tag{4.3}$$

The above is an alternative to the notation (from Chapter 2)

$$E\{X(t_1)X(t_2)\} = R_{X(t_1)X(t_2)}. \tag{4.4}$$

4.1.3 Autocovariance

Analogous to the autocorrelation, the *autocovariance of the random process* X, at times t_1 and t_2, is given by

$$E\{[X(t_1) - m_X(t_1)][X(t_2) - m_X(t_2)]\} \triangleq K_X(t_1, t_2). \tag{4.5}$$

It follows from (4.1) to (4.5) that

$$K_X(t_1, t_2) \equiv R_X(t_1, t_2) - m_X(t_1)m_X(t_2). \tag{4.6}$$

4.1.4 Cross-Correlation and Cross-Covariance

Similar to the definitions of autocorrelation and autocovariance, the *cross-correlation* and *cross-covariance* for two random processes X and Y are defined by

$$R_{XY}(t_1, t_2) \triangleq E\{X(t_1)Y(t_2)\}, \tag{4.7}$$

$$K_{XY}(t_1, t_2) \triangleq E\{[X(t_1) - m_X(t_1)][Y(t_2) - m_Y(t_2)]\} \tag{4.8}$$

$$\equiv R_{XY}(t_1, t_2) - m_X(t_1)m_Y(t_2). \tag{4.9}$$

Analogous notation is used for discrete-time processes, for example,

$$m_X(n) \triangleq E\{X_n\}, \tag{4.10}$$

$$R_X(n_1, n_2) \triangleq E\{X_{n_1} X_{n_2}\}. \tag{4.11}$$

4.2 Examples of Random Processes and Autocorrelations

4.2.1 Bernoulli Process

Consider an infinite sequence of independent Bernoulli trials of a binary experiment, such as flipping a coin. The resultant sequence of event indicators,

$$X_n \triangleq \begin{cases} 1 & \text{for } success \text{ in } n\text{th trial}, \\ 0 & \text{for } failure \text{ in } n\text{th trial}, \end{cases}$$

can be interpreted as one statistical sample, denoted by $\{x_n\} = \{X_n(s)\}$, of the single composite experiment of performing an infinite sequence of Bernoulli trials. This is an example of a discrete-amplitude, discrete-time random process. The probability of success is denoted by

$$P\{X_n = 1\} = p.$$

It is easily shown (exercise 1) that

$$m_X(n) = p, \qquad (4.12)$$

$$K_X(n_1, n_2) = \begin{cases} p(1-p), & n_1 - n_2 = 0, \\ 0, & n_1 - n_2 \neq 0. \end{cases} \qquad (4.13)$$

It is noted that $m_X(n)$ is independent of the time n, and $K_X(n_1, n_2)$ depends on only the time difference $n_1 - n_2$.

4.2.2 Binomial Counting Process

Consider counting the number of successes in the Bernoulli process, with the count at time n denoted by Y_n:

$$Y_n \triangleq \sum_{i=1}^{n} X_i. \qquad (4.14)$$

The infinite sequence $\{Y_n\}$ of binomial random variables is an example of a discrete-amplitude, discrete-time random process. Use of (4.13) and (4.14) and the linearity property of expectation (Section 2.2) yields (exercise 1)

$$m_Y(n) = np \qquad (4.15)$$

$$K_Y(n_1, n_2) = p(1-p)\min\{n_1, n_2\}, \qquad (4.16)$$

where

$$\min\{n_1, n_2\} \triangleq \begin{cases} n_1, & n_1 - n_2 \leq 0, \\ n_2, & n_1 - n_2 \geq 0. \end{cases} \qquad (4.17)$$

Unlike the Bernoulli process, the mean of this binomial counting process depends on the time n, and the autocovariance depends on more than just the time difference $n_1 - n_2$.

In addition to the mean and covariance, the probability distribution for the binomial process can be determined. Specifically, the probability of a *particular* sequence of n Bernoulli trials yielding k successes is

$$p^k(1-p)^{n-k}.$$

But this is the probability of only one of many sequences that will yield k

successes in n trials. The total number of such sequences is given by the *binomial coefficient*

$$\frac{n!}{k!(n-k)!}.$$

Consequently, the probability of the event $Y_n = k$ is

$$P_{Y_n}(k) = \frac{n!}{k!(n-k)!} p^k (1-p)^{n-k}. \tag{4.18}$$

This result is used in the derivation of the Wiener and Poisson random processes in Chapter 6.

4.2.3 Random-Walk Process

Consider the modified (± 1) Bernoulli process for which failure is represented by -1 instead of 0:

$$Z_n \triangleq \begin{cases} +1 & \text{for success in } n\text{th trial,} \\ -1 & \text{for failure in } n\text{th trial.} \end{cases}$$

Consider now the sum of these binary variables [analogous to (4.14)], which is denoted by W_n:

$$W_n \triangleq \sum_{i=1}^{n} Z_i. \tag{4.19}$$

A typical statistical sample of this process is shown in Figure 4.1. The process W is referred to as the *one-dimensional random walk*. It is easily shown that the underlying process Z is related to the Bernoulli process X (Section 4.2.1) by

$$Z_i = 2(X_i - \tfrac{1}{2}), \tag{4.20}$$

from which it follows that the random-walk process W is related to the binomial counting process Y by

$$W_n = 2Y_n - n. \tag{4.21}$$

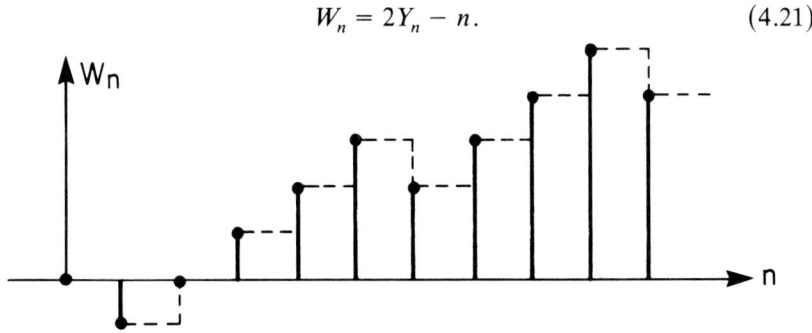

Figure 4.1 Statistical Sample of the Discrete-Time Random-Walk Process

Use of (4.15), (4.16), (4.21), and the linearity property of expectation yields (exercise 1)

$$m_W(n) = n(2p - 1), \quad (4.22)$$

$$K_W(n_1, n_2) = 4p(1 - p)\min\{n_1, n_2\}. \quad (4.23)$$

It is noted that the underlying Bernoulli process Z can be recovered from the random walk W by differencing:

$$Z_n = W_n - W_{n-1}. \quad (4.24)$$

Since this sequence of increments (differences) is a sequence of independent random variables, the random walk is called an *independent-increment process*. The random walk plays a crucial role in the development of continuous-time processes for modeling noise, as explained in Chapter 6.

4.2.4 Random-Amplitude Sine Wave

As a particularly simple example of a continuous-time process, consider a sine wave with a random amplitude A,

$$X(t) = A\sin(\omega_o t + \theta),$$

for which ω_o and θ are nonrandom. It is easily shown (exercise 8) that

$$m_X(t) = m_A \sin(\omega_o t + \theta), \quad (4.25)$$

$$R_X(t_1, t_2) = E\{A^2\} \sin(\omega_o t_1 + \theta)\sin(\omega_o t_2 + \theta)$$
$$= E\{A^2\}[\tfrac{1}{2}\cos[\omega_o(t_1 - t_2)] - \tfrac{1}{2}\cos[\omega_o(t_1 + t_2) + 2\theta]]. \quad (4.26)$$

4.2.5 Random-Amplitude-and-Phase Sine Wave

As a modification to the previous example, consider a sine wave $Y(t)$ with random phase as well as random amplitude. It is assumed that the random phase Θ is independent of the random amplitude A and is uniformly distributed on the interval $[-\pi, \pi)$. Through use of the concept of conditional expectation, and the fundamental theorem of expectation (2.8), the mean and autocorrelation for $Y(t)$ can be obtained* (exercise 8) from the

---

*In (4.27) and (4.28), m_X and R_X are interpreted as random variables because of their dependence on the random phase Θ.

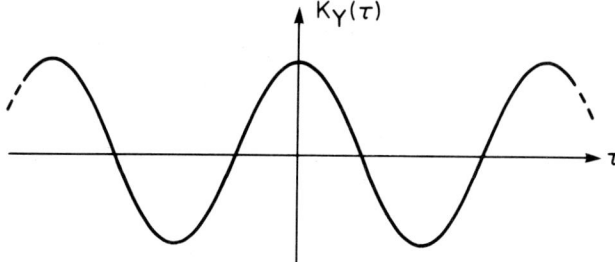

Figure 4.2 Autocovariance of the Random-Amplitude-and-Phase Sine-Wave Process

mean and autocorrelation for $X(t)$ (Section 4.2.4):

$$\begin{aligned}
m_Y(t) &= E\{E\{Y(t)|\Theta\}\} \\
&= E\{m_X(t)\} \\
&= m_A \int_{-\infty}^{\infty} \sin(\omega_o t + \theta) f_\Theta(\theta)\, d\theta \\
&= m_A \int_{-\pi}^{\pi} \sin(\omega_o t + \theta) \frac{1}{2\pi}\, d\theta \\
&= 0, \quad (4.27)
\end{aligned}$$

$$\begin{aligned}
R_Y(t_1, t_2) &= E\{E\{Y(t_1)Y(t_2)|\Theta\}\} \\
&= E\{R_X(t_1, t_2)\} \\
&= \tfrac{1}{2} E\{A^2\} \cos[\omega_o(t_1 - t_2)] \quad (4.28) \\
&= K_Y(t_1, t_2).
\end{aligned}$$

It is noted that $m_Y(t)$ is independent of t, and $K_Y(t_1, t_2)$ depends on only the time difference $t_1 - t_2$. A graph of the autocovariance, as a function of the time difference (denoted by $\tau = t_1 - t_2$), is shown in Figure 4.2.

4.2.6 Sampled-and-Held Noise Process

Another simple example of a continuous-time random process is a sampled-and-held zero-mean noise voltage, denoted by $X(t)$. A typical statistical sample of this process is shown in Figure 4.3. This process can be expressed analytically by

$$X(t) = \sum_{n=-\infty}^{\infty} V(nT) h(t - nT), \quad (4.29)$$

where T is the sampling period, and $h(\cdot)$ is the holding pulse,

$$h(t) \triangleq \begin{cases} 1, & 0 \le t < T, \\ 0, & \text{otherwise.} \end{cases} \quad (4.30)$$

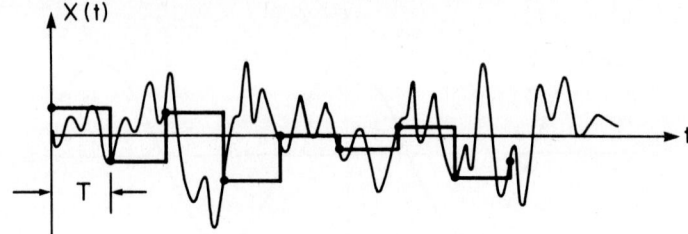

Figure 4.3 Statistical Sample of the Sampled-and-Held Noise Process

If T is sufficiently large to render adjacent time samples, $V(nT)$ and $V([n + 1]T)$, statistically independent, and if these noise samples have equal variance, then it can be shown (exercise 10) that

$$K_X(t_1, t_2) = \sum_{n=-\infty}^{\infty} \sigma_V^2 h(t_1 - nT) h(t_2 - nT). \qquad (4.31)$$

4.2.7 Phase-Randomized Sampled-and-Held Process

As a modification to the previous example, consider a sampled-and-held noise voltage for which the phase Θ of the periodic sampling clock is random and uniformly distributed over one clock period:

$$Y(t) = \sum_{n=-\infty}^{\infty} V(nT + \Theta) h(t - nT - \Theta), \qquad (4.32)$$

$$f_\Theta(\theta) = \begin{cases} \dfrac{1}{T}, & -\dfrac{T}{2} \le \theta < \dfrac{T}{2}, \\ 0, & \text{otherwise.} \end{cases} \qquad (4.33)$$

Use of the conditional-expectation approach illustrated in Section 4.2.5 yields (exercise 10), for $t_1 = t + \tau$ and $t_2 = t$,

$$K_Y(t + \tau, t) = \frac{1}{T} \sum_{n=-\infty}^{\infty} \sigma_V^2 \int_{-T/2}^{T/2} h(t + \tau - nT - \theta) h(t - nT - \theta) \, d\theta. \qquad (4.34)$$

With the change of variables $\phi = t - nT - \theta$, (4.34) reduces to

$$K_Y(t + \tau, t) = \frac{\sigma_V^2}{T} \int_{-\infty}^{\infty} h(\phi + \tau) h(\phi) \, d\phi.$$

Thus, $K_Y(t_1, t_2)$ depends on only the time difference $t_1 - t_2 = t + \tau - t = \tau$:

$$K_Y(t_1, t_2) = \frac{\sigma_V^2}{T} r_h(t_1 - t_2), \qquad (4.35)$$

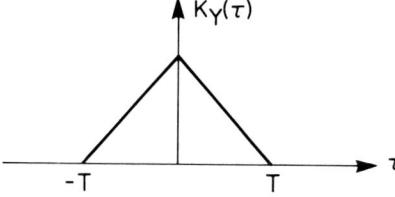

Figure 4.4 Autocovariance of the Phase-Randomized Sample-and-Held Noise Process

where

$$r_h(\tau) \triangleq \int_{-\infty}^{\infty} h(\phi + \tau)h(\phi)\, d\phi. \tag{4.36}$$

The function $r_h(\cdot)$ is called the *finite autocorrelation* of the pulse $h(\cdot)$. A graph of the autocovariance K_Y as a function of the time separation $\tau = t_1 - t_2$ is shown in Figure 4.4.

4.2.8 Amplitude-Modulated Sine Wave

Consider a sine wave with random phase and randomly modulated amplitude

$$Z(t) = Y(t)\sin(\omega_o t + \Phi). \tag{4.37}$$

It is assumed that $Y(t)$ and Φ are independent. It can be shown by the method of Section 4.2.5 (exercise 8) that the mean and autocorrelation are given by

$$m_Z(t) = m_Y(t) E\{\sin(\omega_o t + \Phi)\}, \tag{4.38}$$

$$R_Z(t_1, t_2) = R_Y(t_1, t_2)\left[\tfrac{1}{2}\cos[\omega_o(t_1 - t_2)]\right.$$
$$\left. - \tfrac{1}{2}E\{\cos[\omega_o(t_1 + t_2) + 2\Phi]\}\right], \tag{4.39}$$

and if Φ is uniformly distributed on the interval $[-\pi, \pi)$, then (4.38) and (4.39) yield

$$m_Z(t) = 0, \tag{4.40}$$

$$R_Z(t_1, t_2) = \tfrac{1}{2} R_Y(t_1, t_2)\cos[\omega_o(t_1 - t_2)]. \tag{4.41}$$

For example, if the amplitude $Y(t)$ is the phase-randomized sampled-and-held noise process discussed in Section 4.2.7, then $R_Y(t_1, t_2)$ is given by (4.35) and (4.36).

4.3 Summary

The *second-order theory* of random processes focused on in this book (with the exception of Chapter 5) deals with the *mean function* and the *autocovariance* and *autocorrelation functions*. Explicit formulas for these functions are

90 • **Random Processes**

calculated for various specific random process models in this chapter. The first three examples, the *Bernoulli*, *binomial*, and *random-walk processes*, are quite elementary, but form the bases for more sophisticated models, such as the *Wiener* and *Poisson processes*, which are studied in Chapter 6. The *sine-wave* and *sampled-and-held processes* are also quite elementary, but are introductory to more sophisticated models such as the *modulated-sine-wave* processes and *modulated-pulse-train processes* discussed in the exercises and Chapters 11 and 12.

In addition to introducing several basic models of random processes in this chapter, some important techniques for calculating mean and autocovariance functions are illustrated. These include exploitation of the *linearity property of expectation* and the *fundamental theorem of expectation*, and judicious use of *conditional expectation* to decompose a calculation into more elementary parts.

EXERCISES

★**1.** Verify Equations (4.12), (4.13), (4.15), (4.16), (4.22), and (4.23).

2. Consider the generalized random-walk process in which the step sizes Z_i in (4.19) are continuous random variables (independent and identically distributed). Determine how the formulas (4.22) and (4.23) for the mean and autocovariance must be generalized for this process. Express the final result in terms of only m_Z and σ_Z^2 (and the time parameters).

3. Let W_n be the random-walk process defined in Section 4.2.3, and let $U_n \triangleq W_n - W_{n-m}$ for some positive integer m. Determine explicit formulas for the mean and autocovariance of U_n.

Hint: Express U_n as

$$U_n = \sum_{i=-\infty}^{\infty} h(n-i) Z_i,$$

where

$$h(j) = \begin{cases} 1, & 0 \leq j \leq m, \\ 0, & \text{otherwise.} \end{cases}$$

4. Let Y_n be the *moving average process* $Y_n = X_n + X_{n-1} + X_{n-2} + \cdots + X_{n-N+1}$, where $\{X_n\}$ are independent identically distributed random variables. Determine the mean and autocorrelation for Y_n.

5. The autocorrelations $R_X(k)$ and $R_Y(k)$ for two discrete-time processes X_n and Y_n are both of the form $r^{|k|}$, $k = 0, \pm 1, \pm 2, \ldots$, except that $r > 0$ for X_n and $r < 0$ for Y_n. Explain how the sample sequences from these two processes differ from one another.

Hint: Consider $Y_n = X_n \cos(\pi n + \Theta)$, where Θ is independent of $\{X_n\}$ and uniformly distributed over $[0, 2\pi)$.

6. Consider the sine wave process $X(t) = A \cos \Omega t + B \sin \Omega t$ in which A, B, and Ω are mutually independent random variables, and A and B have zero-mean values and equal variances. Determine the autocovariance function for $X(t)$ solely in terms of the characteristic function for Ω and t_1 and t_2.

7. Consider the random periodic process

$$X(t) = \sum_{p=-P}^{P} C_p e^{ip(\omega_0 t + \Theta)},$$

where $\{C_p\}$ have arbitrary means and variances, except that $C_{-p} = C_p^*$, and Θ is independent of $\{C_p\}$ and uniformly distributed over $[0, 2\pi)$. Show that the mean of $X(t)$ is independent of t and the autocovariance depends on only the time differences $t_1 - t_2$.

★8. Verify Equations (4.38) to (4.41). Compare this result with (4.25) and (4.26).

Hint: To obtain (4.39), use the identity $\sin a \sin b = \frac{1}{2} \cos(a - b) - \frac{1}{2} \cos(a + b)$.

9. As a generalization of the result in exercise 8, verify that the autocorrelation function for the product of two independent processes, $Y(t) = X(t)Z(t)$, is the product of their autocorrelations, $R_Y = R_X R_Z$.

★10. Verify Equations (4.31) and (4.35). Then draw graphs [as surfaces above the (t_1, t_2) plane] of the functions $K_X(t_1, t_2)$ in (4.31), and $K_Y(t_1 - t_2)$ in (4.35). Describe the paths, in the (t_1, t_2) plane, along which K_X is periodic and K_Y is constant.

Hint: To obtain (4.35), use the identity

$$\sum_{n=-\infty}^{\infty} \int_{t-nT-T/2}^{t-nT+T/2} f(\phi) \, d\phi = \int_{-\infty}^{\infty} f(\phi) \, d\phi,$$

which holds for any integrable function $f(\cdot)$.

11. (a) Consider the idealized pulse-code-modulation process defined by

$$X(t) = \sum_{n=-\infty}^{\infty} A_n p(t - nT),$$

where $\{A_n\}$ are independent identically distributed zero-mean binary random variables with values ± 1, and $p(t)$ is equal to unity for $0 \leq t < T/2$ and equal to zero otherwise. Determine the autocovariance for $X(t)$ and sketch it as the height of a surface above the plane with coordinates t_1, t_2.

(b) Consider the randomly delayed version $Y(t) = X(t - \Theta)$ of the process from part (a), where Θ is uniformly distributed over $[0, T)$ and is statistically independent of $\{A_n\}$. Determine the autocovariance for $Y(t)$ and sketch it as the height of a surface above the plane with coordinates t_1, t_2.

12. Consider the quadrature-amplitude-modulated signal

$$Y(t) = X(t) \cos(\omega_o t) - Z(t) \sin(\omega_o t), \qquad (4.42)$$

where $X(t)$ and $Z(t)$ are zero-mean independent processes with identical autocorrelation functions, $R_X = R_Z$. Determine $R_Y(t_1, t_2)$, and show that if $R_X(t_1, t_2) = R_X(t_1 - t_2)$, then $R_Y(t_1, t_2) = R_Y(t_1 - t_2)$.

Hint: Use the hint in exercise 8.

13. Let $X(t)$ and $Z(t)$ be independent processes. Determine the autocorrelation functions (in terms of m_X, m_Z, R_X, R_Z) of

$$Y_1(t) = X(t) + Z(t) \quad \text{and} \quad Y_2(t) = X(t) - Z(t).$$

Then determine the cross-correlation function of $Y_1(t)$ and $Y_2(t)$. Finally, consider the special case for which $R_X = R_Z$ and $m_X = m_Z$, and simplify the preceding results.

14. Determine the autocorrelation function for the process

$$Y(t) = X(t) - X(t - T),$$

for the case in which

$$R_X(t_1, t_2) = R_X(t_1 - t_2).$$

⋆15. (a) Determine an infinite-series formula [analogous to (4.29)] for the pulse-position-modulated signal described graphically in Figure 4.5, where the random positions $\{P_n\}$ are each independent and identically distributed, and the zero-position pulse is denoted by $p(t)$.

(b) Determine the mean function for this signal, and verify that it is periodic. Then let P_n be uniformly distributed on $[0, T - \Delta)$ and draw a graph of the mean function.

(c) Modify the model of the signal by introducing a random phase variable, uniformly distributed over $[-T/2, T/2)$, and show that the autocorrelation function is given by

$$R_X(\tau) = \frac{1}{T}\left[r_p(\tau) + \sum_{n \neq 0} r_p(\tau - nT) \otimes r_f(\tau - nT)\right], \qquad (4.43)$$

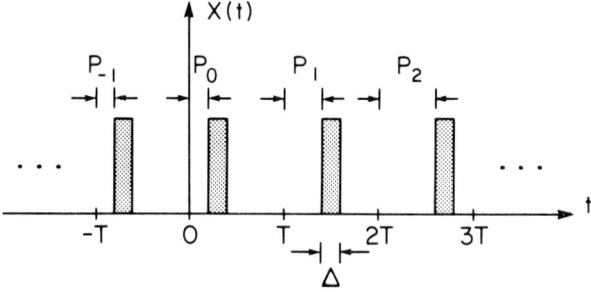

Figure 4.5 Statistical Sample of a Pulse-Position-Modulated Signal

where r_p and r_q are the finite autocorrelations for the pulse p and the expected pulse

Hint: First verify that

$$R_X(t + \tau, t) = E\left\{\sum_{n=-\infty}^{\infty} E\{p(t - nT + \tau - P_n - \Theta)\right.$$

$$\times p(t - nT - P_n - \Theta)|\Theta\}\bigg\}$$

$$+ E\left\{\sum_{r \neq 0} \sum_{n=-\infty}^{\infty} E\{p(t - nT + \tau - P_n - \Theta)|\Theta\}\right.$$

$$\times E\{p(t - nT - rT - P_{n+r} - \Theta)|\Theta\}\bigg\}.$$

(4.44)

Then use changes of variables of integration, analogous to that used to obtain (4.35) from (4.34). See the hint in exercise 10.

16. (a) Determine an infinite-series formula [analogous to (4.29)] for the pulse-width-modulated signal described graphically in Figure 4.6, where the random widths $\{W_n\}$ are each independent and identically distributed, and the unity-width pulse is denoted by $p(t)$.
    (b) Determine the mean function and verify that it is periodic.
    (c) Modify the model of the signal by introducing a random phase variable, uniformly distributed over $[-T/2, T/2)$, and show that the autocorrelation function is given by

$$R_X(\tau) = \frac{1}{T} E\left\{Wr_p\left(\frac{\tau}{W}\right)\right\} + \frac{1}{T} \sum_{n \neq 0} r_q(\tau - nT), \quad (4.45)$$

where r_p and r_q are the finite autocorrelations for the pulse p and the expected pulse

$$q(t) \triangleq E\{p(t/W)\}.$$

Hint: Consider an analog of (4.44), and proceed as in exercise 15.

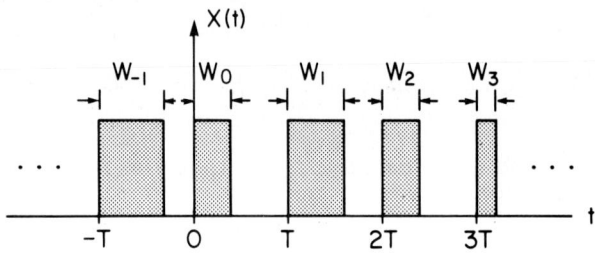

Figure 4.6 Statistical Sample of a Pulse-Width-Modulated Signal

5

Classes of Random Processes

IN THIS CHAPTER the wide variety of classes of random processes, for which considerable theory has been developed, are briefly surveyed. Section 5.2 introduces the class of processes most widely exploited in the analysis and design of signals and systems: the Gaussian processes. Section 5.3 on Markov processes is a brief introduction to an important class of processes, the extensive theory of which is only touched on in this book. Only the examples presented are reconsidered in subsequent chapters, particularly Chapter 6, where the Wiener and Poisson processes are studied, and Chapter 9, where dynamical systems are studied. Section 5.4, on the other hand, introduces the kinds of processes that are studied in some depth in this book: the stationary processes and various classes of related processes, such as cyclostationary and asymptotically stationary processes.

5.1 Specification of Random Processes

Although calculation of the mean and autocorrelation functions in the examples in Chapter 4 is straightforward, it can be considerably more involved for more interesting types of processes. In general, calculation of the mean requires specification of the univariate probability densities $f_{X(t)}(x)$ for all values of t:

$$m_X(t) = \int_{-\infty}^{\infty} x f_{X(t)}(x)\, dx. \qquad (5.1)$$

96 • **Random Processes**

Similarly, calculation of the autocorrelation requires specification of the bivariate probability densities $f_{X(t_1)X(t_2)}(x_1, x_2)$ for all pairs of times t_1 and t_2:

$$R_X(t_1, t_2) = \int_{-\infty}^{\infty} \int_{-\infty}^{\infty} x_1 x_2 f_{X(t_1)X(t_2)}(x_1, x_2) \, dx_1 \, dx_2. \tag{5.2}$$

Explicit determination of these densities was avoided in examples in Chapter 4, by use of the fundamental theorem of expectation [Equation (2.8)], regarding the expectation of a function of a random variable. This theorem applies when the random process is a deterministic function of time and some random parameters, such as the amplitude, position, and width of a pulse, and the amplitude, phase, and frequency of a sine wave. Nevertheless, it can be seen from (5.1) and (5.2) that, in general, determination of the mean and autocorrelation requires specification of the probability densities for the entire process. In fact, a random process is in general not completely specified unless the multivariate probability densities

$$f_{X(t_1)X(t_2)\cdots X(t_n)}(x_1, x_2, \ldots, x_n) \tag{5.3}$$

are specified for all n-tuples of times $\{t_1, t_2, \ldots, t_n\}$, for all positive integers n. However, since second-order properties of random processes are focused on in this book, only the bivariate probability densities are needed typically.

5.2 Gaussian Processes

Although complete specification of a random process can be a formidable task in practice, it is unusually simple for Gaussian random processes. A process X is defined to be a *Gaussian process* if and only if every n-tuple of time samples $X = \{X(t_i) : i = 1, 2, \ldots, n\}$ is a set of jointly Gaussian random variables (cf. Section 2.3.4) for every positive integer n. The joint probability densities are given by (2.37) provided that the covariance matrix K_X, with (i, j)th element defined by $K_X(t_i, t_j)$, is nonsingular. This will be the case provided that every set of distinct time samples $\{X(t_i)\}$ is a linearly independent set, as discussed in Chapter 2. Otherwise, the joint probability densities can be obtained by inverse Fourier transformation of the joint characteristic functions, given by (2.38). Moreover, simple formulas for joint moments of sets of more than two time samples $\{X(t_i)\}$, in terms of sums of products of the means and covariances, can be obtained from the joint characteristic function, as shown in exercise 3.

Example 1: It is easily shown that if the random amplitude in Section 4.2.4 is a Gaussian variable, then the sine-wave process is a Gaussian process. However, its covariance matrix is singular (and has rank 1). ∎

Example 2: If the time samples of the noise voltage in Section 4.2.6 are each Gaussian, and are mutually independent, then they are jointly Gaussian. As a result, the sampled-and-held noise process can be shown to be a Gaussian process. ∎

It is emphasized that although both of these examples are Gaussian processes, typical statistical samples of each appear to have little in common with each other. This is reflected in the fact that the autocorrelations of the two processes (cf. Figures 4.2 and 4.4) have little in common.

An especially important example of a Gaussian process is the *Wiener process*, which is obtained from the random walk by a limiting procedure in which the time increment between steps and the step size approach zero. This process is derived in Chapter 6.

5.3 Markov Processes*

An alternative to specifying all of the joint densities (5.3), which is completely equivalent, is to specify all of the first-order conditional densities

$$f_{X(t_1)}(x_1),$$

$$f_{X(t_2)|X(t_1)}(x_2|x_1),$$

$$f_{X(t_3)|X(t_2)X(t_1)}(x_3|x_2, x_1),$$

$$\vdots$$

$$f_{X(t_n)|X(t_{n-1})\cdots X(t_1)}(x_n|x_{n-1},\ldots, x_1) \qquad (5.4)$$

for every set of ordered time points $t_1 < t_2 < t_3 < \cdots < t_n$, for every positive integer n. Repeated use of the definition of conditional probability density, (1.29), yields

$$f_{X(t_1)\cdots X(t_n)}(x_1,\ldots, x_n) = f_{X(t_n)|X(t_{n-1})\cdots X(t_1)}(x_n|x_{n-1},\ldots, x_1)$$
$$\times f_{X(t_{n-1})|X(t_{n-2})\cdots X(t_1)}(x_{n-1}|x_{n-2},\ldots, x_1) \cdots f_{X(t_2)|X(t_1)}(x_2|x_1) f_{X(t_1)}(x_1),$$

$$(5.5)$$

which reveals that (5.3) can be obtained from (5.4). This alternative approach to specification immediately suggests a class of processes for which

---

*This section is a brief introduction to a highly developed subject; see [Parzen, 1962; Blanc-Lapierre and Fortet, 1965; Feller, 1968,1971; Breiman, 1968; Karlin and Taylor, 1975, 1981; Larson and Shubert, 1979; Chung 1982; Ross, 1983].

the problem of specification is greatly simplified, and this is the class that exhibits the property

$$f_{X(t_n)|X(t_{n-1})\cdots X(t_1)} \equiv f_{X(t_n)|X(t_{n-1})}, \qquad (5.6)$$

where $t_1 < t_2 < \cdots < t_n$. If we interpret t_n as the present time, then this property indicates that the probabilistic behavior at the present, given all of the past behavior, depends on only the most recent past. Or, if we interpret t_{n-1} as the present, then the future probabilistic behavior, given the present and all past behavior, depends on only the present. Equation (5.6) is called the *Markov property*, and a process with this property is called a *Markov process*. Repeated application of (5.6) to (5.5) reveals that

$$f_{X(t_1)\cdots X(t_n)} = f_{X(t_n)|X(t_{n-1})} f_{X(t_{n-1})|X(t_{n-2})} \cdots f_{X(t_2)|X(t_1)} f_{X(t_1)}. \qquad (5.7)$$

It follows that a Markov process $X(t)$, $t \geq t_0$ is completely specified by the initial marginal density $f_{X(t_0)}$ and the set of first-order conditional densities

$$f_{X(t)|X(u)}, \qquad t > u \geq t_0. \qquad (5.8)$$

The conditional density (5.8) is called the *transition density*. The set of transition densities for a Markov process cannot be specified completely arbitrarily because, for consistency, they must satisfy

$$f_{X(t_3)|X(t_1)}(x_3|x_1) = \int_{-\infty}^{\infty} f_{X(t_3)|X(t_2)}(x_3|x_2) f_{X(t_2)|X(t_1)}(x_2|x_1)\, dx_2 \quad (5.9)$$

for all $t_1 < t_2 < t_3$ and all x_1 and x_3 (see Figure 5.1). Equation (5.9) is called the *Chapman–Kolmogorov equation*. This equation can be derived by

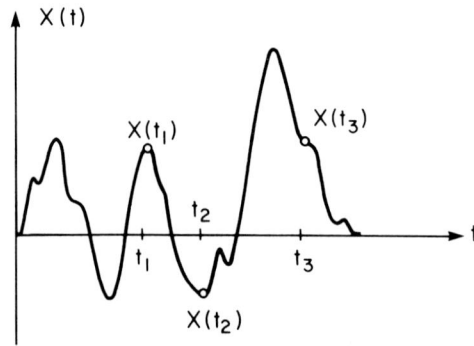

Figure 5.1 Pictorialization of Three Ordered Time Samples for the Chapman-Kolmogorov Equation

applying the definition of conditional probability density and the Markov property to $f_{X(t_3)X(t_2)|X(t_1)}$, and integrating out $X(t_2)$ (exercise 15). By using the definition of conditional probability density

$$f_{X(t)|X(u)} = \frac{f_{X(t)X(u)}}{f_{X(u)}} \qquad (5.10)$$

in (5.7), it can be seen that a Markov process is completely specified by the first- and second-order densities only.

The preceding discussion applies equally well to discrete-time processes. Moreover, if the process has discrete distributions, then the Markov property (5.6) is expressed in terms of probability mass functions (Section 1.8.3) as

$$P_{X(t_n)|X(t_{n-1})\cdots X(t_1)} = P_{X(t_n)|X(t_{n-1})}. \qquad (5.11)$$

A discrete-amplitude Markov process, regardless of whether it is discrete-time or continuous-time, is called a *Markov chain*. The specific values $x_1, x_2, x_3, \ldots$ that can be taken on by the discrete random variables in a Markov chain are called the *states*. A continuous-time Markov chain is necessarily piecewise constant with step discontinuities. Continuous-amplitude Markov processes are sometimes called *continuous-state* processes.

Example 3 (Binary digital system): As an example of a two-state, discrete-time Markov chain, we consider a digital system that can change state, from a 0 to a 1 or a 1 to a 0, once every unit of time. We assume that the transition probabilities do not change with time, so that

$$P_{X_{n+1}|X_n}(1|1) = P_{11},$$

$$P_{X_{n+1}|X_n}(1|0) = P_{10},$$

$$P_{X_{n+1}|X_n}(0|0) = P_{00},$$

$$P_{X_{n+1}|X_n}(0|1) = P_{01}, \qquad (5.12)$$

for all n. This probabilistic model can be described by the schematic shown in Figure 5.2. Given the initial state probability distribution [e.g., $P_{X_0}(1) = p$ and $P_{X_0}(0) = 1 - p$] and the transition probabilities (5.12), the state probability distributions for any other times can be derived. For example, it can be shown (exercise 16) that the marginal distribution at time n is

$$\boldsymbol{P}_n = \boldsymbol{P}^n \boldsymbol{P}_0, \qquad (5.13)$$

where $\boldsymbol{P}_0$ is the vector composed of the two initial state probabilities

$$\boldsymbol{P}_0 = [P_{X_0}(1), P_{X_0}(0)]^T, \qquad (5.14)$$

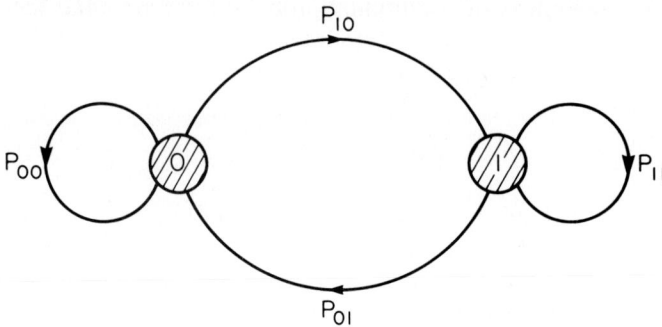

Figure 5.2   Schematic of a Two-State Markov Chain

P^n is the nth power of the matrix composed of the four state transition probabilities,

$$P^n = \begin{bmatrix} P_{11} & P_{10} \\ P_{01} & P_{00} \end{bmatrix}^n, \qquad (5.15)$$

and P_n is composed of the two marginal state probabilities at time n. ∎

Example 4 (Birth-death processes):   As an example of a continuous-time Markov chain, we consider a *birth-death process*, which is a nonnegative-integer-valued Markov chain, with step discontinuities of only ±1. The value of the process represents the size of an evolving population. The process is fully specified by the initial state probability mass function, a sequence of three state-transition probabilities

$$P(X_n = q | X_{n-1} = q),$$
$$P(X_n = q + 1 | X_{n-1} = q),$$
$$P(X_n = q - 1 | X_{n-1} = q),$$
$$n = 1, 2, 3, \ldots, \qquad q = 1, 2, 3, \ldots,$$

and a probabilistic model for the sequence of state-transition times (say) $T_1, T_2, T_3, \ldots$, such as the Poisson process (cf. Section 5.3.5). ∎

Example 5 (Dynamical systems):   A particularly important application area in which Markov processes arise is the area of finite-order dynamical systems. For example, a first-order linear discrete-time system is described by the difference equation

$$X_{n+1} = a_n X_n + Z_n, \qquad (5.16)$$

where $\{Z_n\}$ is the excitation sequence, $\{X_n\}$ is the sequence of system states, and $\{a_n\}$ characterizes internal feedback. It is clear from this equation that the future state X_{n+1} depends on only the current state X_n; given the current state, past states are irrelevant. Thus, if the excitation $\{Z_n\}$ has no memory, that is, if it is a sequence of independent random variables, then the sequence of states is a Markov process. In fact, this example includes the binomial counting process and the random walk, introduced in Chapter 4, as the special case for which $a_n = 1$. Similar remarks apply to continuous-time dynamical systems, such as

$$\frac{dX}{dt} = a(t)X(t) + Z(t). \tag{5.17}$$

This can be seen by reexpressing (5.17) in the differential form

$$\begin{aligned} dX(t) &= a(t)X(t)\,dt + Z(t)\,dt \\ &= a(t)X(t)\,dt + dW(t), \end{aligned} \tag{5.18}$$

where

$$W(t) \triangleq \int_0^t Z(u)\,du. \tag{5.19}$$

It follows from (5.18) that the change in state, $dX(t)$, depends on only the current state $X(t)$. Thus, if the current increment $dW(t)$ due to the excitation $Z(t)$ is independent of its past, then $X(t)$ is a Markov process. Dynamical system models, such as (5.16) and (5.17), are studied in Chapter 9. ∎

A Markov process is said to be *homogeneous* if the transition densities (or probability mass functions) are *translation-invariant*, in the sense that

$$f_{X(t_1+t)|X(t_2+t)}(x_1|x_2) = f_{X(t_1)|X(t_2)}(x_1|x_2) \tag{5.20}$$

for all time translations t for which the process X is defined. It does not necessarily follow that the joint densities are translation-invariant. The binomial counting process and the random-walk process are both homogeneous Markov chains (exercise 18).

5.3.1 Independent Increments

A process is said to have *independent increments* if the incremental random variables

$$\begin{aligned} Y_1 &= X(t_1) - X(t_0), \\ Y_2 &= X(t_2) - X(t_1) \\ &\;\;\vdots \\ Y_n &= X(t_n) - X(t_{n-1}) \end{aligned}$$

are independent for all ordered times $t_0 < t_1 < \cdots < t_n$ for all positive integers n. It can be shown (exercise 17) that an independent-increment process with known starting value $X(0) = x_0$ is completely specified by its first- and second-order distributions alone. Moreover, it can be shown (exercise 17) that every independent-increment process with known starting value $X(0) = x_0$ is a Markov process, although there are many Markov processes that do not have independent increments. As examples, the binomial counting process and the random-walk process both have independent increments (exercise 18).

5.3.2 Gauss-Markov Processes

If a process is both Gaussian and Markov, then its autocovariance function satisfies the equation (exercise 19)

$$K_X(t_3, t_1) = \frac{K_X(t_3, t_2) K_X(t_2, t_1)}{K_X(t_2, t_2)} \tag{5.21}$$

for all $t_1 \le t_2 \le t_3$. In fact, it can be shown that if the autocovariance function of a Gaussian process satisfies (5.21), then it is a Markov process.

Example 6 (Ornstein-Uhlenbeck process*): If $X(t)$, $t \ge 0$, is a zero-mean Gaussian process with autocovariance function

$$K_X(t_1, t_2) = \sigma^2 \exp(-\alpha|t_1 - t_2|), \tag{5.22}$$

then $X(t)$ is a Markov process. Moreover, if the autocovariance function of a Gauss-Markov process depends on only $t_1 - t_2$, then it can be shown that it must be given by (5.22). This process is known as the *Ornstein-Uhlenbeck process* [Uhlenbeck and Ornstein, 1930]. ∎

Example 7 (Wiener process*): If $X(t)$ is a zero-mean Gaussian process for which $K_X(t_1, t_2) > 0$ for all t_1 and t_2 for which $X(t)$ is defined, then it can be shown that (5.21) implies that K_X must have the form

$$K_X(t_1, t_2) = g[\max\{t_1, t_2\}] h[\min\{t_1, t_2\}] \tag{5.23}$$

for some functions $g(\cdot)$ and $h(\cdot)$. Moreover, it can be shown that (5.23) implies that $X(t)$ is of the form

$$X(t) = g(t) W\left[\frac{h(t)}{g(t)}\right], \tag{5.24}$$

---

*The Wiener process (sometimes called the *Wiener-Lévy process* or the *Wiener-Einstein process*) and the Ornstein-Uhlenbeck process are particularly important examples from the class of continuous-state Markov processes called *diffusion processes* (cf. exercise 21), which have important applications in a variety of fields such as physics, population dynamics, genetics, epidemiology, ecology, and neurology [Breiman, 1968; Feller, 1971; Arnold 1974; Fleming and Rishel, 1975; Larson and Shubert, 1979; Schuss, 1980; Karlin and Taylor, 1981; Knight, 1981].

where $W(t)$ is a Gauss-Markov process for which $K_W(t_1, t_2) = \alpha^2 \min\{t_1, t_2\}$. The process $W(t)$ is the Wiener process, which is studied in Chapter 6. The process $X(t)$ is simply an amplitude-scaled and time-warped Wiener process. For example, if $g(t) = e^{-t}$ and $h(t) = e^t$, then $X(t)$ is the Ornstein-Uhlenbeck process (exercise 19), and

$$X(t) = e^{-t} W(e^{2t}). \quad \blacksquare$$

5.3.3 Mth-Order Markov Processes

If a discrete-time process X_n, $n \geq 0$, has the property

$$f_{X_{n+1}|X_n X_{n-1} \cdots X_0} \equiv f_{X_{n+1}|X_n X_{n-1} \cdots X_{n-m+1}} \qquad (5.25)$$

for all n, and for some particular m, then it is said to be an *m*th-*order Markov process*. The *memory* of this process extends m units of time into the past. Such a process can be reinterpreted as a vector-valued first-order Markov process. Specifically, if we define vectors by

$$\mathbf{X}_n \triangleq [X_n, X_{n-1}, \ldots, X_{n-m+1}]^T,$$

then it follows from (5.25) that $\mathbf{X}_n$ satisfies the Markov property

$$f_{\mathbf{X}_n|\mathbf{X}_{n-1} \cdots \mathbf{X}_0} \equiv f_{\mathbf{X}_n|\mathbf{X}_{n-1}}. \qquad (5.26a)$$

Moreover, it can be shown (exercise 17) that (5.26a) guarantees that

$$f_{\mathbf{X}_{n_m}|\mathbf{X}_{n_{m-1}} \cdots \mathbf{X}_{n_1}} \equiv f_{\mathbf{X}_{n_m}|\mathbf{X}_{n_{m-1}}}, \qquad (5.26b)$$

for all $n_1 < n_2 < \cdots < n_m$, which is the discrete-time counterpart of (5.6). In addition, it can be shown that if (5.26a) is translation invariant, then so too is (5.26b), in which case $\mathbf{X}_n$ is a homogeneous Markov process.

A particularly important setting in which mth-order Markov processes arise is in the study of deterministic dynamical systems with memoryless random excitation. When such a system is modeled as an mth-order difference equation with an excitation sequence consisting of independent random variables, then the system response is an mth-order Markov process. This can be seen by analogy with Example 5 since such models can be put into the form of an m-dimensional vector-valued first-order difference equation. Such discrete-time dynamical systems and their continuous-time counterparts are studied in Chapter 9. The continuous-time counterpart is modeled as an mth-order differential equation, which can be put into the form of an m-dimensional vector-valued first-order differential equation in which the vector at time t is given by the process and its first $m - 1$ derivatives at time t, $\mathbf{X}(t) = [X(t), dX(t)/dt, \ldots, d^{m-1}X(t)/dt^{m-1}]^T$. Again, by analogy with Example 5, if the vector-valued process $\mathbf{X}(t)$ is Markov, then the scalar-valued process $X(t)$ is mth-order Markov.

5.3.4 Martingale Processes*

An independent increment process with zero mean has the property

$$E\{X(t_n) - X(t_{n-1}) | X(t_1), X(t_2), \ldots, X(t_{n-1})\} = 0 \quad (5.27)$$

for $t_1 < t_2 < t_3 < \cdots < t_n$ for all integers $n \geq 2$. This property can be reexpressed as (exercise 23)

$$E\{X(t_n) | X(t_1), X(t_2), \ldots, X(t_{n-1})\} = X(t_{n-1}), \quad (5.28)$$

and is called the *martingale property*. A process with this property is called a *martingale process*, or just a *martingale*. Not all martingales have zero mean or independent increments. However, the increments of every martingale are *unpredictable* in the sense that the random variable Y that best predicts the value of the current increment $Z = X(t_n) - X(t_{n-1})$, given the past, $X(t_1), X(t_2), \ldots, X(t_{n-1})$, is $Y = 0$, in the sense that the mean squared prediction error

$$E\{(Z - Y)^2 | X(t_1), X(t_2), \ldots, X(t_{n-1})\} \quad (5.29)$$

is minimized by $Y = 0$. This is proved in Chapter 13, where it is shown that the best Y is given by (5.27). If a martingale is also a Gaussian process, then it can be shown that it has independent increments. If, in addition, the process has zero starting value $X(0) = 0$, then the martingale property reduces to (exercise 23)

$$E\{X(t_n) | X(t_{n-1})\} = X(t_{n-1}). \quad (5.30)$$

Example 8: Let X_n be an arbitrary random sequence, and let Z be an arbitrary finite-variance random variable. Then the properties of conditional expectation (Chapter 2) can be used to verify that the sequence of random conditional expectations

$$Y_n \triangleq E\{Z | X_1, X_2, \ldots, X_n\} \quad (5.31)$$

is a martingale (exercise 24). ∎

---

*The term *martingale* comes from the French, and means, among other things, a certain gambling strategy. The theory of this broad class of processes is considerable. See [Doob, 1953; Breiman, 1968; Karlin and Taylor, 1975; Lipster and Shiryayev, 1977; Hall and Heyde, 1980; Brémaud, 1981].

5.3.5 Point Processes*

A sequence of ordered random variables, $X_1 \le X_2 \le X_3 \le \cdots$, when interpreted as time points $T_1 = X_1$, $T_2 = X_2$, $T_3 = X_3, \ldots$, is called a *point process*. For example, if the times at which a Markov chain changes state are random, as they would be for example in a birth-death process, then the set of state transition times is a point process. Point processes are useful for modeling a wide range of phenomena involving the occurrence of events at random points in time, such as the emission of subatomic particles from radioactive material, the detection of photons by the receiver of a weak optical signal, neural discharges, lightning discharges, seismic events, and the arrival of customers in a service queue (waiting line). Every point process defines a counting process $X(t)$, whose value at time t is the number of events that have occurred since some starting time, say $t = 0$; that is,

$$X(t) = 0, \quad 0 \le t < T_1,$$
$$X(t) = 1, \quad T_1 \le t < T_2$$
$$\vdots$$
$$X(t) = n, \quad T_n \le t < T_{n+1}. \qquad (5.32)$$

Depending on the probabilistic model for the random time points, the counting process can be a martingale, a Markov process, or an independent-increment process. A particularly important example of a counting process is the *Poisson process*, which is studied in Chapter 6. One situation in which the Poisson process arises as an appropriate model is that of a service queue. For example, if the customers arrive at the queue independently of one another, and at a fixed average rate, say λ, and the arrival of two customers at precisely the same time is impossible, then a complete probabilistic model for the point process is specified, and is in fact the Poisson process. The corresponding counting process has independent increments and is therefore a Markov process. The mean of this process is λt, and if subtracted from the count yields a martingale (exercise 24).†

5.4 Stationary Processes

With reference to random thermal noise, it seems intuitive that if the physical parameters of this phenomenon (e.g., temperature, chemical com-

---
*The highly developed subject of point processes [Snyder, 1975; Larson and Shubert, 1979; Karlin and Taylor, 1981; Brémaud, 1981] is briefly studied in Chapter 6.

†The highly developed subject of queueing [Feller, 1971; Kleinrock, 1975; Karlin and Taylor, 1981; Ross, 1983] is not treated in this book, except for exercise 24 in Chapter 6.

position of the resistor, electromagnetic environment of the resistor, etc.) do not change with time, then the parameters of the probabilistic model should be time-invariant also. That is, if time is translated by any amount, say u, then the time-translated process,

$$Y(t) \triangleq X(t - u),$$

should have the same multivariate probability densities as the process $X(t)$; for example,

$$f_{X(t_1)X(t_2)}(y_1, y_2) = f_{Y(t_1)Y(t_2)}(y_1, y_2) \equiv f_{X(t_1-u)X(t_2-u)}(y_1, y_2)$$

(5.33)

for all t_1, t_2, u, y_1, y_2.

As a result, the mean and autocorrelation functions should be *translation-invariant* also:

$$m_X(t) = m_Y(t) \equiv m_X(t - u), \qquad (5.34)$$

$$R_X(t_1, t_2) = R_Y(t_1, t_2) \equiv R_X(t_1 - u, t_2 - u) \qquad (5.35)$$

for all t, t_1, t_2, u.

In general, a process X for which (5.33) is valid is said to be *second-order stationary*. If the nth order multivariate densities for X are translation-invariant, then X is said to be nth-*order stationary*. And if, for every positive integer n (no matter how large), X is nth-order stationary, then X is said to be *stationary in the strict sense*. On the other hand, if it is known only that (5.34) and (5.35) are valid, then X is said to be *stationary in the wide sense* (wide-sense stationary, abbreviated WSS).

Similarly, two processes X and Y are said to be *jointly WSS* if and only if every linear combination of them is a WSS process. That is, the process

$$Z(t) = aX(t) + bY(t)$$

must be WSS for all real numbers a and b. It follows (exercise 25) that two processes are jointly WSS if and only if both means, both autocorrelations, *and the cross-correlation* are translation-invariant.

It follows directly from the defining properties (5.34) and (5.35) that for a WSS process X, the mean waveform $m_X(t)$ is independent of time; it is therefore abbreviated to

$$m_X(t) = m_X. \qquad (5.36)$$

Likewise the autocorrelation (and autocovariance) depends on only the time difference $t_1 - t_2$, and is therefore abbreviated to

$$R_X(t_1, t_2) = R_X(t_1 - t_2). \qquad (5.37)$$

Hence, for a WSS process X the autocorrelation is a function of the single variable $\tau \triangleq t_1 - t_2$, and

$$R_X(\tau) = E\{X(t+\tau)X(t)\}, \qquad (5.38)$$

for every value of t. Since τ is the amount by which the time t lags behind (for $\tau > 0$) the time $t + \tau$, the variable τ is referred to as the *lag variable*. Examples in Sections 4.2.1, 4.2.5, and 4.2.7 are all WSS processes.

5.4.1 Stationary Gaussian Processes

Although wide-sense stationarity is, in general, the weakest form of stationarity of practical interest, it can be shown that for the exceptional case of all Gaussian processes, wide-sense stationarity guarantees strict-sense stationarity. This follows directly from the formula for the probability density (or characteristic function) of jointly Gaussian random variables (2.37) [or (2.38)]. This formula depends on only the mean vector

$$\boldsymbol{m}_X = [m_X(t_1), m_X(t_2), \ldots, m_X(t_n)]^T, \qquad (5.39)$$

and the covariance matrix, with (i, j)th element

$$[\boldsymbol{K}_X]_{ij} = K_X(t_i, t_j). \qquad (5.40)$$

Thus, if the mean and covariance are translation-invariant, the probability density (and characteristic function) must be translation-invariant.

Application (Thermal noise). As discussed in Section 10.7, a thermal-noise voltage is appropriately modeled as a stationary Gaussian process (assuming the environment of the noisy resistance is time-invariant). Furthermore, it is shown there that the autocovariance is given by

$$K_X(\tau) = N_o \int_{-\infty}^{\infty} h(t+\tau) h(t) \, dt,$$

where $h(t)$ is the impulse response of the voltmeter used to measure the noise voltage waveform, and $N_o = 2KTR$ [cf. (1.24)]. This result is based on the practical assumption that the response time, denoted by τ_*, of the voltmeter is much larger than the mean relaxation time of free electrons within the resistance (i.e., $\tau_* > 10^{-12}$ seconds at room temperature). If

$$h(t) = \begin{cases} e^{-\alpha t}, & t \geq 0, \\ 0, & t < 0, \end{cases}$$

then K_X is given by (5.22), and $X(t)$ is the Ornstein-Uhlenbeck process.

■

5.4.2 Autocorrelation Properties

The autocorrelation function R_X for every WSS process possesses the following properties (exercise 30):

1. R_X is even:
$$R_X(\tau) = R_X(-\tau). \tag{5.41}$$

2. R_X has its maximum at the origin*:
$$|R_X(\tau)| \le R_X(0). \tag{5.42}$$

3. $R_X(\tau)$ is continuous for all τ if it is continuous at $\tau = 0$.

There are only two ways in which an even function can be discontinuous at the origin. Either $R_X(\tau)$ contains a singularity such as an impulse at the origin, or it contains an additive null function at the origin. The first case is discussed in Chapters 6 and 7. For the second case, $R_X(\tau)$ can be expressed as
$$R_X(\tau) = a(\tau) + b(\tau),$$
for which $a(\tau)$ is continuous at $\tau = 0$, and $b(\tau)$ is the null function
$$b(\tau) = \begin{cases} b_0, & \tau = 0, \\ 0, & \tau \ne 0, \end{cases}$$
for some positive constant b_0.

Example 9 (Linear FM): To gain some insight into the kind of phenomenon that could give rise to a model for which the autocorrelation contains a null-function discontinuity, consider the linear frequency-modulated sine wave
$$X(t) = \sin(\beta t^2), \tag{5.43}$$
which has instantaneous frequency
$$\omega(t) \triangleq \frac{d(\beta t^2)}{dt} = 2\beta t.$$

The empirical autocorrelation (3.2) for this waveform can be shown to be
$$\hat{R}_X(\tau) = \begin{cases} \tfrac{1}{2}, & \tau = 0, \\ 0, & \tau \ne 0, \end{cases} \tag{5.44}$$
which is a null function. Similarly, a sum of delayed replicas of this waveform would yield null-function discontinuities at various values of τ determined by the delays. Furthermore, the sum of this waveform and

---
*Equality in (5.42) can hold if and only if the process X is periodic in the mean-square sense (cf. Section 4.2.5 and exercise 30(b)).

any other unrelated waveform would yield these same discontinuities in the autocorrelation. ∎

In this book, we shall not consider any stationary random processes whose autocorrelations contain null-function discontinuities. Such random processes are rarely (if ever) used as models.

5.4.3 Stationary Increments

A process $X(t)$ is said to have *stationary increments* if the joint densities of increments are translation-invariant; for example, for

$$Y_{12}(t) = X(t_2 + t) - X(t_1 + t),$$
$$Y_{34}(t) = X(t_4 + t) - X(t_3 + t), \qquad (5.45a)$$

we have

$$f_{Y_{12}(t-u)Y_{34}(t-u)}(y_1, y_2) = f_{Y_{12}(t)Y_{34}(t)}(y_1, y_2) \qquad (5.45b)$$

for all t and u for which the stationary-increment process X is defined. A process with stationary increments need not be a stationary process. For example, the Wiener process has stationary increments, but is a nonstationary process. The same is true for the Poisson counting process, the binomial counting process, and the random-walk process (exercise 18).

5.4.4 Asymptotically Stationary Processes

With reference to thermal noise, it seems intuitive that if some physical parameters of this phenomenon (e.g., temperature) undergo a sudden transition from their old values to new values, and all other physical parameters remain fixed at their old values, then at least some of the parameters of the probabilistic model (e.g., variance) should undergo similar transitions to new values; however, because of the nonzero response time of a physical system, there will result a transient phenomenon in the thermal noise that should be reflected in transients in the probabilistic parameters. That is, the probabilistic parameters should not change instantaneously to their new fixed values, but should *approach* new fixed values as time progresses.

If the time at which some change occurs is identified as $t = 0$, then the old values of probabilistic parameters can be referred to as *initial parameters*, and the new values that are approached as time progresses can be referred to as *steady-state parameters*.

Many transient phenomena are accurately modeled by decaying exponentials, which do not reach their steady state at any finite time, but approach a steady state as $t \to \infty$. In such cases, probabilistic parameters are never time-invariant, but approach steady-state values as $t \to \infty$. Random processes with such asymptotically time-invariant probabilistic parameters are called *asymptotically stationary processes*. A specific example of an asymptotically stationary process is the asynchronous telegraph signal discussed in Chapter 6. Its mean is given by $m_X(t) = e^{-2\lambda t}$, $t \geq 0$, and therefore approaches the asymptote 0, $m_X(t) \to 0$.

5.4.5 Cyclostationary Processes

With reference to random thermal noise, it seems intuitive that if some of the physical parameters of this phenomenon (e.g., temperature) vary periodically with time, and all other physical parameters are time-invariant, then at least some of the parameters of the probabilistic model (e.g., the variance) should vary periodically with time. That is, the multivariate probability densities for the translated process $Y_u(t) = X(t - u)$ should be periodic functions of the translation variable u. For example, if the period of such periodicity is denoted by T, then

$$m_X(t - u) \equiv m_{Y_u}(t) = m_{Y_{u+T}}(t) \equiv m_X(t - u - T) \qquad (5.46)$$

and

$$R_X(t_1 - u, t_2 - u) \equiv R_{Y_u}(t_1, t_2) = R_{Y_{u+T}}(t_1, t_2)$$
$$\equiv R_X(t_1 - u - T, t_2 - u - T). \qquad (5.47)$$

It follows from (5.46) and (5.47) (with the change of variables $t' = t - u$, $t'_1 = t_1 - u$, $t'_2 = t_2 - u$) that

$$m_X(t') = m_X(t' - T), \qquad (5.48)$$

$$R_X(t'_1, t'_2) = R_X(t'_1 - T, t'_2 - T). \qquad (5.49)$$

A process X for which the mean and autocorrelation exhibit the periodicity properties (5.48) and (5.49) is said to be *cyclostationary* in the wide sense.

Cyclostationary processes are particularly appropriate probabilistic models for many signals and noises encountered in signal-processing systems, such as communication, telemetry, radar, and sonar systems, where the periodicity arises from sampling, scanning, modulating, multiplexing, and coding operations [Gardner and Franks, 1975; Gardner, 1987a]. For example, the sine-wave signal process of Section 4.2.4 and the sampled-and-held noise process of Section 4.2.6 are both cyclostationary. Similarly, the pulse-position-modulated and pulse-width-modulated signals in exercises 15 and 16 in Chapter 4 are both cyclostationary. It can be shown that amplitude-modulated and phase-modulated sine waves and all pulse-modulated periodic pulse trains are cyclostationary processes if the modulating signals are stationary. This class of processes is studied in Chapter 12.

Phase Randomization. It can easily be shown (exercise 31) that if X is any cyclostationary process with period T, then the phase-randomized process

$$Y(t) = X(t - \Theta), \tag{5.50}$$

for which the random phase variable Θ is uniformly distributed over one period, say $[-T/2, T/2)$, and is independent of the process $X(t)$, is a stationary process, with mean and autocorrelation given by the time averages

$$m_Y = \frac{1}{T} \int_{-T/2}^{T/2} m_X(t)\, dt, \tag{5.51}$$

$$R_Y(\tau) = \frac{1}{T} \int_{-T/2}^{T/2} R_X(t + \tau, t)\, dt. \tag{5.52}$$

The phase-randomized sine-wave process (Section 4.2.5), the phase-randomized sampled-and-held noise process (Section 4.2.7), and the phase-randomized pulse-position- and pulse-width-modulated processes (exercises 15 and 16 in Chapter 4) provide examples. In situations for which the phase of a cyclostationary process is not known or is not of interest, the phase-randomized model is an appropriate model of the associated physical phenomenon. However, in other situations, the random phase can be considered a mathematical artifice that masks the periodicity of interest that is present in the associated physical phenomenon. Hence, the cyclostationary model without random phase is more appropriate in such situations. This is pursued in Chapter 12.

Vector-Stationary Processes. A scalar-valued cyclostationary process can always be treated as a vector-valued discrete-time stationary process, by reinterpreting the set of values over each period to be a single value of a vector

$$X_n \triangleq \{ X(s) : (n-1)T < s \leq nT \}. \tag{5.53a}$$

For continuous-time processes the vector is infinite-dimensional and therefore requires relatively sophisticated mathematical techniques. However, for discrete-time processes, the vector is finite-dimensional. For example, if X_n is cyclostationary with period N, then

$$X_n \triangleq \{ X_m : (n-1)N < m \leq nN \} \tag{5.53b}$$

is stationary (n, m, and N are all integers). Nevertheless, the theory and

5.4.6 Transient and Persistent Processes

A *transient* process is a process that dies out as time progresses, and therefore cannot be stationary or asymptotically stationary. An example of a transient process is the natural response of a deterministic, stable dynamical system to random initial stored energy, with no external excitation. Transient processes are not of major concern in this book, but are briefly studied in terms of dynamical systems in Chapter 9. The processes of major concern in this book are those that can be described as *persistent*. This includes, in addition to stationary, asymptotically stationary, and cyclostationary processes, other processes that are nonstationary, but that do not die out or blow up as time progresses. In particular, the nonstationary processes of primary concern in this book exhibit time-variant distributions whose asymptotic time averages (means) exist. Such processes are called *asymptotically mean stationary* (AMS) [Gray, 1987]. For example, for a process that is AMS in the *wide sense*, the limits

$$\lim_{T \to \infty} \frac{1}{T} \int_{-T/2}^{T/2} m_X(t)\, dt,$$

$$\lim_{T \to \infty} \frac{1}{T} \int_{-T/2}^{T/2} R_X(t_1 + t, t_2 + t)\, dt \qquad (5.54)$$

exist and the latter does not vanish. The class of AMS processes includes all stationary, cyclostationary, and asymptotically stationary processes.

5.5 Summary

In general, the complete specification of a random process requires specification of the joint probability densities for every set of n time points for every positive integer n. But certain subclasses of processes admit a much more economical specification. A *Gaussian process* is completely specified by its mean function and autocovariance function alone. A Markov process, which is defined by a finite-memory property (5.6) called the *Markov property*, is completely specified by its initial first-order density and the set of first-order conditional densities (5.8), called *transition densities*, for all pairs of time points. However, a valid process model cannot be constructed from an arbitrary set of transition densities; it is required that they satisfy a consistency condition defined by the *Chapman-Kolmogorov equation* (5.9).

In principle, the transition densities for a continuous-state Markov process can be obtained by solving the *Fokker-Planck equation* (5.65), which is specified by a differential mean function and a differential variance function. However, this approach is not always viable in practice. A Markov process that is discrete-valued is called a *Markov chain*. A Markov process is said to be *homogeneous* if the transition densities are translation-invariant.

A necessary and sufficient condition for a Gaussian process to be a Markov process can be specified in terms of only the autocovariance function (5.21). There is only one process that is Gaussian, Markov, and stationary, and it is called the *Ornstein-Uhlenbeck process* (5.22). An important example of a nonstationary Gauss-Markov process is the *Wiener process*.

The definition of a Markov process can be generalized by modifying the defining finite-memory property to increase the memory length beyond unity to (say) m (5.25). This yields the *mth-order Markov process*. An mth-order Markov process can be reinterpreted as a m-vector-valued first-order Markov process.

A more specific class of processes are those with the *independent-increment property*. Every independent-increment process is a Markov process. A more general class of processes are those with the *martingale property* (5.28). Every independent increment process with zero mean is a martingale process, and every Gaussian martingale process has independent increments.

A different type of process from those preceding is the *point process*, which consists of a discrete set of random points in time or some other dimension such as space. Associated with every point process is a *counting process* defined by (5.32). Depending on the probabilistic model for the point process, the counting process can be a martingale, a Markov process, or an independent-increment process. A particularly important point process is the Poisson process (Chapter 6).

Stationary processes are defined by a *translation-invariance property* exhibited by their probabilistic parameters. If all probability densities of order n are translation-invariant, the process is said to be *nth-order stationary*. A process that is nth-order stationary for $n \to \infty$ is said to be *strict-sense stationary*. If only the mean and autocovariance are translation-invariant, the process is said to be *wide-sense stationary*. A more general class of processes are those with *stationary increments*. Every stationary process has stationary increments. Another more general class of processes are those with probabilistic parameters that are translation-invariant asymptotically as time approaches infinity. Every stationary process is trivially an *asymptotically stationary process*. A third, more general class of processes are those with probabilistic parameters that vary periodically with translation. These are called *cyclostationary processes*. Every stationary process is trivially a cyclostationary process. Also every cyclostationary process can be reinterpreted as a *vector-valued stationary process*. Furthermore, every cyclostationary process can be made stationary by *phase randomization*. A more general class of nonstationary processes are those that *exhibit cyclostationarity* with possibly more than one period (Chapter 12).

The most general class of processes of interest in this book that exhibit some form of stationarity (but not stationarity associated with only the increments) are the *asymptotically mean stationary processes*. These are processes that are persistent in a sufficiently regular fashion that the asymptotic time averages of certain time-variant probabilistic parameters, such as the mean and autocorrelation (5.54), exist and are not all identically zero. Every cyclostationary process is an asymptotically mean stationary process.

EXERCISES

1. Consider the random-amplitude sine wave process in Section 4.2.4, and let the amplitude be a Gaussian random variable.
   (a) Show that this is a Gaussian random process.
   (b) Show that the covariance matrix for any set of n time points has rank equal to unity.
   (c) Derive an explicit expression for the joint probability density for the random variables occurring at an arbitrary pair of times t_1 and t_2.

   Hint: See (1.34).

★2. Consider the sampled and held noise process $X(t)$ in Section 4.2.6, and let the samples of the noise voltage $\{V(nT)\}$ be jointly Gaussian random variables. Prove that $X(t)$ is a Gaussian process.

3. *Isserlis's formula for Gaussian moments* [Isserlis, 1918]. The *rth-order joint moment* of the n random variables $X_1, X_2, \ldots, X_n$ is defined by

$$m(k_1, k_2, \ldots, k_n) \triangleq E\{X_1^{k_1} X_2^{k_2} \cdots X_n^{k_n}\}, \quad (5.55)$$

where

$$r = k_1 + k_2 + \cdots + k_n,$$

and the *nth-order joint characteristic function* of these n random variables is defined by

$$\Phi(\omega_1, \omega_2, \ldots, \omega_n) \triangleq E\{e^{i(\omega_1 X_1 + \omega_2 X_2 + \cdots + \omega_n X_n)}\}. \quad (5.56)$$

By expanding the exponential in an n-dimensional power series, the following identity can be obtained:

$$\left.\frac{\partial^r \Phi(\omega_1, \ldots, \omega_n)}{\partial \omega_1^{k_1} \cdots \partial \omega_n^{k_n}}\right|_{\omega_1 = \omega_2 = \cdots = \omega_n = 0} = i^r m(k_1, \ldots, k_n). \quad (5.57)$$

If we now use the formula (2.38) for Φ for jointly Gaussian zero-mean variables, we can show that all odd-order joint moments are zero, and we can derive *Isserlis's formula* for the even-order joint moments of a set of jointly Gaussian zero-mean variables,

$$E\{X_1 X_2 \cdots X_r\} = \sum E\{X_{j_1} X_{j_2}\} \cdots E\{X_{j_{r-1}} X_{j_r}\}, \quad r \text{ even}, \quad (5.58)$$

in which the sum is taken over all possible ways of dividing the r integers into $r/2$ combinations of pairs. The number of terms in the summation is $(1)(3)(5) \cdots (r - 3)(r - 1)$. Note that some of the X_j can be identical.

(a) Apply the result (5.58) to show that for a zero-mean Gaussian process $X(t)$, the fourth joint moment is given by

$$E\{X(t_1)X(t_2)X(t_3)X(t_4)\} = K_X(t_1, t_2)K_X(t_3, t_4) \\ + K_X(t_1, t_3)K_X(t_2, t_4) \\ + K_X(t_2, t_3)K_X(t_1, t_4). \quad (5.59)$$

(b) Derive the result (5.59) by starting with the identity (5.57).

4. To generalize Isserlis's formula for fourth moments, substitute $X(t) = X'(t) - m_{X'}(t)$, where $X'(t)$ is a nonzero-mean Gaussian process, into (5.59) and solve for the fourth moment of $X'(t)$ in terms of *only* first and second moments. Show that the result is

$$E\{X(t_1)X(t_2)X(t_3)X(t_4)\} = R_X(t_1, t_2)R_X(t_3, t_4) + R_X(t_1, t_3)R_X(t_2, t_4) \\ + R_X(t_1, t_4)R_X(t_2, t_3) - 2m_X(t_1)m_X(t_2)m_X(t_3)m_X(t_4). \quad (5.60)$$

5. Let $Y(t) = X(t)X(t - \Delta)$, where $X(t)$ is a stationary Gaussian process. Determine formulas for the mean and autocorrelation of $Y(t)$ that depend only on the mean and autocovariance of $X(t)$.

6. To illustrate that the result of exercise 9 in Chapter 4 is invalid if the processes $X(t)$ and $Z(t)$ are dependent, let $Z(t) = X(t)$ be a Gaussian process, and apply (5.59) to determine R_Y, for $Y(t) = X(t)Z(t) = X^2(t)$.

7. (a) If the processes in Sections 4.2.4 and 4.2.6 are Gaussian, then their phase-randomized versions in Sections 4.2.5 and 4.2.7 cannot be Gaussian. Verify this.

Hint: Express $f_{Y(t)}(y)$ in terms of $f_{X(t)}(x)$ and $f_\Theta(\theta)$ using the definition of conditional probability density, and use the fact that the sum of two or more Gaussian functions, with different widths or locations, cannot be a Gaussian function.

(b) Let the amplitude of the process $Y(t)$ in Section 4.2.5 be Rayleigh-distributed, and use the fact that the two random variables
$$W \triangleq A\cos\Theta \quad \text{and} \quad Z \triangleq A\sin\Theta$$
are independent, identically distributed, and Gaussian (see exercise 15 in Chapter 1) to prove that $Y(t)$ is a Gaussian process (not just a Gaussian random variable for each t).

⋆8. For a Gaussian process $X(t)$, solve for the conditional expectation
$$E\{X(t_2)|X(t_1)\}$$
in terms of $X(t_1)$, $m_X(t_1)$, $m_X(t_2)$, $\sigma_X^2(t_1)$ and $K_X(t_1, t_2)$ [cf. exercise 12(d) of Chapter 1].

9. *Price's theorem* [Price, 1958]. Let $X(t)$ be a zero-mean stationary Gaussian process, and let $X_1 = X(t_1)$ and $X_2 = X(t_2)$ be two time samples with correlation coefficient
$$\rho = \frac{E\{X_1 X_2\}}{\sigma_{X_1}\sigma_{X_2}} = \frac{R_X(t_1 - t_2)}{R_X(0)}.$$
Also, let Y_1 and Y_2 be obtained from X_1 and X_2 by two deterministic functions
$$Y_1 = g_1(X_1), \quad Y_2 = g_2(X_2).$$
Then it can be shown that the correlation
$$R_Y = E\{Y_1 Y_2\}$$
satisfies
$$\frac{d^k R_Y}{d\rho^k} = R_X^k(0) E\left\{ \frac{d^k g_1(X_1)}{dX_1^k} \frac{d^k g_2(X_2)}{dX_2^k} \right\}. \tag{5.61}$$
This formula can be used to evaluate R_Y by differentiating g_1 and g_2 enough times to simplify evaluation of the expectation in the right member of (5.61). Then R_Y is obtained by integration. As an illustration, consider the clipper
$$g_1(x) = g_2(x) = g(x) \triangleq \begin{cases} +1, & x > 0, \\ -1, & x < 0, \end{cases}$$
and use $k = 1$ to obtain
$$\frac{dR_Y}{d\rho} = \frac{2}{\pi} \frac{1}{(1-\rho^2)^{1/2}}.$$

Then integration yields

$$R_Y = \frac{2}{\pi} \sin^{-1} \rho.$$

For example, if $t_1 - t_2 = \tau$, then

$$R_Y(\tau) = \frac{2}{\pi} \sin^{-1}\left[\frac{R_X(\tau)}{R_X(0)}\right], \qquad (5.62)$$

which is the autocovariance of a clipped zero-mean Gaussian random process. It can also be shown (although not as easily) that if

$$g(x) = \left(\frac{2}{\pi \alpha^2}\right)^{1/2} \int_0^x e^{-z^2/2\alpha^2} \, dz,$$

then

$$\rho_Y(\tau) = \frac{\sin^{-1}[\rho_X(\tau)/(1+\eta)]}{\sin^{-1}[1/(1+\eta)]}, \qquad (5.63)$$

where

$$\eta \triangleq \frac{\alpha^2}{R_X(0)}$$

and

$$\rho_X(\tau) \triangleq \frac{R_X(\tau)}{R_X(0)},$$

$$\rho_Y(\tau) \triangleq \frac{R_Y(\tau)}{R_Y(0)}.$$

For $\eta \to 0$, g approaches the clipper, and (5.63) becomes (5.62). For $\eta = 1$, $g(\cdot)$ makes a smoother transition from -1 to $+1$, and (5.63) reduces to

$$R_Y(\tau) = \frac{4}{\pi} \sin^{-1}\left[\frac{R_X(\tau)}{2R_X(0)}\right]. \qquad (5.64)$$

Show that in this case $Y = g(X)$ has a uniform density on $[-1, 1)$.

Hint: Use (1.36).

10. (a) Consider the clipping function

$$g(X) = \begin{cases} +1, & X \geq 0, \\ -1, & X < 0. \end{cases}$$

The product $X(t + \tau)g[X(t)]$ is simply $\pm X(t + \tau)$, depending on the sign of $X(t)$. The time average of this product can therefore be evaluated without performing any multiplication operations. To gain some insight into how this empirical cross-correlation is related to the empirical autocorrelation obtained without the clipper, evaluate the probabilistic cross-correlation $E\{X(t + \tau)g[X(t)]\}$ for a stationary zero-mean Gaussian process.

Hint: Use Price's theorem. The result is $E\{X(t + \tau)g[X(t)]\} = cR_X(\tau)$, where $c = \sqrt{2}[\pi R_X(0)]^{-1/2}$.

(b) Repeat part (a) for an arbitrary nonlinearity $g(\cdot)$, and show that the result is of the same form, but the constant c is given by $c = E\{h[X(t)]\}$, where $h(x) = dg(x)/dx$.

11. A control device has the property that its position at time $n + 1$ is a linear combination of its position at time n and a random error that is independent of the positions at times prior to $n + 1$. Prove that the position process X_n is a Markov process.

12. Consider the Markov process model for the binary digital system in Example 3. Assume that $P_{10} = P_{01}$ so that the matrix $\mathbf{P}$ of transition probabilities is symmetric $\mathbf{P}^T = \mathbf{P}$. Study the eigenvalues of this matrix and classify the asymptotic ($n \to \infty$) behavior of the state-probability vector $\mathbf{P}_n$ in (5.13) in terms of these eigenvalues. Also, explain in terms of the eigenvectors of $\mathbf{P}$ the different types of behavior of $\mathbf{P}_n$ that can be obtained by the choice of the initial state vector $\mathbf{P}_0$.

13. Consider the following *queueing model*. A device serves customers one at a time. If the service of a customer is not complete at any integer time n, the probability of service being completed before time $n + 1$ is p. When service is complete, the next customer in the queue begins being served at the next integer time. The number of customers arriving in the queue for service in the time interval between n and $n + 1$ is a Poisson random variable Z_n, with probability mass function

$$P(k) = \frac{\lambda^k}{k!} e^{-\lambda}$$

for some fixed value of the parameter λ. $\{Z_n\}$ is an independent sequence. Let X_n denote the total number of customers in the queue at time n. Show that this is a Markov process and determine its transition probabilities.

14. Consider the following model for a *renewal process*. Each one of a certain type of device has a particular sequence of probabilities $p_1, p_2, p_3, \ldots$ of failing during the 1st, 2nd, 3rd, $\ldots$ periods in service. One device starts in service at time 0. If it fails between times n and $n + 1$, it is replaced at time $n + 1$ by a new device of the same type. This process continues on indefinitely. The age in periods of the device in

service at time n is denoted by X_n. Show that this is a Markov process and determine its transition probabilities.

⋆**15.** (a) Derive the Chapman-Kolmogorov equation (5.9). Write down the discrete-distribution counterpart [corresponding to (5.11)] of (5.9).

(b) Show that the n-stage transition density for a discrete-time Markov process can be obtained from its one-stage transition densities by the formula

$$f_{X_n|X_m}(x_n|x_m) = \int_{-\infty}^{\infty} \cdots \int_{-\infty}^{\infty} f_{X_n|X_{n-1}}(x_n|x_{n-1})$$
$$\times f_{X_{n-1}|X_{n-2}}(x_{n-1}|x_{n-2}) \times \cdots$$
$$\times f_{X_{m+1}|X_m}(x_{m+1}|x_m)\, dx_{n-1}\, dx_{n-2} \cdots dx_{m+1}.$$
(5.65)

16. (a) Derive the formula (5.13) for the evolution of the marginal distribution of a binary digital system modeled by a Markov chain. Show that if $P_{10} = P_{01} = r$ and $P_{11} = P_{00} = q = 1 - r$, and $\boldsymbol{P}_0 = \{\tfrac{1}{2}, \tfrac{1}{2}\}$, then $\boldsymbol{P}_n = \{\tfrac{1}{2}, \tfrac{1}{2}\}$ for all n.

(b) Draw a schematic diagram, analogous to that shown in Figure 5.2, for a birth-death process. This reveals why the term *chain* is appropriate for such processes.

⋆**17.** (a) Show that if $f_{X_n|X_{n-1}\cdots X_1} = f_{X_n|X_{n-1}}$, then

$$f_{X_{n_m}|X_{n_{m-1}}\cdots X_{n_1}} = f_{X_{n_m}|X_{n_{m-1}}} \quad \text{for } n_m \geq n_{m-1} \geq \cdots \geq n_1.$$

(b) For an independent-increment process with known starting value $X(0) = X_0$, derive a formula for the nth-order distribution in terms of the first- and second-order distributions.

(c) Show that an independent-increment process with known starting value $X(0) = X_0$ is a Markov process.

18. (a) Show that the binomial counting process and the random-walk process (Chapter 4) both have independent stationary increments and are homogeneous Markov processes.

(b) Is the sampled-and-held process described in Section 4.2.6 a Markov process? Determine if this process has independent increments.

19. (a) Use the Chapman-Kolmogorov equation to prove that for a Markov process

$$E\{E\{X(t_3)|X(t_2)\}|X(t_1)\} = E\{X(t_3)|X(t_1)\}$$

for $t_1 \leq t_2 \leq t_3$. Then use the result of exercise 8 to show that the covariance of a Gauss-Markov process satisfies (5.21). Then show that (5.22) satisfies (5.21).

(b) Let $g(t) = e^{-t}$ and $h(t) = e^t$, and show that (5.23) reduces to (5.22) with $\sigma = \alpha = 1$.

20. (a) It is shown in Chapter 13 that the best (minimum mean-squared prediction error) qth order p-step predictor for a process X_n is the conditional mean

$$\hat{X}_{n+p} = E\{X_{n+p}|X_n, X_{n-1}, X_{n-2}, \ldots, X_{n-q+1}\}.$$

Let X_n be a first-order Markov process. Show that $\hat{X}_{n+p}$ depends only on X_n. What if X_n were an mth-order Markov process with $m < q$?

(b) Show that if X_n is a Gaussian process, then the best predictor from part (a) is simply a constant plus a linear combination of the available data

$$\hat{X}_{n+p} = c_0 X_n + c_1 X_{n-1} + \cdots + c_{q-1} X_{n-q+1} + c.$$

21. *Fokker-Planck equation.* The objective of this exercise is to point out that the transition density for every diffusion process satisfies the following partial differential equation known as the Fokker-Planck equation*:

$$\frac{\partial}{\partial t} f(x, t) + \frac{\partial}{\partial x} [m(x, t) f(x, t)] - \frac{1}{2} \frac{\partial^2}{\partial x^2} [\sigma^2(x, t) f(x, t)] = 0,$$

(5.66a)

where f is the transition density,

$$f(x, t) \triangleq f_{X(t)|X(t_0)}(x|x_0),$$

(5.66b)

and m and σ^2 are the conditional mean and variance of the differential $dx(t)$, normalized by dt:

$$m(x, t) \triangleq \frac{E\{dX(t)|X(t) = x\}}{dt},$$

(5.66c)

$$\sigma^2(x, t) \triangleq \frac{E\{[dX(t) - m(x, t) \, dt]^2|X(t) = x\}}{dt}.$$

(5.66d)

This equation (5.66a) can be derived from the Chapman-Kolmogorov

---

*The class of diffusion processes consists of those continuous-state Markov processes that have transition densities possessing certain smoothness properties that guarantee the existence of the derivatives in (5.66).

equation with the use of the following characterizations of $m(x, t)$ and $\sigma^2(x, t)$:

$$m(x, t) = \frac{\partial}{\partial s} m_{X(s)|X(t) = x} \Big|_{s=t+}, \quad (5.66e)$$

$$\sigma^2(x, t) = \frac{\partial}{\partial s} \sigma^2_{X(s)|X(t)=x} \Big|_{s=t+}. \quad (5.66f)$$

Derive (5.66e) and (5.66f) from (5.66c) and (5.66d).

Hint: Use $dX(t) = X(t^+) - X(t)$ to express (5.66c) as

$$m(x, t) \, dt = m_{X(t^+)|X(t)=x} - m_{X(t)|X(t)=x},$$

and to express (5.66d) as

$$\sigma^2(x, t) \, dt = \sigma^2_{X(t^+) - X(t)|X(t)=x}$$

$$= \sigma^2_{X(t^+)|X(t)=x} - \sigma^2_{X(t)|X(t)=x};$$

then divide through by dt to obtain (5.66e) and (5.66f). (The Fokker-Planck equation for the Wiener process is derived in exercise 7 of Chapter 6.)

22. Consider a diffusion process with transition density satisfying the Fokker-Planck equation (5.66a) with $m(x, t) \equiv 0$ and $\sigma^2(x, t) \equiv \alpha^2$. Show that

$$f(x, t) = [2\pi\alpha^2(t - t_0)]^{-1/2} \exp\left[\frac{(x - x_0)^2}{2\alpha^2(t - t_0)}\right]$$

is the transition density for this process. Then let $x_0 = t_0 = 0$, and show that for $t \geq 0$, this is a Gaussian process. (It is shown in Chapter 6 that this is the Wiener process.)

★23. Derive (5.28) from (5.27). Then show that, for an independent-increment process with known initial value, (5.28) reduces to (5.30).

24. (a) Show that (5.31) is a martingale.
    (b) Show that if $X(t)$ has independent increments and known initial value, then $Y(t) \stackrel{\Delta}{=} X(t) - m_X(t)$ is a martingale.

25. (a) Show that the mean and autocorrelation of $Z(t) = aX(t) + bY(t)$ are translation-invariant for all deterministic a and b if and only if both means, both autocorrelations, and the cross-correlation of $X(t)$ and $Y(t)$ are translation-invariant.
    (b) A complex process is defined to be stationary if and only if its real and imaginary parts are jointly stationary. The autocorrelation of

a complex process is defined by

$$R_X(t_1, t_2) \triangleq E\{X(t_1)X^*(t_2)\}.$$

Show that a zero-mean complex process is WSS if and only if both

$$R_X(t_1, t_2) = R_X(t_1 - t_2)$$

and

$$E\{X(t_1)X(t_2)\} \triangleq R_{XX^*}(t_1, t_2) = R_{XX^*}(t_1 - t_2).$$

Are any of the processes $A(t)\cos\omega t$, $A(t)\sin\omega t$, $A(t)e^{i\omega t}$ WSS if $A(t)$ is WSS?

26. Let $\{X_n\}$ be a sequence of independent random variables. Under what condition will this be a stationary process? In this case what must be known to have a complete probabilistic specification of this process? Is this a Markov process?

★27. (a) Show that the second-order autoregressive process

$$X_n = aX_{n-1} + bX_{n-2} + Z_n,$$

where $\{Z_n\}$ is an independent sequence of random variables for which Z_n is independent of $X_{n-1}, X_{n-2}, X_{n-3}, \ldots$, is a second-order Markov process.

(b) Show that if $\{Z_n\}$ are identically distributed and $\{X_n\}$ is WSS for all $n \leq n_o$ for any n_o in part (a), then it is WSS for all $n > n_o$.

28. Consider the first-order autoregressive process (5.16), and let Z_n be stationary and $a_n = a$, a constant. Discuss the role of the value of a in determining the stationarity of X_n.

29. (a) A square matrix R is said to be *nonnegative definite* if

$$v^T R v \geq 0$$

for all vectors v. Show that every autocorrelation matrix and every autocovariance matrix is nonnegative definite.

Hint: Consider the mean-squared value of the linear combination
$$Y \triangleq v_1 X(t_1) + v_2 X(t_2) + \cdots + v_n X(t_n).$$

Show that every autocorrelation matrix R is symmetric, $R^T = R$.

(c) A square matrix is said to be *Toeplitz* if the elements along its diagonal are all identical, and the elements along each line parallel to the diagonal are all identical. Show that the autocorrelation matrix for every uniformly time sampled ($t_i = i\Delta$) WSS process is Toeplitz.

★30. (a) Verify properties 1 to 3 in Section 5.4.2.

Hint: For any two random variables, the *Cauchy-Schwarz inequality*

$$|E\{XY\}| \leq (E\{X^2\}E\{Y^2\})^{1/2} \qquad (5.67)$$

holds (Section 13.3). To verify property 3, let $X = X(t)$ and $Y = X(t + \tau + \varepsilon) - X(t + \tau)$ in (5.67). To verify property 2, make appropriate choices for X and Y in (5.67).

(b) Show that if $R_X(\tau_0) = R_X(0)$ for some $\tau_0 \neq 0$, then $X(t)$ is mean-square equivalent to a periodic process in the sense that

$$E\{[X(t) - X(t - \tau_0)]^2\} = 0 \quad \text{for all } t.$$

31. Verify (5.51) and (5.52). Then show that $K_Y(\tau)$ is not given by

$$\frac{1}{T}\int_{-T/2}^{T/2} K_X(t + \tau, t)\, dt$$

unless $m_X(t)$ is constant.

32. Verify that the processes in Sections 4.2.6 and 4.2.8 [assuming $Y(t)$ in Section 4.2.8 is WSS] are cyclostationary in the wide sense.

33. Verify that the process in exercise 15 of Chapter 4 is cyclostationary in the wide sense. (The same can be shown for the process in exercise 16 of Chapter 4.)

34. Show that if the process $X(t)$ dies out in the mean-square sense,

$$E\{X^2(t)\} \to 0 \quad \text{as } t \to \infty,$$

then the autocorrelation dies out in every radial direction from the origin of the (t_1, t_2) plane, that is,

$$R_X(at, bt) \to 0, \quad t \to \infty, \quad \text{for all } a, b.$$

Then conclude that R_X cannot depend on only the difference of its arguments, $R_X(t_1, t_2) \neq R_X(t_1 - t_2)$. Include appropriate sketches.

35. (a) Draw a Venn diagram representing the following classes of processes: Gaussian, Markov, stationary, and independent increment (with known initial value).
    (b) Indicate the appropriate location in this diagram of the two points corresponding to the following two processes: Wiener process and Ornstein-Uhlenbeck process.

# 6

The Wiener and Poisson Processes

TWO OF THE MOST FUNDAMENTALLY IMPORTANT MODELS of random processes, the Wiener process and the Poisson process, are introduced in this chapter.

6.1 Derivation of the Wiener Process*

A particular random process that is of fundamental importance, not only as an appropriate model for a variety of physical phenomena, but also as the core of advanced theories of calculus for random processes,[†] is the Wiener process. Among other applications, the Wiener process provides a model for Brownian motion in gases and liquids, thermal noise in electrical conductors, and various diffusions. A particle on the order of, say, a micron in diameter, immersed in a liquid or gas, exhibits persistent erratic motion, called *Brownian motion*, which can be observed with a microscope. This erratic motion is explained in terms of bombardment by molecules of the surrounding medium. We let $W(t)$ denote the displacement in one dimension, after time t, of a particle in Brownian motion from its initial position at $t = 0$. Thus $W(0) = 0$. Since the impacts of molecules upon the particle

---

*The Wiener process is a special case of a class of Markov processes called *diffusion processes*, for which there is a well-developed theory that has important applications in a variety of fields as briefly mentioned in Section 5.3.2, where references are given.

[†]See the references in the first footnote in Chapter 7.

are completely random, we do not expect the particle to drift off in one particular direction. Thus $E\{W(t)\} = 0$. Due to the continual impacts of molecules upon the particle, the displacement of the particle over a time interval (s, t) that is long compared with the time between impacts can be regarded as the sum of a large number of small random displacements. Consequently the central limit theorem of probability theory can be used to argue that $W(t) - W(s)$ is a Gaussian random variable. For a medium in equilibrium the parameters of this Gaussian random variable should be independent of time translation. Since the particle's motion is due to very frequent and irregular molecular impacts, it can be argued that the random incremental displacements, $W(t_1) - W(s_1)$ and $W(t_2) - W(s_2)$, corresponding to disjoint time-intervals, (s_1, t_1) and (s_2, t_2), are statistically independent, since the numbers and strengths of impacts in these intervals are independent. We are thus led to define the *Wiener process* as the unique process exhibiting the following three axiomatic properties:

1. The initial position is zero:
$$W(0) = 0. \tag{6.1a}$$

2. The mean is zero:
$$E\{W(t)\} = 0, \qquad t \geq 0. \tag{6.1b}$$

3. The increments of $W(t)$ are independent, stationary, and Gaussian.

As an alternative to this axiomatic definition, the Wiener process will now be *derived* as a limiting form of the random-walk process (Section 4.2.3). We begin with the following continuous-time version of the discrete-time ± 1 Bernoulli process Z_n with parameter $p = \frac{1}{2}$:

$$Z_\Delta(t) = \sum_{n=1}^{\infty} Z_n \Delta_w \delta(t - n\Delta_t), \qquad t \geq 0, \tag{6.2}$$

$$\text{Prob}\{Z_n = +1\} = \text{Prob}\{Z_n = -1\} = \tfrac{1}{2},$$

where Δ_t and Δ_w represent time and amplitude step sizes, and $\delta(\cdot)$ is the impulse function. Analogous to (4.19) for the discrete-time random walk W_n, the continuous-time random walk is defined by

$$W_\Delta(t) \triangleq \int_0^t Z_\Delta(v)\, dv \tag{6.3}$$

$$= \sum_{n=1}^{\infty} Z_n \Delta_w u(t - n\Delta_t), \qquad t \geq 0, \tag{6.4}$$

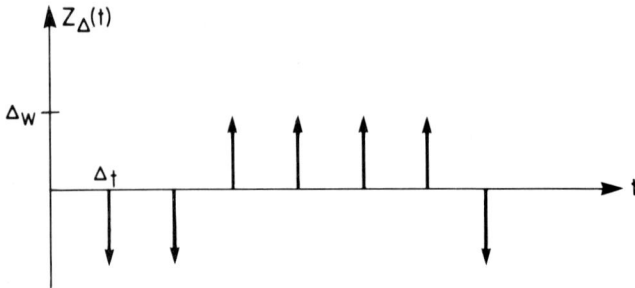

Figure 6.1 Statistical Sample of the Continuous-Time (Impulsive) ±1 Bernoulli Process

where $u(\cdot)$ is the unit step function. Typical statistical samples of $Z_\Delta(t)$ and $W_\Delta(t)$ are shown in Figures 6.1 and 6.2.

In order to obtain the Wiener process, denoted by $W(t)$, the limit of $W_\Delta(t)$ as the step sizes Δ_t and Δ_w approach zero is taken. In order to obtain a limiting process with appropriate properties, the relative sizes of Δ_w and Δ_t are constrained so that $\Delta_t \to 0$ twice as fast as $\Delta_w \to 0$:

$$\Delta_w = \alpha\sqrt{\Delta_t} \tag{6.5}$$

for some $\alpha > 0$. The limit of $W_\Delta(t)$ subject to (6.5) is the Wiener process,

$$W(t) \triangleq \lim_{\Delta_t \to 0} W_\Delta(t). \tag{6.6}$$

Therefore, we can heuristically interpret $W(t)$ as an infinitely dense collection of infinitesimal steps. Two typical statistical samples of $W(t)$ are shown in Figure 6.3.

An alternative interpretation of the Wiener process is, from (6.3), the indefinite integral of $Z(t)$, which is defined by

$$Z(t) \triangleq \lim_{\Delta_t \to 0} Z_\Delta(t), \tag{6.7}$$

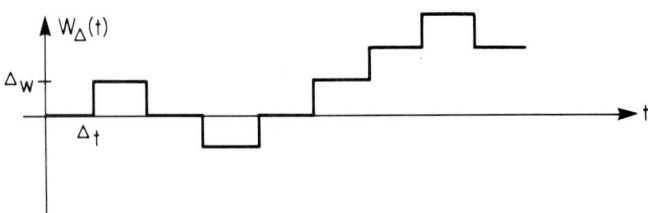

Figure 6.2 Statistical Sample of the Continuous-Time Random-Walk Process

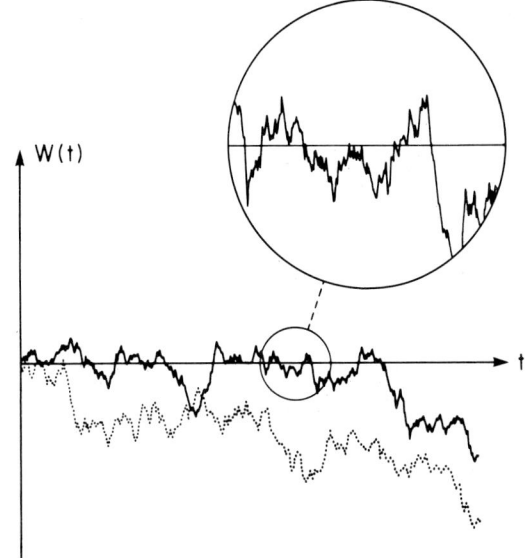

Figure 6.3 Two Statistical Samples of the Wiener Process

where the limit is subject to the constraint (6.5). Although there are mathematical difficulties associated with the existence of the limit (6.7), it provides a useful heuristic interpretation of $W(t)$ as the integral of $Z(t)$, which is an infinitely dense collection of impulses with infinitesimal areas.

The mean and covariance of $W(t)$ can be obtained as the limits of the mean and covariance of $W_\Delta(t)$. Since the number of Bernoulli trials made in the time interval $[0, t_i]$ is

$$n_i = \frac{[t_i]}{\Delta_t}, \tag{6.8}$$

where $[t_i]$ denotes the largest integer multiple of Δ_t that is less than or equal to t_i, then it follows directly from (4.22) and (4.23) that

$$m_{W_\Delta}(t) = 0, \tag{6.9}$$

$$K_{W_\Delta}(t_1, t_2) = \alpha^2 \Delta_t \min\{n_1, n_2\}. \tag{6.10}$$

Substitution of (6.8) into (6.10) yields

$$K_{W_\Delta}(t_1, t_2) = \alpha^2 \min\{[t_1], [t_2]\}. \tag{6.11}$$

Hence, the mean and covariance for $W(t)$ are the limits

$$m_W(t) = \lim_{\Delta_t \to 0} m_{W_\Delta}(t) = 0, \qquad t \geq 0, \qquad (6.12)$$

$$K_W(t_1, t_2) = \lim_{\Delta_t \to 0} K_{W_\Delta}(t_1, t_2)$$

$$= \alpha^2 \min\{t_1, t_2\}, \qquad t_1 \geq 0, t_2 \geq 0, \qquad (6.13)$$

where the simple fact

$$\lim_{\Delta_t \to 0} [t_i] = t_i \qquad (6.14)$$

has been used.

Since it can be shown that the Wiener process is a Gaussian process (exercise 6), then the mean and covariance, (6.12) and (6.13), completely specify the process.

It can be shown by using (6.13) that $W(t)$ has uncorrelated (and therefore independent) stationary increments (exercise 3).

6.2 The Derivative of the Wiener Process

It follows from (6.3) that

$$Z_\Delta(t) = \frac{dW_\Delta(t)}{dt}, \qquad (6.15)$$

and therefore

$$Z(t) = \frac{dW(t)}{dt}, \qquad (6.16)$$

provided that $W(t)$ is differentiable [i.e., provided that the limit (6.7) exists]. As a matter of fact, $W(t)$ is not differentiable (recall that it has infinitesimal step discontinuities infinitely close together), as we shall see in Chapter 7. Nevertheless, the *symbolic process* $Z(t)$ suggested by (6.16) is a very useful idealized heuristic model for broadband continuous-waveform noise, such as thermal noise. We can obtain the mean and covariance for this symbolic process by formal interchange of the linear operations of expectation and differentiation. We shall justify this formal manipulation in Chapter 7. The result is

$$E\{Z(t)\} = E\left\{\frac{dW(t)}{dt}\right\} = \frac{d}{dt} E\{W(t)\} = 0, \qquad t \geq 0, \qquad (6.17)$$

$$E\{Z(t_1)Z(t_2)\} = E\left\{\left(\frac{dW(t_1)}{dt_1}\right)\left(\frac{dW(t_2)}{dt_2}\right)\right\} = \frac{d}{dt_1}\frac{d}{dt_2} E\{W(t_1)W(t_2)\},$$

$$(6.18)$$

which can be expressed as

$$R_Z(t_1, t_2) = \frac{\partial^2}{\partial t_1 \, \partial t_2} R_W(t_1, t_2). \tag{6.19}$$

Substitution of (6.13) into (6.19) yields (exercise 4)

$$R_Z(t_1, t_2) = \alpha^2 \delta(t_1 - t_2), \qquad t_1, t_2 \geq 0, \tag{6.20}$$

where $\delta(\cdot)$ is the impulse function. This impulsive autocorrelation appears to be consistent with the heuristic interpretation of $Z(t)$ as an infinitely dense collectional of infinitesimal impulses. Since only for $t_1 \geq 0$ and $t_2 \geq 0$ is $R_Z(t_1, t_2)$ a function of $t_1 - t_2$, then $Z(t)$ is stationary for $t > 0$ only.

Application (Thermal noise). The process $Z(t)$ is an idealized model (for $t > 0$) for a thermal-noise voltage, for which each infinitesimal impulse corresponds to the tiny pulse of voltage caused by a single ion core or free electron. This is discussed in more detail in Section 10.7. ∎

6.3 Derivation of the Poisson Process*

Like the Wiener process, the Poisson process defined in this section is of fundamental importance, not only as an appropriate model for a variety of physical phenomena (such as shot noise in electronic devices, radioactive decay, and photon detection), but also as the core of advanced theories of point processes. Also like the Wiener process, which was originally derived as a limiting form of the discrete-time discrete-amplitude random-walk process, the Poisson counting process was originally derived by a limiting procedure, as explained in the following.

We consider the placement, at random, of m points in the time interval $[0, T]$, and we seek the probability $P_t(n)$ of the event that n of these points will lie in the subinterval $[0, t]$, $t < T$. We can reinterpret this as the probability of getting n successes out of m independent trials of the experiment of placing one point at random in the interval $[0, T]$. Consequently, the probability we seek is given (Section 4.2.2) by the binomial distribution,

$$P_t(n) = \frac{m!}{n!(m-n)!} p^n (1-p)^{m-n}, \tag{6.21}$$

where $p = t/T$, which is the probability of success in one trial. Let us now

---

*The Poisson process is a special type of point process. The theory of more general point processes is well developed and has a wide variety of applications, as briefly mentioned in Section 5.3.5, where references are given.

consider a very long interval $T \gg t$, with many ($m \gg 1$) points. Then for n on the order of mp (a reasonable requirement, since n/m converges to p as $m \to \infty$, by the law of large numbers), we can apply the Poisson theorem [Feller, 1968] to closely approximate the binomial distribution,

$$P_t(n) \simeq e^{-mp}\frac{(mp)^n}{n!}. \tag{6.22}$$

We can arrive at this approximation by simply noting that since n is on the order of mp and $p \ll 1$, then $n \ll m$, and therefore

$$\frac{m!p^n}{(m-n)!} = m(m-1)\cdots(m-n+1)p^n \simeq m^n p^n = (mp)^n \tag{6.23}$$

and

$$(1-p)^{m-n} \simeq (1-p)^m \simeq (e^{-p})^m = e^{-mp}. \tag{6.24}$$

In the limit $T \to \infty$, with $m = \lambda T$ for some fixed parameter λ, which represents the average rate of occurrence or time density of points, the Poisson approximation becomes exact:

$$P_t(n) = e^{-\lambda t}\frac{(\lambda t)^n}{n!}, \tag{6.25}$$

(for which $0! \triangleq 1$). From this expression, we can determine that for $t \to 0$

$$P_t(1) \to \lambda t,$$
$$P_t(0) \to 1 - \lambda t. \tag{6.26}$$

This implies that there can be at most one point in an infinitesimal interval — there can be no coincidences. The preceding results hold for any time interval of length t, say $[\tau, \tau + t]$, not just the interval $[0, t]$. Moreover, it can be shown that (for $T \to \infty$), the numbers of points in each of two (or more) disjoint intervals, say $[\tau_1, \tau_1 + t_1]$ and $[\tau_2, \tau_2 + t_2]$, are statistically independent. Consequently the probability that n_1 points lie in $[\tau_1, \tau_1 + t_1]$ and n_2 points lie in $[\tau_2, \tau_2 + t_2]$ is simply $P_{t_1}(n_1)P_{t_2}(n_2)$.

Now let us reverse our approach, by taking these derived properties as axioms, and thereby obtain an axiomatic definition of what we shall call the *Poisson counting process*. We consider the random occurrence of arbitrary events (e.g., detection of photons) with the passage of time, and we denote the random number of events (the count) at time t, assuming a zero count at $t = 0$, by $N(t)$. Unlike the binomial counting process (Section 4.2.2), for which time is quantized so that counts can occur only at equal time intervals, the counting process that we consider here allows counts to occur

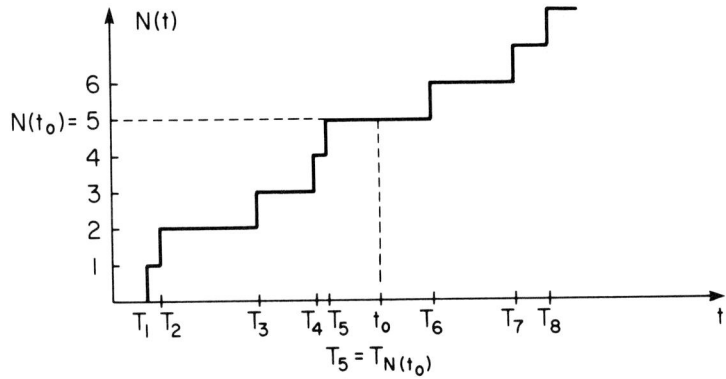

Figure 6.4 Statistical Sample of the Poisson Counting Process

at random times. We can express the counting process $N(t)$ by

$$N(t) = \sum_{i=1}^{\infty} u(t - T_i), \qquad t \geq 0 \tag{6.27}$$

$$= \sum_{i=1}^{N(t)} u(t - T_i) = \sum_{i=1}^{N(t)} 1, \qquad t \geq 0, \tag{6.28}$$

where $u(\cdot)$ is the unit step function (here defined to be continuous from the right),

$$u(t) \triangleq \begin{cases} 1, & t \geq 0, \\ 0, & t < 0. \end{cases} \tag{6.29}$$

Equation (6.28) follows from (6.27) because

$$u(t - T_i) = \begin{cases} 1, & i \leq N(t), \\ 0, & i > N(t), \end{cases} \tag{6.30}$$

since $T_i \leq t$ for $i \leq N(t)$, and $T_i > t$ for $i > N(t)$, where T_i is the random time of occurrence of the ith event. This can be verified pictorially using the typical statistical sample of this counting process shown in Figure 6.4.

The probability of the event $N(t) = n$ is denoted by

$$\text{Prob}\{N(t) = n\} = P_t(n) \tag{6.31}$$

for all nonnegative times t and all nonnegative integers n. It is assumed that the event occurrence times $\{T_i\}$ possess the following two axiomatic properties:

1. For any time interval $[s, t)$, the random number of events that occur in $[s, t)$, which is $N(t) - N(s)$, depends on only the length of the

interval, $t - s$, and is independent of the random number of events that occur in any disjoint interval, say $[u, v)$.
2. The probability of exactly one event occurring in an interval of length Δ_t becomes proportional to Δ_t, say $\lambda \Delta_t$, as Δ_t approaches zero, and the probability of zero events occurring in an interval of length Δ_t becomes approximately $1 - \lambda \Delta_t$ as Δ_t approaches zero.

These two axiomatic properties can be expressed respectively by

$$\text{Prob}\{N(t) - N(s) = n, N(v) - N(u) = m\} = P_{t-s}(n) P_{v-u}(m) \tag{6.32}$$

for $v \geq u \geq t \geq s \geq 0$, and

$$P_{\Delta_t}(1) \to \lambda \Delta_t$$
$$P_{\Delta_t}(0) \to 1 - \lambda \Delta_t \tag{6.33}$$

for $\Delta_t \to 0$, for some $\lambda > 0$. Using nothing more than (6.32) and (6.33), it can be shown (exercise 17) that

$$P_t(n) = \frac{(\lambda t)^n e^{-\lambda t}}{n!}, \qquad t \geq 0, \quad n \geq 0. \tag{6.34}$$

Furthermore, for any set of times $0 \leq t_1 \leq t_2 \leq \cdots \leq t_m$ (not necessarily samples of the random variables $\{T_i\}$), it follows from (6.32) that

$$\text{Prob}\{N(t_1) = n_1, N(t_2) = n_2, \ldots, N(t_m) = n_m\}$$
$$= P_{t_1}(n_1) P_{t_2 - t_1}(n_2 - n_1) \cdots P_{t_m - t_{m-1}}(n_m - n_{m-1}). \tag{6.35}$$

Therefore, all joint probabilities for the process $N(t)$ are completely specified by (6.34). The sequence of random event times $\{T_i\}$ in the Poisson counting process $N(t)$ is called the *Poisson point process*.

Before continuing, a slight generalization of this counting process is introduced. It is assumed that each event point T_i has associated with it a random weight Y_i, so that the *weighted count* is

$$W(t) \stackrel{\Delta}{=} \sum_{i=1}^{\infty} Y_i u(t - T_i), \qquad t \geq 0 \tag{6.36}$$

$$= \sum_{i=1}^{N(t)} Y_i u(t - T_i) = \sum_{i=1}^{N(t)} Y_i, \qquad t \geq 0, \tag{6.37}$$

[which should be compared with (6.4)]. It is assumed that the weights Y_i are independent and have identical distributions, and that the sequence $\{Y_i\}$ is independent of the sequence of occurrence times, $\{T_i\}$.

The mean and covariance of $W(t)$ can be obtained, with the use of (6.32) and (6.34), as follows:

$$E\{W(t)\} = E\{E\{W(t)|N(t)\}\}$$

$$= \sum_{n=1}^{\infty} E\left\{\sum_{i=1}^{N(t)=n} Y_i\right\} P_t(n)$$

$$= m_Y \sum_{n=1}^{\infty} n P_t(n)$$

$$= m_Y \sum_{n=1}^{\infty} \frac{(\lambda t)^n e^{-\lambda t}}{(n-1)!}$$

$$= m_Y \lambda t \left\{\sum_{m=0}^{\infty} \frac{(\lambda t)^m}{m!}\right\} e^{-\lambda t}$$

$$= m_Y \lambda t = m_W(t), \qquad t \geq 0. \qquad (6.38)$$

Similarly, it can be shown (exercise 10) that

$$K_W(t_1, t_2) = \lambda(\sigma_Y^2 + m_Y^2) \min\{t_1, t_2\}, \qquad t_1, t_2 \geq 0, \qquad (6.39)$$

which is of the same form as the autocovariance of the Wiener process. This is directly related to the fact that both processes have stationary uncorrelated increments.

6.4 The Derivative of the Poisson Counting Process

Analogous to our heuristic interpretation of the Wiener process as the integral of a symbolic process $Z(t)$, which is an infinitely dense collection of infinitesimal impulses, we have the heuristic interpretation of the randomly weighted Poisson counting process $W(t)$ as the integral of a symbolic process $Z(t)$, which is a finitely dense collection of finite impulses:

$$W(t) = \int_0^t Z(u)\,du, \qquad t \geq 0, \qquad (6.40)$$

$$Z(t) = \frac{dW(t)}{dt}, \qquad t \geq 0, \qquad (6.41)$$

where, from (6.36),

$$Z(t) = \sum_{i=1}^{\infty} Y_i \delta(t - T_i), \qquad t \geq 0. \qquad (6.42)$$

A typical statistical sample of $Z(t)$ is shown in Figure 6.5.

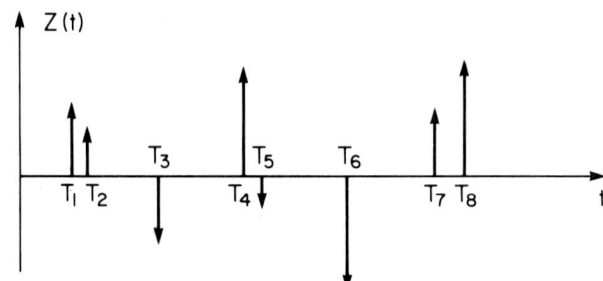

Figure 6.5 Statistical Sample of a Randomly Weighted Poisson Impulse Process

The mean and covariance of $Z(t)$ can be obtained from the mean and covariance of $W(t)$ by formal interchange of the linear operations of expectation and differentiation:

$$m_Z(t) = \frac{d}{dt} m_W(t) = m_Y \lambda, \qquad t \geq 0, \tag{6.43}$$

$$K_Z(t_1, t_2) = \frac{\partial^2}{\partial t_1 \partial t_2} K_W(t_1, t_2) = \lambda(\sigma_Y^2 + m_Y^2)\delta(t_1 - t_2), \qquad t_1, t_2 \geq 0. \tag{6.44}$$

This formal manipulation is justified in Chapter 7. We see that, like the derivative of the Wiener process, the derivative of the Poisson counting process is stationary for $t \geq 0$.

6.5 Marked and Filtered Poisson Processes

There are many physical phenomena that give rise to random waveforms consisting of trains of randomly occurring pulses, and in many situations the Poisson point process $\{T_i\}$ is an appropriate model for the occurrence times of the random pulses. Frequently, the randomness of each pulse can be modeled in terms of a few random parameters, such as amplitude or width. Thus, we consider the model

$$X(t) = \sum_{i=1}^{\infty} h(t - T_i, Y_i), \tag{6.45}$$

where $\{Y_i\}$ is a sequence of independent and identically distributed vectors, whose elements are the random parameters of the pulses. If the pulses are causal in the sense that

$$h(t, Y) = 0, \qquad t < 0, \tag{6.46}$$

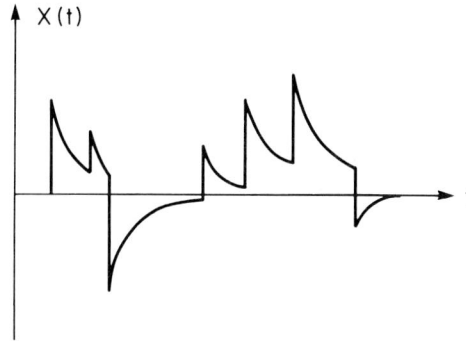

Figure 6.6 Statistical Sample of a Marked and Filtered Poisson Process with Causal Pulses

as illustrated in Figure 6.6, then (6.45) can be reexpressed in terms of the Poisson counting process $N(t)$:

$$X(t) = \sum_{i=1}^{N(t)} h(t - T_i, Y_i). \qquad (6.47)$$

The parameter vectors Y_i are called *marks*, since each vector marks the corresponding event time with a set of event characteristics. The process $X(t)$ is said to be a *marked and filtered Poisson process*, because it can be envisioned as having been generated by passing the impulses of a Poisson impulse process through randomly marked filters with impulse response functions $h(t, Y)$.

Example 1 (Shot noise): Consider a vacuum-tube diode in which electrons that are emitted from the heated cathode are attracted to the anode. If the electron emission rate is temperature-limited so that space-charge effects are negligible, then the emission times are well modeled by the Poisson process. Furthermore, the current through the diode that results from these emissions can be modeled as a filtered Poisson process

$$X(t) = \sum_{i=1}^{N(t)} h(t - T_i), \qquad t \geq 0,$$

where $t = 0$ is the time at which the diode is energized, $N(t)$ is the number of emissions during $[0, t)$, $\{T_i\}$ are the emission times, and the form of the pulse h is a function of cathode-anode geometry, cathode temperature, and anode voltage. The area of h is the charge on one electron, and the duration of h is approximately the anode-to-cathode transit time. Shot noise occurs also in other electronic devices: it results from generation recombination noise in bulk semiconductors, emission noise in *pn* junction semiconductor devices, microwave tube noise, Barkhausen noise in magnetic tapes, and so on. A detailed study of shot noise in vacuum tubes is presented in [Davenport and Root, 1958, Chapter 7]. This is discussed in more detail in Section 10.7. ∎

Example 2 (Photon detection): Electron-hole pairs created in a bulk semiconductor by the absorption of light photons increase the conductivity of the material during the lifetime of the pairs. This effect, called *photoconductivity*, provides a mechanism for the detection of optical signals. Upon application of a bias voltage across the semiconductor, the change in conductivity due to the absorption of a photon results in a measurable change in current flowing through the device. For a constant light intensity, the current can be modeled by a marked and filtered Poisson process

$$X(t) = \sum_{i=1}^{N(t)} \frac{1}{Y_i} h\left(\frac{t - T_i}{Y_i}\right), \quad t \geq 0,$$

where $t = 0$ is the time of application of the light, $N(t)$ is the number of photons absorbed during $[0, t)$, $\{T_i\}$ are the absorption times, and Y_i is the duration of the incremental contribution to $X(t)$, which depends on the lifetime of the electron-hole pair created by the photon absorption. The area of $(1/Y_i)h(t/Y_i)$ is the charge of one electron. If the light intensity varies with time, the model $\{T_i\}$ can be generalized to what is called an *inhomogeneous Poisson process*, for which the rate parameter λ varies with time. Furthermore, if the light intensity is modeled as a random process, then the time-variant rate parameter $\lambda(t)$ is a random process, and the model for the point process $\{T_i\}$ is called a *doubly stochastic Poisson process* [Snyder, 1975]. ∎

Example 3 (ELF-VLF atmospheric noise): Atmospheric noise in frequency bands below 30 kHz (i.e., the very-low-frequency and extremely-low-frequency bands) is due mainly to lightning discharges. The effect of such noise in a radio receiver can be modeled by a marked and filtered Poisson process

$$X(t) = \sum_{i=1}^{N(t)} Y_i h(t - T_i), \quad t \geq 0,$$

where $t = 0$ is the time at which the receiver is energized, $N(t)$ is the number of received discharges during $[0, t)$, $\{T_i\}$ are the occurrence times of the discharges, h is the response of the receiver to a discharge (which appears to be an impulse in the ELF-VLF band), and Y_i is the strength of the discharge. ∎

Example 4 (*Radar clutter and scattering communication channels*): Consider a signal $s(t)$ propagated through a radio channel consisting of small scatterers distributed in space. Examples are radar signals reflected from water droplets in rain, and communication signals reflected from a cloud of small dipoles. Let the signal be given by

$$s(t) = a(t)\cos[\omega_o t + \phi(t)], \quad t \geq 0,$$

where $a(t)$ and $\phi(t)$ are deterministic amplitude and phase modulation, respectively, of the sine-wave carrier. The received signal can be modeled by the marked and filtered Poisson process

$$X(t) = \sum_{i=1}^{N(t)} Y_{1i} a(t - T_i) \cos[\omega_o(t - T_i) - \phi(t - T_i) + Y_{2i}],$$

$$= \sum_{i=1}^{N(t)} h(t - T_i, Y_{1i}, Y_{2i}), \quad t \geq 0,$$

where $N(t)$ is the number of scattered signals received during $[0, t)$, $\{T_i\}$ are their arrival times at the receiver, and Y_{1i} and Y_{2i} are random attenuation and phase variables associated with each scatterer. ∎

6.6 Summary

The *Wiener process* is a nonstationary Gaussian process with zero mean, independent stationary increments, and autocovariance given by (6.13). Its formal derivative is a stationary symbolic process with impulsive autocovariance, (6.20). This process can be either defined axiomatically as in (6.1), or constructed as the limiting form of an elementary random walk process.

The *Poisson counting process* is analogous to the Wiener process; it is a nonstationary process with independent stationary increments, and autocovariance given by (6.39), but it is non-Gaussian and has nonzero mean. It also can be either defined axiomatically, as in (6.32)–(6.33), or constructed as the limiting form of an elementary binomial process. Its formal derivative is a stationary symbolic process with impulsive autocovariance (6.44).

The *Poisson point process* underlying the Poisson counting process forms the basis for models of a wide variety of physical phenomena involving randomly occurring events at discrete points in time. The models are called *marked and filtered Poisson processes*.

The *homogeneous Poisson process* with constant rate parameter λ can be generalized to an *inhomogeneous Poisson process* with time-variant rate parameter $\lambda(t)$, and this can be further generalized to a *doubly stochastic Poisson process* with random time-variant rate parameter.

EXERCISES

1. Investigate the consequences on the variance of the Wiener process model of changing the relationship (6.5) between step sizes to
$$\Delta_w = \alpha \Delta_t^x$$
for $x > 1/2$ and also for $x < 1/2$.

2. Let $W(t)$ be the Wiener process and define $X(t)$ by
$$X(t) = e^{-t}W(e^{2t}).$$
Show that $X(t)$ is a stationary Gauss Markov process and is therefore the Ornstein-Uhlenbeck process.

3. Use (6.13) to verify that the Wiener process has uncorrelated increments, that is,
$$E\{[W(t_1) - W(t_2)][W(t_3) - W(t_4)]\} = 0, \qquad (6.48)$$
provided that the time intervals $[t_1, t_2)$ and $[t_3, t_4)$ are disjoint.

★4. Use the representation
$$\min\{t_1, t_2\} = t_1 u(t_2 - t_1) + t_2 u(t_1 - t_2), \qquad (6.49)$$
where $u(\cdot)$ is the unit step function (6.29), and
$$\delta(t) = \frac{d}{dt}u(t), \qquad (6.50)$$
to derive (6.20) from (6.19).

Hint: Use the property $f(u)\delta(t - u) = f(t)\delta(t - u)$ for any function f continuous at t.

Then draw a graph of the function $\min\{t_1, t_2\}$ as a surface above the (t_1, t_2) plane. Describe the paths in the (t_1, t_2) plane along which $\min\{t_1, t_2\}$ is constant.

5. Let $Y(t) = X(t)Z(t)$, where $Z(t)$ is the symbolic process (6.16) obtained by formal differentiation of the Wiener process $W(t)$, and where $X(t)$ is independent of $W(t)$. Show that the autocorrelation of $Y(t)$ is given by
$$R_Y(t_1, t_2) = \alpha^2[m_X^2(t_1) + \sigma_X^2(t_1)]\delta(t_1 - t_2).$$

★6. *Gaussian property of Wiener process:* In order to show that the Wiener process is Gaussian, we proceed as follows.

   (a) In order for the random-walk process $W_\Delta(t)$ to take on the value $r\Delta_w$ at time $n\Delta_t$, there must have been k positive steps (successes) and $n - k$ negative steps, for which $r = k - (n - k) = 2k - n$. Thus, the event $W_\Delta(n\Delta_t) = r\Delta_w$ is identical to the event of k successes out of n Bernoulli trials, where $k = (r + n)/2$. It follows from (4.18) (with $p = \frac{1}{2}$) that the probability of this event is

$$P_{W_\Delta(n\Delta_t)}(r\Delta_w) = \frac{n!}{k!(n-k)!} \left(\frac{1}{2}\right)^n. \qquad (6.51)$$

Moreover, it follows from the DeMoivre-Laplace theorem [Feller, 1968] that for large n we have the close approximation

$$P_{W_\Delta(n\Delta_t)}(r\Delta_w) \simeq (2\pi n)^{-1/2} \exp\left[-\frac{2(k - n/2)^2}{n}\right], \qquad (6.52)$$

with $k = (n + r)/2$, which becomes exact in the limit $n \to \infty$, provided that r is on the order of $\sqrt{n}$. Use this result to verify that the first-order probability density for the Wiener process is

$$f_{W(t)}(w) = (2\pi \alpha^2 t)^{-1/2} \exp\left[-\frac{w^2}{2\alpha^2 t}\right]. \qquad (6.53)$$

Hint: Let $w = r\Delta_w$, $t = n\Delta_t$; hold w and t fixed while letting $\Delta_t \to 0$, subject to (6.5); and define $f_{W(t)}$ to be the limit of $(1/\Delta_w) P_{W_\Delta}$ as $\Delta_t \to 0$.

Verify that r is indeed on the order of $\sqrt{n}$ as $n \to \infty$.
(b) Since the random-walk process (Section 4.2.3) has stationary independent increments, then so too does the Wiener process (although this needs a proof). It follows from the fact that $W(t)$ has stationary increments and $W(0) = 0$ that

$$f_{W(t)}(w) = f_{W(t+s) - W(s)}(w),$$

and therefore (a) reveals that $W(t)$ has Gaussian increments. Use this and the fact that $W(t)$ has independent increments to show that $W(t)$ is a Gaussian process.

★7. (a) Verify that the Wiener process is a homogeneous Markov process and is nonstationary.
(b) Show that the transition density $f_{W(t)|W(t_o)}(w|w_o)$ for the Wiener process is given by (6.53) with the replacements $t \to t - t_o$ and $w \to w - w_o$ for $t_o \geq 0$. Then verify that this transition density satisfies the following *diffusion equation*:

$$\frac{\partial}{\partial t} f_{W(t)|W(t_o)}(w|w_o) = D^2 \frac{\partial^2}{\partial w^2} f_{W(t)|W(t_o)}(w|w_o),$$

where $D = \alpha/\sqrt{2}$ is the *diffusion constant*. Show also that this diffusion equation follows from the Fokker-Planck equation (5.66).

8. *Level crossings of Wiener process:* Use the symmetry property (from Desiré André's *reflection principle* [Doob, 1953, Chapter VIII])

$$\text{Prob}\{W(t) \le w \mid W(s) = w, s \le t\}$$
$$= \text{Prob}\{W(t) \ge w \mid W(s) = w, s \le t\} = \frac{1}{2}, \quad (6.54)$$

exhibited by the Wiener process for $w > 0$, to prove that the probability $P_w(t)$ that the Wiener process crosses the amplitude level w in the interval $[0, t]$ is

$$P_w(t) = 2 - 2F_{W(t)}(w). \quad (6.55)$$

Then use the mean and variance expressions, from (6.12) and (6.13), together with the fact that $W(t)$ is Gaussian, to obtain an explicit formula for $P_w(t)$ in terms of the error function (cf. Section 1.8). Show that as $t \to \infty$, $P_w(t) \to 1$, regardless of w.

Hint: To establish (6.55), use the definition of conditional probability together with the fact that the probability of the level-crossing event, say A, satisfies

$$P(A) = P(A, W(t) \le w) + P(A, W(t) \ge w).$$

9. Verify that for the Poisson counting process,

$$E\{\text{number of events in } (t, t + T]\} = E\{N(t + T) - N(t)\} = \lambda T,$$
$$(6.56)$$

and therefore that λ is the *mean rate of occurrence of events.*

★**10.** Use a procedure similar to that used to derive (6.38) to derive (6.39).

★**11.** Verify that the Poisson counting process is a homogeneous Markov process.

12. *Asynchronous telegraph signal:* Consider the process

$$Y(t) = (-1)^{N(t)}, \quad t \ge 0,$$

where $N(t)$ is the Poisson counting process with rate parameter λ. This process starts at $Y(0) = 1$ and switches back and forth from $+1$ to -1 at random Poisson times $\{T_i\}$ as illustrated in Figure 6.7. Verify the following derivations of the mean and autocorrelation for $Y(t)$.
(a) Since $(-1)^i = 1$ for i even and $(-1)^i = -1$ for i odd, then for $t \ge 0$

$$m_Y(t) = \sum_{\substack{n=0 \\ n \text{ even}}}^{\infty} P_t(n) - \sum_{\substack{n=1 \\ n \text{ odd}}}^{\infty} P_t(n)$$

$$= e^{-\lambda t} \cosh \lambda t - e^{-\lambda t} \sinh \lambda t$$

$$= e^{-2\lambda t}. \quad (6.57)$$

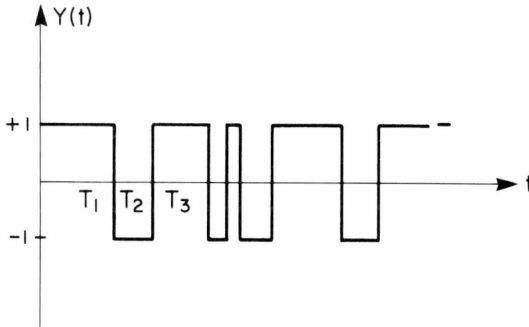

Figure 6.7 Statistical Sample of the Asynchronous Telegraph Signal

(b) Similarly, since $Y(t + \tau)Y(t) = 1$ if there are an even number of events in $(t, t + \tau]$ for $\tau > 0$ (or in $(t + \tau, t]$ for $\tau < 0$), and $Y(t + \tau)Y(t) = -1$ if there are an odd number of events, then for $t \geq 0$ and $t + \tau \geq 0$,

$$R_Y(\tau) = \sum_{\substack{n=0 \\ n \text{ even}}}^{\infty} P_{|\tau|}(n) - \sum_{\substack{n=1 \\ n \text{ odd}}}^{\infty} P_{|\tau|}(n) \qquad (6.58)$$
$$= e^{-2\lambda|\tau|}.$$

It follows from (a) and (b) that the asynchronous telegraph signal is asymptotically WSS.

(c) The preceding process is sometimes referred to as the *semirandom asynchronous telegraph signal* because its initial value $Y(0) = 1$ is nonrandom. This bias is removed in the model

$$U(t) = AY(t),$$

where A is independent of $Y(t)$ and takes on the values ± 1 with equal probability. Show that $U(t)$ is WSS for $t \geq 0$.

13. On the basis of the number k of photons counted in a symbol interval of length T, a binary digital receiver for a fiber optic communication system must decide which of two weak light intensities has occurred. Use Poisson process models with rate parameters λ_1 and λ_2 ($\lambda_2 < \lambda_1$) for the photon counts resulting from each of the two possible light intensities to show that the most probable light intensity is determined by the decision rule: Decide λ_1 if

$$k > \frac{(\lambda_1 - \lambda_2)T - \ln[P(\lambda_1)/P(\lambda_2)]}{\ln(\lambda_1/\lambda_2)},$$

otherwise decide λ_2.

Hint: Determine $P[k|\lambda_1]$ and $P[k|\lambda_2]$ and then use Bayes' law to determine $P[\lambda_1|k]$ and $P[\lambda_2|k]$. (Note: If on-off keying is used, then λ_2 is simply the rate parameter for background radiation. Also, if equiprobable symbols are transmitted over the optical fiber, then $P(\lambda_1) = P(\lambda_2)$.)

14. One approach to calculating the steady-state mean and autocovariance of marked and filtered Poisson processes is developed in this exercise. We consider, as an example, the shot noise process described in Section 6.5,

$$X(t) = \sum_{i=1}^{N(t)} h(t - T_i), \qquad t \geq 0.$$

We begin by letting the starting time of the process approach $-\infty$ and then approximating the process by

$$\hat{X}(t) \triangleq \sum_{n=-\infty}^{\infty} X_n h(t - n\Delta_t),$$

where

$$X_n \triangleq \begin{cases} 1, & \text{if } n\Delta_t < T_i < (n+1)\Delta_t, \\ 0, & \text{otherwise} \end{cases}$$

for any event time T_i, and X_n and X_m are independent for $n \neq m$. By considering a sufficiently small time increment Δ_t, we obtain the close approximations

$$P[X_n = 0] \simeq 1 - \lambda\Delta_t$$

$$P[X_n = 1] \simeq \lambda\Delta_t.$$

Use this approximate model and the straightforward techniques illustrated in Chapter 4 to evaluate the mean and autocovariance for $\hat{X}(t)$. Then take the limit as $\Delta_t \to 0$, and use the definition

$$\lim_{\Delta_t \to 0} \sum_{n=-\infty}^{\infty} h(t - n\Delta_t)\Delta_t = \int_{-\infty}^{\infty} h(t - u)\, du$$

of the Riemann integral, to show that the steady state parameters are given by

$$m_X = \lim_{\Delta_t \to 0} m_{\hat{X}} = \lambda \int_{-\infty}^{\infty} h(u)\, du$$

$$K_X(\tau) = \lim_{\Delta_t \to 0} K_{\hat{X}}(\tau) = \lambda \int_{-\infty}^{\infty} h(\tau + u)h(u)\, du.$$

15. (a) Use the approach developed in the preceding exercise to derive the steady-state mean and autocovariance for the ELF-VLF atmospheric noise model described in Section 6.5.
(b) Repeat part (a) for the more general model

$$X(t) = \sum_{i=1}^{N(t)} h(t - T_i, Y_i).$$

16. Consider the Poisson-impulse-sampled process

$$U(t) = Z(t)X(t) = \sum_{i=1}^{\infty} X(T_i)\delta(t - T_i), \qquad (6.59)$$

for which $Z(t)$ is the derivative of the Poisson counting process with unit counts, $Y_i = 1$, and $X(t)$ is a mean-square continuous stationary process that is independent of $Z(t)$. Determine the mean and autocorrelation functions for $U(t)$. Compare (6.59) with (6.42), and explain why the result for K_U is different from (6.44).

Hint: Use the result of exercise 9 in Chapter 4.

★17. *Count probability:*
(a) Use (6.32) and (6.33) to obtain

$$\frac{1}{\Delta_t}[P_{t+\Delta_t}(0) - P_t(0)] = -\lambda P_t(0)$$

for $\Delta_t \to 0$. Then use the limiting ($\Delta_t \to 0$) form of this equation,

$$\frac{dP_t(0)}{dt} = -\lambda P_t(0),$$

to obtain the solution

$$P_t(0) = e^{-\lambda t}. \qquad (6.60)$$

(b) There are $n + 1$ distinct ways to obtain n counts in the interval $[0, t + \Delta_t)$ according to the numbers of counts in the disjoint intervals $[0, t)$ and $[t, t + \Delta_t)$; for example, there can be 0 counts in $[0, t)$ and n counts in $[t, t + \Delta_t)$, or there can be 1 count in $[0, t)$ and $n - 1$ counts in $[t, t + \Delta_t)$, etc. Since these $n + 1$ different ways are mutually exclusive events, and since [from (6.32)] each pair of events making up each one of these $n + 1$ ways consists of two independent events, we have

$$P_{t+\Delta_t}(n) = \sum_{k=0}^{n} P_t(n - k)P_{\Delta_t}(k).$$

Use (6.33) to show that, with $\Delta_t \to 0$, this equation yields

$$\frac{dP_t(n)}{dt} + \lambda P_t(n) = \lambda P_t(n-1). \tag{6.61}$$

Hint: Terms corresponding to $k > 1$ vanish in the limit. (Why?) Use (6.60) and (6.61) to obtain

$$P_t(n) = e^{-\lambda t} \frac{(\lambda t)^n}{n!}. \tag{6.62}$$

18. *Event-point probability density:*
    (a) The probability that the kth event point T_k occurs before time t is
    $$F_{T_k}(t) = \text{Prob}\{T_k < t\} = 1 - \text{Prob}\{N(t) < k\}.$$
    Verify that this can be expressed as
    $$F_{T_k}(t) = 1 - \sum_{i=0}^{k-1} P_t(i),$$
    and then differentiate this expression for the probability distribution to obtain the following formula for the probability density:
    $$f_{T_k}(t) = \begin{cases} \lambda e^{-\lambda t} \dfrac{(\lambda t)^{k-1}}{(k-1)!}, & t \geq 0, \\ 0, & t < 0. \end{cases} \tag{6.63}$$
    Sketch this function for $k = 1, 2, 3$.
    (b) Use (6.62) to show that
    $$E\{T_k\} = k/\lambda,$$
    $$\text{Var}\{T_k\} = k/\lambda^2. \tag{6.64}$$

19. *Interarrival-time probability density:* Denote the interarrival time between the $k - 1$ and k event points by
    $$S_k \triangleq T_k - T_{k-1}.$$
    The probability that S_k is less than s is
    $$F_{S_k}(s) = 1 - \text{Prob}\{S_k > s\}.$$
    Verify that this can be expressed as
    $$F_{S_k}(s) = 1 - P_s(0),$$
    and then differentiate this expression for the probability distribution to obtain the following formula for the probability density:
    $$f_{S_k}(s) = \begin{cases} \lambda e^{-\lambda s}, & s \geq 0 \\ 0, & s < 0. \end{cases} \tag{6.65}$$

20. Let $\{S_j: j = 1, 2, 3, \ldots\}$ be a sequence of independent, identically distributed random variables with probability density

$$f_S(s) = \begin{cases} \lambda e^{-\lambda s}, & s \geq 0, \\ 0, & s < 0, \end{cases}$$

and form the series

$$T_k = \sum_{j=1}^{k} S_j.$$

It can be shown that $\{T_k: k = 1, 2, 3, \ldots\}$ is the Poisson point process with rate parameter λ. Verify that the probability density of T_k is indeed given by (6.63).

21. The photon count from a photo detector is modeled as a Poisson process with rate parameter proportional to the intensity of light impinging on the detector. Determine how the mean and variance of the interarrival time, and the mean and variance of the time of the kth count, change when the light intensity is doubled.

22. *Reliability:* In reliability studies, the *conditional failure rate* $\lambda(t)$ plays a central role. This rate is defined so that $\lambda(t) \, dt$ is the probability of failure in the infinitesimal time interval $(t, t + dt)$, given that failure did not occur before time t. Thus, if $f_{T|T \geq t}(t)$ is the probability density for this event, then

$$\lambda(t) = f_{T|T \geq t}(t).$$

Show that

$$\lambda(t) = \frac{dF_T(t)/dt}{1 - F_T(t)}. \tag{6.66}$$

Assuming that operation commences at $t = 0$, so that $F_T(0) = 0$, show that

$$F_T(t) = 1 - \exp\left\{-\int_0^t \lambda(\tau) \, d\tau\right\}. \tag{6.67}$$

Then show that for a constant failure rate,

$$f_T(t) = \lambda e^{-\lambda t}, \quad t \geq 0. \tag{6.68}$$

Compare this with the probability density for the first event point in a Poisson point process.

23. (a) Consider a device consisting of a large number of small parts that wear out continually. An appropriate failure rate for such a device would increase linearly with time, $\lambda(t) = kt$. Use this model together with the reliability relationship (6.66) to show that the probability density function for the random failure time is the Rayleigh density.

(b) Consider the failure rate $\lambda(t) = \gamma t^\beta$. Show that the probability density of the failure time is the *Weibull density*

$$f_T(t) = \gamma t^\beta \exp\left(-\frac{\gamma}{\beta + 1} t^{\beta+1}\right), \quad t \geq 0. \quad (6.69)$$

24. *Single-server queue:* Consider a service queue into which the ith unit to be served enters at the *arrival time* A_i, and is served with a *service time* of S_i, and leaves from the queue at the *departure time* D_i. Assume that there is a single server, so that the *waiting time* for the ith unit in the queue is

$$W_i = D_{i-1} - A_i \quad (6.70a)$$

and the service time is the interdeparture time,

$$S_i = D_i - D_{i-1}, \quad (6.70b)$$

as shown in Figure 6.8. It follows that the total time that the ith unit is in the queue,

$$Q_i \triangleq D_i - A_i, \quad (6.71)$$

is the sum of its waiting time and service time,

$$Q_i = W_i + S_i. \quad (6.72)$$

This total time in the queue is called the *queue time*. Assume that the arrival times $\{A_i\}$ constitute a Poisson point process with associated counting process $N(t)$ and rate parameter λ, and assume that the service times $\{S_i\}$ are independent and identically distributed positive random variables.

(a) Show that the expected number of arrivals during the ith service time is given by

$$E\{N(t + S_i) - N(t)\} = \lambda m_S.$$

This reveals that the expected queue size will not grow without bound only if $\lambda < 1/m_S$.

Hint: Evaluate the expectation, conditional on $S_i = s$, and use (6.56).

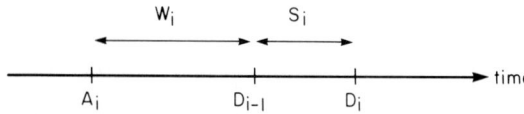

Figure 6.8 Relationship Among Arrival Times, Departure Times, Service Times, and Waiting Times for a Single-Server Queue.

(b) Let the number of arrivals during the ith unit's stay in the queue be denoted by N_i,

$$N_i \triangleq N(D_i) - N(A_i). \qquad (6.73)$$

Since all units that arrived in the queue prior to the ith unit have departed by time D_i, and the ith unit departs at time D_i, then N_i is the total number of units in the queue immediately following time D_i. Consequently, N_i is referred to as the *state of the system* at time D_i. Show that the mean state is given by

$$m_N = \lambda m_Q = \lambda(m_W + m_S). \qquad (6.74)$$

Hint: Evaluate the expectation of N_i conditional on D_i and A_i, and use (6.56).

(c) It follows from the celebrated *Khinchin-Pollaczek formula* for the characteristic function for the waiting time W_i [Feller, 1971] that

$$m_W = \frac{\lambda E\{S^2\}}{2(1 - \lambda E\{S\})}. \qquad (6.75)$$

Use (6.74) and (6.75) to graph the mean state m_N as a function of the ratio of the mean arrival rate λ to the reciprocal $1/m_S$ of the mean service time, which can be thought of as the mean service rate.

Hint: Use $E\{S^2\} = \sigma_S^2 + m_S^2$ and let $\lambda \sigma_S$ be fixed.

7

Stochastic Calculus

SIGNAL-PROCESSING SYSTEMS can be broadly described in terms of the operations they perform on postulated classes of inputs— for example, linear operations such as differentiation, integration, multiplication, and convolution of input waveforms, and the counterparts of differencing, summing, and so on, of discrete-time input sequences. Thus, the probabilistic study of signals and systems is, in a general sense, the study of averages of ensembles of waveforms and sequences subjected to these various operations. The underlying probability theory is, in essence, a calculus of averages. In this chapter, the concept of a stochastic calculus, a calculus for random processes, is introduced, and the technicalities of a particular stochastic calculus are briefly studied.

7.1 *The Notion of a Calculus for Random Functions*

In order to introduce the concept of a calculus for random functions, we consider the following inquiry regarding thermal noise. Can we in some sense predict, in advance of actual measurements, the squared root-mean-square (rms) value of the thermal noise voltage $V(t)$, given by

$$V^2_{\text{rms}_T} = \frac{1}{T}\int_0^T V^2(t)\,dt\,? \qquad (7.1)$$

We shall model the random noise process $V(t)$ by the convolution of the symbolic broadband Gaussian noise process $Z(t)$ (Section 6.2) with the nonrandom impulse-response function, denoted by $h(t)$, of the instantaneous voltmeter used to measure $V(t)$:

$$V(t) = \int_{-\infty}^{t} h(t-u)Z(u)\,du, \qquad (7.2)$$

where $Z(t)$ is the formal derivative of the Wiener process $W(t)$,

$$Z(t) = \frac{dW(t)}{dt}. \qquad (7.3)$$

Since the waveform $V(t)$ is random, then the number $V_{\text{rms}_T}^2$ is random. So for now we shall settle for determining the mean and the variance of this random variable, but we shall also consider the problem of determining the behavior of these probabilistic parameters as the length T of the time interval over which $V_{\text{rms}_T}^2$ is measured is increased. Hence, we want to evaluate the quantities

$$m_{Y_T} \quad \text{and} \quad \sigma_{Y_T}^2,$$

$$\lim_{T \to \infty} m_{Y_T} \quad \text{and} \quad \lim_{T \to \infty} \sigma_{Y_T}^2, \qquad (7.4)$$

where

$$Y_T \triangleq V_{\text{rms}_T}^2. \qquad (7.5)$$

We see from (7.1) to (7.3) that the random variable Y_T involves derivatives and integrals of random processes. It would appear, therefore, that such a problem requires a calculus for random functions, that is, a *stochastic calculus*.

A natural response to this statement is: "Since a random process is a collection of nonrandom functions, why not simply define the derivative of a process to be the collection of derivatives of the nonrandom functions, and likewise for the integral of a process?" But recall that the derivative of a function $X(\cdot, s)$ is defined to be the limit of the sequence of differences

$$\frac{dX(t,s)}{dt} \triangleq \lim_{\varepsilon \to 0} \frac{1}{\varepsilon} [X(t,s) - X(t-\varepsilon, s)], \qquad (7.6)$$

and the Riemann integral is defined to be the limit of the sequence of areas of rectangles

$$\int_0^t X(u,s)\,du \triangleq \lim_{\varepsilon \to 0} \varepsilon \sum_{i=1}^{t/\varepsilon} X(i\varepsilon, s), \qquad (7.7)$$

where the sequence of values of ε that approach zero are chosen so that the corresponding sequence of values of t/ε are integers. We see from (7.6) and

(7.7) that if we are to define the derivative and the integral of a process $X(t)$ in terms of the conventional derivative and integral of each sample function $X(t, s)$, then we must require the existence of limits of samples of sequences of random variables for every sample point s in the sample space. But, as we have already discussed in Section 2.5, such a requirement is too stringent; for many random processes of interest these limits do not exist for every sample point. [In fact, for some processes, such as the Wiener process, the limit (7.6) fails to exist for almost every sample point.] Therefore, we must use one of the weaker types of convergence of sequences of random variables, defined in Section 2.5, in order to obtain useful definitions of the derivative and integral of a random process. In other words, we do indeed need a stochastic calculus, which is a calculus based on stochastic convergence.

A stochastic calculus based on almost sure convergence is quite involved* and is beyond the scope of this book. However, a stochastic calculus based on mean-square convergence, to be called *mean-square calculus*, is quite tractable, and is appropriate for the study of second-order properties of random processes. This is especially true for the study of random processes subjected to linear transformations, for example, linear dynamical systems with random excitation, and random signals and noise subjected to filtering operations, which are the application areas of primary concern in this book. Nevertheless, it should be understood that for the study of nonlinear dynamical systems with random excitation, mean-square calculus is often inadequate, and the much more technical stochastic calculus based on almost sure convergence must be resorted to. A primary application of this more technical calculus is to studies of the evolution of the probability distribution of the state of a dynamical system.

7.2 Mean-Square Continuity

Definition. A process X is *mean-square continuous at t* if and only if the limit

$$\lim_{\varepsilon \to 0} E\{[X(t) - X(t - \varepsilon)]^2\} = 0 \tag{7.8}$$

exists.

It is easily shown [Loève, 1978, Chapter XI] that (7.8) is satisfied if and only if $R_X(t_1, t_2)$ is continuous in t_1 and in t_2 at the point $t_1 = t_2 = t$; that is, the limit

$$\lim_{\varepsilon_1, \varepsilon_2 \to 0} [R_X(t - \varepsilon_1, t - \varepsilon_2) - R_X(t, t)] = 0 \tag{7.9}$$

must exist.

---

*See, for example, [Jazwinski, 1970], [Gihman and Skorohod, 1974], [Arnold, 1974], [Fleming and Rishel, 1975], [Lipster and Shiryayev, 1977], and [Larson and Shubert, 1979].

This is *not*, in general, satisfied by the property

$$\lim_{\varepsilon \to 0} \left[R_X(t - \varepsilon, t - \varepsilon) - R_X(t, t) \right] = 0. \tag{7.10}$$

In fact, for the sampled-and-held noise process (Section 4.2.6) it can be shown that (7.10) is satisfied but (7.9) is violated (exercise 1). This simply means that $R_X(\cdot, \cdot)$ is continuous *along* the line $t_1 = t_2 = t$ in the (t_1, t_2) plane, but discontinuous *across* this line, at least at some points $t_1 = t_2 = t$. (See exercise 10 of Chapter 4.)

If the process X is WSS, then it follows immediately that (7.9) is satisfied if and only if $R_X(\tau)$ is continuous at $\tau = 0$; that is, the limit

$$\lim_{\varepsilon \to 0} \left[R_X(\varepsilon) - R_X(0) \right] = 0 \tag{7.11}$$

must exist. Hence, if a WSS process is mean-square continuous at one time t, it is at all times. Moreover, since the only models considered in this book for random WSS processes whose autocorrelations are discontinuous at the origin are those for which the autocorrelation has an impulse at the origin (as discussed in Section 5.4.2), these are the only mean-square discontinuous WSS processes of interest to us.

Example 1: For the asynchronous telegraph signal Y (Chapter 6, exercise 12), the autocorrelation is (for $t > 0$ and $t + \tau > 0$)

$$R_Y(\tau) = e^{-2\lambda |\tau|}, \tag{7.12}$$

from which it can be shown that (7.11) is satisfied. Hence Y is mean-square continuous for all time ($t > 0$). We note however that *every* sample function of Y has many step discontinuities. This illustrates the fact that mean-square continuity is a weak form of continuity. ∎

Example 2: For the Wiener process (Section 6.1), the autocorrelation is (for $t_1 > 0$, $t_2 > 0$)

$$R_W(t_1, t_2) = \alpha^2 \min\{t_1, t_2\}, \tag{7.13}$$

from which it can be shown that (7.9) is satisfied for all $t > 0$. Hence, W is mean-square continuous for all time ($t > 0$). This seems in agreement with the heuristic interpretation of W as an infinitely dense collection of infinitesimal steps. ∎

Example 3: For the symbolic broadband Gaussian noise process Z, described in Chapter 6 as the formal derivative of the Wiener process W, the autocorrelation is

$$R_Z(\tau) = \alpha^2 \delta(\tau), \tag{7.14}$$

from which it can be shown that (7.11) is violated. Hence Z is not mean-square continuous. This seems in agreement with the heuristic interpretation of Z as an infinitely dense collection of infinitesimal impulses. ∎

It can be shown (exercise 5) that if $X(\cdot)$ is mean-square continuous at t, then the mean function $m_X(\cdot)$ is continuous in the ordinary sense at t. This reveals that *for a mean-square continuous process, the order of execution of the operations of expectation and limiting can be interchanged*, since then

$$\lim_{\varepsilon \to 0} E\{X(t+\varepsilon)\} = \lim_{\varepsilon \to 0} m_X(t+\varepsilon) = m_X(t)$$

$$= E\{X(t)\} = E\{\operatorname*{l.i.m.}_{\varepsilon \to 0} X(t+\varepsilon)\}, \quad (7.15)$$

in which l.i.m. denotes *limit in mean* (square).

7.3 Mean-Square Differentiability

Definition. A process X is *mean-square differentiable* at t if and only if there exists a random variable, denoted by $X^{(1)}(t)$ [or alternatively by $dX(t)/dt$], such that the limit

$$\lim_{\varepsilon \to 0} E\left\{\left(\frac{1}{\varepsilon}[X(t) - X(t-\varepsilon)] - X^{(1)}(t)\right)^2\right\} = 0 \quad (7.16)$$

exists.

It is easily shown [Loève, 1978, Chapter XI] that (7.16) is satisfied if and only if $R_X(t_1, t_2)$ is differentiable jointly in t_1 and t_2 at the point $t_1 = t_2 = t$; that is, the limit

$$\left.\frac{\partial^2 R_X(t_1, t_2)}{\partial t_1 \partial t_2}\right|_{t_1=t_2=t} \triangleq \lim_{\varepsilon_1, \varepsilon_2 \to 0} \frac{1}{\varepsilon_1 \varepsilon_2} \{R_X(t-\varepsilon_1, t-\varepsilon_2)$$
$$- R_X(t, t-\varepsilon_2) - R_X(t-\varepsilon_1, t) + R_X(t,t)\}$$
$$(7.17)$$

must exist. This is not, in general, satisfied by the existence of the limit

$$\frac{d^2 R_X(t,t)}{dt^2} \triangleq$$
$$\lim_{\varepsilon \to 0} \frac{1}{\varepsilon^2} [R_X(t,t) - 2R_X(t-\varepsilon, t-\varepsilon) + R_X(t-2\varepsilon, t-2\varepsilon)].$$
$$(7.18)$$

In fact, for the sampled-and-held noise process, it can be shown (exercise 1) that (7.18) exists, but (7.17) does not. (See Chapter 4, exercise 10.)

If the process X is WSS, then it follows immediately that (7.17) exists if and only if $R_X(\tau)$ is twice differentiable at $\tau = 0$; that is, the limit

$$\left.\frac{d^2 R_X(\tau)}{d\tau^2}\right|_{\tau=0} \triangleq \lim_{\varepsilon \to 0} \frac{1}{\varepsilon^2}[R_X(\varepsilon) - 2R_X(0) + R_X(-\varepsilon)] \quad (7.19)$$

must exist. Hence if a WSS process is mean-square differentiable at one time t, it is at all times.

If a process X is mean-square differentiable at $t = t_1$ and $t = t_2$, then it can be shown that the mean and autocorrelation of the differentiated process $X^{(1)}(t)$ are

$$m_{X^{(1)}}(t_1) = \frac{d}{dt_1} m_X(t_1), \quad (7.20)$$

$$R_{X^{(1)}}(t_1, t_2) = \frac{\partial^2}{\partial t_1 \, \partial t_2} R_X(t_1, t_2), \quad (7.21)$$

and if X is WSS, then

$$m_{X^{(1)}} = 0, \quad (7.22)$$

$$R_{X^{(1)}}(\tau) = -\frac{d^2}{d\tau^2} R_X(\tau). \quad (7.23)$$

Equations (7.20) to (7.23) are what one would obtain by formal interchange of the operations of differentiation and expectation.

Example 4: For the asynchronous telegraph signal, the autocorrelation (7.12) violates (7.19). Hence, this process is not mean-square differentiable, although it is mean-square continuous. ∎

Example 5: For the Wiener process, the autocorrelation (7.13) violates (7.17). Hence, this process is not mean-square differentiable, although it is mean-square continuous. This means that the symbolic broadband Gaussian noise process $Z(t)$, formally defined to be the derivative of the Wiener process, does not actually exist. ∎

7.3.1 *Broadened Interpretation of Differentiability*

Because the processes in the preceding examples are not mean-square differentiable, their autocorrelations are not twice differentiable. If we were to reverse this reasoning, we would say that because the autocorrelations are not twice differentiable, these processes are not mean-square differentiable.

Moreover, we can see that for these examples the autocorrelations are not twice differentiable because they possess step discontinuities in slope (see, for example, Chapter 6, exercise 4), and the derivative at a step discontinuity does not exist. However, in engineering we are quite familiar with the broadened interpretation of differentiation for which the derivative of a step discontinuity is defined to be an impulse. With this broadened definition, the previous two processes can be interpreted as being mean-square differentiable. Heuristically, in both cases the differentiated sample functions consist of impulses (as discussed in Chapter 6) and their autocorrelations consist of impulses. We shall henceforth interpret (7.21) and (7.23) as valid provided that, in any finite interval, R_X contains (at the worst) a finite number of step discontinuities in slope. The formal interchange of differentiation and expectation employed in Chapter 6 is thereby justified, and the broadband Gaussian and Poisson processes $Z(t)$ introduced in Chapter 6 do exist in the broader sense, that is, as *symbolic* processes. It turns out that as long as these symbolic processes are filtered before possibly entering a nonlinear signal-processing operation, typical calculations will be consistent and unambiguous. However, if one of these symbolic processes is directly processed by a zero-memory nonlinearity (e.g., a squaring device), calculations will typically break down, and one must either resort to less idealized mathematical models for such broadband noise processes (see Section 10.7), or employ more sophisticated mathematical techniques. The most fruitful way to proceed in many such signal-processing applications is to abandon the continuous-time model for a discrete-time model of broadband noise, for which calculus is irrelevant. Also, as illustrated in exercise 3, ambiguities can arise with these symbolic processes when subjected to signal-processing operations that are linear but discontinuous.*

7.4 Mean-Square Integrability

Definition. A process X is *mean-square integrable* on the interval $(0, t)$ if and only if there exists a random variable, denoted by $X^{(-1)}(t)$ [or alternatively by $\int_0^t X(u)\, du$], such that the limit

$$\lim_{\varepsilon \to 0} E\left\{ \left[\varepsilon \sum_{i=1}^{t/\varepsilon} X(i\varepsilon) - X^{(-1)}(t) \right]^2 \right\} = 0 \qquad (7.24)$$

exists (where the sequence of values of ε is chosen such that t/ε are integers).

---

*For mathematical rigor, these symbolic processes $Z(t)$ can be treated as generalized random functions, in the same sense that the Dirac delta is a generalized function. Strictly speaking, a generalized function can be specified only by the values taken on by the integral of its product with arbitrary continuous functions; for example, $\delta(t)$ is specified by $\int_{-\infty}^{\infty} \delta(t) f(t)\, dt = f(0)$ for all continuous functions $f(\cdot)$.

It can be shown [Loève, 1978, Chapter XI] that (7.24) exists if and only if $R_X(t_1, t_2)$ is Riemann-integrable on the square $(0, t) \times (0, t)$; that is, the limit

$$\int_0^t \int_0^t R_X(t_1, t_2)\, dt_1\, dt_2 \triangleq \lim_{\varepsilon \to 0} \left[\varepsilon^2 \sum_{i,j=1}^{t/\varepsilon} R_X(i\varepsilon, j\varepsilon) \right] \quad (7.25)$$

must exist.

If the process X is WSS, then it can be shown that (7.25) exists if and only if $R_X(\tau)$ is Riemann-integrable on $(0, t)$; that is, the limit

$$\int_0^t R_X(u)\, du \triangleq \lim_{\varepsilon \to 0} \left[\varepsilon \sum_{i=1}^{t/\varepsilon} R_X(i\varepsilon) \right] \quad (7.26)$$

must exist.

If X is mean-square integrable on $(0, t_1)$ and on $(0, t_2)$, then it can be shown that the mean and autocorrelation of the integrated process $X^{(-1)}(t)$ are

$$m_{X^{(-1)}}(t_1) = \int_0^{t_1} m_X(u)\, du, \qquad t_1 > 0 \quad (7.27)$$

$$R_{X^{(-1)}}(t_1, t_2) = \int_0^{t_1} \int_0^{t_2} R_X(u_1, u_2)\, du_1\, du_2, \qquad t_1, t_2 > 0. \quad (7.28)$$

Equations (7.27) and (7.28) are what one would obtain by formal interchange of the operations of integration and expectation.

7.4.1 Broadened Interpretation of Integrability

For both the symbolic Gaussian and Poisson broadband noise processes, and for the symbolic derivative of the asynchronous telegraph signal, the autocorrelations are impulsive, and therefore are not Riemann-integrable. In general, a function is Riemann integrable only if it contains (at the worst) a finite number of step discontinuities in a finite interval. But in engineering we are quite familiar with the broadened interpretation of integration for which the integral of an impulse is defined to be a step function. With this broadened definition, the preceding three processes can be interpreted as being mean-square integrable processes.

Although stationarity is preserved by differentiation, it is not necessarily preserved by integration.

Example 6: For the symbolic broadband Gaussian noise process $Z(t)$, which is WSS, its integral is the Wiener process $W(t)$, which has autocorrelation given by (7.28),

$$R_W(t_1, t_2) = \int_0^{t_1} \int_0^{t_2} \alpha^2 \delta(u_1 - u_2)\, du_1\, du_2 = \alpha^2 \min\{t_1, t_2\}, \quad (7.29)$$

and is clearly nonstationary. ∎

Example 7: For the differentiated asynchronous telegraph signal $V(t) = dY(t)/dt$, the autocorrelation is (for $t > 0$ and $t + \tau > 0$)

$$R_V(\tau) = 4\lambda\delta(\tau) - 4\lambda^2 e^{-2\lambda|\tau|}, \qquad (7.30)$$

and the mean is asymptotically zero. Therefore the process is asymptotically WSS. The integral of this process, Y, has zero mean (asymptotically) and autocorrelation (7.12), and is therefore asymptotically WSS also. ∎

Now that we have developed a meaning for differentiation and integration of random processes, we can evaluate the mean and the variance of the squared rms thermal-noise voltage described in Section 7.1. We complete this task in the next chapter.

7.5 *Summary*

The stochastic calculus introduced in this chapter is analogous to the more elementary calculus of real-valued functions of real variables, which is based on the usual Euclidean measure of distance. In terms of the Euclidean distance measure the concept of convergence is defined, and in terms of convergence the properties of continuity, differentiability, and integrability are defined. The stochastic calculus described here is based on the *root-mean-square measure of distance*. In terms of this distance measure the concept of *mean-square convergence* is defined (in Chapter 2), and in terms of mean-square convergence the properties of *mean-square continuity*, *mean-square differentiability*, and *mean-square integrability* are defined. For each of these three latter definitions, there is a necessary and sufficient condition that is completely specified in terms of a related property of the autocorrelation function of the process. That is, $X(t)$ is mean-square continuous at t if and only if $R_X(t_1, t_2)$ is continuous in t_1 and in t_2 at the point $t_1 = t_2 = t$; $X(t)$ is mean-square differentiable at t if and only if $R_X(t_1, t_2)$ is differentiable jointly in t_1 and t_2 at the point $t_1 = t_2 = t$; and $X(t)$ is mean-square integrable on $(0, t)$ if and only if $R_X(t_1, t_2)$ is integrable on $(0, t) \times (0, t)$. For WSS processes, the differentiability and integrability conditions on $R_X(t_1, t_2)$ simplify to twice-differentiability of $R_X(\tau)$ and integrability of $R_X(\tau)$, respectively.

For some applications these definitions of differentiability and integrability are too strict, in the same sense that in the calculus of real-valued functions of real variables, the nondifferentiability of step functions and the nonintegrability of impulse (Dirac delta) functions are often undesirable. By introducing symbolic functions, such as the impulse function, the definitions of differentiability and integrability can be broadened. Similarly, by introducing symbolic processes, such as the formal derivatives of the Wiener process and Poisson counting process, the definitions of mean-square differentiability and mean-square integrability can be broadened. Nevertheless, the same sorts of limitations that apply to the use of symbolic functions such as the impulse, also apply to the use of symbolic processes.

When a process is mean-square continuous, then the order of the limiting and expectation operations can be interchanged: see (7.15). When a process is mean-square differentiable [integrable], then the order of the differentiation [integration] and expectation operations can be interchanged: see (7.22) and (7.23) [(7.27) and (7.28)]. This interchangeability of operations is found to be very useful in the study of linear systems in Chapter 9.

EXERCISES

1. Show that the sampled-and-held noise process (Section 4.2.6) is neither mean-square continuous nor mean-square differentiable in the narrow sense, but that it is mean-square differentiable in the broad sense. [You need not actually evaluate the limits (7.9) and (7.17); graphical descriptions are adequate—see Chapter 4, exercise 10.]

2. Show that the phase-randomized sampled-and-held noise process (Section 4.2.7), is mean-square continuous, but not mean-square differentiable in the narrow sense, although it is mean-square differentiable in the broad sense. (Observe that the sample functions of the processes in this and the preceding exercise are of identical form except for time shift; yet the latter is mean-square continuous, and the former is not. This is a result of the smoothing effect—over the ensemble, not over time—of a random phase.)

3. (a) Show that, formally,

$$E\left\{W(t)\frac{dW(t)}{dt}\right\} = \alpha^2 \int_0^t \delta(\tau - t)\, d\tau, \qquad (7.31)$$

where

$$W(t) = \int_{-\infty}^{\infty} h(t - \tau)Z(\tau)\, d\tau,$$

$$h(t - \tau) = \begin{cases} 1, & \tau < t, \\ 0, & \tau > t, \end{cases} \qquad (7.32)$$

and $Z(t)$ is the symbolic broadband noise process defined in Chapter 6. It is not clear whether the integral (7.31) has value 1 or 0 or something between 0 and 1. Thus, our formal manipulation of the derivative of a process that is not differentiable in the narrow sense, and the integral of a process that is not integrable in the narrow sense, has led us to an ambiguous situation. The reason is that the factor h in the integrand in (7.32) is discontinuous. A deeper study of stochastic calculus reveals that the value of $\frac{1}{2}$ for (7.31) is appropriate when $W(t)$ is the Wiener process.

(b) Let $X(t)$ be a mean-square differentiable WSS process. Show that

$$E\left\{X(t)\frac{dX(t)}{dt}\right\} = 0$$

for all t.

Hint: Verify that this is equivalent to $dR_X(\tau)/d\tau = 0$ at $\tau = 0$. Then differentiate the identity $R_X(\tau) = R_X(-\tau)$, and set $\tau = 0$.

★4. Verify that for a WSS integrable process X, the integrated process $Y = X^{(-1)}$ has mean and variance given by

$$m_Y(t) = m_X t, \qquad t > 0 \tag{7.33}$$

$$\sigma_Y^2(t) = t\int_{-t}^{t}\left(1 - \frac{|\tau|}{t}\right)K_X(\tau)\,d\tau. \tag{7.34a}$$

Hint: Show that

$$\sigma_Y^2(t) = \int_0^t\int_0^t K_X(t_1 - t_2)\,dt_1\,dt_2, \tag{7.34b}$$

and then make the change of variables $\tau = t_1 - t_2$, after introducing the window function

$$h(t_i) = \begin{cases} 1, & 0 \le t_i \le t, \\ 0, & \text{otherwise,} \end{cases}$$

for $i = 1, 2$, so that the limits of integration can be changed to $\pm \infty$.

5. Use the inequality

$$E\{[X(t) - X(t - \varepsilon)]^2\} \ge [E\{X(t) - X(t - \varepsilon)\}]^2, \tag{7.35}$$

together with (7.8), to prove that mean-square continuity of X at t guarantees continuity, in the ordinary sense, of the mean function m_X at t.

6. Show that if $X(t)$ is mean-square differentiable, then it must be mean-square continuous.

7. Consider the phase-randomized sampled-and-held process $Y(t)$ described in Section 4.2.7. Determine the autocorrelation of the symbolic process $Y^{(1)}(t)$ obtained by differentiating $Y(t)$.

8. Consider the random-amplitude-and-phase sine wave process $Y(t)$ de-

scribed in Section 4.2.5. Determine the autocorrelation of the process $Y^{(-1)}(t)$ obtained by integrating $Y(t)$.

9. Let $Z(t)$ be the symbolic broadband noise process with intensity $\alpha^2 = 1$. Show that

$$E\{Y_1 Y_2^*\} = \int_0^T \phi_1(t)\phi_2^*(t)\, dt, \qquad (7.36)$$

where

$$Y_1 \triangleq \int_0^T \phi_1(t) Z(t)\, dt,$$

$$Y_2 \triangleq \int_0^T \phi_2(t) Z(t)\, dt. \qquad (7.37)$$

It follows that if ϕ_1 and ϕ_2 are *orthogonal* in the sense that the integral in (7.36) is zero (see Chapter 13), then Y_1 and Y_2 are orthogonal random variables.

10. (a) Let $X(t)$ be a mean-square differentiable process that satisfies the equation

$$\frac{dX(t)}{dt} = aX(t) + Z(t), \qquad t \geq 0,$$

where $Z(t)$ is WSS. Obtain an explicit formula in terms of m_Z for the mean function $m_X(t)$ for $t \geq 0$, given that $m_X(0) = 1$.

(b) Let $X(t)$ be a mean-square integrable process that satisfies the equation

$$\int_0^t X(u)\, du = bX(t) + W(t), \qquad t \geq 0,$$

where $W(t)$ is a zero-mean process. Obtain an explicit formula for the mean function $m_X(t)$ for $t \geq 0$.

11. Let $X(t)$ be a voltage applied across a series connection of a resistor, capacitor, and inductor, and let $Y(t)$ be the resultant current flow. Determine a differential equation for the mean $m_Y(t)$.

12. Consider a sliding-interval smoother with input $W(t)$ and output

$$V(t) = \int_{t-T}^{t} W(s)\, ds.$$

Determine formulas for the mean and autocorrelation for $V(t)$ when $W(t)$ is the Wiener process.

13. (a) Let $X(t)$ be the Ornstein-Uhlenbeck process (Section 5.3.2). De-

termine the autocorrelation function for the smoothed process

$$Y(t) \triangleq \frac{1}{\Delta} \int_{t-\Delta}^{t} X(\tau)\, d\tau. \tag{7.38}$$

Hint: Use a change of variables so that the identity (7.34) can be used.

(b) Determine the effect of this smoothing on the mean-square value of $Y(t)$ as Δ is increased.

14. The velocity of an aircraft is measured using an inertial navigational device, and the error $Y(t)$ in the measurement after t seconds of flight is given by

$$Y(t) = g \int_{0}^{t} \sin X(\tau)\, d\tau, \tag{7.39}$$

where $g = 980$ cm/s² is the gravitational acceleration and $X(t)$ is the angular error of the gyro axis. Let $X(t)$ be a zero-mean Ornstein-Uhlenbeck process with dimensions of radians and with autocorrelation

$$R_X(\tau) = 10^{-8} e^{-|\tau|/100},$$

and determine the mean-square error in the measurement after 12 hours of flight.

Hint: Justify and use the approximation $\sin x \simeq x$.

★15. (a) Consider the infinite-interval integrated process

$$X^{(-1)}(t) \triangleq \int_{-\infty}^{t} X(u)\, du,$$

where $X(t)$ is WSS. Show that in this case the formula (7.28) for a finite-interval integrated process becomes

$$R_{X^{(-1)}}(\tau) = \int_{-\infty}^{\tau} \int_{\sigma}^{\infty} R_X(v)\, dv\, d\sigma. \tag{7.40}$$

(b) Show that for processes with nonoscillatory autocorrelations like $R_X(\tau) = \delta(\tau)$ or $R_X(\tau) = e^{-|\tau|}$, $R_{X^{(-1)}}(\tau)$ does not exist and, therefore, $X^{-1}(t)$ does not exist (even in the broadened sense). However, for an autocorrelation such as $R_X(\tau) = 2\delta(\tau) - e^{-|\tau|}$, $R_{X^{(-1)}}(\tau)$ does indeed exist. Similarly, for

$$R_X(\tau) = \left[\frac{\sin \pi\tau b}{\pi t}\right]^2 \cos(2\pi f_o \tau)$$

with $f_o > b$, $R_{X^{(-1)}}(\tau)$ does indeed exist (although for this latter case we must wait until Chapter 10 [exercise 26b] to see how to show this).

16. (a) Let $X(t)$ be a mean-square differentiable process and let $Y(t)$ be the process

$$Y(t) \triangleq \left|\frac{dX(t)}{dt}\right|.$$

Let $N_x(u, v)$ denote the number of times $X(t)$ crosses the level x [$X(t) = x$] in the time interval $[u, v]$. Show that the mean number of level crossings is given by

$$E\{N_x(u, v)\} = \int_u^v f_{X(t)}(x)\, E\{Y(t)|X(t) = x\}\, dt. \qquad (7.41)$$

Hint: A formal method can be based on the following identity for the Dirac delta:

$$\delta[X(t) - x] = \left|\frac{dX(t)}{dt}\right|^{-1} \sum_i \delta(t - t_i),$$

where $X(t)$ is any differentiable function and $\{t_i\}$ is the set of values for which $X(t) = x$. First show that

$$N_x(u, v) = \int_u^v \sum_i \delta(t - t_i)\, dt,$$

and then substitute the preceding identity into the integrand of this equation, and finally use the relation

$$f_{Y(t)X(t)}(y, x) = f_{Y(t)|X(t)}(y|x) f_{X(t)}(x)$$

to evaluate the expected value of the resultant expression for $N_x(u, v)$ (treating the integrand as if it were mean-square integrable.)

(b) Show that if $X(t)$ is stationary, then $E\{N_x(u, v)\}$ is proportional to $(v - u)$.

(c) Let $X(t)$ be a zero-mean stationary Gaussian process and show that

$$E\{N_x(u, v)\} = \frac{(v - u)}{\pi \sigma_X} \exp\left(\frac{-x^2}{2\sigma_X^2}\right) \left[\frac{-d^2 R_X(\tau)}{d\tau^2}\right]_{\tau=0}^{1/2}. \qquad (7.42)$$

Hint: Use the fact that $E\{X(t)\, dX(t)/dt\} = 0$ [cf. part (b) of exercise 3] to show that $X(t)$ and $dX(t)/dt$ are independent and therefore $E\{Y(t)|X(t)\} = E\{Y(t)\}$.

17. To gain some intuitive insight into the concept of two stationary processes $X(t)$ and $Y(t)$ being mean-square equivalent, consider the time-average counterpart to the ensemble average (recall Section 2.5.2), for which mean-square equivalence is defined by

$$\lim_{T \to \infty} \frac{1}{T} \int_{-T/2}^{T/2} [X(t) - Y(t)]^2\, dt = 0. \qquad (7.43)$$

Now, let $Y(t) = X(t) + V(t)$, where $V(t)$ is any function with finite energy,
$$\int_{-\infty}^{\infty} V^2(t)\, dt < \infty,$$
and verify that (7.43) is valid. If we envision $V(t)$ as a stationary process with ensemble members
$$V(t, \sigma) = V(t - \sigma),$$
then the mean squared value of $V(t)$ is zero, which reveals why $X(t)$ and $Y(t)$ are mean-square equivalent.

8

Ergodicity and Duality

AS DISCUSSED in the preface, and again in the introductory Chapter 3, a theme of this book is the exploitation of the duality that exists between the deterministic theory of random processes, based on time averages, and the probabilistic theory, based on expected values (ensemble averages). A fundamental part of the duality is the fact that under certain conditions, called *ergodicity*, the two types of averages are not only dual, but also *equivalent*. This notion of ergodicity is explored in this chapter, and the more general notion of duality is explained.*

8.1 The Notion of Ergodicity

With reference to our inquiry regarding prediction of the rms value of thermal noise, initiated in Chapter 7, we now consider the problem of evaluating the mean and variance of the squared rms thermal-noise voltage

$$V_{\text{rms}}^2(T) \triangleq \frac{1}{T} \int_0^T V^2(t)\, dt, \tag{8.1}$$

and the limits of these quantities as the integration time T approaches infinity. For simplicity, we consider an idealized rectangular impulse-

*This chapter provides a brief, practically oriented introduction to a highly developed subject that can be treated more deeply only within the mathematical framework of measure theory; see [Gray, 1987].

response function h for the voltmeter:

$$h(t) = \begin{cases} 1/T_o, & 0 \leq t \leq T_o, \\ 0, & \text{otherwise,} \end{cases} \quad (8.2)$$

in which case the measured voltage is modeled by [see (7.2)]

$$V(t) = \frac{1}{T_o} \int_{t-T_o}^{t} Z(u) \, du, \quad (8.3)$$

where Z is the symbolic broadband Gaussian noise process. Since Z is mean-square integrable (in the broad sense), we obtain

$$V(t) = \frac{1}{T_o} [W(t) - W(t - T_o)], \quad (8.4)$$

where $W(t)$ is the Wiener process. By using (8.4) and the formula for the autocorrelation of the Wiener process, [or, more simply, (8.3) and (7.27)–(7.28)] we obtain

$$m_V = 0, \qquad \sigma_V^2 = \alpha^2/T_o, \quad (8.5)$$

$$K_V(\tau) = \begin{cases} \dfrac{\alpha^2}{T_o}\left(1 - \dfrac{|\tau|}{T_o}\right), & |\tau| \leq T_o, \\ 0, & \text{otherwise.} \end{cases} \quad (8.6)$$

The mean and variance of $V_{\text{rms}}^2(T)$ are

$$E\{V_{\text{rms}}^2(T)\} = \frac{1}{T} \int_0^T R_V(0) \, dt = \sigma_V^2, \quad (8.7)$$

$$\text{Var}\{V_{\text{rms}}^2(T)\} = E\{V_{\text{rms}}^4(T)\} - \sigma_V^4, \quad (8.8)$$

where

$$E\{V_{\text{rms}}^4(T)\} = \frac{1}{T^2} \int_0^T \int_0^T E\{V^2(t) V^2(u)\} \, dt \, du. \quad (8.9)$$

For zero-mean jointly Gaussian variables $V(t)$ and $V(u)$, we have the identity (Chapter 5, exercise 3)

$$E\{V^2(t) V^2(u)\} = E\{V^2(t)\} E\{V^2(u)\} + 2(E\{V(t)V(u)\})^2$$
$$= \sigma_V^4 + 2K_V^2(t - u). \quad (8.10)$$

Substitution of (8.10) into (8.9) into (8.8) yields (cf. Chapter 7, exercise 4)

$$\text{Var}\{V_{\text{rms}}^2(T)\} = \frac{4}{T} \int_0^T \left(1 - \frac{|\tau|}{T}\right) K_V^2(\tau) \, d\tau, \quad (8.11)$$

which, upon substitution of (8.6), becomes

$$\text{Var}\{V_{\text{rms}}^2(T)\} = \frac{4\sigma_V^4}{T} \int_0^{T_o} \left(1 - \frac{|\tau|}{T}\right)\left(1 - \frac{|\tau|}{T_o}\right)^2 d\tau, \quad (8.12)$$

where it is assumed that $T > T_o$. For long integration times in (8.1), $T \gg T_o$, (8.12) reduces to

$$\text{Var}\{V_{\text{rms}}^2(T)\} \simeq \frac{4\sigma_V^4}{T} \int_0^{T_o} \left(1 - \frac{|\tau|}{T_o}\right)^2 d\tau \quad (8.13)$$

with increasing accuracy of approximation as T is increased. It follows directly from (8.7) and (8.13) that

$$E\{V_{\text{rms}}^2(T)\} = \sigma_V^2, \quad (8.14)$$

$$\lim_{T \to \infty} \text{Var}\{V_{\text{rms}}^2(T)\} = 0, \quad (8.15)$$

from which we obtain

$$\lim_{T \to \infty} E\{(V_{\text{rms}}^2(T) - \sigma_V^2)^2\} = 0. \quad (8.16)$$

Equation (8.14) implies that $V_{\text{rms}}^2(T) = \sigma_V^2$, on the average, regardless of the integration time T. Equation (8.15) implies that, on the average, the squared deviation of $V_{\text{rms}}^2(T)$ about its average value approaches zero as $T \to \infty$. Equation (8.16) states that, in the limit as $T \to \infty$, the random variable $V_{\text{rms}}^2(T)$ is equal, in the mean-square sense, to the nonrandom variable σ_V^2. We conclude that for a sufficiently long integration time T, $V_{\text{rms}}(T)$ is a close approximation to the standard deviation σ_V. This is conceptually striking, since the probabilistic parameter σ_V is defined to be an average over the sample space of waveforms comprising the random process $V(t, s)$, that is, an average over s with t fixed (which can be visualized as a *vertical* average in Figure 3.3); whereas the empirical quantity $V_{\text{rms}}(T)$ is defined to be an average over time for one statistical sample, that is, an average over t with s fixed (which can be visualized as a *horizontal* average in Figure 3.3).

In general, the property of equality between an ensemble average and an infinite-time average is referred to as an *ergodic property*.* Many random

---

*The word *ergodic* derives from the Greek words meaning *work path*. Historically, the concept of equivalence between ensemble averages and time averages first arose in statistical mechanics, where *work path* means the path of motions followed by a system of particles (such as a gas) whose total energy (potential work) remains stationary (constant). The property of ergodicity guarantees that time averages over a stationary work path for a single system equal ensemble averages over the stationary work paths of many systems at one time instant. Thus, a sample path of a stationary random process in our treatment is the counterpart of a stationary work path in statistical mechanics.

processes of practical interest exhibit ergodic properties; however, some important random processes do not. This motivates a careful discussion of these properties and methods for determining whether or not a particular mathematical model for a process possesses them. For this purpose, it is assumed throughout the chapter that the processes $X(t)$ and $Y_\tau(t) \triangleq X(t+\tau)X(t)$ are mean-square integrable on every finite interval.

8.2 Discrete and Continuous Time Averages

Consider the problem of estimating the mean of a WSS process $X(t)$ using an average of equally spaced discrete time samples of one sample function, as illustrated in Figure 8.1. Thus, the estimate of

$$m_X \triangleq E\{X(t)\} \tag{8.17}$$

is

$$\hat{m}_X^N \triangleq \frac{1}{N} \sum_{i=1}^{N} X(i\Delta). \tag{8.18}$$

It is easily shown (exercise 1) that

$$E\{\hat{m}_X^N\} = m_X,$$

$$\text{Var}\{\hat{m}_X^N\} = \frac{1}{N^2} \sum_{i=1}^{N} \sum_{j=1}^{N} K_X([i-j]\Delta). \tag{8.19}$$

Consider now the two extremes of very sparse sampling and very dense sampling.

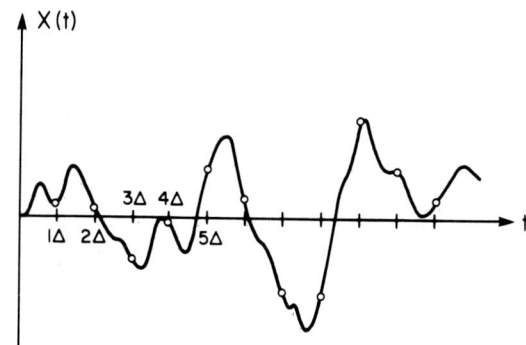

Figure 8.1 Equally Spaced Time Samples of a WSS Process

8.2.1 Sparse Sampling

It is assumed that Δ is larger than the *correlation time* τ_o of the process [i.e., $2\tau_o$ is the approximate width of the covariance function $K_X(\tau)$]. Then

$$K_X([i-j]\Delta) \begin{cases} = \sigma_X^2, & i = j, \\ \ll \sigma_X^2, & i \neq j, \end{cases} \quad (8.20)$$

and (8.19) reduces to (exercise 1)

$$\text{Var}\{\hat{m}_X^N\} \simeq \sigma_X^2/N, \quad (8.21)$$

in which case the mean squared error between m_X and the estimate $\hat{m}_X^N$ is (exercise 1)

$$\text{MSE}_N \triangleq E\{[m_X - \hat{m}_X^N]^2\} \simeq \sigma_X^2/N. \quad (8.22)$$

Thus, the mean squared error can be made as small as desired by choosing the number N of time samples sufficiently large.

In practice N cannot be made arbitrarily large because of the unavailability of an arbitrarily long record of the process X. Thus, if only a T-second record is available, the maximum value for N is T/Δ. In the next subsection, the procedure of letting N get large by decreasing the sampling increment Δ is considered.

8.2.2 Dense Sampling

Let $N = T/\Delta$, and consider what happens as $\Delta \to 0$ [in which case (8.20) no longer holds]. By using a change of variables ($k = i - j$), (8.19) can be reexpressed as (exercise 2)

$$\text{Var}\{\hat{m}_X^{T/\Delta}\} = \frac{1}{N} K_X(0) + \frac{2}{N^2} \sum_{k=1}^{N} (N-k) K_X(k\Delta)$$

$$= \frac{\sigma_X^2}{N} + \frac{2}{T} \sum_{k=1}^{N} \left(1 - \frac{k\Delta}{T}\right) K_X(k\Delta) \Delta. \quad (8.23)$$

The second term here is a Riemann approximating sum, and in the limit as $\Delta \to 0$ becomes a Riemann integral. Also, in the limit as $\Delta \to 0$ ($N \to \infty$) the first term vanishes. Thus,

$$\lim_{\Delta \to 0} \text{Var}\{\hat{m}_X^{T/\Delta}\} = \frac{2}{T} \int_0^T \left(1 - \frac{\tau}{T}\right) K_X(\tau) \, d\tau \triangleq \text{MSE}(T). \quad (8.24)$$

For $T \gg \tau_0$, (8.24) is closely approximated by (exercise 3)

$$\text{MSE}(T) \simeq \frac{2}{T} \int_0^T K_X(\tau) \, d\tau \qquad (8.25)$$

$$\simeq \frac{\sigma_X^2}{T/\tau_0}, \qquad (8.26)$$

where the second approximation follows from the fact that a reasonable measure of the width $2\tau_0$ of an even function K_X satisfies

$$\int_0^T K_X(\tau) \, d\tau \simeq \frac{\tau_0}{2} K_X(0) \qquad (8.27)$$

for $T \gg \tau_0$ [e.g., (8.27) is exact for a triangle-shaped $K_X(\tau)$].

Comparison of (8.22) and (8.26) reveals that the *effective number of uncorrelated (sparse) samples* in a dense set of samples is

$$N = T/\tau_0. \qquad (8.28)$$

Hence, the MSE cannot be reduced below that for $\Delta = \tau_0$ by further increase in the density of sampling (decrease in Δ). However, the implementation of the averaging operation (8.18) is sensitive to how small Δ is (how large N is). Specifically, by reexpressing (8.18) as

$$\hat{m}_X^{T/\Delta} = \frac{1}{T} \sum_{i=1}^{N} X(i\Delta) \Delta, \qquad (8.29)$$

it can be seen that it is a Riemann approximating sum, and in the limit as $\Delta \to 0$ becomes the mean-square Riemann integral

$$\underset{\Delta \to 0}{\text{l.i.m.}} \, \hat{m}_X^{T/\Delta} = \frac{1}{T} \int_0^T X(t) \, dt \triangleq \hat{m}_X(T). \qquad (8.30)$$

Hence, for $\Delta \ll \tau_0$, the discrete-time average (8.18) is closely approximated by the continuous-time average (8.30), which can be implemented with an integrating device.

8.3 Mean-Square Ergodicity of the Mean

In the preceding discussion it was tacitly assumed that the covariance function $K_X(\tau)$ decays to zero as $\tau \to \infty$, so that it does indeed possess an effective finite width $2\tau_0$. In this case, it follows [from (8.26)] that

$$\lim_{T \to \infty} \text{MSE}(T) = 0. \qquad (8.31)$$

That is, the MSE can be made as small as desired if the record length T can be made sufficiently long. Specifically, for the estimate

$$\hat{m}_X(T) \triangleq \frac{1}{T}\int_0^T X(t)\,dt, \tag{8.32}$$

it can be shown directly (without resort to Riemann approximating sums) that

$$E\{\hat{m}_X(T)\} = m_X \tag{8.33}$$

$$\text{Var}\{\hat{m}_X(T)\} = \frac{2}{T}\int_0^T\left(1 - \frac{\tau}{T}\right)K_X(\tau)\,d\tau = \text{MSE}(T), \tag{8.34}$$

from which it follows [cf. (8.25)–(8.27)] that

$$\text{MSE}(T) \triangleq E\{[m_X - \hat{m}_X(T)]^2\} \simeq \frac{\sigma_X^2}{T/\tau_0}, \tag{8.35}$$

provided that a finite τ_0 does indeed exist. Nevertheless, it follows from (8.33) and (8.34) that (8.31) holds even if $K_X(\tau)$ does not decay to zero as $\tau \to \infty$, provided that $K_X(\tau)$ oscillates with appropriate symmetry about zero.

Example 1: For the phase-randomized sine-wave process Y of Section 4.2.5, the covariance is given by

$$K_Y(\tau) = \tfrac{1}{2}E\{A^2\}\cos\omega_0\tau, \tag{8.36}$$

and this oscillatory covariance, when substituted into (8.34), satisfies (8.31) (exercise 6). ∎

Example 2: Consider the random periodic waveform

$$X(t) = Z + A\cos(\omega_0 t + \Theta), \tag{8.37}$$

which is the sum of a random d-c component Z and the random sine wave from example 1. With the assumption that Z, A, and Θ are independent, it can be shown that

$$K_X(\tau) = \sigma_Z^2 + \tfrac{1}{2}E\{A^2\}\cos\omega_0\tau. \tag{8.38}$$

Substitution of (8.38) into (8.34) yields (exercise 6)

$$\lim_{T\to\infty}\text{MSE}(T) = \sigma_Z^2. \tag{8.39}$$

Thus, the MSE cannot be made small, no matter how long the record length T is. It is not difficult to see why $\hat{m}_X(T)$ cannot be an accurate estimate of m_X, since

$$\underset{T\to\infty}{\text{l.i.m.}} \hat{m}_X(T) = Z, \tag{8.40}$$

whereas

$$m_X = m_Z. \tag{8.41}$$

∎

Necessary and Sufficient Condition. Based on the preceding discussion, it is said that a WSS process X has *mean-square ergodicity of the mean* for $t \geq 0$ if and only if

$$\lim_{T\to\infty} E\{[m_X - \hat{m}_X(T)]^2\} = 0, \tag{8.42a}$$

and a necessary and sufficient condition for this is

$$\lim_{T\to\infty} \frac{1}{T}\int_0^T \left(1 - \frac{\tau}{T}\right) K_X(\tau)\, d\tau = 0. \tag{8.42b}$$

It can be shown (exercise 4) that (8.42b) is equivalent to

$$\lim_{T\to\infty} \frac{1}{T}\int_0^T K_X(\tau)\, d\tau = 0. \tag{8.42c}$$

In (8.42a), $\hat{m}_X(T)$ is defined by (8.32). However, if the interval of integration in (8.32) is changed from $[0, T]$ to $[-T/2, T/2]$, then (8.42a) is the definition of mean-square ergodicity of the mean for *all* time t, and it can be shown that (8.42c) is a necessary and sufficient condition.

The limit in mean square, denoted by

$$\hat{m}_X \triangleq \underset{T\to\infty}{\text{l.i.m.}} \hat{m}_X(T) \tag{8.43}$$

and defined with the use of (8.32), is called the *one-sided empirical mean* of the process $X(t)$. The *two-sided empirical mean*, also denoted by $\hat{m}_X$ and defined with the use of (8.32) with the interval of integration changed from $[0, T]$ to $[-T/2, T/2]$, will be referred to simply as the *empirical mean*. From this point forward, the discussion focuses on two-sided averages.

8.4 Mean-Square Ergodicity of the Autocorrelation

Consider now the problem of estimating the autocorrelation for a WSS process X,

$$R_X(\tau) \triangleq E\{X(t+\tau)X(t)\}, \tag{8.44}$$

using a time average. The results of Sections 8.2 and 8.3 can be applied by identifying the product

$$Y_\tau(t) \triangleq X(t+\tau)X(t) \tag{8.45}$$

as a process $Y_\tau(t)$ depending on a parameter τ. Since the mean of this process,

$$m_{Y_\tau} \triangleq E\{Y_\tau(t)\} \equiv R_X(\tau), \tag{8.46}$$

is the autocorrelation for the underlying process, then estimation of the mean m_{Y_τ} is identical to estimation of the autocorrelation $R_X(\tau)$. Hence, the estimate is defined by

$$\hat{R}_X(\tau)_T \triangleq \frac{1}{T}\int_{-T/2}^{T/2} X(t+\tau)X(t)\,dt \equiv \frac{1}{T}\int_{-T/2}^{T/2} Y_\tau(t)\,dt \triangleq m_{Y_\tau}(T). \tag{8.47}$$

It follows from (8.42) that

$$\lim_{T\to\infty} E\{[R_X(\tau) - \hat{R}_X(\tau)_T]^2\} = 0 \tag{8.48a}$$

if and only if

$$\lim_{T\to\infty} \frac{1}{T}\int_0^T K_{Y_\tau}(u)\,du = 0, \tag{8.48b}$$

where K_{Y_τ} is the autocovariance for the process Y_τ, and is given by [using (8.45)]

$$K_{Y_\tau}(u) = E\{X(t+\tau+u)X(t+u)X(t+\tau)X(t)\} - R_X^2(\tau), \tag{8.49}$$

where it is assumed that $X(t)$ has stationary fourth moments. Hence, the stationary process X is said to possess *mean-square ergodicity of the autocorrelation* if and only if (8.48a) holds for every τ, and a necessary and sufficient condition is that (8.48b) holds* for every τ. The limit in mean square, denoted by

$$\hat{R}_X(\tau) \triangleq \operatorname*{l.i.m.}_{T\to\infty} \hat{R}_X(\tau)_T \tag{8.50}$$

and defined with the use of (8.47), is called the *empirical autocorrelation* of the process $X(t)$.

*A discussion of the conditions on $X(t)$ under which the discrete-time counterpart of (8.48b) holds is given in [Hannan, 1970]. For example, it holds for all stable MA, AR, and ARMA processes driven by independent, identically distributed sequences with finite fourth moments (cf. Section 9.4.1 herein).

Gaussian Processes. It is known from Isserlis's formula (Chapter 5, exercise 3) that for a zero-mean Gaussian process,

$$E\{(X(t_1)X(t_2)X(t_3)X(t_4)\} = K_X(t_1,t_2)K_X(t_3,t_4)$$
$$+ K_X(t_1,t_3)K_X(t_2,t_4)$$
$$+ K_X(t_1,t_4)K_X(t_2,t_3). \quad (8.51)$$

Substitution of (8.51) into (8.49) yields

$$K_{Y_\tau}(u) = K_X^2(u) + K_X(u+\tau)K_X(u-\tau), \quad (8.52)$$

which when substituted into (8.48b) yields an equation that can be put into the form

$$\lim_{T\to\infty} \left| \frac{1}{2T}\int_{-T}^T K_X^2(u)\,du + \frac{1}{2T}\int_{-T}^T K_X(u+\tau)K_X(u-\tau)\,du \right| = 0. \quad (8.53)$$

The condition (8.53) can be reexpressed as

$$\hat{R}_K(0) + \hat{R}_K(2\tau) = 0, \quad (8.54)$$

where $\hat{R}_K$ is the empirical autocorrelation of the nonrandom function $K_X(\cdot)$. It can be shown that the empirical autocorrelation possesses the property

$$|\hat{R}_K(2\tau)| \leq \hat{R}_K(0), \quad (8.55)$$

which is analogous to a property of probabilistic autocorrelations. Now, (8.54) cannot be valid for all τ unless it is valid for $\tau = 0$:

$$\hat{R}_K(0) = 0. \quad (8.56)$$

This necessary condition and the above inequality require that

$$\hat{R}_K(2\tau) \equiv 0, \quad (8.57)$$

which renders (8.54) valid for all τ. Thus, *the necessary and sufficient condition for a zero-mean stationary Gaussian process to have mean-square ergodicity of the autocorrelation for all τ is that it has it for $\tau = 0$*; hence, the condition is, from (8.56),

$$\hat{R}_K(0) = \lim_{T\to\infty} \frac{1}{2T}\int_{-T}^T K_X^2(u)\,du = 0. \quad (8.58)$$

This can be interpreted as the condition that the function $K_X(\cdot)$ has zero average power. It is easily shown that a necessary condition for this is that K_X does not contain any finite additive periodic components, which in turn requires that $X(\cdot)$ not contain any random finite additive periodic components. In fact, it can be shown that this condition is sufficient as well as necessary (cf. [Koopmans, 1974]).

Higher-Order Ergodicity. To illustrate that a process can exhibit ergodicity of the autocorrelation without exhibiting ergodicity of higher-order moments, consider the *mixture process* defined by

$$X(t) \triangleq \begin{cases} U(t), & \text{probability} = p > 0, \\ V(t), & \text{probability} = 1 - p > 0, \end{cases} \quad (8.59)$$

where the processes $U(t)$ and $V(t)$ have identical autocorrelations and exhibit ergodicity of the autocorrelation, but have different higher-order moments. For example, let

$$U(t) = Z_1(t) \otimes h(t),$$
$$V(t) = Z_2(t) \otimes h(t), \quad (8.60)$$

where $Z_1(t)$ is the derivative of the Wiener process, and $Z_2(t)$ is the derivative of the Poisson counting process, such that both of these processes have the autocorrelation

$$R_Z(\tau) = \delta(\tau) \quad (8.61)$$

(for $t > 0$ and $t + \tau > 0$). Then it can be shown that

$$R_U(\tau) = R_V(\tau) = r_h(\tau), \quad (8.62)$$

and therefore

$$R_X(\tau) = pR_U(\tau) + (1 - p)R_V(\tau) = r_h(\tau). \quad (8.63)$$

Hence, regardless of whether a random sample from the Wiener process or the Poisson process is obtained, the empirical autocorrelation will be the same as the probabilistic autocorrelation, since both $U(t)$ and $V(t)$ have ergodicity of the autocorrelation. However, since $U(t)$ is Gaussian and $V(t)$ is Poisson, then all nonzero moments higher than second-order are different. For example, the one-sided time average of $X^4(t)$ is given by

$$\langle X^4 \rangle = \begin{cases} \langle U^4 \rangle = E\{U^4\}, & \text{probability} = p, \\ \langle V^4 \rangle = E\{V^4\}, & \text{probability} = 1 - p, \end{cases} \quad (8.64)$$

and the expected value of X^4 is given by

$$E\{X^4\} = pE\{U^4\} + (1 - p)E\{V^4\}. \quad (8.65)$$

Therefore, since $E\{V^4\} \neq E\{U^4\}$, then

$$\langle X^4 \rangle \neq E\{X^4\}, \qquad (8.66)$$

and $X(t)$ does not exhibit ergodicity of the fourth moment. (As a matter of fact, the difference between $E\{V^4\}$ and $E\{U^4\}$ depends on the rate parameter λ for the Poisson process, and this difference vanishes in the limit $\lambda \to \infty$, as explained in Section 10.7.3.)

8.5 Regular Processes*

In order to properly discuss ergodic properties of nonstationary processes, it is necessary to introduce terminology for the property that guarantees existence of time-average limits such as the empirical mean and empirical autocorrelation,

$$\hat{m}_X \triangleq \underset{T \to \infty}{\text{l.i.m.}} \frac{1}{T} \int_{-T/2}^{T/2} X(t) \, dt, \qquad (8.67a)$$

$$\hat{R}_X(\tau) \triangleq \underset{T \to \infty}{\text{l.i.m.}} \frac{1}{T} \int_{-T/2}^{T/2} X(t + \tau) X(t) \, dt, \qquad (8.67b)$$

regardless of whether or not these limits are mean-square equivalent to the corresponding probabilistic parameters m_X and $R_X(\tau)$, that is, regardless of whether or not $X(t)$ exhibits mean-square ergodicity. Thus, $X(t)$ will be said to exhibit *mean-square regularity of the mean* if and only if $\hat{m}_X$ exists (as a mean-square limit), and similarly, $X(t)$ will be said to exhibit *mean-square regularity of the autocorrelation* if and only if $\hat{R}_X(\tau)$ exists (as a mean-square limit) for every τ. It can be shown (by interchanging the order of expectation, limit, and integration) that mean-square regularity guarantees the equalities

$$E\{\hat{m}_X\} = \langle m_X \rangle \triangleq \lim_{T \to \infty} \frac{1}{T} \int_{-T/2}^{T/2} m_X(t) \, dt, \qquad (8.68a)$$

$$E\{\hat{R}_X(\tau)\} = \langle R_X \rangle(\tau) \triangleq \lim_{T \to \infty} \frac{1}{T} \int_{-T/2}^{T/2} R_X(t + \tau, t) \, dt. \qquad (8.68b)$$

If the function $\langle R_X \rangle(\tau)$ is not identically zero, then $X(t)$ is said to be *asymptotically mean stationary in the wide sense* (Section 5.4.6).

Let us consider the conditions under which a regular asymptotically mean stationary process exhibits mean-square ergodicity of the mean. If

---

*The term *regular* has several loosely related meanings in mathematics. It is sometimes used in probability theory to denote the residual left after subtracting from a process its deterministic part. A *deterministic process* is defined to be a process whose future can be perfectly predicted from its past. This is explained in Section 11.9. Also, some authors have used the term *regular* to denote the property of ergodicity, which is a stronger property than regularity as defined herein.

$X(t)$ has time-variant mean $m_X(t)$, then since $\hat{m}_X$ is time-invariant it is impossible to have $m_X(t) = \hat{m}_X$ even in the mean-square sense. Nevertheless, it is possible to have $\langle m_X \rangle = \hat{m}_X$. It follows from (8.68a) that this equality holds in the mean-square sense if and only if $\hat{m}_X = E\{\hat{m}_X\}$ in the mean-square sense. But this can be so if and only if the variance of $\hat{m}_X$ is zero.

The expected value of the estimate of the mean is

$$E\left\{\frac{1}{T}\int_{-T/2}^{T/2} X(t)\,dt\right\} = \frac{1}{T}\int_{-T/2}^{T/2} m_X(t)\,dt, \qquad (8.69)$$

and the variance is

$$\mathrm{Var}\left\{\frac{1}{T}\int_{-T/2}^{T/2} X(t)\,dt\right\} = \frac{1}{T^2}\int_{-T/2}^{T/2}\int_{-T/2}^{T/2} K_X(t,u)\,dt\,du. \qquad (8.70)$$

Therefore, if

$$\lim_{T\to\infty} \frac{1}{T^2}\int_{-T/2}^{T/2}\int_{-T/2}^{T/2} K_X(t,u)\,dt\,du = 0, \qquad (8.71)$$

then (and only then), in the limit $T \to \infty$, the estimate, which becomes the empirical mean

$$\hat{m}_X \triangleq \underset{T\to\infty}{\mathrm{l.i.m.}} \frac{1}{T}\int_{-T/2}^{T/2} X(t)\,dt, \qquad (8.72)$$

equals (in the mean-square sense) its expected value

$$\hat{m}_X = E\{\hat{m}_X\}. \qquad (8.73a)$$

This equality is taken to be the definition of *mean-square ergodicity of the mean* of a possibly nonstationary process, $X(t)$. Consequently, (8.71) *is the necessary and sufficient condition for mean-square ergodicity of the mean of a nonstationary process*. Equation (8.71) reduces to (8.42c) for a stationary process. When necessary for clarity, a process that exhibits both stationarity and ergodicity can be said to exhibit *strong ergodicity*.

By analogy with the preceding, a nonstationary process is said to exhibit *mean square ergodicity of the autocorrelation* if and only if the equality

$$\hat{R}_X(\tau) = E\{\hat{R}_X(\tau)\} \qquad (8.73b)$$

holds in the mean-square sense. It follows that (8.71), with $X(t)$ replaced by $Y_\tau(t) \triangleq X(t+\tau)X(t)$, is the necessary and sufficient condition for mean-square ergodicity of the autocorrelation of a nonstationary process $X(t)$, and this reduces to (8.48b) for a stationary process.

Use of (8.68) and (8.73) reveals that if $X(t)$ exhibits ergodicity, then

$$\hat{m}_X = \langle m_X \rangle \qquad (8.74a)$$

$$\hat{R}_X(\tau) = \langle R_X \rangle(\tau), \qquad (8.74b)$$

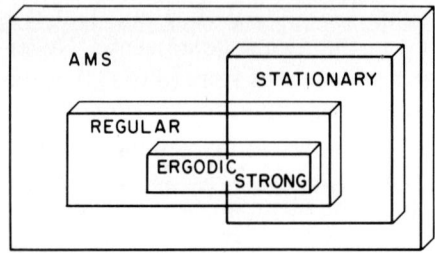

Figure 8.2 Venn Diagram of Classes of Asymptotically Mean Stationary Processes

whereas if $X(t)$ exhibits both ergodicity and stationarity (strong ergodicity), then

$$\hat{m}_X = m_X \tag{8.75a}$$

$$\hat{R}_X(\tau) = R_X(\tau). \tag{8.75b}$$

As a conceptual aid, the classes of processes of interest in this discussion are depicted in a Venn diagram in Figure 8.2.

A simple example that illustrates that not all regular processes are ergodic is the process $X(t) = Z$, a random d-c signal. Clearly, $\hat{m}_X = Z$ exists, but $E\{\hat{m}_X\} = E\{Z\} \neq Z = \hat{m}_X$. A simple example that illustrates that not all ergodic processes are stationary is the sine-wave process $X(t) = \sin \omega_o t$. Clearly, $E\{\hat{m}_X\} = E\{0\} = 0 = \hat{m}_X$, but $m_X(t) = \sin \omega_o t$ is not time-invariant.

8.5.1 Autocorrelation for Ergodic and Nonergodic Regular Stationary Processes

For a WSS regular process the *empirical autocorrelation function*,

$$\hat{R}_X(\tau) \triangleq \lim_{T \to \infty} \frac{1}{T} \int_{-T/2}^{T/2} X(t+\tau) X(t) \, dt \equiv \langle X(t+\tau) X(t) \rangle, \tag{8.76}$$

is in general a random function, since each statistical sample of the process $X(t)$ gives rise to a sample of $\hat{R}_X(\tau)$. The mean value of $\hat{R}_X(\tau)$ is, from (8.68b),

$$E\{\hat{R}_X(\tau)\} = \lim_{T \to \infty} \frac{1}{T} \int_{-T/2}^{T/2} R_X(\tau) \, dt = R_X(\tau), \tag{8.77}$$

which is the *probabilistic autocorrelation*.

In the special case for which $X(t)$ has mean-square ergodicity of the autocorrelation, $\hat{R}_X(\tau)$ equals its mean value (i.e., its variance is zero), and therefore

$$\langle X(t+\tau) X(t) \rangle \equiv \hat{R}_X(\tau) = R_X(\tau) \equiv E\{X(t+\tau) X(t)\}. \tag{8.78}$$

Example 3: For the random-amplitude and -phase sine-wave process of Section 4.2.5, the empirical autocorrelation is

$$\hat{R}_X(\tau) = \tfrac{1}{2}A^2\cos\omega_o\tau, \tag{8.79}$$

where A is the random amplitude parameter. Since the probabilistic autocorrelation is

$$R_X(\tau) = \tfrac{1}{2}E\{A^2\}\cos\omega_o\tau, \tag{8.80}$$

it can be seen that (8.77) holds, but (8.78) does not, because $X(t)$ does not have mean-square ergodicity of the autocorrelation (exercise 7), unless A^2 is nonrandom. ∎

An important type of nonergodic regular stationary processes is that which results from applying a random transformation, such as filtering, to an otherwise ergodic process. Random filters are discussed in Section 9.3.2.

8.5.2 *Autocorrelation for Regular Nonstationary Processes*

The mean of the empirical autocorrelation (8.76) for a regular nonstationary process is given by (8.68b). $E\{\hat{R}_X(\tau)\}$ is therefore equal to the *time average of the instantaneous* (*or time-variant*) *probabilistic autocorrelation* $R_X(t + \tau, t)$. Furthermore, if the nonstationary process $X(t)$ has mean-square ergodicity of the autocorrelation, the empirical autocorrelation equals its mean, and therefore

$$\langle X(t+\tau)X(t)\rangle \equiv \hat{R}_X(\tau) = \langle R_X\rangle(\tau) \equiv \langle E\{X(t+\tau)X(t)\}\rangle. \tag{8.81}$$

Example 4: For the random-amplitude sine-wave process of Section 4.2.4 the empirical autocorrelation is

$$\hat{R}_X(\tau) = \tfrac{1}{2}A^2\cos\omega_o\tau, \tag{8.82}$$

and the probabilistic autocorrelation is

$$R_X(t+\tau,t) = \tfrac{1}{2}E\{A^2\}\cos\omega_o\tau - \tfrac{1}{2}E\{A^2\}\cos(\omega_o[2t+\tau]+2\theta). \tag{8.83}$$

Since the time average of the second term in (8.83) is zero, then it can be seen that (8.68b) holds. However, since $X(t)$ does not have mean-square ergodicity of the autocorrelation (unless A^2 is nonrandom), then (8.81) does not hold. ∎

178 • **Random Processes**

Example 5 (*Cyclostationary process*): For the amplitude-modulated sine-wave process

$$X(t) = A(t)\cos(\omega_o t + \Theta), \qquad (8.84)$$

for which $A(t)$ and Θ are independent, the empirical autocorrelation is (making use of trigonometric identities)

$$\hat{R}_X(\tau) = \lim_{T \to \infty} \frac{1}{T} \int_{-T/2}^{T/2} A(t+\tau)A(t)$$
$$\times \left[\tfrac{1}{2}\cos\omega_o\tau + \tfrac{1}{2}\cos(\omega_o[2t+\tau] + 2\Theta)\right] dt \qquad (8.85)$$
$$= \tfrac{1}{2}\hat{R}_A(\tau)\cos\omega_o\tau + \tfrac{1}{2}\operatorname{Re}\{\hat{R}_A^{2\omega_o}(\tau)e^{-i\omega_o\tau}e^{-i2\Theta}\}, \qquad (8.86)$$

where $\operatorname{Re}\{\cdot\}$ denotes the real-part operation, and

$$\hat{R}_A^{2\omega_o}(\tau) \triangleq \lim_{T \to \infty} \frac{1}{T} \int_{-T/2}^{T/2} A(t+\tau)A(t)e^{-i2\omega_o t}\,dt. \qquad (8.87)$$

The probabilistic autocorrelation is

$$R_X(t+\tau,t) = \tfrac{1}{2}R_A(t+\tau,t)\cos\omega_o\tau$$
$$+ \tfrac{1}{2}\operatorname{Re}\left[R_A(t+\tau,t)e^{-i2\omega_o t}e^{-i\omega_o\tau}E\{e^{-i2\Theta}\}\right]. \qquad (8.88)$$

The time-averaged value of R_X is (exercise 17)

$$\langle R_X \rangle(\tau) = \tfrac{1}{2}E\{\hat{R}_A(\tau)\}\cos\omega_o\tau + \tfrac{1}{2}\operatorname{Re}\left[E\{\hat{R}_A^{2\omega_o}(\tau)\}e^{-i\omega_o\tau}E\{e^{-i2\Theta}\}\right]$$
$$= E\{\hat{R}_X(\tau)\}. \qquad (8.89)$$

Thus, (8.68b) holds. However, (8.81) does not hold unless

$$\hat{R}_A(\tau) = E\{\hat{R}_A(\tau)\} \qquad (8.90)$$

and

$$\hat{R}_A^{2\omega_o}(\tau)e^{-i2\Theta} = E\{\hat{R}_A^{2\omega_o}(\tau)\}E\{e^{-i2\Theta}\}. \qquad (8.91)$$

If $A(t)$ has mean-square ergodicity of the autocorrelation, then (8.90) is satisfied. However, (8.91) will not be satisfied unless

$$\hat{R}_A^{2\omega_o}(\tau) = E\{\hat{R}_A^{2\omega_o}(\tau)\} \qquad (8.92)$$

and Θ is nonrandom, so that

$$e^{i2\Theta} = E\{e^{i2\Theta}\}, \qquad (8.93)$$

or unless

$$\hat{R}_A^{2\omega_o}(\tau) \equiv 0. \qquad (8.94)$$

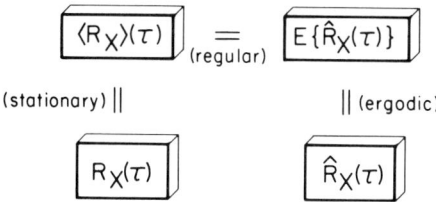

Figure 8.3 Relationships among Empirical and Probabilistic Autocorrelations

The condition (8.94) will be satisfied if $A(t)A(t + \tau)$ contains no additive periodicity with radian frequency $2\omega_o$. For example, if $A(t)$ contains no components at frequencies greater than or equal to $f = \omega_o/2\pi$, then (8.94) holds. If $A(t)A(t + \tau)$ does exhibit such periodicity, then (8.92) must be satisfied and Θ must be nonrandom. Equation (8.92) is an ergodic property associated with the periodicity and is therefore called a *cycloergodic property* [Boyles and Gardner, 1983]. This generalization of the concept of ergodicity is treated in Chapter 12. ∎

In summary, the relationships among empirical and probabilistic autocorrelations can be displayed in the form of a diagram, as shown in Figure 8.3.

8.6 Duality and the Role of Ergodicity

Before leaving the subject of ergodicity, its role in the application of random-process theory to practical problems should be clarified. The basic question regarding ergodicity is: when we choose or construct a mathematical model of a random process, do we care if it has ergodic properties or not? The answer depends on the purpose to which the model is to be put. If it is important that ensemble averages obtained from the model equal time averages obtained from the model, then clearly we want a model with ergodic properties. However, in applications for which there is no desire to relate time averages to ensemble averages, or even no interest in the existence of time averages, ergodic properties are irrelevant. Returning to the former situation, when it is important that time averages equal ensemble averages, it is entirely possible that the most appropriate model is a nonprobabilistic model for which all statistical parameters of interest are defined directly in terms of time averages, such as

$$\hat{m}_X \triangleq \lim_{T \to \infty} \frac{1}{T} \int_{-T/2}^{T/2} X(t)\, dt \qquad (8.95)$$

$$\hat{R}_X(\tau) \triangleq \lim_{T \to \infty} \frac{1}{T} \int_{-T/2}^{T/2} X(t + \tau) X(t)\, dt, \qquad (8.96)$$

as discussed in Section 3.2. As a matter of fact, these statistical parameters can be given a probabilistic interpretation in terms of *fraction-of-time distributions* [Hofstetter, 1964; Brillinger, 1975]. For example, let us define the probability of the amplitude of a specific waveform $x(t)$ (e.g., a statistical sample of a random process) being less than some value y to be the fraction of time that this occurs. Specifically,

$$\hat{F}_X(y) \triangleq [\text{probability that } x(t) < y]$$

$$\triangleq \lim_{T \to \infty} \frac{1}{T} \int_{-T/2}^{T/2} u[y - x(t)] \, dt, \qquad (8.97)$$

where the integrand is the event indicator

$$u[y - x(t)] \triangleq \begin{cases} 1, & x(t) < y, \\ 0, & x(t) \geq y. \end{cases} \qquad (8.98)$$

Then the mean value of the waveform is

$$\hat{m}_X \triangleq \int_{-\infty}^{\infty} y \hat{f}_X(y) \, dy, \qquad (8.99)$$

where

$$\hat{f}_X(y) \triangleq \frac{d\hat{F}_X(y)}{dy}. \qquad (8.100)$$

Substitution of (8.97) and (8.98) into (8.100), and (8.100) into (8.99), yields (8.95) (exercise 23). The analogous result for $\hat{R}_X$ can be obtained from the definition of the joint fraction-of-time distribution

$$\hat{F}_{X(t_1)X(t_2)}(y_1, y_2) \triangleq \lim_{T \to \infty} \frac{1}{T} \int_{-T/2}^{T/2} u[y_1 - x(t_1 + t)] u[y_2 - x(t_2 + t)] \, dt, \qquad (8.101)$$

by using

$$\hat{R}_X(t_1 - t_2) \triangleq \int_{-\infty}^{\infty} \int_{-\infty}^{\infty} y_1 y_2 \hat{f}_{X(t_1)X(t_2)}(y_1, y_2) \, dy_1 \, dy_2, \qquad (8.102)$$

where

$$\hat{f}_{X(t_1)X(t_2)}(y_1, y_2) \triangleq \frac{\partial^2}{\partial y_1 \partial y_2} \hat{F}_{X(t_1)X(t_2)}(y_1, y_2). \qquad (8.103)$$

Substitution of (8.101) into (8.103) into (8.102) yields (8.96). From this point of view, which leads to the identity between (8.95) and (8.99) and between (8.96) and (8.102), it should be clear why there is a duality between the theory based on time averages and the theory based on expected values (ensemble averages). Moreover, we can heuristically identify these two types of averages as one and the same if we define ensemble members by

$$X(t,\sigma) \triangleq x(t - \sigma), \qquad (8.104)$$

where σ denotes a random sample from a sample space, and we apply the law of large numbers (Section 2.5.3). (The discrete-time counterpart of (8.104) is known as *Wold's isomorphism* [Wold, 1948].) Then the fraction-of-time distribution can be given the same relative-frequency interpretation that formed the basis for our definition of probability in Chapter 1. The limitation to this conceptualization of a probabilistic model is that it cannot accommodate nonstationary models. By the very definition of the fraction-of-time distributions, they must be time-invariant.*

8.7 Summary

Ergodic properties of random processes guarantee that probabilistic parameters of the process, such as the mean and autocorrelation, can be determined from their empirical counterparts obtained by time-averaging. *Mean-square ergodicity* guarantees that a probabilistic parameter and its empirical counterpart are mean-square equivalent. The necessary and sufficient condition for a WSS stationary process to exhibit mean-square ergodicity of the mean (8.42a) is specified in terms of the autocovariance function (8.42c). The necessary and sufficient condition for a stationary process to exhibit mean-square ergodicity of the autocorrelation (8.48a) is specified in terms of the fourth joint moment and autocorrelation function, (8.48b) and (8.49). However, for a Gaussian process this reduces to a necessary and sufficient condition specified by only the autocovariance (8.58). Moreover, this condition is equivalent to the condition that the Gaussian process contains no random finite additive sine-wave components. For a Gaussian process, ergodicity of the mean and autocovariance guarantee ergodicity of all higher-order moments. However, for a non-Gaussian process ergodicity of moments up to order n does not in general guarantee ergodicity of moments of order higher than n.

Stationary processes are not the only processes that can exhibit ergodic properties. For example, asymptotically mean stationary processes can exhibit ergodic properties associated with their time-averaged probabilistic parameters such as, $\langle m_X \rangle$ and $\langle R_X \rangle(\tau)$. In this more general setting, it is seen that the appropriate characterization of ergodicity is the equality (in the mean-square sense) of an empirical parameter and its mean, for example $E\{\hat{m}_X\} = \hat{m}_X$ or $E\{\hat{R}_X(\tau)\} = \hat{R}_X(\tau)$. The necessary and sufficient condi-

*One exception to this is the class of cyclostationary models, for which this fraction-of-time-distribution approach can be generalized, as explained in Chapter 12.

tion for mean-square ergodicity of the mean of an asymptotically mean stationary process is specified in terms of the autocovariance (8.71), and this condition reduces to (8.42c) if the process is stationary.

For an asymptotically mean stationary process, $\hat{R}_X(\tau)$ exists if the process is *regular*, and its expected value is given by $E\{\hat{R}_X(\tau)\} = \langle R_X \rangle(\tau)$. If the process is ergodic then $E\{\hat{R}_X(\tau)\} = \hat{R}_X(\tau)$, and if it is stationary then $\langle R_X \rangle(\tau) = R_X(\tau)$.

In some situations where ergodicity is of the utmost importance, it might be that the dual deterministic theory based on time averages is more appropriate than the probabilistic theory. In this case, by adopting the deterministic theory, the concept of ergodicity becomes irrelevant. Nevertheless, in the deterministic theory the various empirical parameters such as the mean $\hat{m}_X$ and autocorrelation $\hat{R}_X(\tau)$ can always be given probabilistic interpretations in terms of fraction-of-time distributions, such as (8.97) and (8.101). From this point of view, it is clear why there is a duality between the theory based on time averages and the theory based on expected values. This method of giving empirical (deterministic) parameters probabilistic interpreations is equivalent to interpreting time-shifted versions $x(t - \sigma)$ of a wareform as random samples $X(t, \sigma)$ from a sample space. This correspondence for discrete time is known as *Wold's isomorphism*.

EXERCISES

1. Verify Equations (8.19), (8.21), and (8.22).

★2. Verify Equation (8.23).

    Hint: Draw a picture of the matrix with (i, j)th element $K_X([i - j]\Delta)$, and indicate on this drawing the paths of summation corresponding to (8.19) and (8.23); note that the kth off-diagonal of an $N \times N$ matrix of this form has $N - k$ elements, all of which are equal. Such matrices are called *Toeplitz matrices*.

3. Draw a graph of the integrand in (8.24) to verify that (8.25) is a close approximation to (8.24) for $T \gg \tau_o$.

4. To prove that (8.42c) is necessary and sufficient for $X(t)$ to exhibit mean-square ergodicity of the mean, proceed as follows:
    (a) Use the Cauchy-Schwarz inequality together with (8.34) to prove that

$$\left|\frac{1}{T}\int_0^T K_X(\tau)\,d\tau\right|^2 = \left|E\left\{[X(0)-m_X]\frac{1}{T}\int_0^T[X(\tau)-m_X]\,d\tau\right\}\right|^2$$
$$\leq 2\sigma_X^2 \frac{1}{T}\int_0^T\left(1-\frac{\tau}{T}\right)K_X(\tau)\,d\tau = \sigma_X^2\,\text{MSE}(T). \quad (8.105)$$

Therefore (8.42c) is necessary for (8.42b).

(b) Show that
$$\text{MSE}(T) = \frac{1}{T^2}\int_0^T\int_0^T K_X(t-u)\,dt\,du$$
$$= \frac{2}{T^2}\int_0^T\int_0^t K_X(\tau)\,d\tau\,dt. \quad (8.106)$$

It can be proved that (8.42c) is sufficient for (8.42b) as follows. Given (8.42c), we know that for any $\varepsilon > 0$ (no matter how small), there exists a T_* such that
$$\left|\frac{1}{t}\int_0^t K_X(\tau)\,d\tau\right| < \varepsilon, \quad t \geq T_*.$$

Therefore,
$$\frac{1}{T^2}\left|\int_{T_*}^T\int_0^t K_X(\tau)\,d\tau\,dt\right| \leq \frac{1}{T^2}\int_{T_*}^T t\left|\frac{1}{t}\int_0^t K_X(\tau)\,d\tau\right|dt$$
$$\leq \frac{1}{T^2}\int_{T_*}^T t\varepsilon\,dt \leq \varepsilon, \quad T > T_*. \quad (8.107)$$

It follows that
$$\text{MSE}(T) = \frac{2}{T^2}\int_0^{T_*}\int_0^t K_X(\tau)\,d\tau\,dt + \frac{2}{T^2}\int_{T_*}^T\int_0^t K_X(\tau)\,d\tau\,dt$$
$$\leq \frac{2}{T^2}\left|\int_0^{T_*}\int_0^t K_X(\tau)\,d\tau\,dt\right| + 2\varepsilon. \quad (8.108)$$

Since T_* is fixed for each ε, then
$$\lim_{T\to\infty}\text{MSE}(T) \leq 2\varepsilon$$

for any ε, no matter how small. This result together with (8.34) establishes the validity of (8.42b). Hence, (8.42b) follows from (8.42c); that is, (8.42c) is sufficient for (8.42b).

5. It follows from (8.23) that the discrete-time counterpart of the necessary and sufficient condition (8.42b) for mean-square ergodicity of the mean is

$$\lim_{N \to \infty} \frac{1}{N} \sum_{k=1}^{N} \left(1 - \frac{k}{N}\right) K_X(k\Delta) = 0. \tag{8.109}$$

If $K_X(\tau) \to 0$ as $\tau \to \infty$, then both (8.109) and (8.42b) are satisfied. However, if $K_X(\tau)$ oscillates asymptotically with a period that is integrally related to Δ, then (8.109) can be violated even though (8.42b) is satisfied. In such a case, the discrete-time estimate (8.18) will not converge to the correct value although the continuous-time estimate (8.32) will. Demonstrate this for the process in Example 1 and explain why (8.18) and (8.32) behave differently for this process.

6. Verify that (8.36) satisfies (8.31), but that (8.38) violates (8.31).

7. Verify that the random-amplitude and -phase sine-wave process (example 3), which is assumed to be a Gaussian process (as will be the case if the amplitude has a Rayleigh distribution), has mean-square ergodicity of the mean, but does not have mean-square ergodicity of the autocorrelation unless $\sigma_A^2 = 0$ ($\Rightarrow E\{A^n\} = A^n$ for all n); that is, verify that (8.48b) is violated.

8. Let $X(t)$ be a nonzero mean WSS process that exhibits ergodicity of both the mean and the autocorrelation, and let

$$Y(t) = AX(t),$$

where A is random and independent of $X(t)$. Prove that $Y(t)$ does not exhibit ergodicity of either the mean or the autocorrelation. Do this first by showing directly that $\hat{m}_Y \neq m_Y$ and $\hat{R}_Y \neq R_Y$. Then show that the necessary and sufficient conditions (8.42c) and (8.48b), with X replaced by Y, are violated.

9. For processes with finite starting times, say $t = 0$, the two-sided time averages for estimating probabilistic parameters will yield half the desired result. Thus, one-sided time averages must be used to define ergodicity of one-sided processes. For example the one-sided stationary process $X(t)$, $t \geq 0$, has strong ergodicity of the mean if and only if

$$\underset{T \to \infty}{\text{l.i.m.}} \frac{1}{T} \int_0^T X(t) \, dt = E\{X(t)\}, \tag{8.110}$$

in which case

$$\underset{T \to \infty}{\text{l.i.m.}} \frac{1}{T} \int_{-T/2}^{T/2} X(t) \, dt = \tfrac{1}{2} E\{X(t)\}. \tag{8.111}$$

(a) Show that (8.111) follows from (8.110) and $X(t) = 0$ for $t < 0$.
(b) Determine if the Ornstein-Uhlenbeck process (Section 5.3.2) has mean-square ergodicity of the autocorrelation.

10. Consider the process

$$X(t) = \sum_n A_n \cos(\omega_n t + \Theta_n) + B(t),$$

where $B(t)$ is a zero-mean stationary Gaussian process which has mean-square ergodicity of the autocorrelation, and $\{A_n\}$ and $\{\Theta_n\}$ are all independent random variables which are also independent of $B(t)$. Let $\{\Theta_n\}$ be uniformly distributed on $[-\pi, \pi)$. Determine the conditions on the $\{A_n\}$ under which $X(t)$ has mean-square ergodicity of the autocorrelation.

11. Consider the radar or sonar range-estimation problem of determining how far away an energy-reflecting object is by estimating the round-trip time lapse τ_* between the transmission of a signal $x(t)$ toward the object and the reception of the randomly corrupted reflected signal $Y(t)$ from the object. In order to ensure that the received signal is not lost in background noise, it is necessary to transmit a signal with sufficient energy. This is often best accomplished by use of a long-duration signal. However, in order to obtain good resolution in range estimation, the long signal must have a narrow autocorrelation function. Given that the signal propagates at a speed v, show that measurement of the cross-correlation of the transmitted and received signals, and determination of the lag value τ at which the cross-correlation peaks, can be used to estimate the range. Do this by evaluating the mean and variance of the measured cross-correlation

$$\hat{R}_{Yx}(\tau)_T \triangleq \frac{1}{T} \int_{-T/2}^{T/2} Y(t + \tau) x(t) \, dt, \qquad (8.112)$$

and showing that the value of τ_* can be obtained from the mean of $\hat{R}_{Yx}(\tau)_T$, and that the variance of $\hat{R}_{Yx}(\tau)_T$ can be made small by choosing T to be sufficiently large. Use the model

$$Y(t) = cx(t - \tau_*) + N(t),$$

for which $x(t)$ is a nonrandom signal whose finite-average autocorrelation [(8.112) with $Y(t)$ replaced by $x(t)$] peaks at $\tau = 0$, c is a constant, and $N(t)$ is the symbolic derivative of the Wiener process. Assume that the energy of $x(t)$,

$$\int_{-T/2}^{T/2} x^2(t) \, dt,$$

grows linearly with T.

12. (a) Draw a block diagram of an instrument that can be used to measure the autocorrelation function, at a fixed lag value τ_*, for an ergodic process. Use only the following components: a delay element with delay $= \tau_*$, a waveform multiplier, and a low-pass filter (LPF)

whose output is given by

$$Z(t) = \int_{-\infty}^{\infty} h(t-u) Y(u)\, du$$

for an input $Y(t)$. (A low-pass filter is, in essence, a sliding-window time-averaging device.)

(b) Then consider the impulse-response function, for the low-pass filter, given by

$$h(t) = \begin{cases} 1/T, & 0 \le t \le T, \\ 0, & \text{otherwise} \end{cases}$$

and show that the measured correlation at time t is given by

$$\hat{R}_X(\tau_*)_T = \frac{1}{T} \int_{t-T}^{t} X(u) X(u - \tau_*)\, du.$$

(c) Let $X(t)$ be a zero-mean Gaussian process with autocovariance

$$K_X(\tau) = \sigma^2 e^{-|\tau|/\tau_0},$$

and evaluate the mean and variance of the measured autocorrelation $\hat{R}_X(\tau_*)_T$. Obtain an explicit formula in terms of σ^2, T, τ_0, and τ_*. Verify that the variance is much smaller than the squared mean when $T \gg \max\{\tau_0, \tau_*\}$.

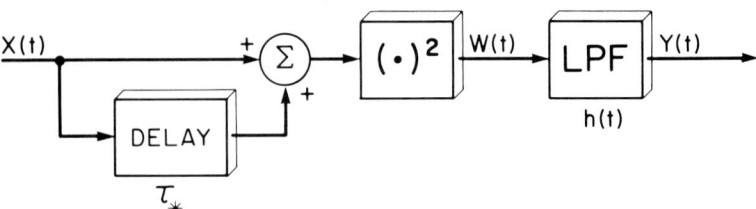

Figure 8.4 A Method for Measuring Autocorrelation

(d) Consider the alternative method for measuring the autocorrelation function depicted in Figure 8.4. Show that

$$E\{Y(t)\} = 2[R_X(0) + R_X(\tau_*)],$$

for $h(t)$ as specified in (b).

13. A stationary process $X(t)$ has a correlation time of $\tau_0 = 1$ s. Determine approximately how long the averaging time must be in order to obtain an estimate of the mean whose variance is no larger than 1/100 of the variance of the process $X(t)$. What is the approximate minimum num-

ber of equally spaced time samples required to implement this estimate as a discrete-time average?

★**14.** Let $X(t)$ be a stationary process for which $X(t)$ and $X(u)$ become statistically independent as $|t - u| \to \infty$. Show that $X(t)$ exhibits mean-square ergodicity of the mean and of the autocorrelation. (In fact, it exhibits mean-square ergodicity of all order moments.)

Hint: If $K(\tau) \to 0$ as $\tau \to \infty$, then

$$\lim_{T \to \infty} \frac{1}{T} \int_0^T K(\tau)\, d\tau = 0.$$

15. Show that a stationary Gaussian process $X(t)$ for which $K_X(\tau) \to 0$ as $\tau \to \infty$ exhibits mean-square ergodicity of the fourth moment. (In fact, it exhibits mean square ergodicity of all order moments.)

16. Use Isserlis's formula from Chapter 5 as a basis for proposing a simplified method for estimating high-order moments of stationary Gaussian processes.

17. Verify (8.86) to (8.88). Then use these equations to verify (8.89) (assuming expectation and limit can be interchanged).

18. Show that the variance formula (8.70) can be reexpressed as

$$\text{Var} = \frac{1}{T} \int_{-T}^{T} \frac{1}{T} \int_{-(T-|\tau|)/2}^{(T-|\tau|)/2} K_X\left(t + \frac{\tau}{2}, t - \frac{\tau}{2}\right) dt\, d\tau. \qquad (8.113)$$

Is the limit, as $T \to \infty$, of this expression the same as the limit of

$$\frac{1}{T} \int_{-T}^{T} \langle K_X \rangle(\tau)\, d\tau, \qquad (8.114)$$

which is the analog of (8.42c)?

19. Consider the Poisson counting process with rate parameter λ. Recall that λ is the expected rate of counts per unit time. Show that the Poisson process exhibits ergodic properties by proceeding as follows. Define a time-average rate parameter by

$$\hat{\lambda}(T) \triangleq \frac{N(T)}{T},$$

and show that

$$E\{\hat{\lambda}(T)\} = \lambda,$$
$$\text{Var}\{\hat{\lambda}(T)\} = \lambda/T,$$

and therefore that
$$\lim_{T\to\infty} E\{[\hat{\lambda}(T) - \lambda]^2\} = 0.$$

★20. Apply the technique used to extend the result (8.42) to (8.48)-(8.49) to develop a necessary and sufficient condition for mean-square ergodicity of the distribution $F_X(x)$ for a stationary process $X(t)$.

Hint: Use the characterization $F_X(x) = E\{u[x - X(t)]\}$. The final result involves only $F_{X(t+\tau)X(t)}(x, x)$ and $F_X(x)$.

★21. Use the result of exercise 20 to show that if $X(t)$ and $X(u)$ become statistically independent as $|t - u| \to \infty$, then $X(t)$ exhibits mean-square ergodicity of the distribution.

Hint: See the hint in the preceding exercise.

22. Let $X(t)$ be a *mixture process* for which the probability law is a mixture of two laws,
$$F_{X(t_1)\cdots X(t_n)} = pF^1_{X(t_1)\cdots X(t_n)} + (1 - p)F^2_{X(t_1)\cdots X(t_n)}, \qquad (8.115)$$
in which $F^1_{X(t_1)\cdots X(t_n)}$ and $F^2_{X(t_1)\cdots X(t_n)}$ are arbitrary distinct distribution functions, and $0 < p < 1$. Show that $X(t)$ cannot, in general, exhibit ergodicity.

Hint: Simply show, as an example, that $X(t)$ cannot exhibit mean-square ergodicity of the mean if $m^1_X \neq m^2_X$. Proceed by analogy with (8.64) to (8.66).

★23. (a) Prove that (8.97) to (8.100) yield (8.95).

Hint: Use $du[y - x(t)]/dy = \delta[y - x(t)]$, where δ is the Dirac delta.

(b) Prove that (8.101) to (8.103) yield (8.96).

(c) Verify that (8.97) and (8.101) are valid distribution functions; that is, $\hat{F}_x$ must be a nondecreasing function for which $\hat{F}_x(-\infty) = 0$ and $\hat{F}_x(+\infty) = 1$.

9

Linear Transformations, Filters, and Dynamical Systems

THE AUTOCORRELATION FUNCTION of a process contains much useful information about the process. In fact, as suggested by the discussion of generalized harmonic analysis in Section 3.2, a complete description of the average frequency content of a WSS process can be obtained from the autocorrelation function. (This is studied in the next chapter.) It is therefore important to know how the autocorrelation function of a process is affected by commonly used signal-processing operations such as filtering and modulation. In this chapter, the effects on the mean and autocorrelation of general linear transformations, such as linear dynamical systems with random excitation, and time-invariant filters with WSS inputs, are studied. Although the effects of a linear transformation of a random process on its probability distributions are not studied in this chapter, it should be pointed out that this is in effect accomplished for Gaussian processes. The reason is that it follows directly from the definition of a Gaussian process (Section 5.2) that a linearly transformed Gaussian process is still a Gaussian process, and such processes are completely specified by their mean functions and autocorrelation functions.

9.1 Linear Transformation of an N-tuple of Random Variables

The fundamental linearity property of expectation (Chapter 1) guarantees that

$$E\{h_1 X_1 + h_2 X_2\} = h_1 E\{X_1\} + h_2 E\{X_2\} \qquad (9.1)$$

for any two random variables X_1 and X_2, and any two numbers h_1 and h_2. More generally, consider the transformation of a 2-tuple (column vector) of random variables, $X = [X_1, X_2]^T$, into another 2-tuple, $Y = [Y_1, Y_2]^T$, according to the transformation matrix

$$H = \begin{bmatrix} h_{11} & h_{12} \\ h_{21} & h_{22} \end{bmatrix}. \tag{9.2}$$

That is,

$$Y_i = \sum_{j=1}^{2} h_{ij} X_j, \quad i = 1, 2, \tag{9.3}$$

which can be reexpressed as

$$Y = HX. \tag{9.4}$$

It follows from (9.1) and (9.3) that the means of X and Y are related by

$$m_{Y_i} \triangleq E\{Y_i\} = \sum_j h_{ij} E\{X_j\} \triangleq \sum_j h_{ij} m_{X_j}. \tag{9.5}$$

If the vector of means is denoted by

$$m_Y = [m_{Y_1}, m_{Y_2}]^T, \tag{9.6}$$

then (9.5) can be reexpressed as

$$m_Y = H m_X. \tag{9.7}$$

This simple result is quite significant, since it reveals that one need not determine the probability densities for Y_1 and Y_2 in order to determine their mean values. One need determine only the mean values for X_1 and X_2, and then perform the linear transformation (9.7). This is quite a short cut, since the determination of the probability densities for Y_1 and Y_2 requires not only knowledge of the joint probability density for X_1 and X_2, but also matrix inversion (assuming H is nonsingular):

$$f_Y(y_1, y_2) = f_X(x_1, x_2) |H|^{-1}, \tag{9.8}$$

where

$$[x_1, x_2]^T = H^{-1}[y_1, y_2]^T, \tag{9.9}$$

and $|H|$ is the magnitude of the determinant of H.

Analogous to (9.5), the correlations of X and Y are related by

$$R_Y(i,j) \triangleq E\{Y_i Y_j\} = \sum_n \sum_m h_{in} h_{jm} E\{X_n X_m\} = \sum_n \sum_m h_{in} R_X(n,m) h_{jm},$$

(9.10)

which can be reexpressed as

$$R_Y = HR_X H^T,$$ (9.11)

where R_Y is the 2×2 correlation matrix with (i,j)th element $R_Y(i,j)$. It is noted that both (9.7) and (9.11) can be obtained directly in matrix notation as follows:

$$m_Y \triangleq E\{Y\} = E\{HX\} = HE\{X\} \triangleq Hm_X,$$ (9.12)

$$R_Y \triangleq E\{YY^T\} = E\{HX(HX)^T\} = E\{HXX^T H^T\}$$

$$= HE\{XX^T\}H^T \triangleq HR_X H^T.$$ (9.13)

Similarly, it follows (exercise 1) that the covariance matrices for X and Y are related by

$$K_Y \triangleq E\{(Y - m_Y)(Y - m_Y)^T\} = R_Y - m_Y m_Y^T = HK_X H^T.$$

(9.14)

The simple formulas (9.11) and (9.14) are quite significant because of the ease with which R_Y and K_Y can be determined from knowledge of only R_X and K_X. Of course, these formulas apply to any n-tuple X and m-tuple Y, in which case H is an $m \times n$ matrix. The case $m = n = 2$ has been focused on simply for the purpose of illustration.

9.2 Linear Discrete-Time Filtering*

Consider a linear, possibly time-variant, discrete-time system with unit-pulse response denoted by $h(i,j)$, which is the response at time i to a unit-pulse sequence applied at time j (exercise 2). With reference to Figure 9.1, the

---

*To emphasize analogies between continuous-time and discrete-time processes, the notation $X(i)$ will be used in place of X_i from this point forward.

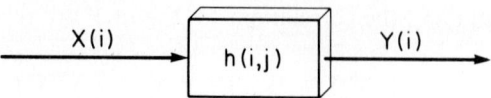

Figure 9.1 Block-Diagram Representation of a Linear Transformation of a Discrete-Time Process $X(i)$ into Another Process $Y(i)$

input-output relation for such a system is the superposition sum

$$Y(i) = \sum_{j=-\infty}^{\infty} h(i, j) X(j), \qquad (9.15)$$

which is assumed to be a mean-square convergent series* for each i. If the system is *causal*, then $h(i, j) = 0$ for $i < j$, and (9.15) reduces to

$$Y(i) = \sum_{j=-\infty}^{i} h(i, j) X(j). \qquad (9.16)$$

If, in addition, the system has *finite memory* with length N, then $h(i, j) = 0$ for $i > j + N$, and (9.16) reduces to

$$Y(i) = \sum_{j=i-N}^{i} h(i, j) X(j). \qquad (9.17)$$

If the system is time-invariant, then $h(i, j) = h(i - j)$ and the system is called a *filter*, and (9.15) can be reexpressed (exercise 2) as

$$Y(i) = \sum_{j=-\infty}^{\infty} h(i - j) X(j) \qquad (9.18)$$

$$= \sum_{k=-\infty}^{\infty} h(k) X(i - k) \qquad (9.19)$$

$$\triangleq h(i) \otimes X(i), \qquad (9.20)$$

where $\otimes$ denotes the *discrete convolution operation*. Similarly, (9.16) and (9.17) can be reexpressed as

$$Y(i) = \sum_{k=0}^{\infty} h(k) X(i - k) \qquad (9.21)$$

and

$$Y(i) = \sum_{k=0}^{N} h(k) X(i - k) \qquad (9.22)$$

for causal and finite-memory causal filters, respectively.

*A sufficient condition is that $X(i)$ is a bounded-mean-square process, and h is a *stable system* in the sense that $\sum_{j=-\infty}^{\infty} h^2(i, j) < \infty$.

The input-output relations for the means and autocorrelations of $X(i)$ and $Y(i)$ are given by (9.5) and (9.10) with $h_{ij} = h(i, j)$ and with limits of summation $\pm \infty$. Furthermore, if the system is time-invariant, these can be reexpressed as

$$m_Y(i) = m_X(i) \otimes h(i), \qquad (9.23)$$

$$R_Y(i, j) = h(i) \otimes R_X(i, j) \otimes h(j). \qquad (9.24)$$

If the input $X(i)$ is WSS, it follows from (9.23) and (9.24) that the output is WSS, and (9.23) and (9.24) reduce (exercise 3) to

$$m_Y = m_X \sum_{i=-\infty}^{\infty} h(i), \qquad (9.25)$$

$$R_Y(k) = R_X(k) \otimes r_h(k), \qquad (9.26)$$

where $r_h(k)$ is the finite autocorrelation of $h(i)$,

$$r_h(k) \triangleq \sum_{i=-\infty}^{\infty} h(i+k)h(i) = \sum_{j=-\infty}^{\infty} h(k-j)h(-j) = h(k) \otimes h(-k).$$

$$(9.27)$$

In summary, (9.26) reveals that when a WSS process is filtered by h, its autocorrelation is, in effect, filtered by the finite autocorrelation of h. By similar means it can be shown (exercise 4) that the cross-correlation of the output of a time-invariant filter with its WSS input is given by the convolutions

$$R_{YX}(k) = R_X(k) \otimes h(k), \qquad (9.28a)$$

$$R_{XY}(k) = R_X(k) \otimes h(-k). \qquad (9.28b)$$

Example 1: The linear discrete-time system shown in Figure 9.2, with WSS input $X(i)$, produces the first difference

$$Y(i) = X(i) - X(i - 1) \qquad (9.29)$$

at its output. To determine the mean and autocorrelation of $Y(i)$, the unit-pulse response of the system is first determined. It is easily seen that

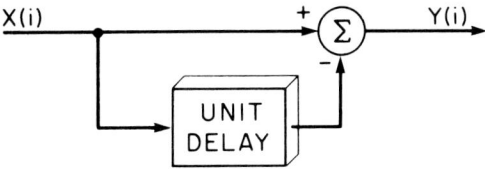

Figure 9.2 Block Diagram for Example 1

this is

$$h(i) = \delta_i - \delta_{i-1}, \qquad (9.30)$$

where δ_i is the unit-pulse sequence

$$\delta_i \triangleq \begin{cases} 1, & i = 0, \\ 0, & i \neq 0, \end{cases} \qquad (9.31)$$

which is called the *Kronecker delta*. Thus,

$$\sum_{i=-\infty}^{\infty} h(i) = 0, \qquad (9.32a)$$

and (9.27) and (9.30) yield the finite autocorrelation

$$r_h(k) = 2\delta_k - \delta_{k-1} - \delta_{k+1}. \qquad (9.32b)$$

It follows from (9.25), (9.26), and (9.32) that $m_Y = 0$ regardless of m_X, and

$$R_Y(k) = 2R_X(k) - R_X(k-1) - R_X(k+1). \qquad (9.33)$$

For this particularly simple example, this result can easily be checked by direct calculation from the system difference equation (9.29). ∎

Example 2: The linear discrete-time system shown in Figure 9.3, with input $X(i)$, produces an output $Y(i)$ governed by the first-order difference equation

$$Y(i+1) = aY(i) + X(i). \qquad (9.34)$$

The unit-pulse response of this system is

$$h(i) = \begin{cases} (1/a)a^i, & i \geq 1, \\ 0, & i \leq 0, \end{cases} \qquad (9.35)$$

where it is assumed that $|a| < 1$ for stability. It can be shown (exercise 5) by using the *geometric-progression* formula

$$\sum_{i=0}^{\infty} a^i = \frac{1}{1-a}, \qquad |a| < 1, \qquad (9.36)$$

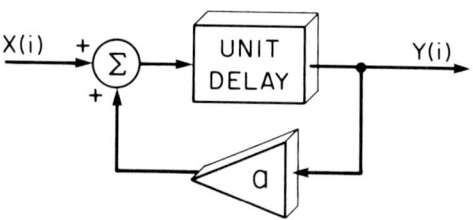

Figure 9.3 Block Diagram for Example 2

that

$$\sum_{i=-\infty}^{\infty} h(i) = \frac{1}{1-a} \tag{9.37}$$

and, using (9.27), that the finite autocorrelation is

$$r_h(k) = \frac{1}{1-a^2} a^{|k|}. \tag{9.38}$$

Consider a WSS input sequence that exhibits zero mean, $m_X = 0$, and zero correlation,

$$R_X(k) = \begin{cases} \sigma^2, & k = 0, \\ 0, & k \neq 0. \end{cases} \tag{9.39}$$

Then it follows from (9.25), (9.26), and (9.37) to (9.39) that the mean and autocorrelation of the output $Y(i)$ are $m_Y = 0$ and

$$R_Y(k) = \frac{\sigma^2}{1-a^2} a^{|k|}. \tag{9.40}$$

∎

9.3 Linear Continuous-Time Filtering

The input-output relations described in Section 9.2 for discrete time yield the input-output relations for continuous time simply by replacement of discrete sums with integrals and replacement of discrete variables i, j, k, n, m with continuous variables t, u, v, r, s. Thus, for the general linear system,

$$Y(t) = \int_{-\infty}^{\infty} h(t, u) X(u) \, du, \tag{9.41}$$

in which $h(t, u)$ denotes the impulse response of the system (exercise 6), Equations (9.5) and (9.10) yield (assuming mean-square integrability for each t)

$$m_Y(t) = \int_{-\infty}^{\infty} h(t, u) m_X(u) \, du, \tag{9.42}$$

$$R_Y(t, u) = \int_{-\infty}^{\infty} \int_{-\infty}^{\infty} h(t, r) h(u, s) R_X(r, s) \, dr \, ds. \tag{9.43}$$

For a WSS process $X(t)$ and a time-invariant system (filter), for which $h(t, u) = h(t - u)$, (9.42) and (9.43) reduce (exercise 6) to

$$m_Y = m_X \int_{-\infty}^{\infty} h(t) \, dt, \tag{9.44}$$

$$R_Y(\tau) = \int_{-\infty}^{\infty} R_X(u) r_h(\tau - u) \, du = \int_{-\infty}^{\infty} r_h(v) R_X(\tau - v) \, dv$$

$$= R_X(\tau) \otimes r_h(\tau), \tag{9.45}$$

where $r_h(\tau)$ is the finite autocorrelation

$$r_h(\tau) \triangleq h(\tau) \otimes h(-\tau), \tag{9.46}$$

and $\otimes$ denotes the continuous convolution operation (cf. Section 3.2). Also,

$$R_{YX}(\tau) = R_X(\tau) \otimes h(\tau), \tag{9.47a}$$
$$R_{XY}(\tau) = R_X(\tau) \otimes h(-\tau). \tag{9.47b}$$

Example 3: Consider subjecting a process $X(t)$ to a sliding-window smoothing operation (see Figure 9.4)

$$Y(t) = \frac{1}{T} \int_{t-T}^{t} X(u)\, du. \tag{9.48}$$

Equation (9.48) can be reexpressed in terms of the rectangular impulse-response function

$$h(t) = \begin{cases} 1/T, & 0 \le t \le T, \\ 0, & \text{otherwise.} \end{cases} \tag{9.49}$$

The finite autocorrelation of h is the triangular function

$$r_h(\tau) = \begin{cases} \dfrac{1}{T}\left(1 - \dfrac{|\tau|}{T}\right), & |\tau| \le T, \\ 0, & \text{otherwise,} \end{cases} \tag{9.50}$$

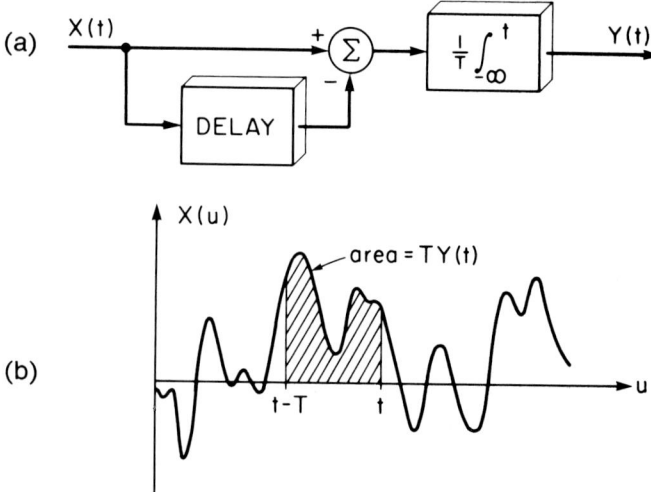

Figure 9.4 (a) Block Diagram of an Implementation of the Sliding-Window Smoother of Example 3; (b) Graphical Illustration of Sliding-Window Smoothing Operation

Linear Transformations, Filters, and Dynamical Systems • 197

and the area of h is

$$\int_{-\infty}^{\infty} h(t)\, dt = 1. \tag{9.51}$$

Therefore, (9.44) and (9.45) yield

$$m_Y = m_X, \tag{9.52}$$

$$R_Y(\tau) = \int_{-T}^{T} \frac{1}{T}\left(1 - \frac{|u|}{T}\right) R_X(\tau - u)\, du. \tag{9.53}$$

Hence, R_X is in effect subjected to a sliding-window smoothing operation with a triangular window. If the process being smoothed is modeled by the symbolic derivative of the Wiener process (with starting time $t_o \to -\infty$), then (9.52) and (9.53) yield

$$m_Y = 0,$$

and

$$R_Y(\tau) = \begin{cases} \dfrac{\alpha^2}{T}\left(1 - \dfrac{|\tau|}{T}\right), & |\tau| \le T, \\ 0, & |\tau| > T, \end{cases} \tag{9.54}$$

which is a triangular autocorrelation. Since the Wiener process is Gaussian, so too is the smoothed process $Y(t)$, and its probability distributions are completely specified by (9.54). ∎

Example 4: Consider the result of measuring a broadband noise voltage through a resistive-capacitive circuit, as shown in Figure 9.5. The noise is modeled by the symbolic derivative of the Wiener process (with starting time $t_o \to -\infty$), so that its mean and autocorrelation are

$$m_Z = 0, \tag{9.55}$$

$$R_Z(\tau) = \alpha^2 \delta(\tau). \tag{9.56}$$

The impulse response function for the circuit is of the form

$$h(t) = \begin{cases} \beta e^{-\gamma t}, & t \ge 0, \\ 0, & t < 0, \end{cases} \tag{9.57}$$

for which $\gamma = 1/RC$. It can be shown (exercise 9) that the finite

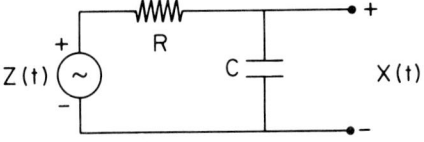

Figure 9.5 Circuit Diagram for Example 4

autocorrelation for h is

$$r_h(\tau) = \frac{\beta^2}{2\gamma} e^{-\gamma|\tau|}. \tag{9.58}$$

Consequently (9.44) and (9.45) yield

$$m_X = 0, \tag{9.59}$$

$$R_X(\tau) = \frac{(\alpha\beta)^2}{2\gamma} e^{-\gamma|\tau|}. \tag{9.60}$$

This autocorrelation is of the same form as that for the Ornstein-Uhlenbeck process (Section 5.3.2), and this process is Gaussian; therefore it *is* the Ornstein-Uhlenbeck process, which is the only stationary Gauss-Markov process. ∎

It should be clarified that the finite autocorrelation defined by

$$r_h(\tau) \triangleq \int_{-\infty}^{\infty} h(t+\tau)h(t)\,dt \tag{9.61}$$

exists for finite-energy functions only (with the exception of impulses), that is, only for transient functions $h(t)$ that die out ($\to 0$) as $|t| \to \infty$ in such a way that

$$\int_{-\infty}^{\infty} h^2(t)\,dt < \infty. \tag{9.62}$$

In contrast, the empirical autocorrelation defined by

$$\hat{R}_h(\tau) \triangleq \lim_{T \to \infty} \frac{1}{T} \int_{-T/2}^{T/2} h(t+\tau)h(t)\,dt \tag{9.63}$$

applies to finite-power functions $h(t)$ (as well as infinitely long trains of impulses) which are persistent and therefore do not die out as $|t| \to \infty$, but for which

$$\lim_{T \to \infty} \frac{1}{T} \int_{-T/2}^{T/2} h^2(t)\,dt < \infty. \tag{9.64}$$

For a finite-energy function, $\hat{R}_h(\tau) = 0$ for all τ. For a finite-power function, $r_h(\tau) = \infty$ for all τ.

9.3.1 Duality for Stationary Processes

Because of the duality between time averages and ensemble averages for stationary processes, discussed in Section 8.6, it can be shown (exercise 10) that the duals to the two discrete-time relations, (9.25) and (9.26), and the two continuous-time relations, (9.44) and (9.45), are valid; specifically

$$\hat{m}_Y = \hat{m}_X \sum_{i=-\infty}^{\infty} h(i), \tag{9.65}$$

$$\hat{R}_Y(k) = \hat{R}_X(k) \otimes r_h(k), \tag{9.66}$$

and

$$\hat{m}_Y = \hat{m}_X \int_{-\infty}^{\infty} h(t)\, dt, \tag{9.67}$$

$$\hat{R}_Y(\tau) = \hat{R}_X(\tau) \otimes r_h(\tau), \tag{9.68}$$

respectively. Moreover, these relations are valid for any regular, possibly nonstationary process. Furthermore, generalizations of these relations exist for asymptotically mean stationary processes. For example, (9.68) generalizes (exercise 10) to

$$\langle R_Y \rangle(\tau) = \langle R_X \rangle(\tau) \otimes r_h(\tau). \tag{9.69}$$

9.3.2 Random Filters

A useful model for the dispersive effects of communication channels is a *random filter* $H(t)$. If $X(t)$ is the input to the channel and $Y(t)$ is the output, then

$$Y(t) = \int_{-\infty}^{\infty} H(t - u) X(u)\, du, \tag{9.70}$$

where H is modeled as statistically independent of X. This random-channel model is appropriate when it is desired to study communication-system performance averaged over all possible channels (e.g., in a communication network). By using the method of conditional expectation (illustrated in Section 4.2.5, in which H represents a random delay Θ), and letting X be WSS, the probabilistic autocorrelation is obtained (exercise 12):

$$R_Y(\tau) = R_X(\tau) \otimes E\{r_H(\tau)\}, \tag{9.71a}$$

where

$$E\{r_H(\tau)\} = E\left\{\int_{-\infty}^{\infty} H(t + \tau) H(t)\, dt\right\} = \int_{-\infty}^{\infty} R_H(t + \tau, t)\, dt.$$

$$\tag{9.71b}$$

Similarly, we can show that the empirical autocorrelation is given by

$$\hat{R}_Y(\tau) = \hat{R}_X(\tau) \otimes r_H(\tau). \qquad (9.72)$$

Because r_H is random,* the empirical and probabilistic autocorrelations $\hat{R}_Y$ and R_Y are distinct, and Y does not have mean-square ergodicity of the autocorrelation. It should be clarified that for a stable channel, $H(t) \to 0$ as $t \to \infty$, and $H(t)$ is therefore a transient process.

9.4 Dynamical Systems

An important problem area in the study of dynamical systems with random excitation is the characterization of the evolution of the mean and variance (and other probabilistic parameters) of the *state* of the system. The *state* describes not only the external response of the system, but its internal behavior as well. Although Equations (9.42) and (9.43) [or (9.5) and (9.10)] do yield the evolution of the mean and variance of any variable in the system in terms of the response h of this variable to an impulse excitation, these formulas can be quite difficult to apply in practice because of difficulties associated with determining the impulse-response function and with carrying out the integrations, which often must be done numerically. An alternative approach, which has many advantages for situations in which the primary objective is to study the evolution of probabilistic parameters, is based on the use of differential (or difference) equations to characterize the dynamical system. In the next two subsections, this alternative approach is pursued, first for discrete-time processes and then for continuous-time processes. As explained in Section 5.3.3, the response of an mth-order system to an excitation process with no memory is an mth-order Markov process, and if the excitation is Gaussian, then the response is an mth-order Gauss-Markov process whose complete probabilistic specification can be obtained from its mean and autocovariance functions. General formulas for these mean and autocovariance functions are derived in this section. An important application area for the results in this section is Kalman filtering, which is briefly treated in Chapter 13.

9.4.1 Discrete-Time State-Variable Analysis

To motivate the general approach and results to be presented, let us consider an example.

---

*r_H is random except in the special case for which $H(t) = h(t - \Theta)$, where $h(\cdot)$ is nonrandom, but the delay Θ is random.

Example 5: We reconsider the first-order system from example 2, but with a finite starting time of, say, $i = 0$:

$$Y(i + 1) = aY(i) + X(i), \quad i \geq 0, \quad (9.73)$$

where

$$E\{X(i + k)X(i)\} \triangleq R_X(k) = \begin{cases} \sigma^2, & k = 0, \\ 0, & k \neq 0, \end{cases} \quad (9.74)$$

$$E\{Y(i)X(j)\} \triangleq R_{YX}(i, j) = 0, \quad j \geq i, \quad (9.75)$$

and

$$E\{X(i)\} \triangleq m_X = 0, \quad i \geq 0. \quad (9.76)$$

Equation (9.75) reveals that the response is orthogonal to the excitation at response times preceding the excitation time. It follows from (9.73), by successive substitution, that the response can be reexpressed as

$$\begin{aligned} Y(i + 1) &= aY(i) + X(i) \\ &= a[aY(i - 1) + X(i - 1)] + X(i) \\ &= a[a\{aY(i - 2) + X(i - 2)\} + X(i - 1)] + X(i) \\ &\vdots \\ &= a^{i-j+1}Y(j) + \sum_{n=j}^{i} a^{n-i}X(n), \quad 0 \leq j \leq i. \end{aligned} \quad (9.77)$$

Equation (9.77) can be interpreted as a family of solutions for $Y(i + 1)$ indexed by the *initial* time j. Now, to determine how the mean evolves with time, we simply take the expected value of (9.73) to obtain

$$m_Y(i + 1) = am_Y(i) + m_X, \quad i \geq 0. \quad (9.78)$$

But to determine how the autocorrelation evolves, we use (9.77) with $i \to i - 1$ to obtain

$$\begin{aligned} R_Y(i, j) &= E\left\{\left[a^{i-j}Y(j) + \sum_{n=j}^{i-1} a^{n-i+1}X(n)\right]Y(j)\right\} \\ &= a^{i-j}R_Y(j, j) + \sum_{n=j}^{i-1} a^{n-i+1}R_{YX}(j, n) \\ &= a^{i-j}R_Y(j, j), \quad i \geq j \geq 0. \end{aligned} \quad (9.79)$$

The last equality in (9.79) follows from (9.75). We see from (9.79) that once we know how the mean square evolves, we can easily determine the evolution of the autocorrelation. To determine the evolution of the mean square, (9.77) is used, with $i \to i-1$, as follows:

$$R_Y(i,i) = E\left\{\left[a^{i-j}Y(j) + \sum_{n=j}^{i-1} a^{n-i+1}X(n)\right]\right.$$
$$\left. \times \left[a^{i-j}Y(j) + \sum_{m=j}^{i-1} a^{m-i+1}X(m)\right]\right\}$$
$$= a^{2(i-j)}R_Y(j,j) + 2a^{i-j}\sum_{n=j}^{i-1} a^{n-i+1}R_{YX}(j,n)$$
$$+ \sum_{n=j}^{i-1}\sum_{m=j}^{i-1} a^{n+m-2i+2}R_X(n,m)$$
$$= a^{2(i-j)}R_Y(j,j) + \sigma^2 \sum_{n=j}^{i-1} a^{2(n-i+1)}, \quad i \geq j \geq 0. \quad (9.80)$$

The last equality follows from (9.74) and (9.75). Now, we let $j = i - 1$ in (9.80) to obtain

$$R_Y(i,i) = a^2 R_Y(i-1, i-1) + \sigma^2, \quad i \geq 1. \quad (9.81)$$

Thus, both the mean (9.78) and the mean square (9.81) evolve according to first-order recursions. Since these are time-invariant recursions, they are easily solved. Specifically, it follows from (9.78) that

$$m_Y(i) = am_Y(i-1) + m_X$$
$$= a[am_Y(i-2) + m_X] + m_X$$
$$\vdots$$
$$= a^i m_Y(0) + a^{i-1} m_X + \cdots + am_X + m_X$$
$$= a^i m_Y(0) + m_X(a^{i-1} + a^{i-2} + \cdots + 1)$$
$$= a^i m_Y(0) + m_X \frac{1-a^i}{1-a}, \quad i \geq 1. \quad (9.82)$$

Similarly, it follows from (9.81) that

$$R_Y(i,i) = a^{2i} R_Y(0,0) + \sigma^2 \frac{1-a^{2i}}{1-a^2}, \quad i \geq 1. \quad (9.83)$$

The steady-state behavior can be obtained in the limit $i \to \infty$, and the results are identical (exercise 14) to the results in example 2. ∎

In order to generalize the method of the preceding example to all discrete-time random processes described by linear difference equations, state-variable techniques can be employed to represent an nth-order difference equation in terms of an n-vector, first-order, difference equation. Specifically, a variety of state-variable methods [Kailath, 1980] can be used to reexpress the nth-order recursion

$$Y(i+n) + a_1(i)Y(i+n-1) + \cdots + a_n(i)Y(i)$$
$$= b_0(i)U(i+n) + b_1(i)U(i+n-1) + \cdots + b_n(i)U(i), \quad i \geq 0, \tag{9.84}$$

with excitation $U(i)$ and response $Y(i)$, in the form of the first-order vector recursion

$$X(i+1) = A(i)X(i) + b(i)U(i), \quad i \geq 0 \tag{9.85a}$$
$$Y(i) = c(i)X(i) + d(i)U(i), \quad i \geq 0, \tag{9.85b}$$

where $X(i)$, $b(i)$, and $c^T(i)$ are n-vectors (columns) and $A(i)$ is an $n \times n$ matrix. The nonrandom parameters $A(i)$, $b(i)$, $c(i)$, and $d(i)$ are specified by the nonrandom parameters $\{a_1(i), \ldots, a_n(i)\}$ and $\{b_0(i), \ldots, b_n(i)\}$. The random vector $X(i)$ is called the *state* of the system. [An example of the state-variable representation for a second-order ($n = 2$) system is given in exercise 18.] In order to generalize the method of example 5 from the time-invariant first-order scalar recursion (9.73) to the time-variant first-order vector recursion (9.85a), the fact that (9.77) can be generalized as follows is used:

$$X(i+1) = \Phi(i+1, j)X(j) + \sum_{m=j}^{i} \Phi(i+1, m+1)b(m)U(m),$$
$$0 \leq j \leq i, \tag{9.86}$$

where $\Phi(i, j)$ is the *state transition matrix*, which is given by

$$\Phi(i, j) = \begin{cases} I, & j = i, \\ A(i-1)A(i-2) \cdots A(j), & j < i, \end{cases} \tag{9.87a}$$

and

$$\Phi(i, j)\Phi(j, i) = I, \tag{9.87b}$$

where I is the identity matrix. For excitations with nonzero means, it is more common for the current and future excitation to be uncorrelated with the current and past state rather than to be orthogonal as in example 5. Thus, it is assumed that

$$K_{XU}(i, j) = 0, \quad j \geq i. \tag{9.88}$$

Similarly, it is assumed that there is no correlation in the excitation,

$$K_U(i, j) = \begin{cases} \sigma^2, & i = j, \\ 0, & i \neq j. \end{cases} \quad (9.89)$$

Now, it can be shown (exercise 16) by methods analogous to those in example 5 that the mean of the state vector evolves according to the recursion

$$m_X(i + 1) = A(i)m_X(i) + b(i)m_U(i), \quad i \geq 0, \quad (9.90)$$

the cross-covariance of the state vector and the scalar excitation evolves according to

$$K_{XU}(i, j) = \sigma^2 \Phi(i, j + 1)b(j), \quad i > j, \quad (9.91)$$

and the autocovariance matrix of the state vector evolves according to the recursion

$$K_X(i + 1, i + 1) = A(i)K_X(i, i)A^T(i) + \sigma^2 b(i)b^T(i), \quad i \geq 0. \quad (9.92)$$

Furthermore, the evolution of the state autocovariance matrix can be determined from (9.92) and the relation

$$K_X(i, j) = \begin{cases} \Phi(i, j)K_X(j, j), & i \geq j, \\ K_X(i, i)\Phi^T(j, i), & i \leq j, \end{cases} \quad (9.93)$$

and the evolution of the state autocorrelation matrix can be determined from (9.90), (9.93), and the relation

$$R_X(i, j) = K_X(i, j) + m_X(i)m_X^T(j). \quad (9.94)$$

Finally, the mean and autocorrelation of the response $Y(i)$ can be obtained directly from (9.90) to (9.94) and the equations

$$m_Y(i) = c(i)m_X(i) + d(i)m_U, \quad (9.95)$$

$$R_Y(i, j) = c(i)R_X(i, j)c^T(j) + c(i)R_{XU}(i, j)d(j)$$
$$+ d(i)R_{UX}(i, j)c^T(i) + d(i)d(j)R_U(i - j), \quad (9.96)$$

which follow directly from the linear equation (9.85b).

This general approach to studying the evolution of the mean and autocorrelation of the response of a system described by a linear difference equation can accommodate excitations that exhibit correlation provided that the excitation can itself be modeled as the response of a linear difference equation with excitation exhibiting no correlation. This is illustrated in the next example.

Example 6: Let us consider the system

$$Z(i + 1) = aZ(i) + bW(i), \tag{9.97}$$

for which the excitation is modeled by

$$W(i + 1) = dW(i) + eU(i), \tag{9.98}$$

for which

$$E\{U(i)U(j)\} = \begin{cases} \sigma^2, & i = j, \\ 0, & i \neq j. \end{cases} \tag{9.99}$$

To put this system of equations into the desired form, we define the state vector

$$X(i) = [Z(i), W(i)]^T \tag{9.100}$$

and response

$$Y(i) = Z(i). \tag{9.101}$$

Then Equations (9.97) and (9.98) can be reexpressed as

$$X(i + 1) = \begin{bmatrix} a & b \\ 0 & d \end{bmatrix} X(i) + \begin{bmatrix} 0 \\ e \end{bmatrix} U(i), \tag{9.102a}$$

$$Y(i) = [1, \ 0] X(i), \tag{9.102b}$$

which is in the desired form (9.85). ∎

ARMA Models. A system described by a linear difference equation of the form (9.84), with an excitation consisting of a zero-mean sequence of statistically independent random variables, gives rise to a response that is called an *autoregressive moving average* (ARMA) process, especially in the case for which all parameters $\{a_1, a_2, \ldots, a_n\}$ and $\{b_0, b_1, \ldots, b_n\}$ are time-invariant. In the special case for which $a_1 = a_2 = \cdots = a_n = 0$, the ARMA model reduces to what is called a *moving-average* (MA) model. This terminology indicates that the present value of the response is simply a finite linear combination (weighted average) of the present and past excitation values. In the special case for which $b_1 = b_2 = \cdots = b_n = 0$, $b_0 \neq 0$, the ARMA model reduces to what is called an *autoregressive* (AR) model. This terminology indicates that the current value of the process depends on only a linear combination of past values of the process and the current value of the excitation. Thus, the process regresses on itself. The process in example 1 is a first-order MA model, and the process in example 2 is a first-order ARMA model, but would be a first-order AR model if the excitation were not delayed by one time unit. AR models are studied in some detail in Section 11.9.

9.4.2 Continuous-Time State-Variable Analysis

For continuous-time systems, described by linear differential equations, the general method of analysis presented in the previous section for difference equations can be mimicked to obtain an analogous method. Let us begin with an example.

Example 7: We consider the first-order time-variant system, with excitation $U(t)$ and response $X(t)$, described by the linear differential equation

$$\dot{X}(t) = a(t)X(t) + b(t)U(t), \qquad t \geq 0, \qquad (9.103)$$

where

$$\dot{X}(t) \triangleq \frac{dX(t)}{dt}, \qquad (9.104)$$

$$E\{U(t)\} \triangleq m_U,$$

$$E\{[U(t+\tau) - m_U][U(t) - m_U]\} \triangleq K_U(\tau) = \delta(\tau), \qquad (9.105)$$

$$E\{[X(t) - m_X(t)][U(v) - m_U]\} \triangleq K_{XU}(t,v) = 0, \qquad t < v, \qquad (9.106)$$

and $U(t)$ is the sum of the symbolic derivative of the Wiener process and a constant m_U. A block diagram representation of this system is shown in Figure 9.6. Since $\dot{X}(t)$ contains an additive component of $U(t)$, it is not mean-square integrable. Although it can be interpreted as being mean-square integrable in the broadened sense described in Chapter 7, we must proceed with care because of the limitations of this interpretation. Using classical methods of solving differential equations, it can be shown that

$$X(t) = \Phi(t,v)X(v) + \int_v^t \Phi(t,w)b(w)U(w)\,dw, \qquad 0 \leq v \leq t, \qquad (9.107)$$

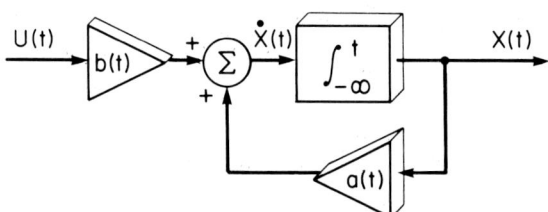

Figure 9.6 Block Diagram for Example 7

where

$$\Phi(t,v) \triangleq \exp\left\{\int_v^t a(s)\,ds\right\}. \tag{9.108}$$

This solution can be verified by substitution into (9.103) (exercise 19). Equations (9.107) and (9.108) can be interpreted as a family of solutions for $X(t)$, indexed by the *initial* time v. Now, the cross-covariance of $X(t)$ and $U(t)$ can be obtained from (9.107), and it is given by (exercise 20)

$$K_{XU}(t,u) = \int_v^t \Phi(t,w)b(w)\delta(u-w)\,dw, \qquad 0 \le v < u \le t. \tag{9.109}$$

For $0 < u < t$, it is clear that

$$K_{XU}(t,u) = \Phi(t,u)b(u), \qquad 0 < u < t; \tag{9.110}$$

however, for $u = t$ the integral in (9.109) is ambiguous. This is precisely the same problem that was investigated in exercise 3 of Chapter 7, and is a result of the fact that $U(t)$ is not mean-square integrable. As claimed in that exercise, the correct value of $K_{XU}(t,t)$ is

$$K_{XU}(t,t) = \tfrac{1}{2}b(t), \qquad t > 0. \tag{9.111}$$

To determine how the mean of $X(t)$ evolves with time, we simply take the expected value of (9.103) to obtain

$$\dot{m}_X(t) = a(t)m_X(t) + b(t)m_U, \qquad t \ge 0. \tag{9.112}$$

But to determine how the autocovariance of $X(t)$ evolves, we use (9.107), with $v = u$, to obtain

$$K_X(t,u) \triangleq \mathrm{Cov}\{X(t), X(u)\}$$

$$= \mathrm{Cov}\left\{\left[\Phi(t,u)X(u) + \int_u^t \Phi(t,w)b(w)U(w)\,dw\right], X(u)\right\}$$

$$= \Phi(t,u)\,\mathrm{Cov}\{X(u), X(u)\}, \qquad t \ge u \ge 0. \tag{9.113}$$

The last equality in (9.113) follows from (9.106). Thus, we have

$$K_X(t,u) = \begin{cases} \Phi(t,u)K_X(u,u), & t \ge u \ge 0, \\ \Phi(u,t)K_X(t,t), & u \ge t \ge 0, \end{cases} \tag{9.114}$$

where the case $u \ge t$ follows by symmetry. We see from (9.114) that once we know how the variance of $X(t)$ evolves, we can easily determine the evolution of the autocovariance. To determine the evolution of the variance, (9.107) is used as follows:

$$K_X(t,t) = \text{Cov}\left\{\left[\Phi(t,v)X(v) + \int_v^t \Phi(t,w)b(w)U(w)\,dw\right],\right.$$
$$\left.\left[\Phi(t,v)X(v) + \int_v^t \Phi(t,u)b(u)U(u)\,du\right]\right\}$$
$$= \Phi^2(t,v)K_X(v,v) + \int_v^t \Phi^2(t,w)b^2(w)\,dw, \qquad t \geq v \geq 0.$$
(9.115)

The last equality follows from (9.105) and (9.106). Now, to obtain a differential equation for the evolution of $K_X(t,t)$, we simply recognize that (9.115) is of precisely the same form as (9.107), with $X(v)$ replaced by $K_X(v,v)$, $b(w)U(w)$ replaced by $b^2(w)$, and $\Phi(t,v)$ replaced by

$$\Phi^2(t,v) = \exp\left\{\int_v^t 2a(s)\,ds\right\}, \qquad (9.116)$$

which corresponds to replacement of $a(t)$ by $2a(t)$. Consequently, $K_X(t,t)$ must satisfy a differential equation of the same form as that satisfied by $X(t)$, namely (9.103), with the preceding replacements, that is,

$$\dot{K}_X(t,t) = 2a(t)K_X(t,t) + b^2(t), \qquad t \geq 0. \qquad (9.117)$$

Equation (9.117) can be verified, somewhat laboriously, by differentiating (9.115). In summary, we see that both the mean (9.112) and the variance (9.117) evolve according to first-order recursions. ∎

As a specific case of the preceding example, the Ornstein-Uhlenbeck process is modeled by (9.103) with $a(t)$ and $b(t)$ equal to constants, and $m_U = 0$. The version of the Ornstein-Uhlenbeck process described in Section 5.3.2 is what one obtains (exercise 21) from (9.103) in the steady state, $t \to \infty$ (or alternatively, the initial time $t = 0$ can be taken to be $t = t_0 \to -\infty$).

In order to generalize the method of the preceding example to all continuous-time random processes described by linear ordinary differential equations, with excitation equal to the symbolic derivative of the Wiener process, state-variable techniques can be employed to represent an nth-order differential equation in terms of an n-vector first-order differential equation. Specifically, a variety of state-variable methods [Kailath, 1980] can be used to reexpress the nth-order differential equation

$$\frac{d^n Y(t)}{dt^n} + a_{n-1}(t)\frac{d^{n-1}Y(t)}{dt^{n-1}} + \cdots + a_0(t)Y(t)$$
$$= b_{n-1}(t)\frac{d^{n-1}U(t)}{dt^{n-1}} + \cdots + b_0(t)U(t), \qquad t \geq 0, \quad (9.118)$$

with excitation $U(t)$ and response $Y(t)$, in the form of the first-order vector differential equation

$$\dot{X}(t) = A(t)X(t) + b(t)U(t), \quad t \geq 0, \quad (9.119a)$$
$$Y(t) = c(t)X(t), \quad t \geq 0, \quad (9.119b)$$

where $X(t)$, $b(t)$, and $c^T(t)$ are n-vectors (columns) and $A(t)$ is an $n \times n$ matrix.* The nonrandom parameters $A(t)$, $b(t)$, and $c(t)$ are specified by the nonrandom parameters $\{a_0(t), \ldots, a_{n-1}(t)\}$ and $\{b_0(t), \ldots, b_{n-1}(t)\}$. The random vector $X(t)$ is called the *state* of the system. [An example of the state-variable representation for a second-order ($n = 2$) differential equation is given in exercise 22.] In order to generalize the method of example 7 from the scalar differential equation (9.103) to the vector differential equation (9.119a), the fact that (9.107) can be generalized as follows is used:

$$X(t) = \Phi(t, v)X(v) + \int_v^t \Phi(t, w)b(w)U(w)\, dw, \quad 0 \leq v \leq t. \quad (9.120)$$

In (9.120), $\Phi(t, v)$ is the *state transition matrix* and is given by the solution to the differential equation

$$\frac{\partial}{\partial t}\Phi(t, u) = A(t)\Phi(t, u), \quad t \geq u, \quad (9.121a)$$

with

$$\Phi(t, u)\Phi(u, t) = I, \quad (9.121b)$$

where I is the identity matrix. If $A(t)$ is independent of t, then we have

$$\Phi(t, v) = \exp\{(t - v)A\}, \quad (9.122)$$

where the matrix exponential is defined by the infinite series

$$\exp\{M\} \triangleq I + M + \frac{1}{2!}M^2 + \frac{1}{3!}M^3 + \cdots \quad (9.123)$$

for any square matrix M.

It is assumed that the future excitation is uncorrelated with the past state,

$$K_{XU}(t, v) = 0, \quad t < v, \quad (9.124)$$

---

*Although (9.118) can be generalized to include the term $b_n(t)d^nU(t)/dt^n$, in which case (9.119b) would include the term $d(t)U(t)$, analogous to (9.84) and (9.85b), respectively, for discrete time, this generalization is rarely needed in practice because the finite response time of all physical systems renders $b_n(t) \equiv d(t) \equiv 0$ in the model.

and that the excitation is the sum of a constant m_U and the symbolic derivative of the Wiener process, in which case

$$K_U(\tau) = \delta(\tau). \qquad (9.125)$$

The results (9.110), (9.111), (9.112), (9.114), and (9.117) from example 7 generalize to the following equations of evolution:

$$\dot{m}_X(t) = A(t)m_X(t) + b(t)m_U, \qquad t \geq 0, \qquad (9.126)$$

$$K_{XU}(t,u) = \begin{cases} \frac{1}{2}b(t), & t = u, \\ \Phi(t,u)b(u), & t > u, \end{cases} \qquad (9.127)$$

$$K_X(t,u) = \begin{cases} \Phi(t,u)K_X(u,u), & t \geq u, \\ K_X(t,t)\Phi^T(u,t), & u \geq t, \end{cases} \qquad (9.128)$$

$$\dot{K}_X(t,t) = A(t)K_X(t,t) + K_X(t,t)A^T(t) + b(t)b^T(t), \qquad t \geq 0. \qquad (9.129)$$

Furthermore, the mean and autocorrelation of the response $Y(t)$ can be obtained directly from (9.126), (9.128), (9.129), and the equations

$$m_Y(t) = c(t)m_X(t), \qquad (9.130a)$$

$$R_Y(t,u) = c(t)R_X(t,u)c^T(u), \qquad (9.130b)$$

which follow directly from the linear equation (9.119b).

This general approach to studying the evolution of the mean and autocorrelation of the response of a system described by a linear differential equation can accommodate excitations that exhibit correlation, provided that the excitation itself can be modeled as the response of a linear differential equation with excitation exhibiting no correlation (the symbolic derivative of the Wiener process). This is illustrated in the next example.

Example 8: Let us consider the system

$$\dot{Z}(t) = a(t)Z(t) + b(t)V(t), \qquad (9.131a)$$

for which the excitation is modeled by

$$\dot{V}(t) = d(t)V(t) + e(t)U(t), \qquad (9.131b)$$

where $U(t)$ is the symbolic derivative of the Wiener process. To put this system of equations into the desired form, we define the state vector

$$X(t) = [Z(t), V(t)]^T \qquad (9.132)$$

and response

$$Y(t) = Z(t). \qquad (9.133)$$

Then Equations (9.131) can be reexpressed as

$$\dot{X}(t) = \begin{bmatrix} a(t) & b(t) \\ 0 & d(t) \end{bmatrix} X(t) + \begin{bmatrix} 0 \\ e(t) \end{bmatrix} U(t), \qquad (9.134a)$$

$$Y(t) = [1, 0] X(t), \qquad (9.134b)$$

which is the desired form (9.119). ∎

Example 9: As an application of the general results obtained in this section, let us consider the problem of determining the evolution of the autocovariance of the random charge on a capacitor, connected in series with a resistor, due to a random initial charge Q (coulombs) on the capacitor. Let the charge be denoted by $X(t)$. This charge satisfies the differential equation

$$\dot{X}(t) = aX(t), \qquad t \geq 0, \qquad (9.135a)$$
$$X(0) = Q, \qquad (9.135b)$$

where

$$a = -1/RC, \qquad (9.136)$$

in which R and C are the values of the resistance and capacitance, in ohms and farads, respectively. It follows from (9.108) that

$$\Phi(t, v) = e^{a(t-v)}, \qquad (9.137)$$

and it follows from (9.117), with $b(t) = 0$, that

$$\dot{K}_X(t, t) = 2aK_X(t, t), \qquad t \geq 0, \qquad (9.138a)$$
$$K_X(0, 0) = \sigma_Q^2. \qquad (9.138b)$$

The solution to (9.138) is given by (9.115) with $v = 0$,

$$K_X(t, t) = \sigma_Q^2 e^{2at}, \qquad t \geq 0. \qquad (9.139)$$

Therefore, (9.114) yields

$$K_X(t, u) = \sigma_Q^2 e^{a(t+u)}, \qquad t, u \geq 0. \qquad (9.140)$$

Hence, the autocovariance of the charge decays uniformly (in τ) exponentially to zero, as time progresses

$$K_X\left(t + \frac{\tau}{2}, t - \frac{\tau}{2}\right) = \sigma_Q^2 e^{-2t/RC}, \qquad |\tau| < 2t \geq 0. \qquad (9.141)$$

∎

9.5 Summary

The means of a set of random variables at the output of a linear transformation can be determined from that transformation and the means of the set of random variables at the input, (9.5) and (9.42). Similarly, the correlations (covariances) at the output can be determined from the transformation and the correlations (covariances) at the input, (9.10) and (9.43). For a linear time-invariant transformation of either a discrete- or a continuous-time WSS process, these input-output relations for the mean and autocorrelation take on particularly simple forms. The output mean is simply a scaled version of the input mean, (9.25) or (9.44), and the output autocorrelation is a convolution of the input autocorrelation with the finite autocorrelation of the impulse-response sequence or function for the transformation, (9.26) or (9.45). Because of the duality between the probabilistic theory of stationary processes and the deterministic theory based on time averages, the input-output relations for the probabilistic mean and probabilistic autocorrelation have duals for the empirical mean and empirical autocorrelation, (9.67) and (9.68). Moreover, these relations can be generalized for asymptotically mean stationary processes. The generalized input-output relations relate the time-averaged probabilistic parameters (mean and autocorrelation) at the input and output of the transformation (9.69).

An important problem area in the study of linear dynamical systems with random excitation is the characterization of the evolution of the mean and covariance of the state of the system. For a linear (possibly time-variant) discrete-time system of order n, the means of the set of n state variables are specified by a set of n simultaneous first-order difference equations (9.90), and the cross-covariances for the n state variables are specified by a set of n^2 simultaneous first-order difference equations, (9.92) and (9.93). Using state-variable techniques, each of these two sets of simultaneous equations is expressed in the form of a single vector matrix first-order difference equation. The same characterizations apply to continuous-time systems, except that the difference equations become differential equations (9.126) to (9.129).

The summation [or integral] equations in terms of impulse-response sequences [or functions] (9.5) [or (9.42)] and (9.10) [or (9.43)] are often preferable for determining steady-state means and covariances for time-invariant systems. However, the difference (or differential) equations in terms of state-variable representations are often preferable for determining transient means and covariances for time-invariant systems, or generally evolving means and covariances for time-variant systems.

EXERCISES

1. Verify (9.14).

2. (a) Use (9.15) to verify that an input consisting of a unit pulse at time $i = k$,
$$X(i) = \delta_{i-k},$$
yields the output sequence
$$Y(i) = h(i, k). \qquad (9.142)$$
   (b) Verify the identity (9.19).

★3. Use (9.23) and (9.24) to verify (9.25) and (9.26) for a WSS process.

4. Derive (9.28) from (9.21).

5. Derive (9.38) from (9.35).

   Hint: use the identity (9.36).

★6. (a) Use (9.41) to verify that an input consisting of an impulse at time v,
$$X(t) = \delta(t - v),$$
yields the output waveform
$$Y(t) = h(t, v). \qquad (9.143)$$

   (b) Derive (9.45) from (9.43) for a time-invariant system with WSS input.

   (c) Derive (9.47) from (9.41) for a time-invariant system with WSS input.

7. Show that for a periodic waveform
$$h(t) = \sum_n h_n e^{in\omega_0 t}, \qquad (9.144)$$
the empirical autocorrelation is periodic and equal to
$$\hat{R}_h(\tau) = \sum_n |h_n|^2 e^{in\omega_0 \tau}. \qquad (9.145)$$

8. The marked and filtered Poisson process discussed in Chapter 6 is given by
$$X(t) = \sum_{i=1}^{\infty} Y_i h(t - T_i)$$
$$= h(t) \otimes Z(t), \qquad (9.146)$$
where $Z(t)$ is the symbolic derivative of the marked counting process
$$Z(t) = \sum_{i=1}^{\infty} Y_i \delta(t - T_i). \qquad (9.147)$$
Determine the asymptotic ($t \to \infty$) mean and autocorrelation function for $X(t)$, using the results (6.43) and (6.44).

9. Derive (9.58) from (9.57). Compare with (9.35) and (9.38).

★10. (a) Derive the results (9.65) and (9.66) for the empirical mean and empirical autocorrelation of a filtered discrete-time process.

   (b) Derive the results (9.67) and (9.68) for a filtered continuous-time process.

(c) Derive generalizations of (9.67) and (9.68), such as (9.69), for asymptotically mean stationary processes.

11. (a) Two separate systems have impulse responses $h_1(t)$ and $h_2(t)$. A process $X_1(t)$ is applied to the first system, and its response is $Y_1(t)$. Similarly, a process $X_2(t)$ invokes a response $Y_2(t)$ from the second system. Find the cross-correlation function of $Y_1(t)$ and $Y_2(t)$ in terms of $h_1(t)$, $h_2(t)$, and the cross-correlation function of $X_1(t)$ and $X_2(t)$. Assume that $X_1(t)$ and $X_2(t)$ are jointly wide-sense stationary.

(b) Two systems are connected in series. A random process $X(t)$ is applied to the input of the first system, which has impulse response $h_1(t)$; its response $W(t)$ is the input to the second system, which has impulse response $h_2(t)$. The second system's output is $Y(t)$. Find the cross-correlation function of $W(t)$ and $Y(t)$ in terms of $h_1(t)$, $h_2(t)$, and the autocorrelation function of $X(t)$. Assume that $X(t)$ is wide-sense stationary.

⋆12. Derive the input-output autocorrelation relation (9.71) for random filters.

13. The block diagram in Figure 9.7 shows a method for measuring the impulse-response function of a linear time-invariant system while it is in operation. The method consists of superimposing a low-level broadband noise process on the system excitation and then measuring the cross-correlation of the system response and this noise process. Let the noise process be modeled as the symbolic derivative of the Wiener process, and show that the cross-correlation $R_{YN}(\tau)$ is proportional to the impulse response $h(\tau)$, provided that $N(t)$ is orthogonal to the system excitation $X(t)$.

14. Show that in the steady state ($i \to \infty$), the results (9.79) and (9.83) in example 5 reduce to the result (9.40) in example 2.

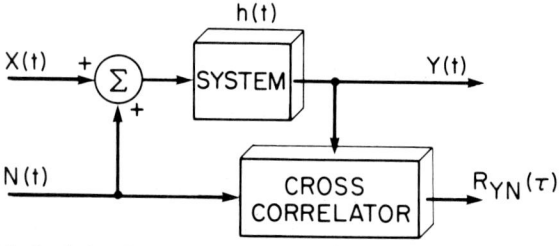

Figure 9.7 Method for Measuring the Impulse-Response Function of a Linear Time-Invariant System While It Is in Operation

15. Since a system described by the state equations (9.85) with $d(i) \equiv 0$ has a unit-pulse response that satisfies

$$h(i, j) = 0, \qquad j > i - 1, \qquad (9.148)$$

and input that satisfies

$$U(j) = 0, \qquad j < 0, \qquad (9.149)$$

then the input-output relation (9.15) reduces to

$$Y(i) = \sum_{j=0}^{i-1} h(i, j) U(j), \qquad i > 0. \qquad (9.150)$$

(a) Use (9.85b) and (9.86) [with $j = 0$, $i \to i - 1$, and $X(0) = 0$ in (9.86)] to show that

$$h(i, j) = c(i) \Phi(i, j + 1) b(j), \qquad 0 \le j < i, \qquad (9.151)$$

for $d(i) \equiv 0$.

(b) Use the result of part (a) together with (9.87) to determine the unit-pulse response for the system

$$Y(i + 1) = \frac{\alpha i}{i + 1} Y(i) + \beta U(i), \qquad i \ge 0. \qquad (9.152)$$

(c) Let $U(i)$ in (9.152) satisfy

$$K_{YU}(i, j) = 0, \qquad j \ge i, \qquad (9.153\text{a})$$

$$K_U(i, j) = \begin{cases} \sigma^2, & i = j, \\ 0, & i \ne j, \end{cases} \qquad (9.153\text{b})$$

and determine explicit difference equations that specify the evolution of the mean and variance of $Y(i)$, in terms of the mean and variance of $U(i)$.

(d) Use the result of part (b) to determine summation formulas for the mean and variance of $Y(i)$, $i > 0$, for the same excitation $U(i)$ as in part (c).

★16. Derive (9.90) to (9.92) by a method analogous to that used in example 5.

17. Use (9.87) to verify that for a time-invariant system,

$$\Phi(i, j) = A^{|i-j|}, \qquad (9.154)$$

where

$$A^0 \triangleq I. \qquad (9.155)$$

18. Consider the linear time-invariant discrete-time system described by the difference equation

$$Y(i+2) + 2Y(i+1) + Y(i) = 3U(i), \qquad i \geq 0. \qquad (9.156)$$

(a) Define the state of this system by

$$X(i) = [Y(i), Y(i+1)]^T, \qquad (9.157)$$

and verify that the state equations (9.85) are specified by $d = 0$ and

$$A = \begin{bmatrix} 0 & 1 \\ -1 & -2 \end{bmatrix}, \qquad b = \begin{bmatrix} 0 \\ 3 \end{bmatrix}, \qquad c = [1, 0]. \qquad (9.158)$$

(b) Let $U(i)$ be a WSS sequence of uncorrelated random variables, with mean $m_U = 1$ and variance $\sigma_U^2 = 4$, for which

$$K_{YU}(i, j) = 0, \qquad j \geq i - 1,$$

and determine difference equations for the evolution of the mean $m_Y(i)$ and the variance $K_Y(i, i)$ of $Y(i)$.

19. Verify the solution (9.107) to the differential equation (9.103).

Hint: Substitute (9.108) into (9.107) into (9.103), and use the relations, obtained from Leibnitz's rule for differentiation,

$$\frac{d\Phi(t, v)}{dt} = a(t)\Phi(t, v) \qquad (9.159)$$

and

$$\frac{d}{dt}\int_v^t \Phi(t, w)b(w)U(w)\,dw = \Phi(t, t)b(t)U(t)$$

$$+ \int_v^t \frac{d}{dt}\Phi(t, w)b(w)U(w)\,dw$$

$$= b(t)U(t) + a(t)\int_v^t \Phi(t, w)b(w)U(w)\,dw; \qquad (9.160)$$

then use (9.107) again to simplify the result, which thereby reduces to (9.103).

★20. Derive (9.109) from (9.107), using (9.105) and (9.106).

21. (a) Show that the model (9.103) with $a(t)$ and $b(t)$ equal to constants, $m_U = 0$, and $K_X(0, 0) = 0$ yields the *finite-starting-time Ornstein-Uhlenbeck process*, which is a zero-mean process with autocovariance of the form

$$K_X(t, u) = k[e^{a|t-u|} - e^{a(t+u)}], \qquad t, u \geq 0, \qquad (9.161)$$

where $a < 0$ and $k = -b^2/2a$.

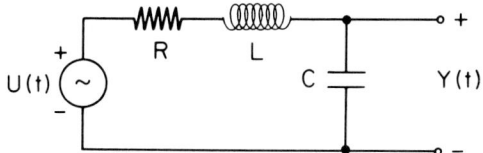

Figure 9.8 Circuit Diagram for Exercise 22

(b) Show that for $t, u \to \infty$, the result of part (a) reduces to the result (5.22) for what is called the *steady-state Ornstein-Uhlenbeck process*.

(c) Show that for $a \to 0$, the Ornstein-Uhlenbeck process of part (a) becomes the Wiener process defined in Chapter 6.

22. Consider the resistive-capacitive-inductive circuit shown in Figure 9.8 for which $R = 1.5\ \Omega$, $L = 0.25$ H, and $C = 0.5$ F. The voltage $Y(t)$ across the capacitor satisfies the differential equation

$$\frac{d^2Y(t)}{dt^2} + \frac{R}{L}\frac{dY(t)}{dt} + \frac{1}{LC}Y(t) = \frac{1}{LC}U(t), \qquad t \geq 0. \quad (9.162)$$

(a) Define the state of this system by

$$X(t) \triangleq [Y(t), \dot{Y}(t)]^T, \qquad (9.163)$$

and verify that the state equations (9.119) are specified by

$$A = \begin{bmatrix} 0 & 1 \\ -8 & -6 \end{bmatrix}, \quad b = \begin{bmatrix} 0 \\ 8 \end{bmatrix}, \quad c = [1, 0]. \quad (9.164)$$

(b) Let $U(t)$ be given by

$$U(t) = Z(t) + 2, \qquad t \geq 0,$$

where $Z(t)$ is the symbolic derivative of the Wiener process, and assume that

$$K_{YU}(t, v) = K_{\dot{Y}U}(t, v) = 0, \qquad t < v.$$

Determine implicit equations for the evolution of the mean $m_Y(t)$ and the variance $K_Y(t, t)$ of $Y(t)$.

23. (a) The *Karhunen-Loève transform* for a finite segment of a discrete-time process, $\{X_1, X_2, \ldots, X_n\}$, is an $n \times n$ matrix, say Ψ, that transforms the n-vector of random variables

$$X = [X_1, X_2, \ldots, X_n]^T$$

into another n-vector, say

$$Y = \Psi X, \qquad (9.165)$$

for which the n random variables $\{Y_1, Y_2, \ldots, Y_n\}$ are mutually orthogonal,

$$E\{Y_i Y_j\} = 0, \quad i \neq j, \qquad (9.166a)$$

and ordered so that

$$E\{Y_1^2\} \geq E\{Y_2^2\} \geq \cdots \geq E\{Y_n^2\}. \qquad (9.166b)$$

Verify that the following matrix Ψ yields the desired result (9.166):

$$\Psi = [\psi_1 \ \psi_2 \ \psi_3 \ \cdots \ \psi_n]^T;$$

that is, the ith row of Ψ is the transpose of the n-vector ψ_i, which is the orthonormal eigenvector of R_X corresponding to the ith largest eigenvalue, say λ_i. More specifically, ψ_i satisfies

$$R_X \psi_i = \lambda_i \psi_i,$$

where

$$\psi_i^T \psi_j = \begin{cases} 1, & i = j, \\ 0, & i \neq j, \end{cases}$$

and

$$\lambda_1 \geq \lambda_2 \geq \cdots \geq \lambda_n.$$

(b) Show that the matrix

$$\Theta \triangleq \Psi^T \Lambda^{-1/2} \Psi, \qquad (9.167)$$

where $\Lambda^{-1/2}$ is a diagonal matrix with diagonal elements $\{1/\sqrt{\lambda_1}, 1/\sqrt{\lambda_2}, \cdots, 1/\sqrt{\lambda_n}\}$, yields the vector

$$Z = \Theta X$$

for which the n random variables $\{Z_1, Z_2, \ldots, Z_n\}$ are mutually orthogonal, and normal in the sense that

$$E\{Z_1^2\} = E\{Z_2^2\} = \cdots = E\{Z_n^2\} = 1. \qquad (9.168)$$

Assume that R_X is nonsingular, so that $\lambda_i > 0$ for $i = 1, 2, \ldots, n$.

10

Spectral Density

IN THIS CHAPTER a means for studying the average frequency composition of stationary and related processes is introduced. It would benefit the reader to return, at this point, to Chapter 3, and reread the material on generalized harmonic analysis, for there is considerable pedagogical value in comparing and contrasting the deterministic and probabilistic notions of average spectral density. Because of the fundamental importance of frequency-domain concepts and methods in the field of signal processing, this chapter can be considered to be one of the most important in the book. This is reflected in the dominant role that the frequency domain plays in the remaining chapters.

10.1 Input-Output Relations

10.1.1 Finite-Energy Waveforms

For a linear time-invariant system with impulse-response function $h(t)$, the input and output waveforms $X(t)$ and $Y(t)$ are related by the convolution operation

$$Y(t) = \int_{-\infty}^{\infty} h(t-u)X(u)\,du \triangleq h(t) \otimes X(t). \quad (10.1)$$

If $X(t)$ and $Y(t)$ are transient finite-energy waveforms, then their Fourier transforms exist, for example,

$$\tilde{X}(f) = \int_{-\infty}^{\infty} X(t) e^{-i2\pi ft} dt, \qquad (10.2)$$

and the convolution theorem for Fourier transforms, applied to (10.1), yields

$$\tilde{Y}(f) = H(f)\tilde{X}(f), \qquad (10.3)$$

where

$$H(f) \triangleq \int_{-\infty}^{\infty} h(t) e^{-i2\pi ft} dt \qquad (10.4)$$

is the *transfer function* of the system. Whereas (10.1) is referred to as a *time-domain* input-output relation, (10.3) is referred to as a *frequency-domain* input-output relation. It is an attractive alternative to (10.1) because multiplication is a simpler operation than convolution. Furthermore (10.3) directly reveals the effect of a linear time-invariant system on the frequency content of a waveform. That is, since the inverse Fourier transform yields

$$Y(t) = \int_{-\infty}^{\infty} \tilde{Y}(f) e^{i2\pi ft} df, \qquad (10.5)$$

then it is clear that $\tilde{Y}(f)$ is the frequency density of the complex sinusoids

$$e^{i2\pi ft} = \cos(2\pi ft) + i\sin(2\pi ft)$$

that compose the waveform $Y(t)$. This frequency density at the output is obtained from the frequency density at the input simply by multiplication with the transfer function $H(f)$ as indicated by (10.3).

10.1.2 Finite-Power Waveforms

If $X(t)$ is a WSS process, then its sample paths are persistent and certainly not finite-energy functions; they are finite-power functions, and their Fourier transforms therefore do not exist. Although a generalized (integrated) Fourier transform exists for persistent functions [Wiener, 1930], when it is applied to a WSS process, it yields a *random* function of frequency.* Nevertheless, the mean squared value of this random Fourier transform is a useful nonrandom measure of the average frequency content of the random process. Expressing this formally in a very simplified way, we

---

*This is the spectral representation developed by Andrei Nikolaevich Kolmogorov, Harald Cramér, Michel Moise Loève, and Joseph Leo Doob; see [Doob, 1953].

have

$$X(t) = \int_{-\infty}^{\infty} \tilde{X}(f) e^{i2\pi ft} df$$

$$\tilde{X}(f) = \int_{-\infty}^{\infty} X(t) e^{-i2\pi ft} dt, \qquad (10.6)$$

from which we obtain

$$E\{\tilde{X}(f_1)\tilde{X}^*(f_2)\} = \int_{-\infty}^{\infty}\int_{-\infty}^{\infty} e^{-i2\pi(f_1 t_1 - f_2 t_2)} R_X(t_1 - t_2) dt_1 dt_2; \qquad (10.7)$$

use of the change of variables $\tau = t_1 - t_2$, followed by integration with respect to τ and t_2 separately yields (exercise 1)

$$E\{\tilde{X}(f_1)\tilde{X}^*(f_2)\} = S_X(f_1)\delta(f_1 - f_2), \qquad (10.8)$$

for which the definition

$$S_X(f) \triangleq \int_{-\infty}^{\infty} R_X(\tau) e^{-i2\pi f\tau} d\tau \qquad (10.9)$$

and the identity

$$\delta(f) = \int_{-\infty}^{\infty} e^{-i2\pi ft} dt \qquad (10.10)$$

have been used. This impulsive autocorrelation of the Fourier transform $\tilde{X}(f)$ reveals that $\tilde{X}(f)$ (with f replaced by t) is similar to the symbolic derivatives of the uncorrelated increment Wiener and Poisson processes described in Chapter 6. Thus, the Fourier transform $\tilde{X}(f)$ is really a symbolic process that can be interpreted heuristically as a very erratic function of frequency f that exhibits zero correlation for all pairs of frequency samples, no matter how close together they are. In fact, it can be treated in a mathematically rigorous way only as a *generalized random function*, analogous to the Dirac delta, which is a (nonrandom) generalized function. Consequently $\tilde{X}(f)$ is more appropriately studied in terms of its integral, which is the *integrated Fourier transform*, denoted by $\tilde{X}^{(-1)}(f)$. Formally, we have

$$\tilde{X}^{(-1)}(f) = \int_{-\infty}^{f} \tilde{X}(\nu) d\nu, \qquad (10.11)$$

$$\tilde{X}(f) = \frac{d}{df}\tilde{X}^{(-1)}(f), \qquad (10.12)$$

although this derivative and integral do not exist in the narrow sense described in Chapter 7. In terms of $\tilde{X}^{(-1)}$, the preceding formal relationships can be replaced with the following rigorous relationships:

$$X(t) = \int_{-\infty}^{\infty} e^{i2\pi ft}\, d\tilde{X}^{(-1)}(f) \tag{10.13}$$

and

$$S_X(f) = \frac{d}{df} S_X^{(-1)}(f), \tag{10.14}$$

where

$$S_X^{(-1)}(f) \triangleq E\{|\tilde{X}^{(-1)}(f)|^2\}. \tag{10.15}$$

Consequently, the mean squared value of the integrated Fourier transform is the integrated Fourier transform of the autocorrelation. This can be seen formally as follows:

$$\begin{aligned} E\{|\tilde{X}^{(-1)}(f)|^2\} &= E\left\{\int_{-\infty}^{f} \tilde{X}(f_1)\, df_1 \int_{-\infty}^{f} \tilde{X}^*(f_2)\, df_2\right\} \\ &= \int_{-\infty}^{f}\int_{-\infty}^{f} E\{\tilde{X}(f_1)\tilde{X}^*(f_2)\}\, df_1\, df_2 \\ &= \int_{-\infty}^{f}\int_{-\infty}^{f} S_X(f_1)\delta(f_1 - f_2)\, df_2\, df_1 = \int_{-\infty}^{f} S_X(f_1)\, df_1, \end{aligned} \tag{10.16}$$

where S_X is given by (10.9). But, rather than pursue this relatively technical approach, we shall derive the function $S_X(f)$, as an average measure of frequency content, by another means in the next subsection.

10.1.3 Spectral Density of Mean Squared Fluctuation

The function S_X arises naturally, without directly considering $\tilde{X}(f)$ or $\tilde{X}^{(-1)}(f)$, in the same way that the Fourier transform for finite-energy functions arises from application of the convolution theorem to the input-output relation (10.1). That is, the autocorrelation function for a WSS process is an average measure of waveform behavior, and the input and output autocorrelations for a linear time-invariant system are related by the convolution operation (from Chapter 9)

$$R_Y(\tau) = r_h(\tau) \otimes R_X(\tau), \tag{10.17}$$

where

$$r_h(\tau) \triangleq h(\tau) \otimes h(-\tau). \tag{10.18}$$

Since the autocorrelation functions for all stationary random-process models of practical interest possess Fourier transforms, the convolution theorem can be applied to (10.17) and (10.18). This yields

$$S_Y(f) = |H(f)|^2 S_X(f), \qquad (10.19)$$

since

$$\int_{-\infty}^{\infty} r_h(\tau) e^{-i2\pi f \tau} d\tau = |H(f)|^2, \qquad (10.20)$$

and since we are here taking S_X to be *defined* as the Fourier transform of R_X, (10.9). The inverse Fourier transformation applied to (10.9) yields

$$R_X(\tau) = \int_{-\infty}^{\infty} S_X(f) e^{i2\pi f \tau} df. \qquad (10.21)$$

Since $R_X(\tau)$ is even, $S_X(f)$ is real and even, and it can be shown to be non-negative. The relation (10.21), which indicates that the autocorrelation of a WSS process can be expressed as the inverse Fourier transform of a real, non-negative, even function S_X with finite area, is called the *Khinchin relation*, and is the probabilistic counterpart of the Wiener relation (3.9). To link $S_X(f)$ to our formal notion of frequency composition $\tilde{X}(f)$, let us see how $S_X(f)$ is related to the output of a narrowbandpass filter with $X(t)$ at the input. Consider an ideal bandpass filter with center frequency f and bandwidth Δ, as depicted in Figure 10.1. The transfer function for this filter has magnitude

$$|H(\nu)| = \begin{cases} 1, & ||\nu| - f| \le \Delta/2, \\ 0, & ||\nu| - f| > \Delta/2, \end{cases} \qquad (10.22)$$

where ν is the frequency variable, and f is a fixed parameter. The mean

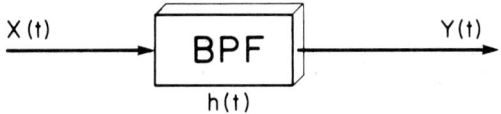

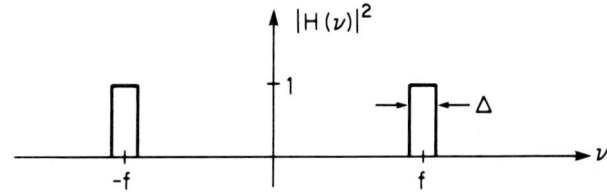

Figure 10.1 Bandpass Filter for Determination of Spectral Density

squared fluctuation of the filtered process $Y(t)$ is, from (10.17), given by

$$E\{Y^2(t)\} \equiv R_Y(0) = \int_{-\infty}^{\infty} r_h(u) R_X(u)\, du. \tag{10.23}$$

Application of Parseval's relation for Fourier transforms to (10.23) yields (exercise 2)

$$E\{Y^2(t)\} = \int_{-\infty}^{\infty} |H(\nu)|^2 S_X(\nu)\, d\nu. \tag{10.24}$$

Substitution of (10.22) into (10.24) yields the close approximation (for sufficiently narrow bandwidth Δ)

$$E\{Y^2(t)\} = \int_{f-\Delta/2}^{f+\Delta/2} S_X(\nu)\, d\nu + \int_{-f-\Delta/2}^{-f+\Delta/2} S_X(\nu)\, d\nu \simeq 2\Delta S_X(f), \tag{10.25}$$

and the limiting relationship

$$\lim_{\Delta \to 0} \frac{E\{Y^2(t)\}}{2\Delta} = S_X(f). \tag{10.26}$$

This reveals that the mean squared fluctuation $E\{Y^2(t)\}$ at the output of an infinitesimal-bandwidth bandpass filter is given by the function S_X evaluated at the center frequency f of the filter, and scaled by the total (positive and negative frequency) bandwidth 2Δ. Furthermore, the mean squared fluctuation of the unfiltered process is, from (10.21) with $\tau = 0$, the integral over all frequencies of this same function,

$$E\{X^2(t)\} = \int_{-\infty}^{\infty} S_X(f)\, df. \tag{10.27}$$

Hence, $S_X(f)$ is the spectral density of the mean squared fluctuation of the process $X(t)$. It will be referred to simply as the *spectral density*. An alternative term that is commonly used for $S_X(f)$ is the *spectrum*.

The frequency-domain input-output relation (10.19) plays a crucial role in the analysis of the average behavior of signal-processing systems involving linear time-invariant transformations. Similar to this relation are the relations obtained by Fourier transformation of both sides of the equations

$$R_{YX}(\tau) = R_X(\tau) \otimes h(\tau),$$
$$R_{XY}(\tau) = R_X(\tau) \otimes h(-\tau), \tag{10.28}$$

namely

$$S_{YX}(f) = S_X(f) H(f),$$
$$S_{XY}(f) = S_X(f) H^*(f), \tag{10.29}$$

where S_{XY} denotes the Fourier transform of R_{XY}. The quantity S_{YX} is called the *cross-spectral density*. Unlike the spectral density, the cross-spectral density is in general not a real-valued function. As explained in Section 10.3, the cross-spectral density plays a crucial role in studies of the degree to which two processes are approximately related by a linear time-invariant transformation.

10.1.4 Discrete-Time Analogs

For discrete-time processes, we have the following analogous input-output relations and formulas:

$$Y(i) = \sum_{j=-\infty}^{\infty} h(i-j)X(j) = h(i) \otimes X(i), \tag{10.30}$$

$$R_Y(k) = R_X(k) \otimes r_h(k), \tag{10.31}$$

$$r_h(k) \triangleq h(k) \otimes h(-k), \tag{10.32}$$

$$S_Y(f) = S_X(f)|H(f)|^2, \tag{10.33}$$

$$S_X(f) \triangleq \sum_{k=-\infty}^{\infty} R_X(k) e^{-i2\pi k f}, \tag{10.34}$$

$$H(f) \triangleq \sum_{k=-\infty}^{\infty} h(k) e^{-i2\pi k f}, \tag{10.35}$$

$$R_{YX}(k) = R_X(k) \otimes h(k), \tag{10.36}$$

$$S_{YX}(f) = S_X(f)H(f). \tag{10.37}$$

The transformation in (10.34) and (10.35) is called the *Fourier series transform* (FST). The *inverse FST* is (exercise 6)

$$R_X(k) = \int_{-1/2}^{1/2} S_X(f) e^{i2\pi f k} df, \tag{10.38}$$

$$h(k) = \int_{-1/2}^{1/2} H(f) e^{i2\pi f k} df. \tag{10.39}$$

10.2 Expected Spectral Density

10.2.1 Expected Power Spectral Density

We can obtain a somewhat more physical interpretation of the spectral density S_X of the mean squared fluctuation of a process X by reinterpreting the mean squared fluctuation. Specifically, if $X(t)$ is the voltage across a 1-Ω

resistance, then the instantaneous rate at which energy is dissipated in the resistance—the instantaneous power dissipation—is (in units of watts)

$$P(t) = X^2(t). \tag{10.40}$$

Therefore the mean squared fluctuation is the *expected power*

$$E\{P(t)\} = R_X(0). \tag{10.41}$$

Hence, $S_X(f)$ is the spectral density of expected power, which is by convention abbreviated to *power spectral density* (PSD). Nevertheless, it must be kept in mind that $S_X(f)$ is really the *expected* PSD. However, for an ergodic process this is the same as the spectral density of the time-averaged power (cf. Section 3.2).

10.2.2 Expected Energy Spectral Density

Another way to obtain the interpretation of $S_X(f)$ as the expected PSD, which avoids the reliance on the notion of bandpass filtering introduced in Section 10.1.3, is as follows. Let $\tilde{X}_T(f)$ denote the Fourier transform of $X_T(t)$, which is the T-second segment of the process $X(t)$ on the interval $[-T/2, T/2]$:

$$\tilde{X}_T(f) \triangleq \int_{-\infty}^{\infty} X_T(t) e^{-i2\pi ft} dt = \int_{-T/2}^{T/2} X(t) e^{-i2\pi ft} dt. \tag{10.42}$$

The total energy dissipated in a 1-Ω resistance by this finite segment of a random voltage waveform is (in units of joules)

$$E_T = \int_{-T/2}^{T/2} X^2(t) dt = \int_{-\infty}^{\infty} X_T^2(t) dt. \tag{10.43}$$

Application of Parseval's relation to (10.43) yields

$$E_T = \int_{-\infty}^{\infty} |\tilde{X}_T(f)|^2 df. \tag{10.44}$$

The integrand in (10.44) will be interpreted* as a *frequency-density of energy*, and its expected value will be referred to as the expected *energy spectral density* (ESD$_T$):

$$\text{ESD}_T \triangleq E\{|\tilde{X}_T(f)|^2\}. \tag{10.45}$$

---

*This interpretation is only heuristic; it cannot be made rigorous. For example, the method used to prove that $S_X(f)$ is the frequency density of the average power does not work for $|\tilde{X}_T(f)|^2$ or for its expected value. However, the method can be used to show that $|\tilde{X}_T(f)|^2$ approximates an energy spectral density with spectral resolution limited to $\Delta \geq 1/T$, by convolution with a rectangle of width Δ. The larger Δ is for a given T, the closer the approximation.

Now the PSD, denoted by S_X', is defined to be the limiting value of ESD_T per unit of time:

$$S_X'(f) \equiv \mathrm{PSD} \triangleq \lim_{T \to \infty} \frac{1}{T} \mathrm{ESD}_T \equiv \lim_{T \to \infty} \frac{1}{T} E\{|\tilde{X}_T(f)|^2\}. \quad (10.46)$$

To verify that (10.46) is precisely the same as (10.9), (10.42) can be substituted into (10.45), and expectation and integration can be interchanged to obtain

$$\mathrm{ESD}_T = \int_{-T/2}^{T/2} \int_{-T/2}^{T/2} R_X(t-u) e^{-i2\pi f(t-u)} \, dt \, du, \quad (10.47)$$

which can be reexpressed as (exercise 7)

$$\mathrm{ESD}_T = T \int_{-T}^{T} \left(1 - \frac{|\tau|}{T}\right) R_X(\tau) e^{-i2\pi f \tau} \, d\tau = T \int_{-\infty}^{\infty} w_T(\tau) R_X(\tau) e^{-i2\pi f \tau} \, d\tau, \quad (10.48)$$

where $w_T(\tau)$ is the triangular window

$$w_T(\tau) \triangleq \begin{cases} 1 - |\tau|/T, & |\tau| \leq T, \\ 0, & \text{otherwise.} \end{cases} \quad (10.49)$$

Application of the convolution theorem to (10.48) yields (exercise 8)

$$\frac{1}{T} \mathrm{ESD}_T = \int_{-\infty}^{\infty} S_X(f-\nu) W_{1/T}(\nu) \, d\nu, \quad (10.50)$$

where

$$W_{1/T}(\nu) = T \left[\frac{\sin(\pi T \nu)}{\pi T \nu}\right]^2. \quad (10.51)$$

Since

$$\lim_{T \to \infty} W_{1/T}(\nu) = \delta(\nu),$$

(10.50) yields

$$S_X'(f) \triangleq \lim_{T \to \infty} \frac{1}{T} \mathrm{ESD}_T = S_X(f). \quad (10.52)$$

It should be emphasized that when the expectation operation is deleted from (10.46), the limit does not exist, in general, in any useful sense (e.g., in mean square). This is discussed further in Section 11.4 of the next chapter.

10.3 Coherence

10.3.1 Definition

A useful measure of the degree to which two WSS processes X and Y are approximately related by a linear time-invariant transformation is the *coherence function* $\rho(f)$ defined by

$$\rho(f) = \frac{S_{XY}(f)}{[S_X(f)S_Y(f)]^{1/2}}. \tag{10.53}$$

Note the similarity between $\rho(f)$ and the correlation coefficient ρ (Chapter 2), which is a measure of the degree to which two random variables are approximately linearly related. In fact, if the output of a one-sided narrow-band filter with center frequency f and bandwidth Δ and with input $X(t)$ is denoted by $X_f^\Delta(t)$ [and similarly for $Y(t)$], then it can be shown that

$$\rho(f) = \lim_{\Delta \to 0} \rho_\Delta(f), \tag{10.54}$$

where $\rho_\Delta(f)$ is the correlation coefficient

$$\rho_\Delta(f) \triangleq \frac{E\{X_f^\Delta(t)Y_f^\Delta(t)^*\}}{\left(E\{|X_f^\Delta(t)|^2\}E\{|Y_f^\Delta(t)|^2\}\right)^{1/2}}. \tag{10.55}$$

It should be noted that $X_f^\Delta(t)$ and $Y_f^\Delta(t)$ are zero-mean random variables for all $f \neq 0$. Since the coherence function is a correlation coefficient, it exhibits the property (Chapter 2)

$$0 \leq |\rho(f)| \leq 1. \tag{10.56}$$

Furthermore, $|\rho(f)| \equiv 1$ if and only if $X(t)$ and $Y(t)$ are exactly linearly related, that is, if and only if (10.1) [or (10.30) for discrete time] holds for some function h (at least in the mean-square sense). A heuristic (but nonrigorous) way to see this is to interpret (10.53) as

$$\rho(f) = \frac{E\{\tilde{X}(f)\tilde{Y}(f)^*\}}{\left(E\{|\tilde{X}(f)|^2\}E\{|\tilde{Y}(f)|^2\}\right)^{1/2}}, \tag{10.57}$$

where $\tilde{X}(f)$ and $\tilde{Y}(f)$ are the symbolic Fourier transforms of $X(t)$ and $Y(t)$. Then $|\rho(f)| = 1$ for some value of f if and only if $\tilde{X}(f)$ and $\tilde{Y}(f)$ are linearly related, as in (10.3). This can hold for all f if and only if $X(t)$ and $Y(t)$ are related by (10.1).

10.3.2 Linear Additive-Noise Model for Two Processes

It is shown in Section 13.4.2 that for *any* two jointly WSS processes X and Y, there exists a linear time-invariant transformation (with impulse response denoted by h) such that Y can be modeled *uniquely* in terms of X by

$$Y(t) = X(t) \otimes h(t) + N(t), \qquad (10.58a)$$

where the error or *noise* process N is WSS and uncorrelated with X:

$$R_{XN}(\tau) \equiv 0. \qquad (10.58b)$$

In terms of this unique representation, the coherence function for X and Y can be expressed as

$$|\rho(f)|^2 = \left\{1 + S_N(f)\left[|H(f)|^2 S_X(f)\right]^{-1}\right\}^{-1} \qquad (10.59)$$

Hence, $|\rho(f)| \simeq 1$ if and only if the mean squared noise fluctuation at f, $S_N(f)$, is small relative to the mean squared fluctuation of $X(t) \otimes h(t)$ at f, $|H(f)|^2 S_X(f)$. The linear time-invariant transformation in this model (10.58) is specified by

$$H(f) = \frac{S_{YX}(f)}{S_X(f)}, \qquad (10.60a)$$

and the noise spectrum is given by

$$S_N(f) = S_Y(f) - \frac{|S_{XY}(f)|^2}{S_X(f)}. \qquad (10.60b)$$

Needless to say, the discrete-time counterparts of (10.58) to (10.60) are valid as well.

10.4 Time-Average Power Spectral Density and Duality

As described in Section 8.6, the probabilistic theory of filtering based on the probabilistic autocorrelation (introduced by Aleksandr Yakovlevich Khinchin [Khinchin, 1934]) is dual to the empirical theory of filtering based on the empirical autocorrelation (introduced by Norbert Wiener [Wiener, 1930]). The latter can be obtained (formally) from the former by replacing the ensemble-average operation (expectation) with the time-average operation. Consequently, the expected PSD has, as an analog, the time-average PSD,

$$\hat{S}_X(f) = \int_{-\infty}^{\infty} \hat{R}_X(\tau) e^{-i2\pi f \tau} d\tau, \qquad (10.61)$$

which, by analogy with the terminology for $\hat{R}_X(\tau)$, is called the *empirical spectral density*. Only slight modification of the arguments in Sections 10.1 to 10.3 yields the properties and interpretations of $\hat{S}_X(f)$ that are the duals of those for $S_X(f)$. Some of these are discussed in Section 3.2.

10.5 Spectral Density for Ergodic and Nonergodic Regular Stationary Processes

As explained in Section 8.5, the probabilistic autocorrelation for a WSS regular process can be interpreted as the expected value of the empirical autocorrelation:

$$R_X(\tau) = E\{\hat{R}_X(\tau)\}. \tag{10.62}$$

Fourier transformation of both sides of (10.62) and interchange of expectation and Fourier transformation operations yields

$$S_X(f) = E\{\hat{S}_X(f)\}. \tag{10.63}$$

Hence, the probabilistic PSD can be interpreted as the expected value of the empirical PSD.

If the process X exhibits mean-square ergodicity of the autocorrelation, then

$$R_X(\tau) = \hat{R}_X(\tau) \tag{10.64}$$

and hence

$$S_X(f) = \hat{S}_X(f); \tag{10.65}$$

that is, $X(t)$ exhibits *mean-square ergodicity of the spectral density*. In this case, the Wiener relation (3.9) and the Khinchin relation (10.21) are one and the same (in the mean square sense), and are therefore called the *Wiener-Khinchin relation*. Nevertheless, it should be emphasized that although the limit

$$\lim_{T \to \infty} \frac{1}{T} \int_{-T/2}^{T/2} X(t+\tau)X(t)\,dt \triangleq \hat{R}_X(\tau) \tag{10.66}$$

exists and its Fourier transform

$$\int_{-\infty}^{\infty} \hat{R}_X(\tau) e^{-i2\pi f \tau}\,d\tau = \hat{S}_X(f) \tag{10.67}$$

exists (in the mean-square sense) for a process that has mean-square ergodicity, and therefore regularity, of the autocorrelation, it can be shown that the limit

$$\lim_{T \to \infty} \frac{1}{T} \int_{-\infty}^{\infty} \int_{-T/2}^{T/2} X(t+\tau)X(t)\,dt\, e^{-i2\pi f \tau}\,d\tau \tag{10.68}$$

does not exist in general (in any useful sense); that is, the Fourier-transformation and limit operations in (10.67) and (10.66) cannot be interchanged. Similarly, the two limits in the alternative formula (cf. Section 3.2)

$$\hat{S}_X(f) = \lim_{\Delta \to 0} \lim_{T \to \infty} \frac{1}{2T\Delta} \int_{-T/2}^{T/2} Y^2(t)\, dt, \tag{10.69}$$

with Y given by (10.1) and (10.22), cannot be interchanged. There are significant practical implications of these restrictions on the definition of the empirical PSD, $\hat{S}_X$. These implications underly the essence of statistical methods of empirical spectral analysis [Gardner, 1987a], which is one of the most useful tools of analysis in empirical science and engineering. The measurement problem of empirical spectral analysis is discussed in Section 11.4.

Example 1 (Nonergodic regular process): Consider the random-frequency and -phase sine-wave process

$$X(t) = \cos(2\pi V t + \Theta), \tag{10.70}$$

where Θ is uniformly distributed on $(-\pi, \pi]$ and is independent of V, which has probability density denoted by $f_V(\cdot)$. The empirical autocorrelation is (exercise 16)

$$\hat{R}_X(\tau) = \tfrac{1}{2}\cos(2\pi V \tau). \tag{10.71}$$

Thus, the empirical PSD is, from (10.61),

$$\hat{S}_X(f) = \tfrac{1}{4}\delta(f - V) + \tfrac{1}{4}\delta(f + V), \tag{10.72}$$

and the probabilistic PSD is, from (10.63) (exercise 16),

$$S_X(f) = \tfrac{1}{4} f_V(f) + \tfrac{1}{4} f_V(-f). \tag{10.73}$$

Whereas $\hat{S}_X$ contains only an impulse at the single (random) frequency V (and its image $-V$), S_X is spread about $E\{V\}$ according to the probability density of V. ∎

10.6 Spectral Density for Regular Nonstationary Processes

The mean of the empirical spectral density (10.63) for a regular process is

$$E\{\hat{S}_X(f)\} = \int_{-\infty}^{\infty} E\{\hat{R}_X(\tau)\} e^{-i2\pi f \tau}\, d\tau. \tag{10.74}$$

For convenience, the time-average probabilistic autocorrelation defined by

232 • **Random Processes**

(8.68b) is reexpressed as (exercise 17)

$$\langle R_X \rangle(\tau) = \lim_{T \to \infty} \frac{1}{T} \int_{-T/2}^{T/2} R_X\left(t + \frac{\tau}{2}, t - \frac{\tau}{2}\right) dt. \tag{10.75}$$

Substitution of (10.75) into (8.68b) into (10.74) yields

$$E\{\hat{S}_X(f)\} = \langle S_X \rangle(f) \triangleq \lim_{T \to \infty} \frac{1}{T} \int_{-T/2}^{T/2} S_X(t, f) \, dt, \tag{10.76}$$

where

$$S_X(t, f) \triangleq \int_{-\infty}^{\infty} R_X\left(t + \frac{\tau}{2}, t - \frac{\tau}{2}\right) e^{-i2\pi f \tau} \, d\tau. \tag{10.77}$$

The quantity $S_X(t, f)$ is referred to as the *instantaneous* (or *time-variant*) *probabilistic spectral density*.* It can be seen that the mean empirical spectral density is the time average of the instantaneous probabilistic spectral density. Furthermore, if the nonstationary process $X(t)$ has ergodicity of the autocorrelation, the empirical spectral density equals its mean, and therefore

$$\hat{S}_X(f) = \langle S_X \rangle(f). \tag{10.78}$$

Example 2 (Cyclostationary process): Consider the amplitude-modulated process

$$X(t) = A(t) \cos(2\pi f_o t + \theta),$$

where f_o and θ are nonrandom, and $A(t)$ is WSS and regular. The empirical autocorrelation is (exercise 19)

$$\hat{R}_X(\tau) = \tfrac{1}{2} \hat{R}_A(\tau) \cos(2\pi f_o \tau), \tag{10.79}$$

assuming that

$$\lim_{T \to \infty} \frac{1}{T} \int_{-T/2}^{T/2} A(t + \tau) A(t) e^{i4\pi f_o t} \, dt \equiv 0$$

(see example 5 in Chapter 8). It follows that the empirical spectral density is

$$\hat{S}_X(f) = \tfrac{1}{4} \hat{S}_A(f - f_o) + \tfrac{1}{4} \hat{S}_A(f + f_o). \tag{10.80}$$

---

*The instantaneous probabilistic spectral density cannot be given a physical interpretation as a time-variant spectral density of power except in the special case of a WSS process (in which case the time variation vanishes) [Gardner, 1987a].

In contrast, the (nonstationary) probabilistic autocorrelation is (exercise 19)

$$R_X\left(t + \frac{\tau}{2}, t - \frac{\tau}{2}\right) = \tfrac{1}{2}R_A(\tau)\cos(2\pi f_o\tau) + \tfrac{1}{2}R_A(\tau)\cos(4\pi f_o t + 2\theta),\tag{10.81}$$

from which (10.77) yields (exercise 19)

$$S_X(t,f) = \tfrac{1}{4}S_A(f - f_o) + \tfrac{1}{4}S_A(f + f_o) + \tfrac{1}{2}S_A(f)\cos(4\pi f_o t + 2\theta).\tag{10.82}$$

Since the time average of the second term in (10.82) is zero, then

$$\langle S_X \rangle(f) = \tfrac{1}{4}S_A(f - f_o) + \tfrac{1}{4}S_A(f + f_o),\tag{10.83}$$

and furthermore, since (10.63) yields

$$E\{\hat{S}_A(f)\} = S_A(f),$$

then it can be seen, from (10.80) and (10.83), that (10.76) is satisfied. Furthermore, if $A(t)$ exhibits mean-square ergodicity of the spectral density, then (10.80) and (10.83) reveal that (10.78) holds. ∎

Examples of spectral densities for other types of cyclostationary signals, such as pulse-amplitude-modulated and pulse-position-modulated pulse trains, and frequency-modulated and phase-modulated sine waves, are presented in the next chapter.

Example 3: Consider the process $X(t)$ formed from the product of a stationary regular process $U(t)$ and one sample path $v(t)$ of another independent, stationary, regular process $V(t)$,

$$X(t) = v(t)U(t).$$

The empirical autocorrelation can be shown to be

$$\hat{R}_X(\tau) = \hat{R}_v(\tau)\hat{R}_U(\tau).$$

Consequently, the empirical spectral density is the convolution

$$\hat{S}_X(f) = \hat{S}_v(f) \otimes \hat{S}_U(f).\tag{10.84}$$

In contrast, the probabilistic autocorrelation is

$$R_X\left(t + \frac{\tau}{2}, t - \frac{\tau}{2}\right) = v\left(t + \frac{\tau}{2}\right)v\left(t - \frac{\tau}{2}\right)R_U(\tau),$$

234 • Random Processes

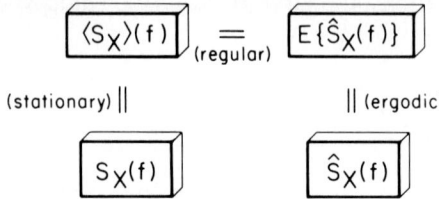

Figure 10.2 Relationships among Probabilistic and Empirical Spectral Densities

and therefore its time-averaged version is

$$\langle R_X \rangle(\tau) = \hat{R}_v(\tau) R_U(\tau).$$

Thus, the time-averaged probabilistic spectral density is the convolution

$$\langle S_X \rangle(f) = \hat{S}_v(f) \otimes S_U(f). \tag{10.85}$$

It follows from (10.84) and (10.85) that (10.76) is satisfied. Furthermore, if $U(t)$ exhibits mean-square ergodicity of the spectral density, then (10.84) and (10.85) reveal that (10.78) holds. ∎

In summary, the relationships among empirical and probabilistic spectral densities can be displayed in the form of a diagram, as shown in Figure 10.2. Note that this diagram can be obtained from the diagram in Figure 8.3 by Fourier transformation of the autocorrelations therein.

10.7 White Noise

10.7.1 Thermal-Noise Model

For the thermal-noise phenomenon in metallic resistors referred to throughout this book, it can be shown* that the PSD is accurately modeled by

$$S_Z(f) = N_0 \left[\frac{|f|/f_o}{\exp(|f|/f_o) - 1} \right]. \tag{10.86}$$

Here the bandwidth parameter f_o (which is approximately the reciprocal of the mean relaxation time of free electrons in the resistance [Middleton, 1960]) is given by

$$f_o \triangleq KT/H \approx 2.1 \times 10^{10} T \text{ Hz}, \tag{10.87}$$

---

*This formula was derived by Harry Nyquist [Nyquist, 1928a], using a thermodynamical argument.

and the *spectral intensity* parameter N_o is defined by

$$N_o \triangleq 2KTR \simeq 2.76 \times 10^{-23} TR \text{ watt-ohm/Hz}, \qquad (10.88)$$

where K and H are Boltzmann's and Planck's constants,

$$K = 1.38 \times 10^{-23} \text{ joule/degree},$$
$$H = 6.63 \times 10^{-34} \text{ joule-second}, \qquad (10.89)$$

and T and R are the temperature in degrees Kelvin and the resistance in ohms. At room temperature, $T = 290$ K, (10.87) and (10.88) yield

$$f_o \simeq 6.0 \times 10^{12} \text{ Hz}, \qquad (10.90)$$
$$N_o \simeq 8.0 \times 10^{-21} R \text{ watt-ohm/Hz}. \qquad (10.91)$$

Since the width of the PSD, (10.86) (which is the bandwidth of the process Z as explained in Section 10.8), is $f_o \simeq 6000$ GHz then the width of the autocorrelation (which is the *correlation width* of the process Z) is $\tau_o \simeq 1/f_o \simeq 1.7 \times 10^{-13}$ second. Thus, the magnitude of $R_Z(\tau)$ is negligible for $|\tau| > \tau_o$. Hence, time samples of thermal noise that are separated by more than 0.17 picosecond are uncorrelated.

It follows from the power-series expansion for exp($\cdot$) that

$$\exp\left(\frac{|f|}{f_o}\right) - 1 \simeq \frac{|f|}{f_o} \quad \text{for} \quad |f| \ll f_o,$$

from which (10.86) yields

$$S_Z(f) \simeq N_o, \quad |f| \le 10^{12} \text{ Hz}. \qquad (10.92)$$

Therefore, when thermal noise enters any linear time-invariant system with one-sided bandwidth $B < 1000$ GHz, the PSD of the system response V is, from (10.19) and (10.92),

$$S_V(f) \simeq N_o |H(f)|^2, \qquad (10.93)$$

where H is the system transfer function, which is assumed to have negligible magnitude for $|f| > B$. Hence, for every system with $B < 1000$ GHZ, thermal noise appears to have a flat PSD for *all* frequencies,

$$S_Z(f) = N_o. \qquad (10.94)$$

The idealized process (10.94) which contains equal amounts of all frequency components (wavelengths) is, by (false) analogy with white light, called *white noise*. The autocorrelation function for the idealized white-noise process is, from (10.21) and (10.94),

$$R_Z(\tau) = N_o \delta(\tau). \qquad (10.95)$$

It follows from (10.41) that white noise has infinite mean power.

236 • **Random Processes**

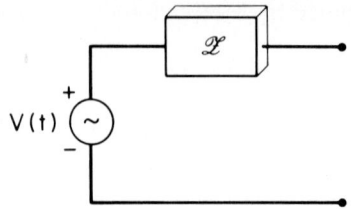

Figure 10.3 Thevenin Equivalent Circuit for Which the Spectral Density $S_V(f)$ of the Voltage V Is Specified by (10.97)

It has been shown* that a thermal-noise process that is bandlimited to $B < f_o$ is accurately modeled by a Gaussian process. Therefore, thermal noise is commonly referred to as *white Gaussian noise* (WGN). Note that the symbolic derivative of the Wiener process (Chapter 6) is a WGN process. Hence, thermal noise is symbolized by the derivative of the Wiener process.

As the time-domain counterpart of the frequency-domain argument leading to (10.93), it is noted that when thermal noise enters a system with response time (rise time) $\tau_* \simeq B^{-1}$ that is much greater than the mean relaxation time $\tau_o \simeq 1/f_o$ (i.e., $\tau_* \gg 0.17$ picosecond), the autocorrelation of the system response V is

$$R_V(\tau) \simeq N_o r_h(\tau). \tag{10.96}$$

This approximation follows from (10.17), since the width of $R_X(\cdot)$ is much less than the width of $r_h(\cdot)$. Fourier transformation of (10.96) yields (10.93).

10.7.2 Thermal Noise in Circuits

Using either (10.93) or (10.96), it can be shown that the thermal-noise voltage developed across the terminals of any linear time-invariant circuit (due to the thermal noise sources associated with all resistances in the circuit) has PSD

$$S_V(f) \simeq 2KT \operatorname{Re}\{\mathscr{Z}(f)\}, \tag{10.97}$$

where $\operatorname{Re}\{\cdot\}$ denotes the real-part operation. Equation (10.97) is referred to as *Nyquist's theorem* [Nyquist, 1928a]. Comparison of (10.97) and (10.88) reveals that the real part of the impedance $\mathscr{Z}(f)$ between the two terminals plays the role of an effective frequency-dependent resistance. The formula (10.97) can be used to characterize the *Thevenin equivalent circuit* shown in Figure 10.3. This circuit, which contains one general frequency-dependent noise-free impedance $\mathscr{Z}(f)$ in series with one general frequency-dependent

---

*This result was derived by G. E. Uhlenbeck and L. S. Ornstein [Uhlenbeck and Ornstein, 1930], using a thermodynamical argument. (See also [Lawson and Uhlenbeck, 1949] and [Middleton, 1960].)

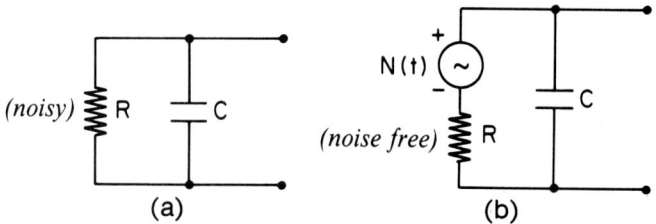

Figure 10.4 (a) Circuit for Example 4; (b) Equivalent Circuit for Example 4

zero-impedance noise voltage source with PSD $S_V(f)$, is equivalent (at the terminals) to the original circuit between the pair of terminals, which can be an arbitrary interconnection of time-invariant inductors, capacitors, and noisy resistors.

Example 4: The resistive-capacitive (RC) circuit shown in Figure 10.4a has impedance given by

$$\mathscr{Z}(f) = \frac{R}{i2\pi fRC + 1},$$

the real part of which is

$$\text{Re}\{\mathscr{Z}(f)\} = \frac{R}{1 + (2\pi fRC)^2}.$$

It follows from the Thevenin equivalent circuit, Figure 10.3, that the open-circuit voltage measured at the terminals of the RC circuit has PSD

$$S_V(f) = \frac{2KTR}{1 + (2\pi fRC)^2}. \tag{10.98}$$

To verify this result, consider the equivalent circuit shown in Figure 10.4b, in which the noisy resistor in Figure 10.4a has been replaced by its Thevenin equivalent, which is a noise-free resistor and a real voltage source $N(t)$ with PSD

$$S_N(f) = 2KTR.$$

Now, the transfer function from this voltage source to the voltage at the terminals of the circuit is

$$H(f) = \frac{1}{i2\pi fRC + 1}.$$

Therefore the PSD of the terminal voltage is

$$S_V(f) = |H(f)|^2 S_N(f) = \frac{2KTR}{1 + (2\pi fRC)^2}, \tag{10.99}$$

which is identical to (10.98), as it must be. ∎

Mean Squared Response. It is established in Section 10.1.3 that the mean squared value of the response of a circuit (or any linear time-invariant system) is given by the formula

$$E\{Y^2(t)\} = \int_{-\infty}^{\infty} |H(f)|^2 S_X(f)\, df, \qquad (10.100)$$

where S_X is the spectral density of the excitation, and H is the transfer function. For many useful models, the integrand in (10.100) is a proper rational function, in which case the integral can be evaluated explicitly. For convenient reference, the value of this integral is given here for $2n$-th order rational functions for $n = 1, 2, 3, 4$. For example, for $n = 4$ the integrand is

$$\left|H\left(\frac{\omega}{2\pi}\right)\right|^2 S_X\left(\frac{\omega}{2\pi}\right) = \left|\frac{-i\omega^3 b_3 - \omega^2 b_2 + i\omega b_1 + b_0}{\omega^4 a_4 - i\omega^3 a_3 - \omega^2 a_2 + i\omega a_1 + a_0}\right|^2. \qquad (10.101)$$

Assuming that $H(f)$ is the transfer function of a stable system, the integral (10.100) is given by the following formulas:

1. for $n = 1$

$$E\{Y^2(t)\} = \frac{1}{2} \frac{b_0^2}{a_0 a_1}, \qquad (10.102\text{a})$$

2. for $n = 2$

$$E\{Y^2(t)\} = \frac{1}{2} \frac{a_2 b_0^2/a_0 + b_1^2}{a_1 a_2}, \qquad (10.102\text{b})$$

3. for $n = 3$

$$E\{Y^2(t)\} = \frac{1}{2} \frac{a_2 a_3 b_0^2/a_0 + a_3(b_1^2 - 2b_0 b_2) + a_1 b_2^2}{a_1 a_2 a_3 - a_0 a_3^2} \qquad (10.102\text{c})$$

4. for $n = 4$

$$E\{Y^2(t)\} = \frac{1/2}{a_1(a_2 a_3 - a_1 a_4) - a_0 a_3^2}$$

$$\times \left[\frac{b_0^2}{a_0}(a_2 a_3 - a_1 a_4) + a_3(b_1^2 - 2b_0 b_2)\right.$$

$$\left. + a_1(b_2^2 - 2b_1 b_3) + \frac{b_3^2}{a_4}(a_1 a_2 - a_0 a_3)\right].$$

$$(10.102\text{d})$$

To use (10.102), the spectral density of the excitation, S_X, must first be factored into the form

$$S_X(f) = G(f)G^*(f), \qquad (10.103)$$

since (10.101) is specified explicitly in terms of $H(f)G(f)$. A general method for accomplishing this spectral factorization is described in Section 13.5.1. For white-noise excitation, the factorization (10.103) is trivial, $G(f) = \sqrt{N_o}$. Another relatively simple example is the spectral density

$$S_X(f) = \frac{S_o}{1 + (f/f_o)^2},$$

which factors into

$$S_X(f) = \frac{\sqrt{S_o}}{1 - i(f/f_o)} \frac{\sqrt{S_o}}{1 + i(f/f_o)}.$$

10.7.3 Shot-Noise Model

As discussed in Section 6.5, the shot-noise voltage in electronic devices can be modeled by

$$Y(t) = \sum_{i=1}^{\infty} A_i g(t - T_i), \qquad t \geq 0, \qquad (10.104)$$

where $\{T_i\}$ is a Poisson point process (with rate parameter λ), $\{A_i\}$ is an independent identically distributed sequence that is independent of $\{T_i\}$, and $g(t)$ is a nonrandom pulse which is zero for $t < 0$. Equation (10.104) is an appropriate steady-state model for shot noise only for $t > \tau_g$, where τ_g is the width of $g(\cdot)$, because of the finite starting time ($t = 0$) of the process. For an exponential pulse, τ_g can be taken to be infinite, in which case steady-state behavior is only approached as $t \to \infty$. In other words, $Y(t)$ is not a stationary process; however, it is *asymptotically stationary*, so that, in the limit $t \to \infty$,

$$E\{Y(t + \tau)Y(t)\} \to R_Y(\tau) \qquad \text{for all } \tau, \qquad (10.105)$$

$$E\{Y(t)\} \to m_Y, \qquad (10.106)$$

where the limits $R_Y(\tau)$ and m_Y are the steady-state (asymptotic) parameters.

It is convenient to interpret the process Y as the result of smoothing an idealized (impulsive) process Z,

$$Z(t) = \sum_{i=1}^{\infty} A_i \delta(t - T_i), \qquad t \geq 0, \qquad (10.107)$$

by passage of Z through a filter with impulse-response function $g(\cdot)$:

$$Y(t) = Z(t) \otimes g(t). \tag{10.108}$$

It is shown in Chapter 6 that (asymptotically)

$$m_Z = m_A \lambda,$$
$$R_Z(\tau) = (m_A \lambda)^2 + \lambda(m_A^2 + \sigma_A^2)\delta(\tau). \tag{10.109}$$

Therefore, the (asymptotic) PSD is

$$S_Z(f) = m_Z^2 \delta(f) + \lambda(m_A^2 + \sigma_A^2). \tag{10.110}$$

It follows from (10.108) and (10.109) that (asymptotically)

$$E\{Y(t)\} = m_Y = m_Z G(0) = m_A \lambda G(0) \tag{10.111}$$
$$E\{Y(t+\tau)Y(t)\} = R_Y(\tau) = R_Z(\tau) \otimes r_g(\tau)$$
$$= [G(0)m_A \lambda]^2 + \lambda(m_A^2 + \sigma_A^2)r_g(\tau) \tag{10.112}$$

(exercise 32). Therefore, the (asymptotic) PSD for the shot-noise process is

$$S_Y(f) = [G(0)m_A \lambda]^2 \delta(f) + \lambda(m_A^2 + \sigma_A^2)|G(f)|^2. \tag{10.113}$$

If either $m_Z = 0$ or $G(0) = 0$, then $m_Y = 0$, and Y contains no d-c component; that is, the PSD S_Y contains no impulse at $f = 0$.

The width of the pulse $g(\cdot)$ is the lifetime of a shot event, for example, the transit time through the shot-noise-producing part of a diode, which is typically fractions of a nanosecond for semiconductor devices. Thus, the one-sided width of $|G(\cdot)|^2$, which is the bandwidth of $Y(t)$, is typically in the high gigahertz range.

If we let f_o be a conservative measure of the bandwidth of $Y(t)$ [such that $|G(f)| \simeq |G(0)|$ for $|f| \leq f_o$], then

$$S_Y(f) \simeq m_Y^2 \delta(f) + N_o, \qquad |f| \leq f_o, \tag{10.114}$$

where

$$N_o \triangleq \lambda(m_A^2 + \sigma_A^2)|G(0)|^2. \tag{10.115}$$

Therefore, when shot noise enters any linear time-invariant system with one-sided bandwidth $B < f_o$, the PSD of the system response V is, from (10.19) and (10.114),

$$S_V(f) \cong m_Y^2 |H(0)|^2 \delta(f) + N_o |H(f)|^2, \tag{10.116}$$

where H is the system transfer function, which is assumed to have negligible magnitude for $|f| > B$. Hence, for every system with $B < f_o$, shot noise *appears* to have a flat PSD for all frequencies, plus a zero-frequency impulsive term

$$S_Y(f) = m_Y^2 \delta(f) + N_o. \qquad (10.117)$$

In the case for which $m_Y = 0$, this idealized process [which is simply (10.107) scaled by $G(0)$] is called *white Poisson impulse noise* (WPIN). Its autocorrelation is

$$R_Y(\tau) = N_o \delta(\tau). \qquad (10.118)$$

Note that WPIN is symbolized by the derivative of the Poisson counting process $W(t)$, with random weights $Z_i = G(0)A_i$ (Chapter 6).

When the rate parameter λ is much greater than the reciprocal of the effective pulse width, then many pulses overlap at each instant in time. As a result, the central limit theorem comes into effect, and the shot-noise process (10.104) can be approximated by a Gaussian process, with the same mean and autocorrelation (10.111)–(10.112) [Snyder, 1975, Section 4.1; Davenport and Root, 1958, Chapter 7].

10.7.4 Other White-Noise Models

It can be shown that the symbolic process obtained by formally differentiating any zero-mean process with stationary and uncorrelated increments has an autocorrelation function that is simply an impulse, and is therefore a white-noise process. As an extension of this, a zero-mean process that has stationary and *independent* increments is said to give rise to a *strict-sense white-noise process* by differentiation. Since both the Wiener process and the Poisson counting process do indeed have independent increments, both WGN and WPIN are strict-sense white-noise processes. Moreover, there is a fundamental theorem concerning the characterization of all stationary independent-increment processes, due to P. Levy and A. N. Kolmogorov [Gnedenko and Kolmogorov, 1954] that reveals that every strict-sense white-noise process is composed of a sum of at most two independent white noises, one of which is WGN and the other of which is a sum (or a limit in distribution of a sum) of independent WPINs. Since the integral of WGN has continuous sample paths and the integral of WPIN has step discontinuities, this theorem also reveals that WGN is the only white-noise process whose integral has continuous sample paths.

10.7.5 Discrete-Time White Noise

By analogy with the terminology for continuous-time processes, a discrete-time process $X(i)$ exhibiting a flat spectral density for all frequencies is referred to as a *white-noise process*. It follows from this definition,

$$S_X(f) = N_o, \tag{10.119}$$

and (10.38) that [analogously to (10.95)]

$$R_X(k) = N_o \delta_k, \tag{10.120}$$

where δ_k is the *Kronecker delta*

$$\delta_k \triangleq \begin{cases} 1, & k = 0, \\ 0, & k \neq 0. \end{cases} \tag{10.121}$$

This reveals that every zero-mean sequence $X(i)$ of equi-mean-square uncorrelated random variables is a white-noise process, regardless of the particular probability distribution. Hence, the random variables $X(i)$ in a discrete-time white-noise process can be discrete (e.g., binary) or continuous (e.g., Gaussian), or mixed. A sequence of zero-mean independent, identically distributed random variables is said to be a *strict-sense white-noise sequence*.

Unlike the idealized continuous-time white-noise models, (WGN and WPIN), which exhibit infinite variance [$\delta(0) = \infty$], the discrete-time white-noise model is quite well behaved [$\delta_0 = 1$].

A link between continuous-time and discrete-time white noise can be established as follows. Let the white-noise process $X(t)$, for which

$$m_X = 0, \tag{10.122}$$
$$R_X(\tau) = N_o \delta(\tau), \tag{10.123}$$

be time-sampled with a sampling switch exhibiting a finite dwell time, and denote the ith time sample by $Y(i)$:

$$Y(i) \triangleq \int_{-\infty}^{\infty} X(t) g(iT - t) \, dt, \tag{10.124}$$

where $g(iT - t)$ represents the transmission characteristic of the switch closed at time iT; the width of $g(\cdot)$ is the dwell time τ_g of the switch. For a switch with unity expected power gain,

$$\int_{-\infty}^{\infty} g^2(t) \, dt = 1. \tag{10.125}$$

If $\tau_g < T$, then it is easily shown that the mean and autocorrelation for $Y(i)$ are (exercise 33)

$$m_Y = 0, \tag{10.126}$$
$$R_Y(k) = N_o \delta_k, \tag{10.127}$$

and therefore $Y(i)$ is a white-noise sequence.

It should also be pointed out that if $X(t)$ is a non-white-noise process with finite correlation width τ_o, so that

$$R_X(\tau) = 0, \qquad |\tau| > \tau_o, \tag{10.128}$$

and if the difference between the sampling increment and the dwell time is greater than the correlation width ($T - \tau_g > \tau_o$), then the sampled process (10.124) is white noise (exercise 33):

$$R_Y(k) = N_o \delta_k, \tag{10.129}$$

where

$$N_o = \int_{-\infty}^{\infty} S_X(f) |G(f)|^2 \, df. \tag{10.130}$$

10.8 Bandwidths

A process is said to be *low-pass* if its spectral density is concentrated around zero frequency, and it is said to be *bandpass* if its spectral density is concentrated around some nonzero frequency, say $f_o > 0$. The bandwidth of a process is defined in terms of the width of its spectral density. There are several useful definitions of width. The *rectangular bandwidth* B_1 of a low-pass process is the width of a rectangle whose height at $f = 0$ and area equal the height at $f = 0$ and area of the spectral density, respectively. Thus,

$$B_1 \triangleq \frac{\int_{-\infty}^{\infty} S_X(f) \, df}{S_X(0)}. \tag{10.131}$$

The *rms bandwidth* B_2 of a low-pass process is defined to be the standard deviation of the density function obtained by normalizing the area of the spectral density to unity. Thus,

$$B_2 \triangleq \left(\frac{\int_{-\infty}^{\infty} f^2 S_X(f) \, df}{\int_{-\infty}^{\infty} S_X(\nu) \, d\nu} \right)^{1/2}. \tag{10.132}$$

The *absolute bandwidth* B_3 of a low-pass process is defined to be the width of the band centered at $f = 0$ over which the spectral density is nonzero.

These three measures of bandwidth are called *two-sided bandwidths*, since they take account of negative as well as positive frequencies. A *one-sided bandwidth* is simply half the value of the corresponding two-sided bandwidth.

For a bandpass process, one-sided bandwidths are defined by analogy to the preceding three definitions of two-sided bandwidths except that $S_X(0)$ is replaced with $S_X(f_o)$, and the lower limit of integration $-\infty$ is changed to 0 in (10.131); f^2 is replaced with $(f - f_o)^2$ and the lower limit of integration $-\infty$ is changed to 0 in (10.132); and in the definition of B_3 $f = 0$ is replaced with $f = f_o$, and only positive frequencies are considered.

10.9 Spectral Lines

It can be shown [Doob, 1953; Priestley, 1981] that a WSS process (with autocorrelation function that is continuous at $\tau = 0$) can be decomposed into two orthogonal components,

$$X(t) = X_c(t) + X_d(t) \tag{10.133}$$

where

$$R_{X_c X_d}(\tau) \equiv 0, \tag{10.134}$$

for which $X_c(t)$ has a continuous spectral density and $X_d(t)$ has a spectral density consisting entirely of Dirac deltas, which are called *spectral lines*. Furthermore, it can be shown that $X_d(t)$ consists entirely of a linear combination of sine waves

$$X_d(t) = \sum_n A_n e^{i 2\pi \alpha_n t}, \tag{10.135}$$

where $\alpha_{-n} = -\alpha_n$, $A_n = A^*_{-n}$, and the spectral density of X_d is given by (exercise 38)

$$S_{X_d}(f) = \sum_n E\{|A_n|^2\} \delta(f - \alpha_n). \tag{10.136}$$

It can be shown (exercise 38) that for the nonzero frequencies $\alpha_n \neq 0$

$$E\{A_n\} = 0, \quad n \neq 0, \tag{10.137a}$$

$$E\{A_n A_m\} = 0, \quad m \neq -n \tag{10.137b}$$

(but it is not necessarily true that $E\{|A_n|^2\} = 0$); otherwise, $X_d(t)$ cannot be WSS. It can also be shown that $X(t)$ can exhibit mean-square ergodicity of the spectral density only if the amplitudes $\{|A_n|\}$ of the sine waves, including $n = 0$, in

$$A_n e^{i 2\pi \alpha_n t} + A_{-n} e^{i 2\pi \alpha_{-n} t} = 2|A_n| \cos(2\pi \alpha_n t + \arg\{A_n\}) \tag{10.138}$$

are nonrandom (cf. Chapter 8, exercise 10). This can be seen from the fact that the empirical spectral density has the spectral lines

$$\hat{S}_{X_d}(f) = 2 \sum_n |A_n|^2 \delta(f - \alpha_n), \tag{10.139}$$

in contrast to (10.136).

It follows from the decomposition (10.133) that the mean of the process $X(t)$ is given (using $A = A_0$) by

$$m_X = E\{A\}. \tag{10.140}$$

Thus, the strength of the spectral line at $f = 0$ is given by

$$E\{A^2\} = m_X^2 + \sigma_A^2, \tag{10.141}$$

and this equals the squared mean if and only if the variance of the zero-frequency component, σ_A^2, is zero (which is necessary for ergodicity).

Results that are analogous to those described here for continuous-time processes hold for discrete-time processes, and are explained in detail in Section 11.9 of the following chapter.

10.10 Summary

The Fourier transform of a stationary process $X(t)$ is a symbolic process analogous to the formal derivative of the Wiener process (Chapter 6) and can therefore be rigorously defined only in terms of its indefinite integral, the *generalized Fourier transform*, analogous to the Wiener process. The mean squared value of the generalized Fourier transform is the indefinite integral (10.16) of the Fourier transform $S_X(f)$ of the autocorrelation function $R_X(\tau)$. The derivative $S_X(f)$ of this mean squared generalized Fourier transform is the *spectral density of mean squared fluctuation* of the process $X(t)$. Since the mean squared fluctuation $E\{X^2(t)\}$ can be interpreted as the expected value of the instantaneous power, then $S_X(f)$ can be interpreted as a *spectral density of expected power*, which is abbreviated to simply *power spectral density*. The power-spectral-density interpretation of the Fourier transform $S_X(f)$ of $R_X(\tau)$ can be obtained from either the limit (10.26) (as the bandwidth approaches zero) of the expected power out of a narrowband filter, or the limit (10.52) (as the segment length approaches infinity) of the expected energy spectral density of a finite segment normalized by the segment length. This Fourier-transform relation between the power spectral density and the autocorrelation is called the *Wiener-Khinchin relation*.

The *cross-spectral density* $S_{XY}(f)$ between two processes $X(t)$ and $Y(t)$ is the *correlation between spectral components* at frequency f from $X(t)$ and

$Y(t)$. For $X(t) \equiv Y(t)$ this correlation becomes a mean squared value, and the cross-spectral density becomes a power spectral density. The cross-spectral density $S_{XY}(f)$ is given by the Fourier transform of the cross-correlation function $R_{XY}(\tau)$. When normalized by the mean squared values of the spectral components, the cross-spectral density becomes a spectrally decomposed correlation coefficient called the *coherence function* (10.53). This is a result of the fact that each spectral component has zero mean and therefore the correlation is a covariance, and the mean squares are variances.

Every pair of jointly stationary processes can be given a unique *linear model* (10.58) in which either process is expressed as a linearly distorted version of the other process plus an uncorrelated noise process. The model is specified by (10.60).

Because of the duality between the probabilistic theory of stationary processes and the deterministic theory based on time averages, the expected power spectral density has a dual, namely, the time-averaged power spectral density or *empirical power spectral density* $\hat{S}_X(f)$, which is the Fourier transform of the empirical autocorrelation $\hat{R}_X(\tau)$. Furthermore, the concept of power spectral density can be generalized for asymptotically mean stationary processes. The time-averaged expected power spectral density $\langle S_X \rangle(f)$ is the Fourier transform of the time-averaged autocorrelation $\langle R_X \rangle(\tau)$. If the process is regular, then $\hat{S}_X(f)$ exists and its expected value is given by $E\{\hat{S}_X(f)\} = \langle S_X \rangle(f)$. If the process is ergodic, then $E\{\hat{S}_X(f)\} = \hat{S}_X(f)$ (in the mean-square sense), and if the process is stationary, then $\langle S_X \rangle(f) = S_X(f)$.

A symbolic process that has a constant power spectral density is called *white noise*. One such symbolic process is the derivative of the Wiener process, and is called *white Gaussian noise* (WGN). WGN is a useful model for thermal noise. Another such symbolic process is the derivative of the Poisson counting process, and is called *white Poisson impulse noise* (WPIN). WPIN is a useful model for shot noise.

Any linear, time-invariant circuit with a multiplicity of thermal noise sources can be modeled with a *Thevenin equivalent circuit* containing one general impedance in series with one idealized noise-voltage source. The power spectral density for this equivalent noise source is given by *Nyquist's theorem* (10.97).

Given the transfer function for any circuit up to fourth order, the mean squared response due to white-noise excitation is given explicitly by (10.102).

All white-noise processes can be characterized as the derivatives of uncorrelated-increment processes. A process that is the derivative of an independent-increment process is called *strict-sense white noise*. Both WGN and WPIN are strict-sense white-noise processes. Moreover, every strict-sense white-noise process is composed of the sum of at most two independent white-noise processes, one of which is WGN and the other of which is a sum of WPINs.

Unlike continuous-time white noise, which can be rigorously defined only in terms of its indefinite integral, discrete-time white noise is quite well behaved. It is simply a sequence of uncorrelated (independent for strict-sense white noise) zero-mean variables.

Two types of processes that arise frequently in practice are the *low-pass* and *bandpass* processes, which are defined in terms of their spectral densities. A low-pass process has spectral density concentrated around $f = 0$, whereas a bandpass process has spectral density concentrated around $f = f_o > 0$. There are various useful alternatives for defining the *bandwidth* of low-pass and bandpass processes.

A stationary process can be decomposed into a sum of two orthogonal components (10.133), one of which has a spectral density consisting entirely of Dirac deltas, called *spectral lines*, and the other of which has a continuous spectral density.

EXERCISES

★1. (a) Derive the formal result (10.8) from the formal result (10.7).

  (b) Substitute (10.6) into (10.11), interchange the order of the two integrations, and formally evaluate the inner integral to obtain

  $$\tilde{X}^{(-1)}(f_1) - \tilde{X}^{(-1)}(f_2) = \int_{-\infty}^{\infty} X(t) \left[\frac{e^{-i2\pi f_2 t} - e^{-i2\pi f_1 t}}{i2\pi t} \right] dt. \tag{10.142}$$

  This integral is well defined for a stationary random process, and therefore provides a sound basis for intuitive interpretation of the integrated Fourier transform. [Note that if $X(t)$ is persistent, then $X(t)/t \to 0$ as $t \to \infty$. Note also that the factor multiplying $X(t)$ in the integrand is finite at $t = 0$.]

  (c) Use (10.142) to prove that

  $$E\{|\tilde{X}^{(-1)}(f_1) - \tilde{X}^{(-1)}(f_2)|^2\} = \int_{f_2}^{f_1} S_X(\nu) \, d\nu. \tag{10.143}$$

  Hint: Use Parseval's relation, exercise 2.

  (d) The rigorous frequency-domain input-output relation for finite-power waveforms is

  $$\tilde{Y}^{-1}(f) = \int_{-\infty}^{f} H(\nu) \, d\tilde{X}^{-1}(\nu). \tag{10.144}$$

  Show by formal manipulation that (10.144) reduces to the formal relation (10.3).

⋆2. (a) Prove Parseval's relation

$$\int_{-\infty}^{\infty} g(t)h^*(t)\,dt = \int_{-\infty}^{\infty} G(f)H^*(f)\,df. \qquad (10.145)$$

(b) Use (10.145) to derive (10.24) from (10.23).

3. Let $X(t)$ be the result of white noise $Z(t)$ interfering with a delayed replica of itself (e.g., due to an echo):

$$X(t) = Z(t) - Z(t - \Delta).$$

Determine the spectral density for $X(t)$. Explain the shape of this spectrum in terms of the effect of a linear filtering operation.

4. (a) Let $W(t)$ be the Wiener process. By expressing $Y(t)$ as a convolution with the derivative of $W(t)$, use filtering methods to show that

$$Y(t) = W(t) - W(t - \Delta)$$

is WSS, and determine the spectral density for $Y(t)$. Compare with the result from exercise 3.

(b) Let $X(t) = dY(t)/dt$, where $Y(t)$ is WSS, and use filtering methods to show that

$$S_X(f) = (2\pi f)^2 S_Y(f).$$

5. (a) Show that the spectral density for the asynchronous telegraph process (Chapter 6, exercise 12), which has autocorrelation

$$R_Y(\tau) = e^{-2\lambda|\tau|},$$

is given by

$$S_Y(f) = \frac{4\lambda}{(2\pi f)^2 + 4\lambda^2}.$$

(b) Show that the spectral density for the Ornstein-Uhlenbeck process (Section 5.3.2), which has autocorrelation

$$R_X(\tau) = \sigma^2 e^{-\alpha|\tau|},$$

is given by

$$S_X(f) = \frac{2\sigma^2 \alpha}{(2\pi f)^2 + \alpha^2},$$

which is sometimes referred to as the *Lorenzian spectrum*.

⋆6. Verify that (10.34) and (10.38) are a transform pair, by substitution of (10.34) into (10.38).

7. Use the change of variables $t = v + \tau/2$ and $u = v - \tau/2$ to derive (10.48) from (10.47).

8. Verify that $W_{1/T}$ in (10.51) is the Fourier transform of w_T in (10.49).

★9. An ergodic stationary process with spectral density

$$S_X(f) = \begin{cases} (1 - |f|/B)S_o, & |f| \leq B, \\ 0, & |f| > B, \end{cases}$$

is passed through an ideal low-pass filter with transfer function

$$H(f) = \begin{cases} 1, & |f| \leq b, \\ 0, & |f| > b. \end{cases}$$

Determine the variance at the input and output of this filter as a function of b and B. What is the mean value at the input and output?

★10. A discrete-time linear time-invariant system has unit-pulse response

$$h(j) = \begin{cases} 1, & 0 \leq j \leq K, \\ 0, & \text{otherwise}, \end{cases}$$

and it is excited by a sequence of zero-mean independent random variables with variance σ^2.
(a) Determine the spectral density and the variance of the response.
(b) Determine the cross-correlation and cross-spectral density of the excitation and response.

11. Consider the discrete-time system consisting of an accumulator followed by a delay-and-subtract device, as depicted in Figure 10.5. The input to the system starts at $i = 0$ and consists of independent, identically distributed, zero-mean random variables. Determine the asymptotic spectral density and the variance of the response.

Hint: First determine the unit-pulse response and then the transfer function.

12. Prove that for a discrete-time WSS process $X(i)$,
(a) $S_X(f) = S_X(-f)$,

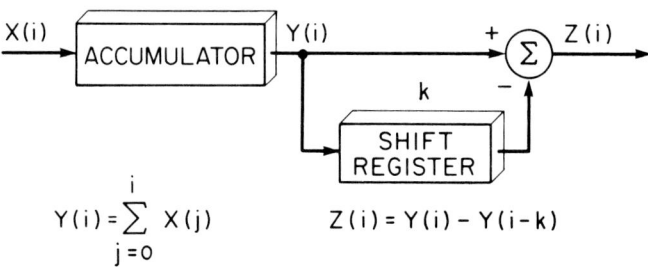

Figure 10.5 System Diagram for Exercise 11

(b) $S_X(f+n) = S_X(f)$, $n = \pm 1, \pm 2, \ldots,$
(c) $S_X(f) \geq 0$.

13. Consider two processes $Y_1(t)$ and $Y_2(t)$ obtained from two jointly WSS processes $X_1(t)$ and $X_2(t)$ by convolution:

$$Y_1(t) = X_1(t) \otimes h_1(t),$$
$$Y_2(t) = X_2(t) \otimes h_2(t).$$

Determine the coherence function for $Y_1(t)$ and $Y_2(t)$, and show that it is independent of $|H_1(f)|$ and $|H_2(f)|$, that is, it depends on only the phases of $H_1(f)$, $H_2(f)$ and the coherence of $X_1(t)$ and $X_2(t)$. [Assume $H_1(f) \neq 0$ and $H_2(f) \neq 0$.] Compare the magnitudes of the coherence functions for $X_1(t)$, $X_2(t)$ and $Y_1(t)$, $Y_2(t)$.

14. On the basis of (10.29), propose a method for identifying the transfer function of an unknown system that utilizes measured cross-spectral densities and spectral densities of the system excitation and response. (It is shown in exercise 19 of Chapter 13 that this method yields a best-fitting linear time-invariant model of the unknown system when it is not linear and/or not time-invariant.)

15. Consider the problem of detecting the presence of a distant disturbance in the ocean (earthquake, storm, etc.) by using local measurements of wave action. Suppose there are two recording stations at which wave amplitudes $Y_1(t)$ and $Y_2(t)$ are measured, and assume that all contributions to these measurements due to the distant source can be attributed to a single waveform $X(t)$, representing an idealized point source. If X, Y_1, and Y_2 are modeled as jointly stationary processes, then we have (from Section 10.3.2) the equivalent linear model

$$Y_1(t) = h_1(t) \otimes X(t) + N_1(t),$$
$$Y_2(t) = h_2(t) \otimes X(t) + N_2(t),$$

for which $N_1(t)$ and $N_2(t)$ are uncorrelated with $X(t)$. Suppose that N_1 and N_2 are due primarily to local disturbances around each of the two recording stations, and assume that these are uncorrelated. Use the fact that locally generated sea waves, such as N_1 and N_2, generally occupy a higher frequency range than that of waves that have traveled a long distance, such as $h_1 \otimes X$ and $h_2 \otimes X$ (because h_1 and h_2 represent low-pass filters), to show that the coherence magnitude between Y_1 and Y_2 can be nearly unity in the low frequency range if the distant source is present, and is much smaller than unity in the high frequency range. Show that if the distant source is absent, the coherence magnitude is much smaller than unity in the low frequency range.

16. (a) Verify the formulas for $\hat{R}_X$, $\hat{S}_X$, and S_X in example 1, that is, Equations (10.71) to (10.73).
 (b) Replace V by $f_o + V$ in (10.70), where f_o is deterministic, and determine S_X. [It is shown in Section 11.3 that if $X(t)$ is made a frequency-modulated sine wave by the replacement

$$Vt \to \int_{-\infty}^{t} V(u)\, du \triangleq \Phi(t),$$

and $R_\Phi(0) \gg 1$, then the spectrum is given to a close approximation by precisely the same formula as that obtained in this exercise. Thus, the probabilistic spectrum for a frequency-modulated sine wave with instantaneous frequency $V(t)$ is the same as that for an unmodulated sine wave with random but constant frequency V, provided that the probability densities of V and $V(t)$ are the same. This is exemplary of how the spectral density of a nonergodic process can be misinterpreted.]

17. Verify the identity

$$\lim_{T \to \infty} \frac{1}{T} \int_{-T/2}^{T/2} R_X(t + \tau, t)\, dt = \lim_{T \to \infty} \frac{1}{T} \int_{-T/2}^{T/2} R_X\left(t + \frac{\tau}{2}, t - \frac{\tau}{2}\right) dt.$$

18. Show that for a regular nonstationary process $X(t)$ in (10.1), the time-average probabilistic spectral density for $Y(t)$ in (10.1) is given by

$$\langle S_Y \rangle(f) = |H(f)|^2 \langle S_X \rangle(f).$$

19. Verify the formulas for $\hat{R}_X$, $\hat{S}_X$, R_X, and S_X in example 2, that is, Equations (10.79), (10.80), (10.81), (10.82).

20. Let $X(t)$ be the asynchronous random telegraph signal (Chapter 6, exercise 12), and consider the amplitude-modulated signal

$$Y(t) = X(t)\cos(2\pi f_o t), \qquad t \geq 0.$$

Determine the time-averaged (over positive time only) probabilistic spectral density for $Y(t)$.

21. A signal with flat spectrum is transmitted through a medium that suppresses high frequencies. The resultant signal spectral density is

$$S_X(f) = \frac{S_o}{1 + (f/f_1)^2}.$$

Show that the circuit depicted in Figure 10.6 can restore the high frequencies up to any desired frequency, say $f_2 \gg f_1$, provided that $f_1 = 1/2\pi R_1 C$ and $f_2 = 1/2\pi R_2 C$.

22. Show that two WSS processes with nonoverlapping spectral densities are orthogonal.

   Hint: Use the fact that the magnitude of the coherence function cannot exceed unity.

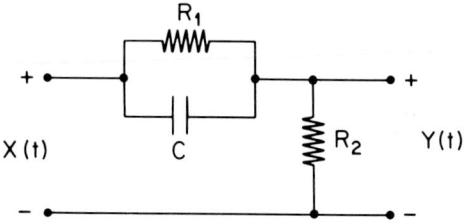

Figure 10.6 Circuit Diagram for Exercise 21

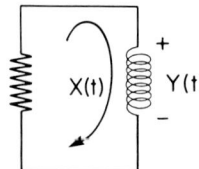

Figure 10.7 Circuit Diagram for Exercise 23

23. Consider the resistive-inductive circuit shown in Figure 10.7. There is a current, say $X(t)$, flowing through the inductor, and a voltage, say $Y(t)$, across the inductor, due to the thermal noise in the resistor. (All losses in the inductor have been lumped with the resistor, so the inductor is noise-free.) Determine the cross-correlation and cross-spectral density of $X(t)$ and $Y(t)$, and the autocorrelation and spectral density of $X(t)$ and of $Y(t)$, assuming that the thermal-noise spectral density is flat far beyond the cutoff frequency $f_o = R/2\pi L$.

24. Consider the problem of using a first-order low-pass filter to attenuate additive white noise $N(t)$ that is corrupting a signal $S(t)$ with spectral density

$$S_S(f) = \frac{S_o}{1 + (f/f_o)^2}.$$

Determine the cutoff frequency f_1 in the transfer-function magnitude

$$|H(f)|^2 = \frac{1}{1 + (f/f_1)^2}$$

that minimizes the mean squared error (MSE) between the noise-free signal $S(t)$ and the filtered signal

$$\hat{S}(t) \triangleq h(t) \otimes [S(t) + N(t)].$$

Assume that $S(t)$ and $N(t)$ are orthogonal.

Hint: Express MSE as

$$\text{MSE} = \int_{-\infty}^{\infty} |1 - H(f)|^2 S_S(f)\,df + N_o \int_{-\infty}^{\infty} |H(f)|^2\,df.$$

Then use the formula (10.102) to show that

$$\text{MSE} = \frac{\pi S_o f_o^2}{(f_o + f_1)} + \pi N_o f_1.$$

Finally, minimize this quantity with respect to f_1. Observe that if the values of S_o and N_o lead to a negative value of f_1, then no optimum exists.

25. It is desired to use rate feedback in a vehicle control system. The rate signal is provided by a permanent-magnet d-c tachometer on a rotating shaft, and is noisy because of commutator brush action. The noisy rate signal cannot simply be low-pass filtered, because the resultant delay incurred by the signal can have serious degrading effects on the dynamical behavior of the feedback control system. However, if an angular accelerometer as well as the d-c tachometer is used on the rotating shaft, then the two signals can be combined and filtered without incurring delay. Let the two signals be modeled by

$$Y_1(t) = S(t) + N_1(t),$$

$$Y_2(t) = \frac{dS(t)}{dt} + N_2(t),$$

where $N_1(t)$ and $N_2(t)$ are independent white noises that are orthogonal to $S(t)$. Show that an estimate of the form

$$\hat{S}(t) = [Y_1(t) + cY_2(t)] \otimes h(t),$$

where

$$H(f) = \frac{1}{1 + i(f/f_1)},$$

results in a perfect replica of the signal $S(t)$ without delay, plus additive filtered noise $\hat{N}(t)$, provided that $c = 1/2\pi f_1$. Then determine the cutoff frequency f_1 that minimizes $E\{\hat{N}^2(t)\}$. **Hint:** Use (10.102).

26. (a) Determine the mean squared error incurred in using the circuit shown in Figure 10.8a to approximate the scaled derivative of a

254 • **Random Processes**

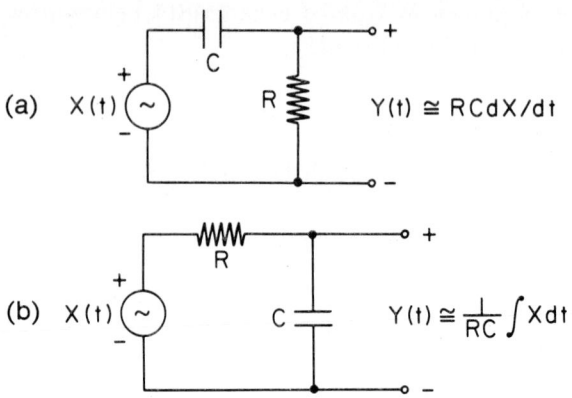

Figure 10.8 Circuit Diagrams for Exercise 10.26(a) and (b)

signal with spectral density $S_X(f)$. Then evaluate the result for $X(t)$ given by the Ornstein-Uhlenbeck process, for which

$$S_X(f) = \frac{S_o}{1+(f/f_o)^2}.$$

Under what conditions on f_o and $f_1 \triangleq 1/2\pi RC$ is the MSE small?
(b) Do the analog of part (a) for approximating the scaled integral of a signal using the circuit shown in Figure 8b, but in this case let the input spectral density be given by

$$S_X(f) = \frac{S_o(f/f_o)^2}{1+(f/f_o)^2}.$$

(Recall that an integrator is a marginally stable system and has transfer function given by $H(f) = 1/i2\pi f + (1/2)\delta(f)$. Consequently, the output spectrum is given by $S_Y(f) = S_X(f)/(2\pi f)^2$ if and only if $S_X(f)/f \to 0$ as $f \to 0$.)

27. (a) Show that the response $Y(t)$ of a system described by the differential equation

$$\frac{d^n Y(t)}{dt^n} + a_{n-1}\frac{d^{n-1}Y(t)}{dt^{n-1}} + \cdots + a_0 Y(t)$$
$$= b_{n-1}\frac{d^{n-1}X(t)}{dt^{n-1}} + \cdots + b_0 X(t),$$

with white-noise excitation $X(t)$, has a spectral density that is a rational function of f^2.
(b) Do the analog of part (a) for a discrete-time ARMA process (cf. Section 9.4.1).

28. The first difference of a process $X(t)$ can be modeled as the response $Y_1(t) = X(t) - X(t - T)$ to the linear time-invariant system with transfer function
$$H(f) = 1 - e^{-i2\pi fT}.$$
Derive the spectral density for the nth difference of $X(t)$. [Recall that the second difference of $X(t)$ is defined by $Y_2(t) \triangleq Y_1(t) - Y_1(t - T)$, and the nth difference is defined by $Y_n(t) \triangleq Y_{n-1}(t) - Y_{n-1}(t - T)$.]

29. (a) The *Hilbert transform* of $X(t)$ is defined by
$$Y(t) = X(t) \otimes h(t),$$
where
$$h(t) = 1/\pi t.$$
The transfer function can be shown to be
$$H(f) = \begin{cases} -i \equiv e^{-i\pi/2}, & f > 0, \\ +i \equiv e^{+i\pi/2}, & f < 0. \end{cases}$$
Consequently, the Hilbert transform simply shifts all frequency components by $\pi/2$ radians (90°). Show that this has no effect on the spectral density.

(b) Determine the cross-spectrum $S_{XY}(f)$, and then the cross-correlation $R_{XY}(\tau)$ [expressed as a convolution with $R_X(\tau)$].

30. In the study of animals, it is often desired to implant a miniature telemetry device so that various body functions can be observed under normal conditions of activity and environment. One possibility for energizing such a device without the use of batteries, which have limited life and can be excessively bulky, is to use a transducer to convert random motion of the animal into electrical energy. In the simplified transducer shown in Figure 10.9, energy conversion occurs in the damping mechanism as simple viscous damping, but in practice this conversion mechanism would have to be replaced with some sort of electromechanical conversion device. Assume that all the power indicated as being dissipated in damping can be converted to electrical form, and determine the optimum value for the spring constant k, that is, the value that maximizes power dissipation. Let the excitation velocity have the autocorrelation function
$$R_V(\tau) = e^{-2|\tau|},$$
and assume that the spring-mass arrangement is critically damped, in which case $b^2/4m = k$.

256 • Random Processes

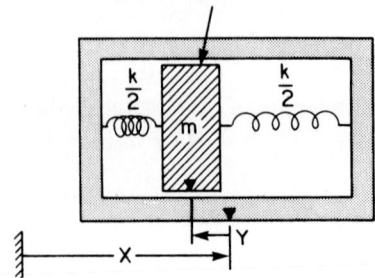

X = device displacement; V = dX / dt
Y = device / mass relative displacement;
W = dY / dt

Fixed
reference

Figure 10.9 Transducer Diagram for Exercise 30

Hint: Use the formula (10.102) to evaluate the response power $P = bE\{W^2(t)\}$ for the equation of motion

$$m\frac{d^2W(t)}{dt^2} + b\frac{dW(t)}{dt} + kW(t) + m\frac{dV(t)}{dt} = 0.$$

31. Consider the resistive-inductive-capacitive circuit shown in Figure 10.10, in which the inductor and capacitor are ideal lossless (noise-free) elements and the resistor is noisy.
    (a) Show that the spectral density of the voltage across the capacitor due to the thermal noise is given by

    $$S_Y(f) = \frac{1}{\left[1 - (f/f_1)^2\right]^2 + (f/f_2)^2},$$

    where

    $$f_1 \triangleq (2\pi\sqrt{LC})^{-1},$$
    $$f_2 \triangleq (2\pi RC)^{-1}.$$

    Assume that the spectral density of the thermal noise is flat for all frequencies of interest, and equal to unity.
    (b) Use the result of part (a) to show that the autocorrelation of the voltage across the capacitor is given by

    $$R_Y(\tau) = ke^{-|\tau/\tau_0|}[\cos(2\pi f_o \tau) + \varepsilon \sin(2\pi f_o |\tau|)],$$

    (10.146a)

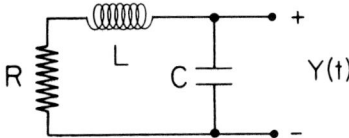

Figure 10.10 Circuit Diagram for Exercise 31

where
$$k \triangleq \pi/2RC,$$
$$\tau_o \triangleq 2L/R,$$
$$f_o \triangleq \left[(f_1)^2 - \left(\frac{1}{2\pi\tau_o}\right)^2\right]^{1/2},$$
$$\varepsilon \triangleq \left[(2\pi\tau_o f_1)^2 - 1\right]^{-1}. \tag{10.146b}$$

(c) Use the result of part (b) to show that when the losses in the circuit are small,
$$R \ll 2\sqrt{L/C},$$
then $\varepsilon \ll 1$ and $f_o \simeq f_1$, and therefore
$$R_X(\tau) \simeq ke^{-|\tau/\tau_o|}\cos(2\pi f_1 \tau). \tag{10.147}$$

32. (a) For the shot-noise process (10.104) derive the formulas (10.111) and (10.112) for the mean and autocorrelation, from the formulas (10.109).
   (b) Use the results of part (a) to verify that the mean and variance for the shot-noise process with unity pulse amplitudes $A_i = 1$ are given by
$$m_Y = \lambda \int_{-\infty}^{\infty} g(t)\,dt,$$
$$\sigma_Y^2 = \lambda \int_{-\infty}^{\infty} g^2(t)\,dt.$$

   This pair of formulas is referred to as *Campbell's theorem*, in honor of Norman Robert Campbell for his pioneering work [Campbell, 1909].

33. (a) Verify that (10.126) and (10.127) follow from (10.122) to (10.125).
    (b) Verify that (10.129) to (10.130) follow from (10.124), (10.128), and $T - \tau_g > \tau_o$.

34. Let $Z(t)$ be a unit-intensity ($N_o = 1$) white-noise process, and consider the process $X(t)$ that is obtained from a time-variant attenuation of $Z(t)$,
$$X(t) = a(t)Z(t),$$
where $a(t)$ is nonrandom.

(a) Show that $X(t)$ is a nonstationary white-noise process, with autocorrelation given by
$$R_X(t+\tau,t) = a^2(t)\delta(\tau).$$
(b) Assume that $a(t)$ is a finite-power waveform, and determine the time-averaged probabilistic autocorrelation and spectral density of $X(t)$.

35. (a) Determine the one-sided bandwidths B_1, B_2, B_3, of the two processes $X(t)$ and $Y(t) = X(t)\cos(2\pi f_o t)$, for which $S_X(f)$ is a triangle centered at $f = 0$ with base $2B < f_o$.
(b) The bandwidth of a function is associated with a limitation on how rapidly the function can fluctuate. As a reflection of this, show that the normalized mean squared incremental fluctuation from time t to time $t + \tau$ is bounded by the product of the time lapse τ and the two-sided absolute bandwidth B_3:

$$\frac{E\{[X(t+\tau) - X(t)]^2\}}{E\{X^2(t)\}} \le (\pi B_3 \tau)^2. \qquad (10.148)$$

Hint: Verify that

$$E\{[X(t+\tau) - X(t)]^2\} = 4\int_{-B_3/2}^{B_3/2} S_X(f)[\sin(\pi f \tau)]^2 \, df, \qquad (10.149)$$

and then use the inequality $|\sin\phi| \le |\phi|$.

36. Show that neither continuous-time white noise nor discrete-time white noise can have a nonzero mean value.

Hint: Refer to Section 10.9.

37. Let $X(t)$ be a zero-mean Gaussian process, and let $Z(t)$ be the lag product
$$Z(t) \triangleq X(t)X(t-\tau).$$
(a) Show that the spectrum of $Z(t)$ is given by
$$S_Z(f) = \int_{-\infty}^{\infty} S_X\left(v - \frac{f}{2}\right) S_X\left(v + \frac{f}{2}\right)$$
$$\times [1 + \cos(4\pi\tau v)] \, dv + R_X^2(\tau)\delta(f). \quad (10.150)$$
(b) Show that for $Z(t) = X^2(t)$,
$$S_Z(f) = 2S_X(f) \otimes S_X(f) + \delta(f)\left[\int_{-\infty}^{\infty} S_X(v) \, dv\right]^2, \quad (10.151)$$

and explain how the bandwidths of $Z(t)$ and $X(t)$ are related.

38. (a) Verify that if (10.137) is violated, then $X(t)$ is not WSS.
   (b) Reexpress (10.137b) for $m = n$ in terms of the magnitude and phase of A_n, and show that if the phase is independent of the magnitude and is uniformly distributed on $[-\pi, \pi)$, then (10.137b) for $m = n$ is satisfied.
   (c) Verify (10.136).

11

Special Topics and Applications

HAVING COMPLETED THE STUDY of autocorrelation and spectral density in the preceding chapters of this part, we are now in a position to apply the theory to some practical problems of fundamental importance in the fields of signals and systems. Thus, in this chapter, various applications are considered, and some extensions of the basic theory, in the form of special topics that are particularly relevant to signal processing, are introduced. Additional applications, extensions, and generalizations of the basic theory are treated in Chapters 12 and 13.

11.1 Sampling and Pulse Modulation

11.1.1 Pulse-Amplitude Modulation

The signals in sample-data systems are typically trains of periodically repeated pulses with amplitudes that are modulated (scaled) by the sequence of data samples, say $\{Z(i)\}$, to yield a *pulse-amplitude-modulated* (PAM) signal of the form

$$X(t) = \sum_{i=-\infty}^{\infty} Z(i) p(t - iT - \Theta). \qquad (11.1)$$

A typical statistical sample of a PAM signal is shown in Figure 11.1. For a fixed value of the phase variable Θ, $X(t)$ is in general cyclostationary if

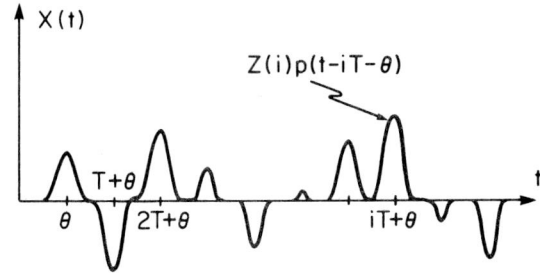

Figure 11.1 Physical Manifestation of a Discrete-Time Process

$Z(i)$ is stationary. However, if Θ is random and uniformly distributed over one pulse period $[-T/2, T/2)$, and independent of $\{Z(i)\}$, then $X(t)$ is stationary. The autocorrelation function for this PAM signal is determined as follows. Straightforward evaluation of the expected value of the lag product yields

$$R_X(t+\tau, t) = \sum_{i=-\infty}^{\infty} \sum_{k=-\infty}^{\infty} R_Z(i-k) \frac{1}{T}$$
$$\times \int_{-T/2}^{T/2} p(t+\tau-iT-\theta) p(t-kT-\theta) \, d\theta. \quad (11.2)$$

Introduction of the changes of variables $i - k = n$ and $t - \theta - kT = \phi$ yields

$$R_X(t+\tau, t) = \frac{1}{T} \sum_{n=-\infty}^{\infty} R_Z(n) \sum_{k=-\infty}^{\infty} \int_{t-kT-T/2}^{t-kT+T/2} p(\tau - nT + \phi) p(\phi) \, d\phi \quad (11.3)$$

$$= \frac{1}{T} \sum_{n=-\infty}^{\infty} R_Z(n) r_p(\tau - nT) = R_X(\tau), \quad (11.4)$$

where r_p is the finite autocorrelation of the pulse,

$$r_p(\tau) \triangleq \int_{-\infty}^{\infty} p(\tau + \phi) p(\phi) \, d\phi. \quad (11.5)$$

The PSD for this PAM signal is

$$S_X(f) = \frac{1}{T} \sum_{n=-\infty}^{\infty} R_Z(n) \int_{-\infty}^{\infty} r_p(\tau + nT) e^{-i2\pi f\tau} \, d\tau \quad (11.6)$$

$$= \frac{1}{T} |P(f)|^2 \sum_{n=-\infty}^{\infty} R_Z(n) e^{i2\pi nTf} \quad (11.7)$$

$$= \frac{1}{T} |P(f)|^2 S_Z(Tf), \quad (11.8)$$

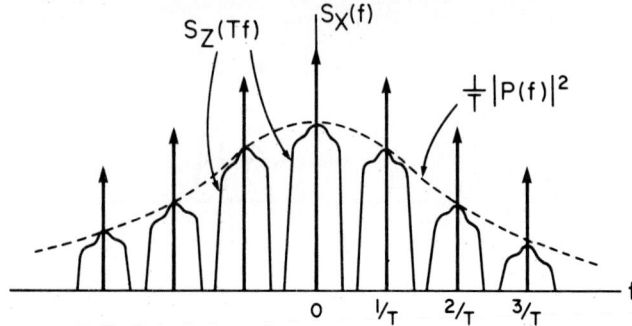

Figure 11.2 Power Spectral Density (11.8) for the Physical Manifestation of a Discrete-Time Process

where S_Z is the spectral density (10.34) for the discrete-time process $Z(i)$. A typical graph of $S_X(f)$ in (11.8) is shown in Figure 11.2.

For applications in which the sample data $\{Z(i)\}$ are time samples of some waveform, say $Y(t)$,

$$Z(i) = Y(iT), \qquad (11.9)$$

the formula (11.8) for the PSD of the PAM signal can be reexpressed in terms of the PSD of the process $Y(t)$. Specifically, by using the sampling property of the impulse δ, the function S_Z can be reexpressed [using $R_Z(n) \equiv R_Y(nT)$] as

$$S_Z(Tf) = \sum_{n=-\infty}^{\infty} R_Z(n) e^{-i2\pi n Tf}$$

$$= \int_{-\infty}^{\infty} R_Y(\tau) \sum_{n=-\infty}^{\infty} \delta(\tau - nT) e^{-i2\pi f\tau} d\tau. \qquad (11.10)$$

Application of Parseval's relation to (11.10), together with the Fourier-transform identity (exercise 1)

$$\int_{-\infty}^{\infty} \sum_{n=-\infty}^{\infty} \delta(\tau - nT) e^{-i2\pi f\tau} d\tau \equiv \frac{1}{T} \sum_{m=-\infty}^{\infty} \delta\left(f - \frac{m}{T}\right), \qquad (11.11)$$

yields

$$S_Z(Tf) = \frac{1}{T} \sum_{m=-\infty}^{\infty} S_Y\left(f - \frac{m}{T}\right), \qquad (11.12)$$

from which (11.8) yields

$$S_X(f) = \left|\frac{P(f)}{T}\right|^2 \sum_{m=-\infty}^{\infty} S_Y\left(f - \frac{m}{T}\right). \qquad (11.13)$$

For a narrow pulse $p(t)$ [i.e., the width of $p(t)$ is much smaller than T], the pulse transform satisfies

$$P(f) \simeq P(0) = \int_{-\infty}^{\infty} p(t)\, dt, \qquad |f| \le \frac{1}{2T}. \tag{11.14}$$

It follows from (11.13) and (11.14) that, if the pulse has area A,

$$S_X(f) \simeq \left[\frac{A}{T}\right]^2 \sum_{m=-\infty}^{\infty} S_Y\!\left(f - \frac{m}{T}\right), \qquad |f| \le \frac{1}{2T}. \tag{11.15}$$

Furthermore, if $Y(t)$ is bandlimited to $|f| \le \frac{1}{2T}$,

$$S_Y(f) = 0, \qquad |f| > \frac{1}{2T}, \tag{11.16}$$

then (11.15) yields

$$S_X(f) \simeq (A/T)^2 S_Y(f), \qquad |f| \le \frac{1}{2T}. \tag{11.17}$$

That is, $S_X(f)$ is simply a scaled version of $S_Y(f)$ within the bandwidth of $S_Y(f)$.

It should be noted that (11.13) reveals that when $Y(t)$ contains a nonzero d-c component, so that $S_Y(f)$ contains a spectral line at $f = 0$, then $S_X(f)$ contains spectral lines at all integer multiples of the pulse repetition frequency, $\{m/T\}$ [assuming $P(m/T) \ne 0$]. This is depicted in Figure 11.2.

11.1.2 Bandlimited Functions and Sampling Theorems

The *sampling theorem* for finite-energy functions establishes that a bandlimited finite-energy function can be perfectly recovered from time samples separated by no more than the reciprocal of twice the bandwidth, $T \le 1/2B$. Furthermore, the recovery is accomplished by interpolation with bandlimited pulses, according to the formula

$$X(t) = \sum_{i=-\infty}^{\infty} X(iT)\, p(t - iT), \tag{11.18}$$

where

$$p(t) = \frac{\sin(\pi t/T)}{\pi t/T}, \tag{11.19}$$

provided that

$$\tilde{X}(f) = 0, \qquad |f| \ge \frac{1}{2T}, \tag{11.20}$$

where $\tilde{X}(f)$ is the Fourier transform of $X(t)$, (10.2). To verify the theorem (11.18)–(11.20), it is simply shown (exercise 2) that the Fourier transform of the series

$$Y(t) \triangleq \sum_{i=-\infty}^{\infty} X(iT) p(t - iT) \qquad (11.21)$$

is given by

$$\tilde{Y}(f) = \frac{1}{T} P(f) \sum_{k=-\infty}^{\infty} \tilde{X}\left(f - \frac{k}{T}\right), \qquad (11.22)$$

where $P(f)$ is the Fourier transform of $p(t)$,

$$P(f) = \begin{cases} T, & |f| \le 1/2T, \\ 0, & |f| > 1/2T. \end{cases} \qquad (11.23)$$

Now, if (11.20) is satisfied, then the translates $\{\tilde{X}(f - k/T) : -\infty < k < \infty\}$ do not overlap each other, and the product (11.22) of the sum of these nonoverlapping translates and the rectangular function (11.23) produces the $k = 0$ term alone:

$$\tilde{Y}(f) = \frac{1}{T} P(f) \tilde{X}(f) = \tilde{X}(f). \qquad (11.24)$$

Therefore, $Y(t) = X(t)$.

On the other hand, if the bandwidth exceeds half the reciprocal of the sampling increment, $B > 1/2T$, then the translates in (11.22) overlap, and therefore the error

$$\tilde{Y}(f) - \tilde{X}(f) = \left[\frac{1}{T} P(f) - 1\right] \tilde{X}(f) + \frac{1}{T} P(f) \sum_{k \ne 0} \tilde{X}\left(f - \frac{k}{T}\right) \qquad (11.25)$$

is not zero. The tails of the translates that overlap into the band passed by the rectangle $P(f)$ cause what is called *aliasing error*. The student should sketch a graph of the right member of (11.22) for the two cases, $B < 1/2T$ and $B > 1/2T$, in order to picture the aliasing phenomenon.

The minimum sampling rate, $1/T = 2B$, for which there is no aliasing error is called the *Nyquist sampling rate*, in honor of Harry Nyquist for his pioneering work [Nyquist, 1928b].

The same result as (11.18)-(11.20) holds true for finite-power functions as well as finite-energy functions. Furthermore, a weaker version of this theorem holds true for finite-power functions that are bandlimited in only the mean-square sense. Before proceeding, the property of bandlimitedness of finite-power functions should be clarified. For a finite-energy function, bandlimitedness simply means that its Fourier transform vanishes outside a finite band of frequencies, as in (11.20). For a finite-power function, bandlimitedness means the same except that the more abstract generalized Fourier transform (Section 10.1.2) must be used. A more concrete way to define (and interpret) a bandlimited function, regardless of whether it has finite energy or finite power, is to observe that a function is bandlimited to the band $(-B, B)$ if and only if the function is invariant when passed through an ideal bandpass filter with transfer function

$$H(f) = \begin{cases} 1, & |f| < B, \\ 0, & |f| \geq B. \end{cases} \qquad (11.26)$$

It is easily shown, using (10.19), that the PSD of a bandlimited WSS process (and the time-averaged probabilistic PSD of a regular nonstationary bandlimited process) is similarly bandlimited:

$$S_X(f) = 0, \qquad |f| \geq B. \qquad (11.27)$$

However, there are also nonbandlimited processes whose PSDs are bandlimited. The reason for this is that the PSD is simply a mean-square measure of the generalized Fourier transform, and a generalized Fourier transform can be bandlimited in mean square without having every sample path bandlimited. This is easily seen for ergodic processes when the ensemble is envisioned as time translations of a single waveform,

$$X(t, \sigma) = X(t - \sigma). \qquad (11.28)$$

For example, if

$$X(t) = Y(t) + z(t), \qquad (11.29)$$

where $z(t)$ is a finite-energy nonbandlimited function, and $Y(t)$ is a finite-power bandlimited function, then every sample $X(t, \sigma)$ is nonbandlimited, but the PSD is bandlimited because it is independent of $z(t)$. This can be seen from the limit

$$\hat{R}_z(\tau) = \lim_{W \to \infty} \frac{1}{W} \int_{-W/2}^{W/2} z\left(t + \frac{\tau}{2}\right) z\left(t - \frac{\tau}{2}\right) dt$$

$$= \lim_{W \to \infty} \frac{1}{W} \int_{-\infty}^{\infty} z\left(t + \frac{\tau}{2}\right) z\left(t - \frac{\tau}{2}\right) dt \equiv 0, \qquad (11.30)$$

which is zero because the integral is upper-bounded by the pulse energy, which is finite; thus

$$\hat{R}_X(\tau) \equiv \hat{R}_Y(\tau), \qquad (11.31)$$

and therefore

$$\hat{S}_X(f) \equiv \hat{S}_Y(f). \tag{11.32}$$

The weaker version of the sampling theorem, which holds for WSS processes that are bandlimited in mean square (i.e., processes whose PSD is bandlimited), states that the mean squared error between the interpolated time-sampled process and the original process is zero,

$$E\left\{\left[X(t) - \sum_{i=-\infty}^{\infty} X(iT)p(t-iT)\right]^2\right\} = 0, \tag{11.33}$$

where

$$p(t) = \frac{\sin(\pi t/T)}{\pi t/T}, \tag{11.34}$$

provided that

$$S_X(f) = 0, \qquad |f| \geq \frac{1}{2T}. \tag{11.35}$$

Furthermore, this theorem holds true for nonstationary regular processes if the mean squared error in (11.33) is time-averaged and the PSD in (11.35) is replaced with the time-averaged instantaneous probabilistic spectral density (10.76). The theorem (11.33)–(11.35) is proved in exercise 3. In addition, a modified sampling theorem is proved in exercise 3 for nonuniformly sampled WSS process. It can also be shown [Gardner, 1972] that the sampling theorem (11.33)–(11.34) holds true for arbitrary mean-square-bandlimited nonstationary processes, which are processes for which the double Fourier transform of the autocorrelation is bandlimited,

$$K_X(f,\nu) \triangleq \int_{-\infty}^{\infty}\int_{-\infty}^{\infty} R_X(t,u) e^{-i2\pi(ft-u\nu)}\,dt\,du = 0 \quad |f| \geq \frac{1}{2T} \text{ or } |\nu| \geq \frac{1}{2T}. \tag{11.36}$$

A comprehensive review of the many variations on the sampling theorem is given in [Jerri, 1977].

11.2 Bandpass Processes

Many signals encountered in signal-processing systems are of the bandpass type; that is, their spectral density is concentrated about some frequency removed from zero. In fact, a major purpose for many types of signal modulations is to convert a low-pass signal into a bandpass signal in order to minimize attenuation and distortion due to propagation through a medium. Nevertheless, for purposes of analysis and computation, it is often more convenient to work with low-pass representations of bandpass signals. These representations are the subject of this section.

11.2.1 Rice's Representation for Finite-Energy Functions

Let us begin by considering the Fourier transform of a finite-energy function. Since every real finite-energy function $X(t)$ has a Fourier transform $\tilde{X}(f)$ with Hermitian (conjugate) symmetry,

$$\tilde{X}(-f) = \tilde{X}^*(f), \tag{11.37}$$

the function $X(t)$ can be completely recovered from the positive-frequency portion of its transform, denoted by

$$\tfrac{1}{2}\tilde{\Psi}(f) \triangleq \begin{cases} \tilde{X}(f), & f > 0, \\ 0, & f < 0, \end{cases} \tag{11.38}$$

as described pictorially in Figures 11.3a, b. Specifically, the transform of the

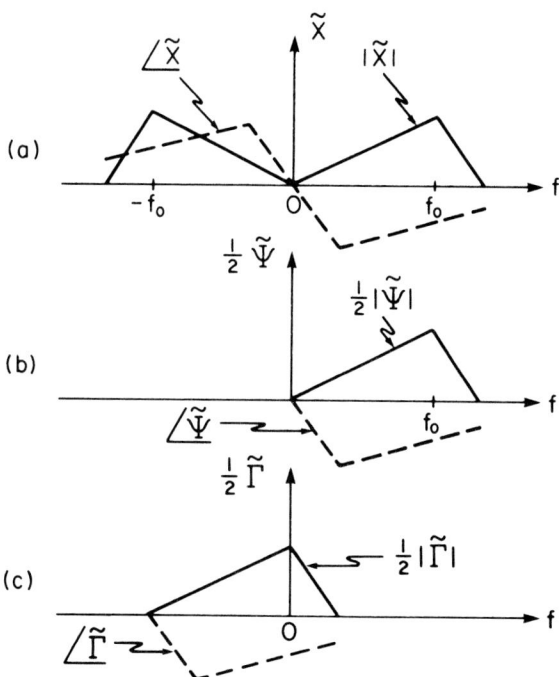

Figure 11.3 Equivalent Signal Representations in the Frequency Domain

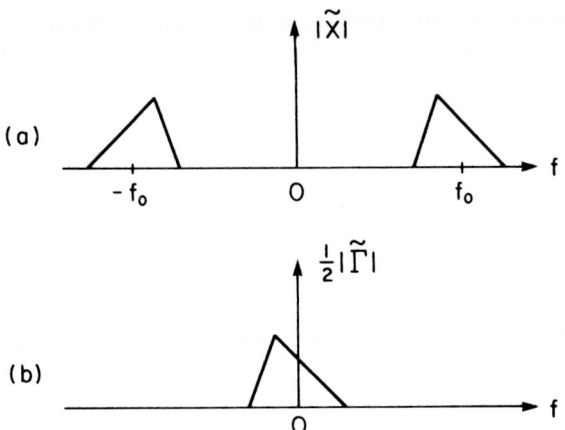

Figure 11.4 Low-Pass Representation of a Bandpass Signal

function $X(t)$ is recovered by

$$\tilde{X}(f) = \tfrac{1}{2}[\tilde{\Psi}(f) + \tilde{\Psi}^*(-f)]. \tag{11.39}$$

Furthermore, $\tilde{X}(f)$ and therefore $X(t)$ can be recovered from the function $\tilde{\Gamma}(f)$ obtained by translating $\tilde{\Psi}(f)$ by any amount (say f_o),

$$\tilde{\Gamma}(f) \triangleq \tilde{\Psi}(f + f_o), \tag{11.40}$$

as depicted in Figure 11.3c. In fact $X(t)$ can be recovered from the inverse transform of $\tilde{\Gamma}(f)$, which is denoted by $\Gamma(t)$. Since $X(t)$ can always be recovered from $\Gamma(t)$, then $\Gamma(t)$ can be interpreted as a representation for $X(t)$. This type of representation is particularly interesting when $X(t)$ is a bandpass function, as depicted in Figure 11.4a, because then $\Gamma(t)$ is a low-pass function if f_o is chosen to be within the passband of $X(t)$, as depicted in Figure 11.4b.

Although $X(t)$ is taken to be a real function, its representation $\Gamma(t)$ is in general complex because its Fourier transform $\tilde{\Gamma}(f)$ does not in general exhibit Hermitian symmetry. But $\Gamma(t)$ can, of course, be represented in terms of two real functions, namely its real and imaginary parts, say $U(t)$ and $V(t)$:

$$\Gamma(t) = U(t) + iV(t). \tag{11.41}$$

By letting $W(t)$ be the imaginary process

$$W(t) = iV(t), \tag{11.42}$$

the transform of $\Gamma(t)$ is represented by the sum of transforms

$$\tilde{\Gamma}(f) = \tilde{U}(f) + \tilde{W}(f), \tag{11.43}$$

and it can be shown that these components exhibit the symmetries

$$\tilde{U}(-f) = \tilde{U}^*(f),$$
$$\tilde{W}(-f) = -\tilde{W}^*(f). \qquad (11.44)$$

Thus, $\tilde{U}(f)$ has even conjugate symmetry, but $\tilde{W}(f)$ has odd conjugate symmetry. Hence $\tilde{\Gamma}(f)$, which has no conjugate symmetry in general, is represented by the sum of its even and odd conjugate-symmetric parts, $\tilde{U}(f)$ and $\tilde{W}(f)$. This is illustrated in Figure 11.5 for the special case in which $\tilde{\Gamma}$, $\tilde{U}$, and $\tilde{W}$ are real.

Now, let us characterize the operations just described for obtaining the real low-pass representors $U(t)$ and $V(t)$ from a real bandpass function $X(t)$, in terms of operations defined in the time domain. It follows from (11.38) that

$$\tilde{\Psi}(f) = \tilde{X}(f) + \text{sgn}(f)\,\tilde{X}(f), \qquad (11.45)$$

where

$$\text{sgn}(f) \triangleq \begin{cases} +1, & f > 0, \\ -1, & f < 0. \end{cases} \qquad (11.46)$$

Furthermore, the function

$$H(f) \triangleq -i\,\text{sgn}(f) \qquad (11.47)$$

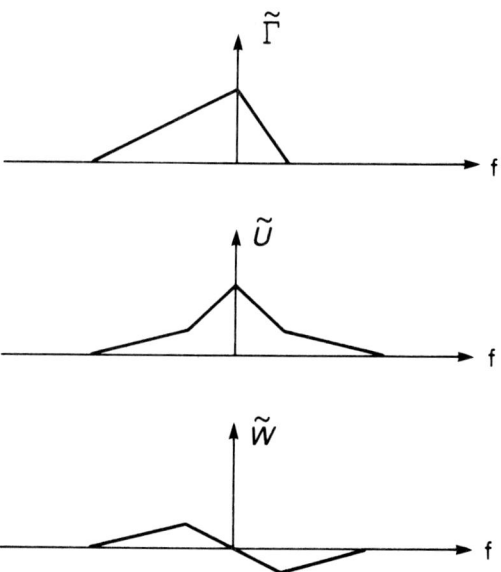

Figure 11.5 Even and Odd Symmetric Components of the Asymmetric Representation for the Case in which $\tilde{\Gamma}$, $\tilde{U}$, and $\tilde{W}$ Are Real

is the transfer function corresponding to the impulse-response function

$$h(t) = 1/\pi t, \qquad (11.48)$$

and the transformation

$$Y(t) = h(t) \otimes X(t) \qquad (11.49)$$

is called the *Hilbert transform*. It follows from the facts

$$|H(f)| \equiv 1,$$

$$\arg\{H(f)\} = \begin{cases} -\pi/2, & f > 0, \\ +\pi/2, & f < 0, \end{cases} \qquad (11.50)$$

that this transformation simply shifts the phase of every positive (negative) frequency component by $-(+)\pi/2$ radians $[-(+)90$ degrees]. It follows from (11.45) to (11.47) that

$$\tilde{\Psi}(f) = \tilde{X}(f) + i\tilde{Y}(f) \qquad (11.51)$$

and therefore

$$\Psi(t) = X(t) + iY(t), \qquad (11.52)$$

which is called the *analytic signal*. Furthermore, it follows from (11.40) that (exercise 5)

$$\Gamma(t) = \Psi(t)e^{-i2\pi f_o t}, \qquad (11.53)$$

which is called the *complex envelope*. Equations (11.52) and (11.53) yield the result

$$\Gamma(t) = [X(t) + iY(t)]e^{-i2\pi f_o t}, \qquad (11.54)$$

where $Y(t)$ is the Hilbert transform of $X(t)$. Now, to obtain the real and imaginary parts of $\Gamma(t)$, Euler's formula

$$e^{-i2\pi f_o t} \equiv \cos(2\pi f_o t) - i\sin(2\pi f_o t) \qquad (11.55)$$

is used in (11.54) to obtain

$$\Gamma(t) = [X(t)\cos(2\pi f_o t) + Y(t)\sin(2\pi f_o t)]$$
$$+ i[Y(t)\cos(2\pi f_o t) - X(t)\sin(2\pi f_o t)]. \qquad (11.56)$$

Thus,

$$U(t) = X(t)\cos(2\pi f_o t) + Y(t)\sin(2\pi f_o t),$$
$$V(t) = Y(t)\cos(2\pi f_o t) - X(t)\sin(2\pi f_o t). \qquad (11.57)$$

In order to recover $X(t)$ from $\Gamma(t)$, (11.54) can be used to obtain

$$X(t) = \text{Re}\{\Gamma(t)e^{i2\pi f_o t}\}, \tag{11.58}$$

where $\text{Re}\{\cdot\}$ denotes the real part. Substitution of (11.41) into (11.58) yields (exercise 5)

$$X(t) = U(t)\cos(2\pi f_o t) - V(t)\sin(2\pi f_o t). \tag{11.59}$$

Equations (11.57) and (11.59) reveal the explicit representation for a real bandpass function $X(t)$ in terms of two real low-pass functions $U(t)$ and $V(t)$, which are called the *in-phase* and *quadrature components* of the *complex representation* $\Gamma(t)$. This is called *Rice's representation* in honor of Stephen O. Rice's pioneering work in signal-representation theory [Rice, 1944, 1945, 1948]. It follows from (11.38) and (11.40) that if $X(t)$ is band-limited to the band $[f_o - B, f_o + B]$ (and its negative-frequency image), then $\Gamma(t)$ and therefore $U(t)$ and $V(t)$ are bandlimited to the band $[-B, B]$. Furthermore, given *any* finite-energy function in the form (11.59) for which $U(t)$ and $V(t)$ are bandlimited to $[-B, B]$ for $B < f_o$, it can be shown that $U(t)$ and $V(t)$ are *uniquely* determined by (11.57) (exercise 5).

11.2.2 Rice's Representation for Stationary Processes

Although we have derived Rice's representation for finite-energy functions, it is valid for finite-power functions as well. It can be shown (exercise 6) that if $X(t)$ is a zero-mean WSS process, then $U(t)$ and $V(t)$ in (11.57) are jointly WSS zero-mean processes. This raises the question of how the spectral densities for $U(t)$, $V(t)$, and $X(t)$ are related. It follows from (11.49) and (11.50) [using (10.19) and (10.29)] that for the Hilbert transformation,

$$S_Y(f) = S_X(f), \tag{11.60}$$

$$S_{YX}(f) = \begin{cases} -iS_X(f), & f > 0, \\ +iS_X(f), & f < 0. \end{cases} \tag{11.61}$$

Also, for any zero-mean WSS process of the form

$$Z(t) = C(t)\cos(2\pi f_o t) - S(t)\sin(2\pi f_o t), \tag{11.62}$$

for which $C(t)$ and $S(t)$ are jointly WSS, it can be shown (exercise 6) that

$$R_Z(\tau) = \tfrac{1}{2}[R_C(\tau) + R_S(\tau)]\cos(2\pi f_o \tau)$$
$$- \tfrac{1}{2}[R_{SC}(\tau) - R_{CS}(\tau)]\sin(2\pi f_o \tau), \tag{11.63}$$

and therefore

$$S_Z(f) = \frac{1}{4}[S_C(f+f_o) + S_S(f+f_o)] - \frac{i}{4}[S_{SC}(f+f_o) - S_{CS}(f+f_o)]$$
$$+ \frac{1}{4}[S_C(f-f_o) + S_S(f-f_o)] + \frac{i}{4}[S_{SC}(f-f_o) - S_{CS}(f-f_o)].$$
(11.64)

Use of $C(t) = U(t)$ and $S(t) = V(t)$ in (11.64) yields the spectrum of $X(t)$ in terms of the spectra and cross-spectrum of $U(t)$ and $V(t)$. Also, use of $C(t) = X(t)$ and $S(t) = -Y(t)$ in (11.64) [and similarly $C(t) = Y(t)$ and $S(t) = X(t)$ in (11.64)], together with (11.60) and (11.61), yields (exercise 6)

$$S_U(f) = S_V(f) = \begin{cases} S_X(f-f_o) + S_X(f+f_o), & |f| < f_o, \\ 0, & |f| > f_o, \end{cases} \quad (11.65)$$

provided that

$$S_X(f) = 0, \quad |f| > 2f_o. \quad (11.66)$$

It can also be shown by similar means (exercise 7) that

$$S_{UV}(f) = -S_{VU}(f) = \begin{cases} i[S_X(f+f_o) - S_X(f-f_o)], & |f| < f_o, \\ 0, & |f| > f_o, \end{cases}$$
(11.67)

provided that (11.66) holds.

It follows from (11.65) that if $X(t)$ is bandlimited in the mean-square sense* to $[f_o - B, f_o + B]$, that is,

$$S_X(f) = 0, \quad ||f| - f_o| > B, \quad (11.68)$$

then $U(t)$ and $V(t)$ are bandlimited in the mean-square sense to $[-B, B]$, that is,

$$S_U(f) = S_V(f) = 0, \quad |f| > B. \quad (11.69)$$

In summary, the spectra of the in-phase and quadrature components $U(t)$ and $V(t)$ are identical, and the cross-spectrum vanishes if and only if the spectrum $S_X(f)$ has even symmetry about the point $f = f_o$ (for $f > 0$). It should be emphasized that these properties result from the stationarity of $X(t)$. For example, for cyclostationary processes the properties of Rice's representation must be generalized (see Section 12.4.2).

*See Section 11.1.2.

Gaussian Processes. Let $X(t)$ be a zero-mean, stationary, Gaussian process. Since $U(t)$ and $V(t)$ are obtained from linear transformations of $X(t)$, they are jointly Gaussian. Furthermore it follows from (11.67) that $U(t)$ and $V(t)$ are uncorrelated if and only if $S_X(f)$ exhibits even symmetry about the point $f = f_o$ (for $f > 0$). In this case, and only in this case, $U(t)$ and $V(t)$ are statistically independent. The representation (11.59), which can be interpreted as being in rectangular-coordinate form, can be converted to polar-coordinate form,

$$X(t) = A(t)\cos[2\pi f_o t + \Phi(t)], \qquad (11.70)$$

by use of trigonomentric identities (exercise 8). The envelope and phase processes, $A(t)$ and $\Phi(t)$, are related to the in-phase and quadrature processes, $U(t)$ and $V(t)$, by

$$A(t) = [U^2(t) + V^2(t)]^{1/2} \qquad (11.71)$$

$$\Phi(t) = \tan^{-1}\left\{\frac{V(t)}{U(t)}\right\}, \qquad (11.72)$$

and are therefore of a low-pass nature if $U(t)$ and $V(t)$ are low-pass processes. It follows from (11.71) and (11.72) and the results of Chapter 1, exercise 15 that for each time instant t, $A(t)$ and $\Phi(t)$ are independent [assuming $S_X(f)$ has even symmetry about $f = f_o$], $A(t)$ is Rayleigh-distributed, and $\Phi(t)$ is uniformly distributed on $[-\pi, \pi)$. However, the random processes $A(t)$ and $\Phi(t)$ can be shown to be dependent [Davenport and Root, 1958, Section 8-5].

11.3 Frequency Modulation and Demodulation

11.3.1 Spectrum of FM

A particularly important type of modulation, because of its noise-immunity properties (which are briefly discussed in Section 11.3.2) is *frequency modulation* (FM). A sinusoid with time-varying phase $\Phi(t)$,

$$X(t) = a\cos[\omega_o t + \Phi(t)], \qquad (11.73)$$

has *instantaneous frequency* defined by

$$\frac{1}{2\pi}\frac{d}{dt}[\omega_o t + \Phi(t)] = f_o + \frac{1}{2\pi}\Psi(t), \qquad (11.74)$$

where

$$\Psi(t) \triangleq \frac{d\Phi(t)}{dt}, \qquad (11.75a)$$

$$f_o \triangleq \frac{\omega_o}{2\pi}. \qquad (11.75b)$$

Thus, a phase-modulated (PM) waveform is also an FM waveform if the modulating waveform $\Psi(t)$ for FM is taken to be the derivative of the modulating waveform $\Phi(t)$ for PM. By use of trigonometric identities, it can be shown (exercise 8) that the autocorrelation for the FM signal (11.73) is given by

$$R_X\!\left(t + \frac{\tau}{2}, t - \frac{\tau}{2}\right) = \frac{a^2}{2}[a(t,\tau)\cos\omega_o\tau - b(t,\tau)\sin\omega_o\tau$$
$$+ c(t,\tau)\cos 2\omega_o t - d(t,\tau)\sin 2\omega_o t], \qquad (11.76)$$

where

$$a(t,\tau) \triangleq E\{\cos[\Phi(t+\tau/2) - \Phi(t-\tau/2)]\},$$
$$b(t,\tau) \triangleq E\{\sin[\Phi(t+\tau/2) - \Phi(t-\tau/2)]\},$$
$$c(t,\tau) \triangleq E\{\cos[\Phi(t+\tau/2) + \Phi(t-\tau/2)]\},$$
$$d(t,\tau) \triangleq E\{\sin[\Phi(t+\tau/2) + \Phi(t-\tau/2)]\}. \qquad (11.77)$$

If $\Phi(t)$ is stationary (in the strict sense), then the functions in (11.77) are independent of time t:

$$a(t,\tau) = a(\tau), \qquad b(t,\tau) = b(\tau),$$
$$c(t,\tau) = c(\tau), \qquad d(t,\tau) = d(\tau), \qquad (11.78)$$

and therefore $R_X(t + \tau/2, t - \tau/2)$ is in general periodic in t with period $T_o = \pi/\omega_o$. In fact, $X(t)$ is in general cyclostationary. The time-averaged value of the probabilistic autocorrelation is (exercise 8)

$$\langle R_X\rangle(\tau) = \frac{a^2}{2}[a(\tau)\cos\omega_o\tau - b(\tau)\sin\omega_o\tau]. \qquad (11.79)$$

If $\Phi(t)$ is a zero-mean Gaussian process, then $\Phi(t + \tau/2) - \Phi(t - \tau/2)$ is a zero-mean Gaussian random variable for each t and τ, and because its probability density is an even function, then $b(\tau) \equiv 0$, and

$$a(\tau) = E\{\exp(i[\Phi(t+\tau/2) - \Phi(t-\tau/2)])\}. \qquad (11.80)$$

It follows from the general formula (2.38) for the characteristic function of jointly Gaussian random variables that $a(\tau)$ is given by (exercise 9)

$$a(\tau) = \exp\{-[R_\Phi(0) - R_\Phi(\tau)]\}, \quad (11.81)$$

and therefore

$$\langle R_X \rangle(\tau) = \frac{a^2}{2} \exp\{-[R_\Phi(0) - R_\Phi(\tau)]\} \cos \omega_o \tau, \quad (11.82)$$

where the quantity

$$2[R_\Phi(0) - R_\Phi(\tau)] = E\{[\Phi(t+\tau) - \Phi(t)]^2\} \quad (11.83)$$

is the *mean squared incremental fluctuation* of the phase process $\Phi(t)$. It follows from (11.82) that the time-averaged probabilistic PSD (or, equivalently, the expected empirical PSD) of the FM signal is given by

$$\langle S_X \rangle(f) = \frac{a^2}{4}[S(f - f_o) + S(f + f_o)], \quad (11.84a)$$

where

$$S(f) \triangleq \int_{-\infty}^{\infty} \exp\{-[R_\Phi(0) - R_\Phi(\tau)]\} e^{-i2\pi f \tau} d\tau. \quad (11.84b)$$

In order to evaluate this Fourier integral, two special cases are considered, corresponding to $R_\Phi(0) \ll 1$ and $R_\Phi(0) \gg 1$. To relate these conditions to spectral parameters, the *modulation index* β, defined by

$$\beta \triangleq \frac{\Delta f}{B_\Psi}, \quad (11.85)$$

is introduced. Here Δf is the *rms frequency deviation*,

$$\Delta f \triangleq \frac{1}{2\pi} \sqrt{R_\Psi(0)} = \frac{1}{2\pi} \left[\int_{-\infty}^{\infty} S_\Psi(f) df\right]^{1/2}, \quad (11.86)$$

and B_Ψ is the *rms bandwidth of the instantaneous frequency*,

$$B_\Psi \triangleq \frac{\left[\int_{-\infty}^{\infty} f^2 S_\Psi(f) df\right]^{1/2}}{\left[\int_{-\infty}^{\infty} S_\Psi(f) df\right]^{1/2}}. \quad (11.87)$$

Substitution of (11.86) and (11.87) into (11.85) yields

$$\beta = \frac{\frac{1}{2\pi} \int_{-\infty}^{\infty} S_\Psi(f) df}{\left[\int_{-\infty}^{\infty} f^2 S_\Psi(f) df\right]^{1/2}}. \quad (11.88)$$

Now, it follows from (11.75) that

$$S_\Psi(f) = (2\pi f)^2 S_\Phi(f), \tag{11.89}$$

and therefore

$$\beta^2 = R_\Phi(0)(B_\Phi/B_\Psi)^2. \tag{11.90}$$

Furthermore, it follows from the Cauchy-Schwarz inequality (which is proved in Section 13.2) that

$$(B_\Phi/B_\Psi)^2 \leq 1, \tag{11.91}$$

and therefore

$$\beta^2 \leq R_\Phi(0). \tag{11.92}$$

Consequently, we have

Narrowband FM: $\quad \beta \ll 1 \quad \Leftarrow \quad R_\Phi(0) \ll 1$

Wideband FM: $\quad \beta \gg 1 \quad \Rightarrow \quad R_\Phi(0) \gg 1. \tag{11.93}$

As we shall see, in the case of narrowband FM, the FM signal has bandwidth equal to the bandwidth of the modulating signal,

$$B_X \simeq B_\Phi, \tag{11.94}$$

whereas for wideband FM, the FM signal has bandwidth greatly exceeding the bandwidth of the modulating signal,*

$$B_X \simeq \Delta f \gg B_\Phi. \tag{11.95}$$

Narrowband FM. Since $R_\Phi(\tau) \leq R_\Phi(0)$, then if $R_\Phi(0) \ll 1$, we have

$$\exp\{-[R_\Phi(0) - R_\Phi(\tau)]\} \simeq 1 - R_\Phi(0) + R_\Phi(\tau), \tag{11.96}$$

and therefore (11.84b) yields

$$S(f) \simeq [1 - R_\Phi(0)]\delta(f) + S_\Phi(f). \tag{11.97}$$

It follows from (11.84a) and (11.97) that a narrowband FM signal has the same spectrum as an AM signal with carrier (cf. example 2 in Chapter 10). As R_Φ is decreased, more of the power resides in the carrier and less in the sidebands. The total power in carrier and sidebands is $a^2/2$, regardless of $\Phi(t)$ (exercise 10).

---

*In some applications, $2\Delta f$ is a more appropriate value for the one-sided (positive-frequency) bandwidth [see (11.100)].

Wideband FM. Using the definitions of β and Δf, it can be shown (exercise 11) that the Taylor-series expansion,*

$$R_\Phi(\tau) = R_\Phi(0) + \tfrac{1}{2}R_\Phi^{(2)}(0)\tau^2 + \tfrac{1}{24}R_\Phi^{(4)}(0)\tau^4 + \cdots \quad (11.98)$$

can be reexpressed as

$$R_\Phi(0) - R_\Phi(\tau) = \frac{1}{2}(2\pi)^2(\Delta f \tau)^2 - \frac{(2\pi)^4}{24}\frac{(\Delta f \tau)^4}{\beta^2} + \cdots . \quad (11.99)$$

By retaining only the first term in the right member of (11.99), (11.84b) yields

$$S(f) \simeq S_o(f) \triangleq \frac{1}{\Delta f \sqrt{2\pi}} \exp\left\{-\frac{1}{2}\left(\frac{f}{\Delta f}\right)^2\right\}. \quad (11.100)$$

Furthermore, it can be shown [Blachman and McAlpine, 1969] that the error in this approximation is approximated by†

$$\frac{|S(f) - S_o(f)|}{\max\{S_o(f)\}} \simeq \frac{1}{8\beta^2}, \quad (11.101)$$

and is therefore small for $\beta \gg 1$. Thus, for wideband FM, the shape of the spectrum is determined not by the spectrum of the modulating signal, but rather by the probability density of the modulating signal, $S_o(f) = f_{\Psi/2\pi}(f)$, which was assumed at the outset to be Gaussian. It can be shown that this result is valid for non-Gaussian modulating-signal probability densities as well, provided that the modulation index is sufficiently large. This more general result is referred to as *Woodward's theorem* in honor of Philip Mayne Woodward's pioneering work [Woodward, 1952; Blachman and McAlpine, 1969].

Cyclostationarity of FM. It can be shown (exercise 12), by using (11.76) and (11.77), that for a zero-mean, stationary, Gaussian phase-process $\Phi(t)$,

$$R_X\left(t + \frac{\tau}{2}, t - \frac{\tau}{2}\right) = \frac{a^2}{2}\left[\exp\{-[R_\Phi(0) - R_\Phi(\tau)]\}\cos\omega_o\tau\right.$$
$$\left. + \exp\{-[R_\Phi(0) + R_\Phi(\tau)]\}\cos 2\omega_o t\right]. \quad (11.102)$$

---
*The superscript in parentheses denotes the order of the derivative.
†Bounds and better approximations than (11.101) are given in [Blachman and McAlpine, 1969].

For $\tau = 0$, (11.102) yields

$$R_X(t,t) = \frac{a^2}{2}\left[1 + \exp\{-2R_\Phi(0)\}\cos 2\omega_o t\right], \qquad (11.103)$$

which reveals that the sinusoidal time-variant component of the mean squared value, which partially reflects the cyclostationarity of the FM signal, is negligible for wideband FM $[R_\Phi(0) \gg 1]$, but is dominant for narrowband FM. However, even some wideband FM signals can exhibit substantial cyclostationarity (although not in the mean squared value). For example, at values of τ, say τ_*, for which $R_\Phi(\tau_*) \simeq -R_\Phi(0)$, the periodically time-variant component in (11.102) is dominant:

$$R_X\left(t + \frac{\tau_*}{2}, t - \frac{\tau_*}{2}\right) \simeq \frac{a^2}{2}\left[\exp\{-2R_\Phi(0)\}\cos\omega_o\tau + \cos 2\omega_o t\right], \qquad (11.104)$$

for $R_\Phi(0) \gg 1$. It should be noted, however, that the property $R_\Phi(\tau_*) \simeq -R_\Phi(0)$ suggests that $\Phi(t)$ is oscillatory (e.g., not a low-pass process).

11.3.2 Noise Immunity of FM

To illustrate the noise-immunity properties of wideband FM, the signal-to-noise ratios at the input and output of an FM receiver can be calculated and compared. For this purpose, the receiver is modeled as a noise-free frequency demodulator with an effective input noise (see Section 11.5) that has flat PSD throughout the spectral band of the signal,

$$S_N(f) = \begin{cases} N_o, & ||f| - f_o| \le \Delta f/2, \\ 0, & \text{otherwise.} \end{cases} \qquad (11.105)$$

The noisy signal at the modulator input is therefore

$$X(t) = a\cos[2\pi f_o t + \Phi(t)] + N(t). \qquad (11.106)$$

The (expected) input signal power is

$$S_{in} = a^2/2, \qquad (11.107)$$

and the input noise power is

$$N_{in} = 2N_o\Delta f. \qquad (11.108)$$

Therefore, the input signal-to-noise ratio is

$$\text{SNR}_{in} = \frac{a^2}{4N_o\Delta f}. \qquad (11.109)$$

Exact determination of the output noise power is complicated by the fact that an FM demodulator is a nonlinear transformation. However, it has been shown that for sufficiently high SNR_{in}, the output signal and noise powers for wideband FM can be determined separately [Middleton, 1960, Chapter 15]. Thus, to determine the output signal power, the noise is set equal to zero, and the fact that the FM demodulator produces the output

$$\Psi(t) = \frac{d\Phi(t)}{dt} \tag{11.110}$$

is used. Since the rms frequency deviation Δf is defined in terms of the power in $\Psi(t)$, (11.86), the output signal power is determined by Δf:

$$S_{\text{out}} = (2\pi\Delta f)^2. \tag{11.111}$$

To determine the output noise power, the input noise is represented in terms of its in-phase and quadrature components, relative to the carrier frequency of the FM signal:

$$N(t) = U(t)\cos(2\pi f_o t) - V(t)\sin(2\pi f_o t). \tag{11.112}$$

Then the modulating signal is set equal to zero to obtain the input

$$\begin{aligned} X(t) &= a\cos(2\pi f_o t) + N(t) \\ &= [a + U(t)]\cos(2\pi f_o t) - V(t)\sin(2\pi f_o t) \\ &= A(t)\cos[2\pi f_o t + \Theta(t)], \end{aligned} \tag{11.113}$$

where

$$A(t) = \{[a + U(t)]^2 + V^2(t)\}^{1/2}, \tag{11.114}$$

$$\Theta(t) = \tan^{-1}\left\{\frac{V(t)}{a + U(t)}\right\}. \tag{11.115}$$

The amplitude fluctuation $A(t)$ is typically removed at the demodulator input by clipping, so that the output is simply the demodulated frequency

$$Y(t) = \frac{d\Theta(t)}{dt}. \tag{11.116}$$

For sufficiently large SNR_{in}, (11.115) yields the close approximation

$$\Theta(t) \simeq \frac{V(t)}{a}, \tag{11.117}$$

and therefore the output noise is

$$Y(t) \simeq \frac{1}{a}\frac{dV(t)}{dt}. \tag{11.118}$$

The spectrum of this output noise is

$$S_Y(f) = (2\pi f/a)^2 S_V(f), \tag{11.119}$$

and it follows from (11.65) and (11.105) that

$$S_V(f) = \begin{cases} 2N_o, & |f| \le \Delta f/2, \\ 0, & \text{otherwise.} \end{cases} \tag{11.120}$$

This output-noise spectrum is shown in Figure 11.6. Since the FM demodulator incorporates a low-pass filter that passes only frequencies within the signal band $[-B_\Psi, B_\Psi]$, the output noise power is

$$N_{\text{out}} = \int_{-B_\Psi}^{B_\Psi} S_Y(f)\, df, \tag{11.121}$$

which, upon substitution of (11.119) and (11.120), yields

$$N_{\text{out}} \simeq \frac{4}{3}\left(\frac{2\pi}{a}\right)^2 B_\Psi^3 N_o. \tag{11.122}$$

Therefore, the output signal-to-noise ratio is, using the definition (11.85),

$$\text{SNR}_{\text{out}} = \frac{3a^2 \beta^3}{4 N_o \Delta f}. \tag{11.123}$$

The fact that the output noise power decreases as the carrier power $a^2/2$ is increased is called the *noise-quieting effect* of FM. Taking the ratio of output-to-input SNR yields the *SNR gain* of the FM demodulator*

$$\frac{\text{SNR}_{\text{out}}}{\text{SNR}_{\text{in}}} = 3\beta^3, \tag{11.124}$$

which reveals that significant improvement in SNR is obtained for wideband FM, $\beta \gg 1$. However, if β is made too large [by increasing Δf in (11.85)], then SNR_{in} is reduced to the point where the FM demodulator fails to perform according to the approximations made in this analysis. In

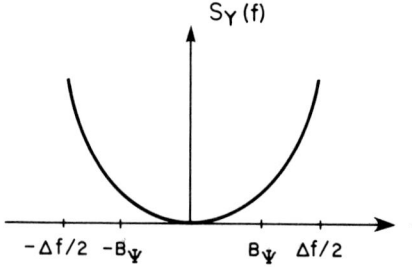

Figure 11.6 Power Spectral Density for Noise at the Output of an FM Demodulator

*If the one-sided bandwidth of $X(t)$ were taken to be $2\Delta f$ rather than Δf, then (11.109) would be half as large and (11.124) would be twice as large.

fact, when SNR_{in} is reduced to between 1 and 10, the FM demodulator is *captured* by the noise and produces an output SNR that is lower than the input SNR. This is called the *threshold effect* of FM.

Pre-emphasis. Additional improvement in the output SNR for FM can be obtained by observing that the output noise is dominated by high frequencies, as revealed by the noise spectrum shown in Figure 11.6. This results from the effective differentiation operation performed by the frequency demodulator. For a signal spectrum $S_\Psi(f)$ that is flat throughout its band, this results in the high frequencies being much more noisy than the low frequencies. This high-frequency-noise effect can be reduced substantially if the high-frequency components of the signal $\Psi(t)$ are emphasized before modulation. Then when they are de-emphasized after demodulation, the high-frequency noise will also be de-emphasized. Specifically, if $\Psi(t)$ is pre-emphasized with a filter having transfer function $H_p(f)$, then the de-emphasis filter should have transfer function,

$$H_d(f) = \frac{1}{H_p(f)}. \tag{11.125}$$

In this case, the output signal power is still given by (11.111), but the output noise power is now given by

$$N_{\text{out},d} = \int_{-B_\Psi}^{B_\Psi} |H_d(f)|^2 S_Y(f)\, df \tag{11.126}$$

rather than by (11.121). Although B_Ψ (after de-emphasis) remains unchanged, Δf is in general affected by the pre-emphasis.

Example 1 (Commercial FM broadcast): If the de-emphasis filter is simply a first-order resistive-capacitive circuit, such as that used for commercial FM broadcasting, then (exercise 14)

$$|H_d(f)|^2 = \frac{1}{1 + (f/B_d)^2}. \tag{11.127}$$

Substitution of (11.119), (11.120), and (11.127) into (11.126) yields (exercise 14) the ratio

$$\frac{N_{\text{out}}}{N_{\text{out},d}} = \frac{(B_\Psi/B_d)^3}{3[B_\Psi/B_d - \tan^{-1}(B_\Psi/B_d)]}. \tag{11.128}$$

Typical values for commercial FM are $B_\Psi = 15$ KHz and $B_d = 2.1$ KHz, in which case

$$\frac{N_{\text{out}}}{N_{\text{out},d}} \approx 20.$$

This is a substantial reduction in output noise power. Note that since Δf is changed by the power loss of the pre-emphasis filter, then N_{in} and therefore N_{out} are changed, and so is S_{out}. Thus (11.128) yields the improvement in SNR_{out} only if the power of Ψ is held constant. ∎

11.4 PSD Measurement Analysis

Consider the problem of measuring the PSD of an ergodic stationary random process $X(t)$ by passing it through a narrowbandpass filter with center frequency $f > 0$ and bandwidth Δf, and then measuring its average power by squaring it and averaging. The averaging can be done using a low-pass filter with bandwidth, say B, on the order of the reciprocal of the desired averaging time Δt: $B = 1/(2\Delta t)$. A block diagram of this measurement system is shown in Figure 11.7. The measurement must be repeated for each value of the frequency f of interest. Now, since the excitation to this measurement system is random, so too is its response. To better understand how to select appropriate values for the two design parameters Δf and Δt, let us determine the mean and variance of the measurement $Z(t)$, which can be interpreted as an estimate, say $\tilde{S}_X(f)$, of the ideal PSD $S_X(f)$:

$$Z(t) \triangleq \tilde{S}_X(f). \qquad (11.129)$$

It follows from the identity

$$E\{[\tilde{S}_X(f) - S_X(f)]^2\} \equiv [E\{\tilde{S}_X(f)\} - S_X(f)]^2 + \text{Var}\{\tilde{S}_X(f)\} \qquad (11.130)$$

that the mean and variance of $Z(t)$ completely determine the mean squared error of the estimate $\tilde{S}_X(f)$.

Now since $Z(t)$ involves a squaring operation on the filtered version $W(t)$ of $X(t)$, the variance of $Z(t)$ depends on fourth joint moments of the process $X(t)$. To simplify the analysis, it is therefore assumed that $X(t)$ is a zero-mean Gaussian stationary process. Then Isserlis's formula can be used to evaluate the variance solely in terms of the PSD of $X(t)$. For this

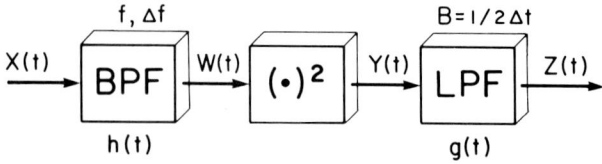

Figure 11.7 PSD Measurement System

purpose, the impulse-response functions of the input bandpass filter and the output low-pass filter are denoted by $h(t)$ and $g(t)$, respectively. It follows from (9.44) that the mean of the measurement is given by

$$m_Z = m_Y G(0), \tag{11.131}$$

where

$$m_Y = R_W(0), \tag{11.132}$$

and it follows from (9.45) that the mean squared value of the filtered excitation $X(t)$ is

$$R_W(0) = \int_{-\infty}^{\infty} R_X(\tau) r_h(\tau) \, d\tau \tag{11.133}$$

$$= \int_{-\infty}^{\infty} S_X(\nu) |H(\nu)|^2 \, d\nu. \tag{11.134}$$

Equations (11.131) to (11.134) yield

$$m_Z = G(0) \int_{-\infty}^{\infty} |H(\nu)|^2 S_X(\nu) \, d\nu. \tag{11.135}$$

For example, if the bandpass filter is ideal, as depicted in Figure 10.1, but with gain $1/\sqrt{2\Delta f}$ and bandwidth Δf, and if $G(0) = 1$, then

$$m_Z = \frac{1}{2\Delta f} \int_{f-\Delta f/2}^{f+\Delta f/2} S_X(\nu) \, d\nu + \frac{1}{2\Delta f} \int_{-f-\Delta f/2}^{-f+\Delta f/2} S_X(\nu) \, d\nu$$

$$= \frac{1}{\Delta f} \int_{f-\Delta f/2}^{f+\Delta f/2} S_X(\nu) \, d\nu. \tag{11.136}$$

Consequently, the bandwidth Δf should be chosen small enough to resolve S_X, in which case

$$m_Z \simeq S_X(f). \tag{11.137}$$

To determine the variance σ_Z^2 of the measurement, the mean squared value is determined, and then the squared mean m_Z^2 is subtracted. It follows from (9.45) that the autocorrelation of the measurement is given by

$$R_Z(\tau) = R_Y(\tau) \otimes r_g(\tau), \tag{11.138}$$

and it follows from Isserlis's formula that

$$R_Y(\tau) = R_W^2(0) + 2R_W^2(\tau), \tag{11.139}$$

where

$$R_W(\tau) = R_X(\tau) \otimes r_h(\tau). \tag{11.140}$$

Equations (11.138) to (11.140), together with (11.131) to (11.133), yield

$$R_Z(\tau) = m_Z^2 + 2[R_X(\tau) \otimes r_h(\tau)]^2 \otimes r_g(\tau). \qquad (11.141)$$

Fourier transformation of (11.141) yields the spectrum of the measurement,

$$S_Z(\nu) = m_Z^2 \delta(\nu) + 2|G(\nu)|^2 \{[S_X(\nu)|H(\nu)|^2] \otimes [S_X(\nu)|H(\nu)|^2]\}. \qquad (11.142)$$

The variance of the measurement is therefore

$$\sigma_Z^2 = R_Z(0) - m_Z^2$$
$$= 2\int_{-\infty}^{\infty} |G(\nu)|^2 \{[S_X(\nu)|H(\nu)|^2] \otimes [S_X(\nu)|H(\nu)|^2]\} d\nu. \qquad (11.143)$$

For example, if the low-pass filter is ideal,

$$G(\nu) = \begin{cases} 1, & |\nu| \leq B = 1/2\Delta t, \\ 0, & \text{otherwise,} \end{cases} \qquad (11.144)$$

and if $S_X(\nu)$ is approximately constant throughout the passband of the ideal bandpass filter, then (11.143) yields (exercise 15)

$$\sigma_Z^2 \simeq \frac{1}{\Delta f} S_X^2(f) \int_{-1/(2\Delta t)}^{1/(2\Delta t)} \left(1 - \frac{|\nu|}{\Delta f}\right) d\nu. \qquad (11.145)$$

In obtaining (11.145) it has been assumed that $|f| > [\Delta f + 1/(2\Delta t)]/2$. Evaluation of the integral in (11.145) yields

$$\sigma_Z^2 \simeq \frac{1}{\Delta t \Delta f} S_X^2(f) \left[1 - \frac{1}{4\Delta t \Delta f}\right]. \qquad (11.146)$$

It follows from the results (11.137) and (11.146) that the variance is small, relative to the squared mean, if and only if the product of the input bandwidth Δf and the output integration time Δt is large. In this case the ratio of the squared mean to the variance (which is a type of signal-to-noise ratio) is given by

$$m_Z^2/\sigma_Z^2 \simeq \Delta t \Delta f, \qquad |f| > \Delta f/2, \quad \Delta t \Delta f \gg 1. \qquad (11.147)$$

Similarly, for $f = 0$, and an ideal input filter defined by

$$H(\nu) = \begin{cases} \frac{1}{\sqrt{\Delta f}}, & |\nu| \leq \Delta f/2, \\ 0, & \text{otherwise,} \end{cases} \qquad (11.148)$$

it can be shown that

$$m_Z \simeq S_X(0) \tag{11.149}$$

and

$$\sigma_Z^2 \simeq \frac{2}{\Delta t \, \Delta f} S_X^2(0) \left[1 - \frac{1}{4 \Delta t \, \Delta f} \right], \tag{11.150}$$

and therefore

$$m_Z^2/\sigma_Z^2 \simeq 2 \Delta t \, \Delta f, \qquad f = 0, \quad \Delta t \, \Delta f \gg 1. \tag{11.151}$$

In summary, to obtain an accurate and reliable measurement of the ideal PSD $S_X(f)$, the bandwidth of the input filter, Δf, must be narrow enough to resolve $S_X(f)$, and the integration time of the output filter, Δt, must be much greater than the reciprocal of the input-filter bandwidth, $\Delta t \gg 1/\Delta f$, in order to obtain small variance.* It can be shown, using (11.143), that with no averaging of the instantaneous power measurement $Y(t)$ (i.e., $\Delta t = 0$), the estimate $Z(t) = \tilde{S}_X(f)$ is extremely unreliable in that its variance is equal to its squared mean (a signal-to-noise ratio of unity). It should be emphasized that Δt can be no longer than the length of the segment of $X(t)$ that is available for processing, and when Δt is relatively small, then the requirement $\Delta t \gg 1/\Delta f$ for high reliability can be met only with a relatively large value for Δf, which limits the resolution; or if Δf is chosen small enough for the desired degree of resolution, then the requirement for high reliability cannot necessarily be met. Thus, there is a fundamental tradeoff between resolution and reliability.

Another approach to measuring the PSD can be based on its interpretation in terms of energy spectral density (Section 10.2.2). Specifically, the squared magnitude of the Fourier transform of a finite segment of the process, located at time t,

$$\tilde{X}_T(t, f) \triangleq \int_{t-T/2}^{t+T/2} X(u) e^{-i 2 \pi f u} \, du, \tag{11.152}$$

normalized by the segment length T,

$$\frac{1}{T} |\tilde{X}_T(t, f)|^2 \triangleq P_T(t, f) \tag{11.153}$$

can be used to obtain an estimate of the PSD $S_X(f)$ by either time averaging or frequency smoothing. The function, $P_T(t, f)$ is called the *time-variant periodogram* of $X(t)$. The two alternative estimates based on

---

*The result (11.147) and (11.151) is valid for a broad class of processes in addition to the Gaussian process.

the periodogram are

$$\tilde{S}'_X(f) \triangleq \frac{1}{\Delta t} \int_{t-\Delta t/2}^{t+\Delta t/2} P_{1/\Delta f}(u, f)\, du, \qquad (11.154)$$

and

$$\tilde{S}''_X(f) \triangleq \frac{1}{\Delta f} \int_{f-\Delta f/2}^{f+\Delta f/2} P_{\Delta t}(t, \nu)\, d\nu, \qquad (11.155)$$

where T is denoted by $T = 1/\Delta f$ for the time-averaged estimate (11.154), and by $T = \Delta t$ for the frequency-smoothed estimate (11.155). In either case, an accurate and reliable estimate is obtained if Δf is small enough to resolve $S_X(f)$, and $\Delta t \gg 1/\Delta f$. In fact formulas for the mean and variance that are quite similar to (11.136) and (11.146) can be derived for each of the estimates (11.154) and (11.155). Furthermore, it can be shown that without time averaging or frequency smoothing, the periodogram (11.153) is an extremely poor estimate with a signal-to-noise ratio of unity. Moreover, as the segment length T is increased, the behavior of the periodogram simply becomes more erratic. For example, it can be shown [e.g., for a Gaussian process, $X(t)$] that the correlation between $P_T(t, f_1)$ and $P_T(t, f_2)$ for two frequencies f_1 and f_2 is negligible for $|f_1 - f_2| \gg 1/T$, and converges to zero as $T \to \infty$ for $|f_1 - f_2| \neq 0$. This is consistent with the formal result (10.8), which indicates the erratic behavior of the symbolic Fourier transform of an unlimited segment of a stationary process.

A heuristic interpretation of the property of reliability (small variance) of a measurement of a PSD can be obtained for ergodic processes by envisioning the ensemble as consisting of time translates of a single waveform,

$$X(t, \sigma) = X(t - \sigma). \qquad (11.156)$$

Then reliability simply means that the PSD measurement is insensitive to the time [t in (11.129), (11.154), and (11.155)] at which it is measured. In fact, the variance of the measurement can be interpreted, using (11.156), as the time-averaged squared deviation of the measurement from its time-averaged (mean) value. For a thorough treatment of PSD measurement based on a deterministic theory of time averages, rather than a probabilistic theory, the reader is referred to [Gardner, 1987a].

11.5 Noise Modeling for Receiving Systems

A signal-receiving system can be thought of as being composed of three major components: a receiving antenna, transmission devices (such as coaxial cables, waveguides, isolators, splitters, etc.) and a receiving amplifier,

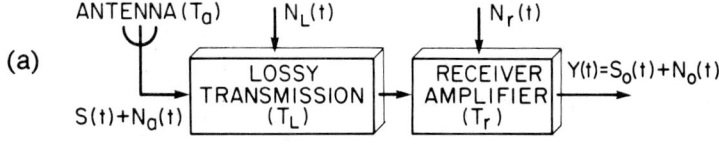

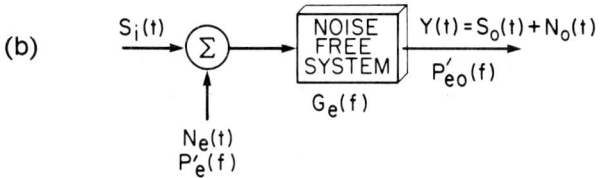

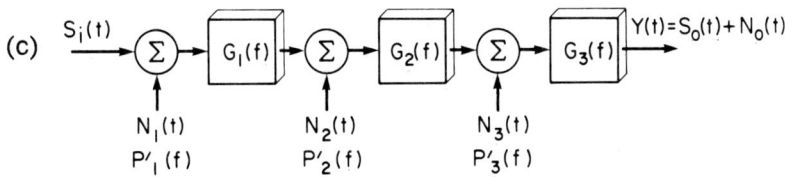

Figure 11.8 (a) Receiving System; (b) Equivalent Model of Receiving System; (c) Intermediate Equivalent Model

as depicted in Figure 11.8a. All three of these components are sources of noise. For example, the antenna output contains thermal noise from the antenna itself; cosmic noise received from outer space (significant below 2 GHz, and often dominant below 300 MHz); radiation noise from the atmosphere (blackbody radiation resulting from energy absorbed by the atmosphere can be significant above 1 GHz); radiation noise from the earth, other planets, and the sun; atmospheric noise from lightning and other electrical phenomena in the atmosphere, and from thermal noise associated with attenuation due to precipitation; and man-made noise such as that from ignition systems and fluorescent lights. The transmission devices are typically passive lossy devices and therefore give rise to primarily thermal noise associated with the loss. The receiving amplifier contributes thermal noise, especially from its initial stages of amplification, and shot noise from its active devices. Thus, a complete description of all noise sources in a receiving system is necessarily quite complicated. Nevertheless, from the point of view of system analysis, for which the receiving system is interpreted as one component in an overall signal-processing system, the three-component receiving system with all its associated noise sources often can be adequately modeled by a single idealized noise-free component, with a single effective noise source at its input, as depicted in Figure 11.8b. The purpose of this section is to describe how to determine the power spectral density of this equivalent noise source.

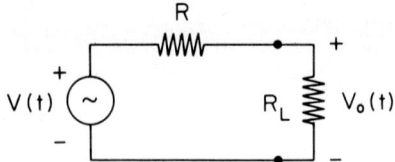

Figure 11.9 Circuit for Determining Available Power

Let us begin by modeling the receiving system as the cascade connection of three subsystems, corresponding to the antenna, transmission devices, and receiving amplifier, described by their available power gains and available noise PSDs, as shown in Figure 11.8c. The equivalent available noise PSD at the input to a component is determined by replacing all noise sources within the component by a single noise source at the input that produces the same available noise PSD at the component output. Moreover, the equivalent input noise source is treated as if it were thermal noise,* in which case the available noise power in an incremental frequency band df is given by

$$dP(f) = \tfrac{1}{2} K T_e \, df \tag{11.157}$$

for some appropriate *effective temperature* T_e, where K is Boltzmann's constant. To see this, consider a thermal-noise source delivering power to a resistive load, as depicted in Figure 11.9. In an incremental band df, the incremental mean squared value of the noise voltage from the source is, from (10.88),

$$S_V(f) \, df = N_o \, df = 2 \, KTR \, df, \tag{11.158}$$

where T is the temperature in degrees Kelvin. It can be shown that the incremental mean squared value of the noise voltage at the load is therefore

$$S_{V_o}(f) \, df = \left[\frac{R_L}{R + R_L}\right]^2 S_V(f) \, df, \tag{11.159}$$

and as a result the incremental noise power dissipated in the load is

$$dP_{V_o}(f) = \frac{S_{V_o}(f) \, df}{R_L} = \frac{R_L}{(R + R_L)^2} S_V(f) \, df, \tag{11.160}$$

and is maximum (exercise 16) when the load resistance matches the source resistance, $R_L = R$, in which case

$$dP_{V_o}(f) = \frac{1}{4R} S_V(f) \, df = \tfrac{1}{2} \, KT \, df. \tag{11.161}$$

---

*This is not necessary for determining the PSD for the equivalent input noise source for the composite system, but it is common practice.

This maximum incremental load power is called the *available incremental power*, and it yields the *available PSD*

$$P'_{V_o}(f) \triangleq \frac{dP_{V_o}(f)}{df} = \tfrac{1}{2}KT. \tag{11.162}$$

Now, given any noise source with any available PSD $P'(f)$, we can treat it as a thermal noise source with an *effective noise temperature*, $T = T_e$, determined by the formula

$$P'(f) = \tfrac{1}{2}KT_e(f). \tag{11.163}$$

The notation indicates that for a frequency-dependent source, this effective temperature is frequency-dependent. However, for an RLC circuit in which all resistors are at the same temperature T, it can be shown that $T_e = T$.

For a cascade connection of three components with available PSD gains* of $G_1(f)$, $G_2(f)$, and $G_3(f)$, and available input PSDs of $P'_1(f)$, $P'_2(f)$, and $P'_3(f)$, as depicted in Figure 11.8c, the available PSD gain of the equivalent composite system (Figure 11.8b) is the product

$$G_e(f) = G_1(f)G_2(f)G_3(f), \tag{11.164}$$

and the available input PSD for the equivalent system is given by

$$P'_e(f) = P'_1(f) + \frac{P'_2(f)}{G_1(f)} + \frac{P'_3(f)}{G_1(f)G_2(f)}. \tag{11.165}$$

The available output PSD is

$$P'_{eo}(f) = P'_e(f)G_e(f). \tag{11.166}$$

It follows from (11.163) and (11.165) that the effective noise temperature for the equivalent system is

$$T_e(f) = T_1(f) + \frac{T_2(f)}{G_1(f)} + \frac{T_3(f)}{G_1(f)G_2(f)}, \tag{11.167}$$

where T_1, T_2, and T_3 are the effective noise temperatures of the three components.[†] In addition, an *effective noise bandwidth* for the equivalent

---

*It follows from (11.162) that the available incremental power gain is identical to the available PSD gain, since the factor df cancels out in the ratio of output and input powers.

[†]A commonly used alternative to the effective noise temperature T_e, as a figure of merit for a noisy system, is the *noise figure* F, given by $F = 1 + T_e/T_s$, where T_s is the noise temperature of the source, typically taken to be 290K in technical specifications.

system can be defined by

$$B_e \triangleq \int_0^\infty \frac{P'_{eo}(f)}{P'_{eo}(f_o)} df, \qquad (11.168)$$

where f_o is the *center frequency* of the system as determined by the function $G_e(f)$. Thus, the total output noise power is given by the product

$$P_o = \int_{-\infty}^\infty P'_{eo}(f) df = 2B_e P'_{eo}(f_o) = 2B_e G_e(f_o) P'_e(f_o). \qquad (11.169)$$

Example 2: It is common practice to lump the losses L_a from the antenna with the losses L_t from the transmission devices (all assumed to be at the same temperature), to obtain a unity-gain antenna ($G_1 \equiv 1$) and a lossy component with gain ($G_2 \equiv 1/L_a L_t$). Now, it can be shown [Mumford and Scheibe, 1968] that the available noise PSD for a purely lossy system at temperature T_L is given by the frequency-independent formula

$$P'_2 = \tfrac{1}{2} K T_L (L - 1), \qquad (11.170\text{a})$$

and the effective noise temperature is therefore also frequency-independent,

$$T_2 = T_L(L - 1) = T_{eL}, \qquad (11.170\text{b})$$

where L is the total loss (e.g., $L = L_a L_t$). The available noise PSD of the antenna is typically constant throughout the passband of the receiver, and the effective antenna noise temperature, $T_1 = T_{ea}$, is therefore also constant,

$$P'_1 = \tfrac{1}{2} K T_1, \qquad (11.171)$$

and similarly for the receiving amplifier,

$$P'_3 = \tfrac{1}{2} K T_3, \qquad (11.172)$$

where $T_3 = T_{er}$ is the effective receiver amplifier noise temperature. Substitution of (11.170) to (11.172) into (11.165) yields the result

$$P'_e(f) = \tfrac{1}{2} K \left[T_{ea} + T_L(L_a L_t - 1) + T_{er} L_a L_t \right], \qquad |f - f_o| \le B_r/2, \qquad (11.173)$$

where B_r is the bandwidth of the receiving amplifier, which is determined by the amplifier gain $G_3(f) = G_r(f)$. It follows from (11.164), (11.166), (11.169), and (11.173) that the total available output noise power from the entire system is

$$P_o = K\left[\frac{T_{ea}}{L_a L_t} + T_L\left(1 - \frac{1}{L_a L_t}\right) + T_{er}\right] G_r(f_o) B_r. \quad (11.174)$$

∎

11.6 Matched Filtering and Signal Detection

Let us consider the problem of detecting the possible presence of a finite-energy signal $s(t)$ of known form buried in additive random noise $N(t)$, based on the noisy measurements,

$$X(t) = \begin{cases} s(t) + N(t), & \text{signal present,} \\ N(t), & \text{signal absent.} \end{cases} \quad (11.175)$$

To try to detect whether or not $s(t)$ is present, the measurements can be filtered to enhance the signal, if present, and attenuate the noise. A time sample of the filtered measurements, taken at an appropriate time, say t_o, determined by the filter characteristics can then be compared with a threshold, say γ. If the measurement sample exceeds the threshold, it is decided that the signal is present:

$$Y(t_o) \begin{cases} > \gamma \Rightarrow & \text{decide } s(t) \text{ present,} \\ < \gamma \Rightarrow & \text{decide } s(t) \text{ absent,} \end{cases} \quad (11.176)$$

where $Y(t)$ is the filtered measurements,

$$Y(t) = X(t) \otimes h(t). \quad (11.177)$$

The simplest way to proceed is to choose the sampling time t_o arbitrarily, and then determine how the filter must accommodate this. To determine a filter that will accomplish the desired task of enhancing the signal while attenuating the noise, the following measure of the signal-to-noise ratio is adopted:

$$\text{SNR} \triangleq \frac{[E\{Y(t_o)|s(t) \text{ present}\}]^2}{\text{Var}\{Y(t_o)|s(t) \text{ present}\}} \quad (11.178)$$

and then the particular filter that maximizes SNR is sought. Given that the noise has zero mean value and the signal is nonrandom, it follows that

$$E\{Y(t_o)|s(t) \text{ present}\} = \int_{-\infty}^{\infty} h(t) s(t_o - t) \, dt \quad (11.179)$$

and
$$\text{Var}\{Y(t_o)|s(t) \text{ present}\} = E\{[N(t_o) \otimes h(t_o)]^2\}. \qquad (11.180)$$

It follows from (9.45) that
$$\text{Var}\{Y(t_o)|s(t) \text{ present}\} = \int_{-\infty}^{\infty}\int_{-\infty}^{\infty} R_N(\tau)h(t+\tau)h(t)\,dt\,d\tau. \qquad (11.181)$$

Application of the convolution theorem and Parseval's relation to (11.179) and (11.181) yields (exercise 18)
$$E\{Y(t_o)|s(t) \text{ present}\} = \int_{-\infty}^{\infty} H(f)S(f)e^{i2\pi f t_o}\,df \qquad (11.182)$$

and
$$\text{Var}\{Y(t_o)|s(t) \text{ present}\} = \int_{-\infty}^{\infty} |H(f)|^2 S_N(f)\,df, \qquad (11.183)$$

where $S(f)$ is the Fourier transform of the signal $s(t)$. Now to put the SNR into a form for which we can recognize the solution, the functions
$$G(f) \triangleq H(f)\sqrt{S_N(f)} \qquad (11.184a)$$

and
$$K(f) \triangleq \frac{S^*(f)e^{-i2\pi f t_o}}{\sqrt{S_N(f)}} \qquad (11.184b)$$

are introduced. Then we have
$$\text{SNR} = \frac{\left[\int_{-\infty}^{\infty} G(f)K^*(f)\,df\right]^2}{\int_{-\infty}^{\infty} |G(f)|^2\,df}. \qquad (11.185)$$

It follows from the Cauchy-Schwarz inequality (which is proved in Section 13.2) that
$$\left[\int_{-\infty}^{\infty} G(f)K^*(f)\,df\right]^2 \leq \int_{-\infty}^{\infty} |G(f)|^2\,df \int_{-\infty}^{\infty} |K(f)|^2\,df, \qquad (11.186)$$

for which equality holds if and only if
$$G(f) = cK(f) \qquad (11.187)$$

for any constant c. Therefore SNR is maximized by the filter obtained from substitution of the definitions (11.184) into (11.187). The result is

$$H(f) = \frac{cS^*(f)e^{-i2\pi f t_o}}{S_N(f)}. \tag{11.188}$$

In view of the threshold test (11.176)–(11.177), in which (11.188) is to be used, the constant c must be chosen appropriately in terms of the threshold level γ, and in any case $c > 0$. For convenience $c = 1$ is used. It can be seen from (11.188) that the filter emphasizes those frequencies for which the signal exceeds the noise power, and de-emphasizes those for which the noise power dominates the signal. Also, the filter includes a delay factor that accommodates the arbitrary choice of sampling time t_o.

The maximum value of the SNR can be determined by substitution of the solution (11.188) into (11.182) and (11.183). The result is (exercise 18)

$$\text{SNR}_{\max} = \int_{-\infty}^{\infty} \frac{|S(f)|^2}{S_N(f)} df. \tag{11.189}$$

As a specific example, if the noise is white, then (11.189) reduces to (exercise 19)

$$\text{SNR}_{\max} = E/N_o, \tag{11.190}$$

where E is the signal energy. Also, in this case, the optimum filter has impulse-response function (exercise 19)

$$h(t) = \frac{1}{N_o} s(t_o - t). \tag{11.191}$$

Substitution of (11.191) into (11.177), evaluated at $t = t_o$, yields

$$Y(t_o) = \frac{1}{N_o} \int_{-\infty}^{\infty} s(t) X(t) \, dt, \tag{11.192}$$

which is recognized as a correlation of the signal to be detected with the noisy measurements. Observe that (11.190) reveals that the performance is evidently independent of the form of the signal; only the signal energy comes into play. Is this true for nonwhite noise?

The solution to this SNR-maximization problem is called a *matched filter*, because the filter is matched to the signal, as revealed by (11.191). Furthermore, when the noise is not white, (11.188) reveals that the filter is inversely matched to the noise spectrum.

It can be shown [Van Trees, 1968] that the matched filter is also an optimum detector in an entirely different sense from that of maximizing

SNR. Specifically, if the noise is Gaussian, then the matched-filter detector minimizes the probability of detection error when the threshold level γ is properly set.

11.7 Wiener Filtering and Signal Extraction

Let us consider the problem of extracting a random WSS signal $S(t)$ from corrupted measurements $X(t)$. It will be assumed that the corruption consists of linear time-invariant signal distortion and additive, zero-mean WSS noise $N(t)$ that is uncorrelated with the signal:

$$X(t) = S(t) \otimes g(t) + N(t), \tag{11.193}$$

where $g(t)$ models the distorting transformation. To extract the signal $S(t)$, the corrupted measurements can be filtered in an attempt to *equalize* (remove) the distortion and attenuate the noise. To determine a filter that will accomplish this, the following measure of extraction error is adopted:

$$\text{MSE} \triangleq E\{[\hat{S}(t) - S(t)]^2\}. \tag{11.194}$$

Then the filter that minimizes MSE is sought, where $\hat{S}(t)$ is the extracted signal,

$$\hat{S}(t) = X(t) \otimes h(t), \tag{11.195}$$

and $h(t)$ is the impulse-response function of the filter being sought.

In Chapter 13, it is shown that a necessary and sufficient condition on the extracted signal $\hat{S}(t)$, in order that it minimize the MSE, is that it yields an error that is orthogonal to the corrupted measurements,

$$E\{[\hat{S}(t) - S(t)] X(u)\} = 0, \tag{11.196}$$

for all values of t and u. This is called the *orthogonality condition*, and is given a geometrical interpretation in Chapter 13. Substitution of (11.195) into (11.196) yields (exercise 22) the condition

$$R_X(\tau) \otimes h(\tau) = R_{SX}(\tau), \tag{11.197}$$

which can be solved by Fourier transformation to obtain

$$H(f) = \frac{S_{SX}(f)}{S_X(f)}. \tag{11.198}$$

This optimum filter is called the noncausal *Wiener filter* in honor of Norbert Wiener's pioneering work in signal extraction theory [Wiener, 1949]. Use of the model (11.193) yields (exercise 22)

$$S_{SX}(f) = G^*(f)S_S(f), \qquad (11.199)$$

$$S_X(f) = |G(f)|^2 S_S(f) + S_N(f). \qquad (11.200)$$

Substitution of (11.199) and (11.200) into (11.198) yields the more explicit solution

$$H(f) = \frac{G^*(f)S_S(f)}{|G(f)|^2 S_S(f) + S_N(f)} \qquad (11.201a)$$

$$= \frac{1}{G(f)}|\rho(f)|^2 \qquad (11.201b)$$

$$= \frac{1/G(f)}{1 + r^{-1}(f)}, \qquad (11.201c)$$

where $|\rho(f)|$ is the coherence-function magnitude (Section 10.3) for the signal and its corrupted version,

$$|\rho(f)| \triangleq \frac{|S_{SX}(f)|}{[S_S(f)S_X(f)]^{1/2}}, \qquad (11.202)$$

and $r(f)$ is the ratio of PSDs of distorted signal and noise,

$$r(f) \triangleq \frac{|G(f)|^2 S_S(f)}{S_N(f)}. \qquad (11.203)$$

We see from (11.201) to (11.203) that at frequencies for which the PSD SNR $r(f)$ is large, the coherence magnitude $|\rho(f)|$ is close to unity and the optimum filter is simply an equalizer that removes the signal distortion. However, at frequencies for which the PSD SNR is low, the coherence magnitude is small compared to unity and the optimum filter is an attenuator.

The minimum value of MSE can be determined as follows. The condition (11.196) can be used to show that for the optimum filter, (11.194) reduces (exercise 23) to

$$\text{MSE}_{\min} = E\{S^2(t)\} - E\{S(t)\hat{S}(t)\}. \qquad (11.204)$$

Substitution of (11.195) into (11.204) yields

$$\text{MSE}_{\min} = R_S(0) - \int_{-\infty}^{\infty} h(\tau) R_{SX}(\tau)\, d\tau. \qquad (11.205)$$

Application of Parseval's relation to (11.205) yields

$$\mathrm{MSE}_{\min} = \int_{-\infty}^{\infty} [S_S(f) - H(f)S_{SX}^*(f)]\, df, \qquad (11.206)$$

which, upon substitution of (11.198) and (11.202), becomes

$$\mathrm{MSE}_{\min} = \int_{-\infty}^{\infty} S_S(f)\left[1 - |\rho(f)|^2\right] df \qquad (11.207\mathrm{a})$$

$$= \int_{-\infty}^{\infty} \frac{S_S(f)}{1 + r(f)}\, df \qquad (11.207\mathrm{b})$$

$$= \int_{-\infty}^{\infty} \frac{S_S(f)S_N(f)}{|G(f)|^2 S_S(f) + S_N(f)}\, df. \qquad (11.207\mathrm{c})$$

Thus, the largest contributions to MSE occur at frequencies for which the signal power $S_S(f)$ is substantial and the PSD SNR $r(f)$ is small, in which case the coherence magnitude is small compared to unity.

An example of the general solution obtained in this section is pursued in exercise 24.

11.8 Random-Signal Detection

Let us consider the problem of detecting the possible presence of a random signal $S(t)$ buried in random noise $N(t)$. Since the form of a random signal is not known, the matched filter solution obtained in Section 11.6 cannot be used. However, the approach of maximizing an appropriate measure of SNR can be generalized to accommodate a random signal. In this case, in place of the linear filter used for detection of a known signal, it is more appropriate to use a quadratic transformation of the noisy measurements,

$$X(t) = \begin{cases} S(t) + N(t), & \text{signal present} \\ N(t), & \text{signal absent} \end{cases}, \quad |t| \le \frac{T}{2}. \qquad (11.208)$$

Thus, $X(t)$ is to be transformed by some quadratic device to obtain a statistic

$$Y = \int_{-T/2}^{T/2} \int_{-T/2}^{T/2} k(u,v) X(u) X(v)\, du\, dv \qquad (11.209)$$

that is to be compared with a threshold γ in order to decide whether or not the signal is present:

$$Y \begin{cases} > \gamma \Rightarrow & \text{decide } S(t) \text{ present,} \\ < \gamma \Rightarrow & \text{decide } S(t) \text{ absent.} \end{cases} \qquad (11.210)$$

Unlike the finite-energy signal of known form, the random signal might be a finite-power signal that continues indefinitely. Thus, we restrict the measurements to a finite time interval (centered at $t = 0$ for convenience), as indicated in (11.208) and reflected in (11.209). The weighting function $k(u, v)$ in (11.209) characterizes the particular quadratic device to be used. Now, as a measure of SNR, the quantity

$$D \triangleq \frac{|E\{Y|S(t) \text{ present}\} - E\{Y|S(t) \text{ absent}\}|}{(\text{Var}\{Y|S(t) \text{ absent}\})^{1/2}} \quad (11.211)$$

is adopted.* D is called the *deflection* because it is related to the normalized difference in the expected deflections of a needle on a power meter that occur when the signal is present and absent. The objective is to seek the particular quadratic device that maximizes D. In order to obtain a tractable problem, it is assumed that the noise is Gaussian. Also, for convenience (but not tractability) it is assumed that the signal is WSS and the noise is white. Then it can be shown (exercise 27) that D reduces to

$$D = \frac{\left|\int_{-T/2}^{T/2}\int_{-T/2}^{T/2} k(u,v) R_S(u-v)\, du\, dv\right|}{\left[2N_o^2 \int_{-T/2}^{T/2}\int_{-T/2}^{T/2} k^2(u,v)\, du\, dv\right]^{1/2}}. \quad (11.212)$$

In obtaining (11.212), it is assumed that the signal and noise processes are orthogonal. A bivariate version of the univariate Cauchy-Schwarz inequality (11.186) yields the solution for the D-maximizing quadratic device,

$$k(u, v) = c R_S(u - v) \quad (11.213)$$

for any constant c. To be consistent with the threshold test (11.209)–(11.210), in which (11.213) is to be used, it is required that $c > 0$, and for convenience $c = 1/N_o^2 T$ is used. Thus, the maximum-D detection statistic is

$$Y = \frac{1}{N_o^2 T}\int_{-T/2}^{T/2}\int_{-T/2}^{T/2} R_S(u-v) X(u) X(v)\, du\, dv. \quad (11.214)$$

By introducing the change of variables

$$u = t + \frac{\tau}{2},$$

$$v = t - \frac{\tau}{2}, \quad (11.215)$$

---

*A comparative study of this and a variety of other maximum-SNR criteria for detector design is reported in [Gardner, 1980].

(11.214) can be reexpressed (exercise 28) as

$$Y = \frac{1}{N_o^2} \int_{-\infty}^{\infty} R_S(\tau) R_X(\tau)_T d\tau, \quad (11.216)$$

where $R_X(\tau)_T$ is defined by

$$R_X(\tau)_T \triangleq \begin{cases} \dfrac{1}{T} \int_{-(T-|\tau|)/2}^{(T-|\tau|)/2} X\left(t + \dfrac{\tau}{2}\right) X\left(t - \dfrac{\tau}{2}\right) dt, & |\tau| \leq T, \\ 0, & |\tau| > T, \end{cases}$$

(11.217)

and is called the *correlogram* of $X(t)$. Application of Parseval's relation to (11.216) yields (exercise 28)

$$Y = \frac{1}{N_o^2} \int_{-\infty}^{\infty} S_S(f) P_T(f) df, \quad (11.218)$$

where $P_T(f)$ is the *periodogram* of $X(t)$,

$$P_T(f) \triangleq \frac{1}{T} |\tilde{X}_T(f)|^2, \quad (11.219)$$

for which

$$\tilde{X}_T(f) \triangleq \int_{-T/2}^{T/2} X(t) e^{-i2\pi ft} dt. \quad (11.220)$$

Thus, this optimum detector calculates the periodogram of the noisy measurements and correlates this with the spectrum of the random signal to be detected. For example, if the signal spectrum is flat throughout its passband, say $[-B, B]$, then

$$Y = \frac{S_o}{N_o^2} \int_{-B}^{B} P_T(f) df, \quad (11.221)$$

which is simply the scaled average power in the measurements within the signal passband. The device that implements (11.221) is typically called a *radiometer*. Thus, the more general optimum detector (11.218) is called an *optimum radiometer*.

The maximum value of D can be determined by substitution of the solution (11.213) into (11.212). The result is (exercise 29)

$$D_{\max} = \left[\frac{1}{2N_o^2} \int_{-T/2}^{T/2} \int_{-T/2}^{T/2} R_S^2(u - v) \, du \, dv \right]^{1/2}. \quad (11.222)$$

Use of the change of variables (11.215) in (11.222) yields

$$D_{max} = \left[\frac{T}{2N_o^2} \int_{-T}^{T} \left(1 - \frac{|\tau|}{T}\right) R_S^2(\tau) \, d\tau \right]^{1/2}. \qquad (11.223)$$

Furthermore, if the measurement time T greatly exceeds the correlation time of $S(t)$ (the width of R_S), then (11.223) yields the close approximation

$$D_{max} \approx \left[\frac{T}{2N_o^2} \int_{-\infty}^{\infty} R_S^2(\tau) \, d\tau \right]^{1/2}$$

$$= \left[\frac{T}{2} \int_{-\infty}^{\infty} \left[\frac{S_S(f)}{N_o} \right]^2 df \right]^{1/2}. \qquad (11.224)$$

Observe the analogy between the formulas (11.189) for SNR_{max} for a known signal and (11.224) for D_{max} for a random signal.

It can be shown [Middleton, 1966] that the optimum radiometer (11.218)–(11.220) [since it is equivalent to (11.214)] is also an optimum detector in an entirely different sense from that of maximizing D. Specifically, if the noise is Gaussian, and the signal is weak in the sense that

$$S_S(f) \ll N_o, \qquad (11.225)$$

then (11.214) minimizes the probability of detection error when the threshold level γ is properly set. It should be understood that (11.225) does not necessarily result in small deflection or poor detection performance. As (11.224) reveals, long measurement times and wide signal bandwidths counteract the low SNR described by (11.225). Moreover, it can be shown [Van Trees, 1971] that for signals that are Gaussian, but not necessarily weak, and long measurement times, the minimum-probability-of-error detector is given by a generalization of (11.214) that can be manipulated into the form

$$Y = \frac{1}{N_o} \int_{-\infty}^{\infty} H(f) P_T(f) \, df, \qquad (11.226)$$

for which

$$H(f) = \frac{S_S(f)}{S_S(f) + N_o}, \qquad (11.227)$$

which is the transfer function for the Wiener filter (Section 11.7). The detection statistic (11.226) can be further manipulated into the form (exercise 30)

$$Y = \frac{1}{N_o T} \int_{-T/2}^{T/2} \hat{S}(t) X(t) \, dt, \qquad (11.228)$$

in which

$$\hat{S}(t) = \int_{-T/2}^{T/2} h(t-u) X(u) \, du \qquad (11.229)$$

is the minimum-MSE estimate of $S(t)$, assuming $S(t)$ is present. Observe the analogy between (11.228) and (11.192). We see by comparing these two results that the optimum detector for a random signal uses the minimum-MSE estimate $\hat{S}(t)$ as if it were a signal of known form, $s(t)$, and correlates it with the noisy measurements. Hence, this optimum detector is called an *estimator-correlator*. Further results on the estimator-correlator characterization of optimum detectors for random signals are reported in [Kailath et al., 1978] and [Gardner, 1982]. Further results on the optimum-radiometer characterization are given in Section 12.8.4 and in [Gardner, 1987a].

11.9 *Autoregressive Models and Linear Prediction*

A particularly useful type of model for discrete-time processes is the *autoregressive* (AR) *model* introduced in Section 9.4. The nth-order AR model for $X(i)$ is the difference equation

$$X(i) + a_1 X(i-1) + a_2 X(i-2) + \cdots + a_n X(i-n) = bZ(i), \qquad (11.230a)$$

where $Z(i)$ is a zero-mean stationary sequence of uncorrelated random variables with unity variance,

$$R_Z(k) = \begin{cases} 1, & k = 0, \\ 0, & k \neq 0. \end{cases} \qquad (11.230b)$$

This model enables one to interpret $X(i)$ as the response of a linear time-invariant system to white-noise excitation $Z(i)$. The transfer function $G(f)$ of this system is obtained by substitution of the excitation $z(j) = e^{i2\pi j f}$ and response $x(j) = G(f) e^{i2\pi j f}$ into (11.230a) to obtain

$$G(f) = \frac{b}{1 + \sum_{p=1}^{n} a_p (e^{-i2\pi f})^p}. \qquad (11.231)$$

Hence, the spectral density for $X(i)$ is given by

$$S_X(f) = \frac{|b|^2}{\left| 1 + \sum_{p=1}^{n} a_p (e^{-i2\pi f})^p \right|^2}. \qquad (11.232)$$

In this section several important properties of this AR model are described, and its role in the theory of linear prediction is explained.

11.9.1 Yule-Walker Equations

The autocorrelation sequence $R_X(k)$ for the AR model satisfies a set of linear equations that can be derived as follows. Multiplication of both sides of (11.230a) (with i replaced with $i+j$) by $X(i)$ and evaluation of the expected value of this product yields the equation (exercise 32)

$$R_X(j) = -a_1 R_X(j-1) - a_2 R_X(j-2) - \cdots - a_n R_X(j-n)$$
$$+ bR_{ZX}(j). \tag{11.233}$$

Since $Z(i)$ has zero mean value, and $X(i)$ depends on only $Z(i)$, $Z(i-1)$, $Z(i-2), \ldots$ [cf. (11.230a)], then

$$R_{ZX}(i) = 0, \quad i \geq 1. \tag{11.234a}$$

Multiplication of both sides of (11.230a) by $Z(i)$ and evaluation of the expected value of this product yields (exercise 32)

$$R_{ZX}(0) = -a_1 R_{ZX}(1) - a_2 R_{ZX}(2) - \cdots - a_n R_{ZX}(n) + b, \tag{11.235}$$

from which (11.234a) yields

$$R_{ZX}(0) = b. \tag{11.234b}$$

It follows from (11.233) to (11.235) that

$$R_X(j) = -\sum_{p=1}^{n} a_p R_X(j-p), \quad j \geq 1 \tag{11.236a}$$

$$R_X(0) = b^2 - \sum_{p=1}^{n} a_p R_X(-p). \tag{11.236b}$$

If the $n+1$ values $\{R_X(j) = R_X(-j): j = 0, 1, 2, 3, \ldots, n\}$ of the autocorrelation of $X(i)$ are known, then these $n+1$ linear equations (for $j = 1, 2, 3, \ldots, n$) can be solved for the $n+1$ model parameters $\{a_p: p = 1, 2, 3, \ldots, n\}$ and b^2. Thus, the AR model is fully specified (except for the sign of b, which is irrelevant) by the first $n+1$ values of $R_X(k)$. Once the model is determined, $R_X(k)$ is fully specified for *all* k. Thus, there must be a way to determine $\{R_X(k): |k| > n\}$ from $\{R_X(k): |k| \leq n\}$. In fact, (11.236a), for $j > n$, is a linear recursion that enables $R_X(k)$ to be determined from $\{R_X(j): |j| < k\}$ for every $k > n$. Of course this *autocorrelation extrapolation* procedure is valid [i.e., the equations (11.236) are valid] only if $R_X(k)$ truly is the autocorrelation sequence for some nth-order AR model. The equations (11.236) are known as the Yule-Walker equations, in honor of George Udny Yule's and Gilbert Walker's pioneering work on AR models [Yule, 1927; Walker, 1931].

11.9.2 Levinson-Durbin Algorithm

The Yule-Walker equations can be solved using a particularly efficient recursive algorithm called the *Levinson-Durbin Algorithm* [Levinson, 1947; Durbin, 1960; Wiggins and Robinson, 1965]. Because of the fact that an AR model of order n_1 is identical to an AR model of order $n_2 > n_1$ if $a_{n_1+1} = a_{n_1+2} = \cdots = a_{n_2} = 0$, then one need not know the model order n in order to use the Yule-Walker equations to determine the model parameters. One can simply solve (11.236a) for $n = 1, 2, 3, \ldots$ until $a_p = 0$ for all $p > n$. Of course, one would in principle have to solve for a_p for $p \to \infty$ to be sure that there are no nonzero a_p for $p > n$, but this problem can often be circumvented in practice. The following Levinson-Durbin algorithm is a computationally efficient recursion for carrying out this solution procedure. Let $\{a_p(n)\}$ and $b(n)$ denote the parameters for an nth-order AR model. Then the algorithm is initialized by

$$a_1(1) = -\frac{R_X(1)}{R_X(0)}, \qquad (11.237a)$$

$$b^2(1) = \left[1 - a_1^2(1)\right] R_X(0), \qquad (11.237b)$$

and the nth step of the recursion is specified by

$$a_n(n) = \frac{-1}{b^2(n-1)} \left[R_X(n) + \sum_{q=1}^{n-1} a_q(n-1) R_X(n-q)\right], \qquad (11.237c)$$

$$b^2(n) = \left[1 - a_n^2(n)\right] b^2(n-1), \qquad (11.237d)$$

$$a_p(n) = a_p(n-1) + a_n(n) a_{n-p}(n-1), \qquad p = 1, 2, 3, \ldots, n-1. \qquad (11.237e)$$

For any value of n, the solution, $\{a_p(n): p = 1, 2, 3, \ldots, n\}$ and $b^2(n)$, provided by (11.237) is identical to the solution provided by (11.236) for $j = 1, 2, 3, \ldots, n$. Thus, at each step, say n, the Levinson-Durbin algorithm provides the solution to the Yule-Walker equations for an AR model of order n (cf. exercise 32).

11.9.3 Linear Prediction

The problem of fitting an AR model to a given autocorrelation sequence $R_X(k)$ is intimately related to the problem of predicting the value $X(i)$ of the process using the previous n values $\{X(i-1), X(i-2), \ldots, X(i-n)\}$ and knowledge of the autocorrelation sequence. This relationship is explained as follows. Since b^2 must be nonnegative, (11.237d) indicates that

$$|a_n(n)| \leq 1 \tag{11.238a}$$

and also that

$$b^2(n) \leq b^2(n-1). \tag{11.238b}$$

Furthermore, it follows from the AR model (11.230) that the *model-error variance* is equal to b^2,

$$E\left\{\left[X(i) - \sum_{p=1}^{n}(-a_p)X(i-p)\right]^2\right\} = b^2(n). \tag{11.239}$$

Hence, the sequence of model-error variances, indexed by n, is nonincreasing. The term *model-error variance* is used because the quantity

$$bZ(i) = X(i) - \sum_{p=1}^{n}(-a_p)X(i-p) \tag{11.240}$$

from (11.230a) is the error in fitting the process $X(i)$ to an nth-order linear regression on its past. The *nth-order regressor* is denoted by

$$\hat{X}(i) \triangleq \sum_{p=1}^{n}(-a_p)X(i-p). \tag{11.241}$$

It can be shown that the solution to the Yule-Walker equations of order n minimizes this error variance for any zero-mean WSS process $X(i)$. This follows from the fact that the variance (11.239) is minimum if and only if the *orthogonality condition* (Chapter 13),

$$E\{[X(i) - \hat{X}(i)]X(i-p)\} = 0, \quad p = 1, 2, 3, \ldots, n, \tag{11.242}$$

is satisfied. This condition can be derived simply by equating to zero the n partial derivatives of the variance (11.239) with respect to the n parameters $\{a_p\}$. Substitution of (11.241) into (11.242) yields the Yule-Walker equations (11.236a) (exercise 32). Furthermore, since the regressor $\hat{X}(i)$ depends on only values of $X(i)$ at times prior to i, then it can be interpreted as a *predictor* of the value $X(i)$. Thus, the solution to the Yule-Walker equations of order n yields the minimum-variance nth-order linear predictor of the process, and the minimum value $b^2(n)$ of the prediction-error variance.

It can be shown (exercise 33) that for any WSS process $X(i)$, the model-error process $bZ(i)$ becomes white, (11.230b), in the limit $n \to \infty$. This reveals that *any* such process can be modeled exactly by an AR model, although the order n can be infinite. Nevertheless, since the model-error variance is a nonincreasing function of the model order n then for any desired degree of accuracy (arbitrarily small difference between the finite-order model-error variance and the minimum error variance $b^2(\infty)$) in the fit of the model to the process, there is a sufficiently large but finite value for n.

11.9.4 Wold-Cramér Decomposition

Any zero-mean, finite-mean-square, WSS process $X(i)$ can be decomposed into two zero-mean processes that are uncorrelated with each other,

$$X(i) = X_r(i) + X_s(i), \tag{11.243a}$$

$$R_{X_r X_s}(k) \equiv 0, \tag{11.243b}$$

for which the component $X_s(i)$ is perfectly predictable (zero prediction-error variance) and is mean-square equivalent to a sum of phase-randomized sine waves, and for which $X_r(i)$ is mean-square equivalent to the response of a linear, time-invariant, causal, stable system, whose inverse is also causal and stable, to a unity-variance white excitation, say $Z(i)$. The component $X_s(i)$ is called *singular* (or *deterministic*), and the component $X_r(i)$ is called *regular*.

The regular component admits a stable AR model [i.e., (11.230) with $X(i) = X_r(i)$] with transfer function $G(f)$ given by (11.231), with n possibly infinite, and the inverse model is also stable and causal, with transfer function denoted by $G^{-1}(f)$,

$$G^{-1}(f) = \frac{1}{G(f)}, \tag{11.244}$$

and corresponding impulse-response sequence denoted by $g^{-1}(i)$. The white excitation,

$$Z(i) = g^{-1}(i) \otimes X_r(i), \tag{11.245a}$$

is called the *innovations representation* for $X_r(i)$,

$$X_r(i) = g(i) \otimes Z(i), \tag{11.245b}$$

because each new value $Z(i)$ of the process is uncorrelated with all prior values $\{Z(j): j < i\}$ of the process, and therefore provides completely *new information*—an *innovation*.

The singular component $X_s(i)$ also admits an AR model, but with $b = 0$. That is, the system is marginally unstable, and produces a response consisting of a sum of sine waves, without any excitation. If there are only m sine waves, then the AR model order is $n = 2m$ and the Levinson-Durbin algorithm terminates [with $b(n) = 0$] at this value of n.

The spectral density for the singular component consists of Dirac deltas only (a *pure line spectrum*), and the spectral density for the regular component is continuous and is equal to $|G(f)|^2$, which can be expressed as (11.232), with n possibly infinite. This representation of $S_{X_r}(f)$ in terms of the product of conjugate factors

$$S_{X_r}(f) = G(f)G^*(f) \tag{11.246}$$

is called a *spectral factorization*. The particular factorization described here, in which both $G(f)$ and $1/G(f)$ are *minimum-phase functions*,* and therefore represent causal stable systems, is called the *canonical spectral factorization*. It follows from (11.245a) that this factorization identifies the *whitening filter* that produces the innovations representation $Z(i)$ from $X_r(i)$.

This decomposition of $X(i)$ and its spectral density $S_X(f)$ is called the *Wold-Cramér decomposition*. However, the Wold-Cramér decomposition is more general than that described here because it applies to nonstationary processes as well. Moreover the Wold-Cramér decomposition is a wide-sense version of a strict-sense decomposition known as the *Doob decomposition* in which the predictor (regressor) is nonlinear (cf. [Larson and Shubert, 1979], [Doob, 1953]). There is a partially analogous decomposition for continuous-time processes (cf. [Larson and Shubert, 1979], [Doob, 1953]). That is, such processes can be decomposed into a sum of singular and regular components that are uncorrelated with each other; however, the singular component need not have a pure line spectrum, and therefore in the decomposition described in Section 10.9, the component $X_c(t)$ with continuous spectral density need not be the regular component. Nevertheless, there is a condition on continuous spectral densities that guarantees that a process with a spectral density satisfying the condition is regular, and this is called the *Paley-Wiener condition* (Chapter 13). If the Paley-Wiener condition is satisfied for the component $X_c(t)$, then $X_c(t)$ and $X_d(t)$ (in Section 10.9) are, respectively, the regular and singular components of $X(t)$.

The innovations representation for discrete-time processes is employed in the derivation of the optimum Kalman filter in Section 13.6. The

---

*If $G(f) = \mathcal{G}(e^{i2\pi f})$, then $G(f)$ is a *minimum-phase function* if $\mathcal{G}(z)$ and $1/\mathcal{G}(z)$ are analytic functions of the complex variable z, for $|z| \geq 1$.

analogous canonical spectral factorization and innovations representation for continuous-time processes are employed in the derivation of the optimum causal Wiener filter in Section 13.5.

11.9.5 Maximum-Entropy Model

An interesting question that arises in the practical application of AR models is what interpretation can be given to the nth-order AR model obtained from the Yule-Walker equations when the only values of the autocorrelation sequence that are known are the $n + 1$ values used in the nth-order Yule-Walker equations. That is, if the values $\{R_X(k): k > n\}$ are unknown (or perhaps cannot be reliably estimated), then it cannot be known whether or not $X(i)$ is truly an nth-order AR process. There is one particularly intriguing answer to this question. Specifically, it is shown in this subsection that among all possible processes that have the same $n + 1$ autocorrelation values $\{R_X(k): |k| \leq n\}$, the nth-order AR process specified by the Yule-Walker equations is the *most random* in the sense of having maximum relative entropy rate for a given relative entropy rate of its innovations representation. The *relative entropy rate*, denoted by $\overline{H}_X$, for a process $X(i)$ is defined by [Shannon and Weaver, 1962]

$$\overline{H}_X \triangleq \lim_{n \to \infty} \frac{1}{n} E\left\{ \ln\left[\frac{1}{f_{X_n}(X_n)}\right] \right\}, \qquad (11.247)$$

where $f_{X_n}(X_n)$ is the nth-order joint probability density for the vector of n random variables $\{X(1), X(2), X(3), \ldots, X(n)\}$, evaluated at the *random* vector X_n. The relative entropy rate is a relative measure of the average uncertainty per time sample of the process.

It can be shown that the model-error variance, for $n \to \infty$, is given by the *Szegö-Kolmogorov formula* [Doob, 1953]

$$b^2(\infty) = \exp\left\{ \int_{-1/2}^{1/2} \ln[S_{X_r}(f)] \, df \right\}. \qquad (11.248)$$

Observe that the singular component has no effect on the prediction-error variance (since it is perfectly predictable). For convenience, it is assumed that the singular component is zero, so that the process is regular, $X(i) \equiv X_r(i)$. It can also be shown (exercise 34) that the difference between the entropy rate $\overline{H}_Z$ of the input $Z(i)$ and the entropy rate $\overline{H}_X$ of the output, $X(i)$, of a minimum-phase linear time-invariant system with transfer function $G(f)$ is given by [Shannon and Weaver, 1962]

$$\overline{H}_X - \overline{H}_Z = \frac{1}{2} \int_{-1/2}^{1/2} \ln\left[|G(f)|^2\right] df. \qquad (11.249)$$

By letting $Z(i)$ be the innovations representation for $X(i)$, it follows from

(11.246), (11.248), and (11.249) that the model-error variance, for $n \to \infty$, can be expressed as

$$b^2(\infty) = \exp\{2[\bar{H}_X - \bar{H}_Z]\}, \qquad (11.250)$$

and it follows from (11.237d) that for finite-order models the model-error variance is nonincreasing,

$$b^2(n) \geq b^2(n+1) \geq b^2(n+2) \geq \cdots \geq b^2(\infty). \qquad (11.251)$$

Consequently, the process with a given nth-order model-error variance $b^2(n)$ will have maximum relative entropy rate, for a given relative entropy rate of its innovations representation, if and only if

$$b^2(\infty) = b^2(n). \qquad (11.252)$$

It follows from (11.252) that a necessary and sufficient condition for a maximum-entropy model for $X(i)$ with given $b^2(n)$ is that the parameters in (11.237) satisfy

$$a_m(m) = 0, \qquad m > n, \qquad (11.253)$$

in which case $X(i)$ is exactly modeled by an nth-order AR process. Furthermore, since $b^2(n)$ is given if $\{R_X(k): |k| \leq n\}$ is given [cf. (11.237)], then the maximum-entropy model for a zero-mean process with given $\{R_X(k): |k| \leq n\}$ is the nth-order AR model specified by the Levinson-Durbin algorithm, or equivalently the Yule-Walker equations.

11.9.6 Lattice Filter

The nth-order minimum-variance linear predictor for a process $X(i)$ can be implemented in the form of a *lattice filter*, which has important properties from an implementation standpoint, especially if the predictor is to be made adaptive (cf. [Friedlander, 1982]). This lattice-filter implementation can be derived from the Levinson-Durbin algorithm as follows. The nth-order linear-prediction-error process, denoted by $E_n(i)$, is given by

$$E_n(i) \triangleq X(i) - \hat{X}_n(i), \qquad (11.254)$$

where $\hat{X}_n(i)$ denotes the nth-order linear predictor, which can be expressed [using (11.241)] as

$$\hat{X}_n(i) = \alpha_n(i) \otimes X(i), \qquad (11.255a)$$

where $\alpha_n(i)$ is the finite impulse-response sequence

$$\alpha_n(i) = \begin{cases} -a_i(n), & i = 1, 2, 3, \ldots, n, \\ 0, & \text{otherwise.} \end{cases} \qquad (11.255b)$$

Consequently, $E_n(i)$ can be expressed as

$$E_n(i) = \eta_n(i) \otimes X(i), \qquad (11.256a)$$

where
$$\eta_n(i) = \delta(i) - \alpha_n(i). \qquad (11.256b)$$

Alternatively, $\hat{X}_n(i)$ can be expressed as

$$\hat{X}_n(i) = [\delta(i) - \eta_n(i)] \otimes X(i). \qquad (11.257)$$

Thus, a realization for the prediction filter $\alpha_n(i)$ can always be obtained from a realization of the prediction-error filter $\eta_n(i)$ and vice versa. A realization of $\eta_n(i)$ can be obtained from the Levinson-Durbin algorithm as follows:

$$E_n(i) = X(i) + \sum_{p=1}^{n} a_p(n) X(i-p) \qquad (11.258a)$$

$$= X(i) + \sum_{p=1}^{n-1} \left[a_p(n-1) + a_n(n) a_{n-p}(n-1) \right] X(i-p)$$

$$+ a_n(n) X(i-n) \qquad (11.258b)$$

$$= E_{n-1}(i) + a_n(n) \check{E}_{n-1}(i-1), \qquad (11.258c)$$

where

$$\check{E}_n(i) \triangleq X(i-n) + \sum_{p=1}^{n} a_p(n) X(i-n+p). \qquad (11.258d)$$

The quantity $\check{E}_n(i)$ is a *backward prediction-error process*, that is, the error in prediction of $X(i-n)$ using $\{X(i-n+p): p = 1, 2, 3, \ldots, n\}$. Because of the stationarity of $X(i)$, the minimum-variance backward-prediction filter coefficients are identical to those for minimum-variance forward prediction. Thus, (11.258d) is the *minimum-variance* backward-prediction-error process. Consequently, the Levinson-Durbin algorithm can again be used to show (exercise 35) that

$$\check{E}_n(i) = \check{E}_{n-1}(i-1) + a_n(n) E_{n-1}(i). \qquad (11.258e)$$

The pair of joint recursions (11.258c) and (11.258e) can be implemented directly with the lattice structure shown in Figure 11.10. One of the great advantages of this lattice structure for implementation of a linear predictor is that as the order n is increased or decreased by the addition or deletion of final stages in the lattice, there is no effect on any of the previous stages. That is, their coefficients need not be readjusted. Moreover, it can be shown that the coefficients

$$\rho_p \triangleq -a_p(p) \qquad (11.259)$$

that completely specify the minimum-variance prediction-error lattice filter

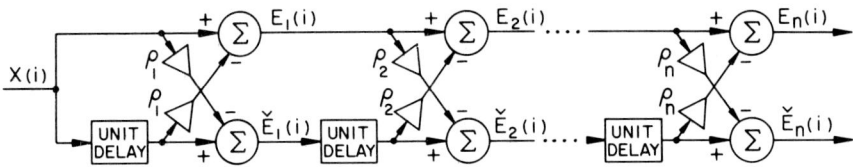

Figure 11.10 Lattice Implementation of Prediction Error Filter

are identical to the correlation coefficients for $E_{p-1}(i)$ and $\check{E}_{p-1}(i-1)$; they are called the partial correlation (PARCOR) coefficients for $X(i)$ and $X(i-p)$, because

$$E_{p-1}(i) = X(i) - \hat{X}_{p-1}(i), \qquad (11.260a)$$

$$\check{E}_{p-1}(i-1) = X(i-p) - \check{X}_{p-1}(i-p), \qquad (11.260b)$$

where $\check{X}_{p-1}(i-p)$ is the $(p-1)$th-order backward linear predictor of $X(i-p)$. Thus ρ_p is the correlation coefficient for $X(i)$ and $X(i-p)$, with the correlation of $\{X(i-1), X(i-2), \ldots, X(i-p+1)\}$ first removed. This explains why $|a_p(p)| \le 1$, as concluded earlier from the Levinson-Durbin algorithm. This and various other properties of lattice models and recursive algorithms for linear prediction can all be given elegant interpretations, as well as derivations, in terms of the geometrical concept of orthogonal projection, which is developed in Chapter 13.

As a final note, it should be clarified that the AR model with input $E(i)$ [or $Z(i)$] and output $X(i)$ is an all-pole model, whereas the lattice model with input $X(i)$ and output $E(i)$ is an all-zero model, and these models are inverses of each other. Furthermore, it can be shown that the AR model is stable if and only if all coefficients $\{\rho_p\}$ in the lattice model have magnitudes less than unity, and it is marginally stable if and only if some coefficients have unity magnitude.

11.10 Summary

This chapter (including the exercises) covers a broad range of special topics and applications in the second-order theory of random processes. In Section 11.1, the power spectral densities for *time-sampled* and *pulse-modulated signals* are derived. Some of these results are then used to explain the mean-square *sampling theorem* for random processes. In Section 11.2, *Rice's representation of a bandpass process* in terms of two low-pass processes is explained, and formulas relating power spectral densities and cross-spectral densities for the process and its representors are derived. In Section 11.3, the cyclostationarity of *frequency-modulated processes* is discussed, and the time-averaged power spectral density is derived for the two cases of high and low modulation index. These results are then used to derive a formula for the demodulation-gain in SNR, which is used to quantify the *noise-immunity* property of frequency-modulated signals. The technique of *pre-emphasis/de-emphasis* for additional SNR gain is also explained. In Section 11.4, several methods for measurement (estimation) of power spectral densities are described. A typical method is analyzed probabilistically in terms of its mean and variance, and a *fundamental tradeoff between resolution and reliability* is explained. In Section 11.5, the many sources of noise in a receiving system are briefly described, and then a general approach to noise modeling is developed. This entails the introduction of the concepts of *available power*, *available power gain*, *effective noise temperature*, and *noise figure*. In Section 11.6, the problem of *detecting the presence of a signal of known form in additive stationary noise* is considered, and an SNR-maximizing filter is derived. It is explained that the optimum filter is a *matched filter*, and (when followed by a sampler) can be implemented as a *correlator*. An explicit formula for the maximized SNR is obtained. In Section 11.7, the problem of *extracting (filtering) a stationary random signal from corrupted jointly stationary measurements* is considered, and a mean-squared-error-minimizing filter is derived. It is explained that the *optimum (Wiener) filter* optimizes a *tradeoff between distortion equalization and noise attenuation*. In Section 11.8, the problem of *detecting the presence of a stationary random signal in additive stationary Gaussian noise* is considered, and an SNR-maximizing quadratic transformation is derived. It is explained that the optimum quadratic transformation can be implemented as a *correlation of the periodogram of the corrupted measurements with the power spectral density of the signal to be detected*. An explicit formula for the maximized SNR is obtained. An alternative interpretation of the optimum quadratic transformation as a *correlation of the corrupted measurements with an optimum linear estimate of the signal* (assumed to be present) is explained. In Section 11.9, the interrelated topics of *autoregressive modeling*, *linear prediction*, and *maximum-entropy modeling* are introduced. The roles of the *innovations representation* and *spectral factorization* are explained; the *Yule-Walker equations*, the *Wold-Cramér decomposition*, and the *Levinson-Durbin algorithm* and its *lattice-filter implementation* are described.

EXERCISES

1. (a) To verify the identity (11.11), show that the left member equals
$$\sum_n e^{-i2\pi nTf},$$
and then expand the right member into a Fourier series
$$\sum_n c_n e^{i2\pi nTf}$$
and show that $c_n = 1$ for all n.

(b) Apply Parseval's relation to (11.10), and then use the identity (11.11) to derive (11.12).

2. Use the same technique as that described by (11.10) to (11.12) to verify that (11.22) is the Fourier transform of (11.21).

Hint: Begin by showing that the Fourier transform of (11.21) is given by

$$P(f)\sum_{j} X(jT)e^{-i2\pi jTf}.$$

3. (a) It follows from the sampling theorem for nonrandom finite-energy functions that

$$R_X(\tau) = \sum_{n=-\infty}^{\infty} R_X(nT) \frac{\sin[(\pi/T)(\tau - nT)]}{(\pi/T)(\tau - nT)} \quad (11.261)$$

provided that $T \leq 1/2B$, where B is the bandwidth of R_X:

$$S_X(f) = 0, \quad |f| > B.$$

Equation (11.261) can be used to prove the sampling theorem for mean-square bandlimited WSS random processes:

$$E\{[X(t) - \hat{X}(t)]^2\} = 0, \quad (11.262)$$

where

$$\hat{X}(t) = \sum_{n=-\infty}^{\infty} X(nT) \frac{\sin[(\pi/T)(t - nT)]}{(\pi/T)(t - nT)}. \quad (11.263)$$

To verify (11.262)–(11.263), use (11.261) to show

$$E\{[X(t) - \hat{X}(t)] X(mT)\} = 0 \quad \text{for all } m, \quad (11.264)$$

from which it follows that

$$E\{[X(t) - \hat{X}(t)] \hat{X}(t)\} = 0. \quad (11.265)$$

Then use (11.261) again to show

$$E\{[X(t) - \hat{X}(t)] X(t)\} = 0. \quad (11.266)$$

Finally, use (11.265) and (11.266) to verify (11.262).

Hint: Use the more general form of (11.261)

$$R_X(\tau - a) = \sum_{n=-\infty}^{\infty} R_X(nT - a - b) \frac{\sin[(\pi/T)(\tau + b - nT)]}{(\pi/T)(\tau + b - nT)}$$

$$(11.267)$$

for any a and b, which corresponds to sampling a function time-shifted by a, at points time-shifted by b.

(b) (Alternative proof.) Show that the mean squared error (MSE) in (11.262) can be expressed by

$$\text{MSE} = \int_{-\infty}^{\infty} [S_X(f) - S_{X\hat{X}}(f) - S_{\hat{X}X}(f) + S_{\hat{X}}(f)] \, df. \tag{11.268}$$

Consider the interpolated time-sampled signal

$$\hat{X}(t) = \sum_{n=-\infty}^{\infty} X(nT + \Theta) \frac{\sin[(\pi/T)(t - nT - \Theta)]}{(\pi/T)(t - nT - \Theta)}, \tag{11.269}$$

where Θ is independent of $X(t)$ and uniformly distributed on $[-T/2, T/2)$. Show that $\hat{X}(t)$ can be modeled by

$$\hat{X}(t) = [X(t)Z(t)] \otimes h(t), \tag{11.270a}$$

where

$$Z(t) = \sum_{n=-\infty}^{\infty} \delta(t - nT - \Theta), \tag{11.270b}$$

and $h(t)$ is an ideal low-pass filter with transfer function given by

$$H(f) = \begin{cases} T, & |f| \leq 1/2T, \\ 0, & |f| > 1/2T. \end{cases} \tag{11.270c}$$

Then show that

$$m_Z = 1/T, \tag{11.271a}$$

$$R_Z(\tau) = \frac{1}{T} \sum_{m=-\infty}^{\infty} \delta(\tau - mT), \tag{11.271b}$$

$$R_{\hat{X}}(\tau) = [R_Z(\tau) R_X(\tau)] \otimes r_h(\tau), \tag{11.272a}$$

$$R_{\hat{X}X}(\tau) = m_Z R_X(\tau) \otimes h(\tau). \tag{11.272b}$$

Use (11.271) and (11.272) together with (11.268) to prove that MSE $= 0$ if

$$S_X(f) = 0 \quad \text{for} \quad |f| > \frac{1}{2T}. \tag{11.273}$$

(c) The method of proof given in (b) is easily adapted to the case of nonuniform sampling, for which

$$\hat{X}(t) = \sum_{n=-\infty}^{\infty} X(T_n) \frac{\sin[2\pi B(t - T_n)]}{\pi(2B + \lambda)(t - T_n)}, \tag{11.274}$$

where $\{T_n\}$ is a Poisson point process with rate parameter λ. Prove that

$$\frac{\text{MSE}}{R_X(0)} = \frac{1}{1 + \lambda/2B}. \qquad (11.275)$$

Hence, no finite sampling rate λ will yield zero MSE. However, the MSE can be made small with a sampling rate that is sufficiently large relative to the Nyquist rate $2B$. [It can be shown that the height $1/(1 + \lambda/2B)$ of the interpolating pulse in (11.274) yields the smallest possible MSE for a given λ.]

4. Consider the pulse-position-modulated signal

$$X(t) = \sum_{j=-\infty}^{\infty} p(t - jT - T_j), \qquad (11.276)$$

where $\{T_j\}$ is a sequence of independent, identically distributed pulse-position variables, with probability density function denoted by $d(u)$. This is a cyclostationary process with period T.

(a) Show that the time-averaged spectrum is given by the formula

$$\langle S_X \rangle(f) = \frac{1}{T} |P(f)|^2 \left[1 - |D(f)|^2 \right]$$

$$+ \left| \frac{P(f)D(f)}{T} \right|^2 \sum_{n=-\infty}^{\infty} \delta\left(f - \frac{n}{T} \right). \qquad (11.277)$$

Hint: Evaluate $R_X(t + \tau, t)$ analogously to (11.2). Then evaluate

$$\langle R_X \rangle(\tau) \triangleq \frac{1}{T} \int_0^T R_X(t + \tau, t) \, dt$$

by a change of variables analogous to that used to obtain (11.3). This requires adding and subtracting the term $i = k$ in the analog of (11.2). The result is

$$\langle R_X \rangle(\tau) = \frac{1}{T} \int_{-\infty}^{\infty} d(u) \int_{-\infty}^{\infty} p(t + \tau - u) p(t - u) \, du \, dt$$

$$- \frac{1}{T} \int_{-\infty}^{\infty} w(t + \tau) w(t) \, dt$$

$$+ \frac{1}{T} \int_{-\infty}^{\infty} \sum_n w(t + \tau) w(t + nT) \, dt,$$

where $w(t) \triangleq d(t) \otimes p(t)$. Show that this result reduces to

$$\langle R_X \rangle(\tau) = \frac{1}{T} \left[r_p(\tau) - r_w(\tau) \right] + \frac{1}{T} \sum_n r_w(\tau - nT),$$

$$(11.278)$$

analogous to (11.4). Show that Fourier transformation yields the final result (11.277) after manipulation analogous to that used to obtain (11.13). Observe that since $D(f)$ is a characteristic function, then $D(0) = 1$.

    (b) Sketch the spectrum $\langle S_X \rangle$, for an exponential pulse $p(t) = e^{-t/T}$, $t \geq 0$, and a uniform probability density d on $[0, a]$ for the two cases $a \gg T$ and $a \ll T$.

5. (a) Derive (11.53) from (11.40) for the complex envelope.
    (b) Use (11.41) and (11.58) to derive the representation (11.59).
    (c) Let $X(t)$ be given by $Z(t)$ in (11.62), and assume that $C(t)$ and $S(t)$ are bandlimited to $(-f_o, f_o)$. Prove that $C(t) = U(t)$ and $S(t) = V(t)$, where $U(t)$ and $V(t)$ are defined by (11.57).

Hint: Let $W(t)$ be bandlimited to $(-f_o, f_o)$, and show that

$$H\{W(t)\cos(2\pi f_o t)\} = W(t)\sin(2\pi f_o t),$$
$$H\{W(t)\sin(2\pi f_o t)\} = -W(t)\cos(2\pi f_o t),$$

where $H\{\cdot\}$ denotes Hilbert transformation. Use this result in (11.57) with (11.62) substituted.

6. (a) Let $X(t)$ be a zero-mean WSS process, and show that $U(t)$ and $V(t)$ in (11.57) are jointly WSS.
    (b) Assume that $C(t)$ and $S(t)$ are WSS, and derive the autocorrelation (11.63) for the WSS process (11.62). (Note that the stationarity of $Z(t)$ constrains the relationship between R_C and R_S, and the form of R_{CS}.)
    (c) Verify that $S_Z(f)$ given by (11.64) is real, as it must be. Does $S_Z(f)$ depend on $\text{Re}\{S_{CS}(f)\}$?
    (d) Use (11.64), (11.60), and (11.61) to show that the spectral densities of the in-phase and quadrature components in Rice's representation are given by

$$S_U(f) = S_V(f) = S_X(f + f_o)u(f_o + f) + S_X(f - f_o)u(f_o - f),$$

(11.279)

where u is the unit step function, and then show that (11.279) reduces to (11.65) under the condition (11.66).
    (e) For the spectral density of $X(t)$ shown in Figure 11.11, sketch the spectral density of $U(t)$ and $V(t)$.

7. (a) Derive a formula for the cross-correlation $R_{UV}(\tau)$ from (11.57).
    (b) Use the result from (a) and (11.61) to derive the cross-spectrum (11.67).

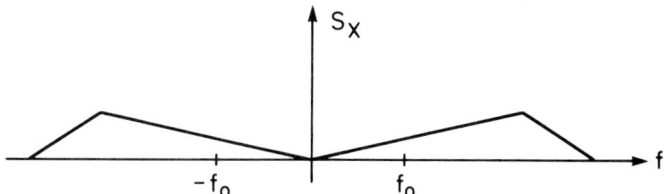

Figure 11.11 Spectral Density of $X(t)$ for Exercise 6(e)

8. (a) Use trigonometric identities to derive (11.59) from (11.70) to (11.72).
   (b) Use the same technique as in part (a) to derive (11.76) from (11.73).
   (c) Substitute (11.78) into (11.76), and then evaluate the asymptotic time average to obtain the formula (11.79) for the time-averaged probabilistic autocorrelation of an FM signal.

9. Show that for a Gaussian process $\Phi(t)$, (11.80) reduces to (11.81).

   Hint: Use (2.38) with $\omega = [1, -1]^T$, $X = [\Phi(t + \tau/2), \Phi(t - \tau/2)]^T$.

10. Evaluate $\langle R_X \rangle(0)$ for the FM signal (11.73) to verify that the average power is $a^2/2$, regardless of the modulation $\Phi(t)$, provided that it is stationary (not necessarily Gaussian).

11. Show that (11.98) can be reexpressed as (11.99).

12. Use (11.76) and (11.77) to derive (11.102) (see hint in exercise 9).

13. An FM signal is transmitted over an additive noise channel with noise spectral density $N_o = 10^{-8}$ Watt/Hz. The rms bandwidth of the modulating signal $\Psi(t)$ is 15 kHz. Determine the rms value of $\Psi(t)$ and the amplitude a of the FM signal needed in order to obtain an input SNR of $\text{SNR}_{in} = 10$ and an output SNR of $\text{SNR}_{out} = 10^4$. Determine the rms bandwidth of the FM signal.

14. Typical pre-emphasis and de-emphasis circuits used in commercial broadcast FM are shown in Figure 11.12, for which $1/2\pi R_2 C \geq B_\Psi$, where B_Ψ is the modulating signal bandwidth.

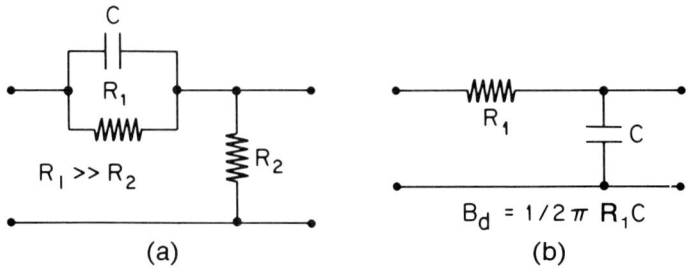

Figure 11.12 (a) Pre-emphasis Circuit for FM; (b) De-emphasis Circuit for FM

(a) Derive the transfer functions, and sketch their magnitudes over the range from $f = 0$ to $f \gg 1/2\pi R_2 C$.
(b) Evaluate the ratio of (11.121) to (11.126), using (11.119), (11.120), and (11.127), in order to verify (11.128).

15. (a) Derive (11.145) from (11.143), under the assumptions that $S_X(\nu)$ is approximately constant throughout the passband of $H(\nu)$, which is specified by

$$|H(\nu)| = \begin{cases} 1/\sqrt{2\Delta f}, & ||\nu| - f_o| < \Delta f/2, \\ 0, & \text{otherwise,} \end{cases}$$

and that $G(\nu)$ is specified by (11.144).

Hint: Draw graphs of the integrand in (11.143).

(b) If a spectral resolution of $\Delta f = 1$ kHz is desired, how long a period of data, Δt, is needed in order to measure the spectral density of a stationary process and obtain a variance no larger than $\frac{1}{100}$ of the squared mean? What assumptions about the process must you make in order to answer this question? Discuss the alternative question where Δt is given and Δf must be determined.

16. (a) Show that the spectra at the input and output of the circuit in Figure 11.9 are related by (11.159).
(b) Use the result of (a) to prove that the incremental power (11.160) is maximized with respect to R_L when $R_L = R$.

17. (a) Two resistors, R_1 and R_2, connected in parallel are at two different temperatures. Determine the effective noise temperature of the parallel connection.
(b) An amplifier has three stages with effective input noise temperatures of $T_1 = 200$ K (first stage), $T_2 = 400$ K (second stage), and $T_3 = 800$ K (final stage). If the available power gain of the second stage is 6, how much gain must the first stage have to guarantee an overall effective input noise temperature of 250 K?
(c) An antenna with effective noise temperature of $T_{ea} = 150$ K is connected to a receiver by a waveguide, both of which are at the physical temperature of 280 K. The power loss factor of the antenna and waveguide together is $L = 1.5$. The receiver has an effective bandwidth of $B_r = 1$ MHz, an effective input noise temperature of $T_{er} = 700$ K, and an available power gain at band center of 10^{12}. Determine the effective input noise temperature of the overall receiving system, and determine the available output noise power.

18. (a) Derive the formulas (11.182) and (11.183) for the mean and variance from (11.179) and (11.181).
(b) Derive the formula (11.189) for the maximum SNR from (11.182), (11.183), and (11.188).

19. Assume that the noise $N(t)$ is white, and derive the simplified formulas (11.190) and (11.191) from (11.189) and (11.188).

20. Derive the discrete-time linear time-invariant filter that maximizes the SNR (11.178) at its output for a signal $s(i)$ of known form in additive stationary noise $N(i)$ at its input.

21. Sketch the output signal from a matched filter for the signal
$$s(t) = \begin{cases} a, & 0 \le t \le T, \\ -a, & T < t \le 2T, \\ 0, & \text{otherwise} \end{cases}$$
when there is no noise at the input.

22. (a) Derive (11.197) from the orthogonality condition (11.196), and then solve to obtain the formula (11.198) for the Wiener filter.
    (b) Verify that for the model (11.193), the general formula (11.198) reduces to (11.201).

23. Use the necessary and sufficient orthogonality condition (11.196) to derive (11.204), and then show that (11.204) can be reexpressed by (11.205). Finally, verify the formula (11.207) for the minimized MSE.

24. Consider the additive noise model
$$X(t) = S(t) + N(t),$$
for which $N(t)$ is white noise, with spectrum N_o, and $S(t)$ has the triangular spectrum
$$S_S(f) = \begin{cases} (1 - |f|/B)S_o, & |f| \le B, \\ 0, & |f| > B. \end{cases}$$

(a) Solve for the Wiener filter for extracting $S(t)$, and sketch the transfer function for $S_o/N_o = 1/10, 1, 10$. Compare this with the ideal low-pass filter,
$$H_*(f) = \begin{cases} 1, & |f| \le B \\ 0, & |f| > B. \end{cases}$$

(b) Show that the minimum MSE is given by
$$\frac{\text{MSE}_{\min}}{E\{S^2(t)\}} = \frac{2N_o}{S_o}\left[1 - \frac{N_o}{S_o}\ln\left(1 + \frac{S_o}{N_o}\right)\right], \quad (11.280)$$

and sketch this as a function of S_o/N_o.

Hint: As an aid in sketching for $S_o/N_o \to 0$, use two terms in the Taylor series expansion of $\ln\{\cdot\}$.

(c) Solve for the MSE resulting from the ideal low-pass filter described in part (a), and sketch this as a function of S_o/N_o. Discuss the relative merits of the Wiener filter and the ideal low-pass filter, in terms of signal distortion as well as MSE.

(d) To gain further insight into the comparison in part (c), for each filter decompose MSE into two components [by similarly decomposing $\hat{S}(t)$], one due to noise and the other due to distortion. Then compare these components for the two filters.

25. As demonstrated in the preceding exercise, the Wiener filter, in general, introduces signal distortion in the process of attenuating noise. For some applications this is not tolerable. However, when there are two sets of measurements available, distortion can be avoided as follows. If one set of measurements, say

$$X_1(t) = S(t) + N_1(t),$$

is filtered with $H(f)$ and the other set, say

$$X_2(t) = S(t) + N_2(t),$$

is filtered with $1 - H(f)$, and the two filtered signals are added, the result is

$$\hat{S}(t) = S(t) - N_1(t) + h(t) \otimes [N_1(t) - N_2(t)].$$

(a) Solve for the filter H that minimizes the MSE between $S(t)$ and $\hat{S}(t)$. [Observe that $S(t)$ need not be WSS; it could be a finite-energy signal.]

Hint: Let $X' \triangleq N_1 - N_2$, $S' \triangleq N_1$, $\hat{S}' \triangleq h \otimes X'$. Then apply the results of Section 11.7, with X, $\hat{S}$, and S there replaced by X' $\hat{S}'$, and S'.

(b) Consider the situation for which $N_1(t)$ is low-pass noise and $N_2(t)$ is high-pass noise, with nonoverlapping spectra. Show that the signal can be recovered perfectly. [This is an idealization of a practical filtering problem that arises in aircraft instrument landing, where $X_1(t)$ is a heading-derived rate signal and $X_2(t)$ is a radio-derived rate signal.]

26. Consider the sampled-and-held process

$$Y(t) = \sum_{j=-\infty}^{\infty} X(jT) p(t - jT),$$

where $p(t)$ is a rectangular pulse of width T. It is desired to reconstruct $X(t)$ from $Y(t)$ using a linear time-invariant filter. $X(t)$ is not necessarily

bandlimited. Determine the filter that minimizes the MSE
$$\text{MSE} = E\{[\hat{X}(t - \Theta) - X(t - \Theta)]^2\},$$
where
$$\hat{X}(t) \triangleq h(t) \otimes Y(t)$$
is the reconstructed process, and Θ is independent of $X(t)$ and uniformly distributed on $[0, T)$. Then determine the minimum MSE. Do this exercise for arbitrary $p(t)$, and show that the minimum MSE is independent of $p(t)$ (assuming that $P(f) \neq 0$ within the passband of the process), and is zero if $X(t)$ is mean-square bandlimited to $(-1/2T, 1/2T)$.

27. Show that for WGN $N(t)$, the deflection (11.211) reduces to (11.212) [with $R_S(u - v)$ replaced by $R_S(u, v)$ for a nonstationary signal].

    Hint: Use Isserlis's formula to evaluate the denominator, and assume, without loss of generality, that $k(u, v) = k(v, u)$.

28. (a) Use the change of variables (11.215) to derive the formula (11.216) from (11.214).
    (b) Derive the formula (11.218) from (11.216). (Recall exercise 3 in Chapter 3.)

29. (a) Derive the formula (11.222) for the maximized deflection from (11.212) and (11.213).
    (b) Derive the approximation (11.224) from (11.222).

30. Derive the alternative formulas (11.228) and (11.229) from (11.226) and (11.227).

31. To compare the signal-to-noise ratios attainable for a known signal and a random signal, we can use the relationship
$$E/T \simeq 2BS_o,$$
which is a close approximation for $BT \gg 1$. E is the expected energy of a segment of a random signal in an interval of length T, and it is reinterpreted as the true energy of the same signal in the same interval when that signal is known. B is the one-sided bandwidth of the signal, and S_o is its spectral density, or effective spectral level if the spectral density is not flat, so that
$$\int_{-B}^{B} S^2(f)\, df = 2BS_o.$$

Show that for a signal in WGN, the maximized SNRs are as follows:

random signal, $\quad$ $\text{SNR}_{\max} \simeq (BT)^{1/2} \dfrac{S_o}{N_o}$; $\quad\quad\quad$ (11.281)

known signal, $\quad$ $\text{SNR}_{\max} \simeq 2BT \dfrac{S_o}{N_o}$. $\quad\quad\quad$ (11.282)

Consequently, the SNR grows linearly in the time-bandwidth product for a known signal, but grows only as the square root of this product for a random signal.

32. (a) Use the AR model (11.230) to verify (11.233) and (11.235).
    (b) Derive the Yule-Walker equations (11.236a) from (11.241) and (11.242).
    (c) Verify that for $n = 2$ the Levinson-Durbin algorithm yields the solution to the Yule-Walker equations.

33. In order to show that the model-error process $bZ(i)$ becomes white in the limit as the model order approaches infinity ($n \to \infty$), proceed as follows. Substitute (11.240) and (11.241) into the necessary and sufficient orthogonality condition (11.242) to show that

$$E\{Z(i)X(i-p)\} = 0, \quad p = 1, 2, 3, \ldots, n, \quad (11.283)$$

and therefore

$$E\{Z(i)Z(i-p)\} = \frac{1}{b(n)} \sum_{q=1}^{n} a_q(n) E\{Z(i)X(i-p-q)\},$$

$$p = 1, 2, 3, \ldots, n. \quad (11.284)$$

Then use (11.283) in (11.284) to obtain

$$R_Z(p) = \frac{1}{b(n)} \sum_{q=n-p+1}^{n} a_q(n) R_{ZX}(p+q), \quad p = 1, 2, 3, \ldots, n.$$

$$(11.285)$$

Finally since the sum in (11.285) contains only p terms, then as long as $R_{ZX}(q)$ is bounded, $1/b(n)$ is bounded, and $a_q(n) \to 0$ as $q \to \infty$ (and $n \to \infty$), this sum must converge to zero as $n \to \infty$, and therefore $R_Z(p) = 0$ for all $p \neq 0$ in the limit $n \to \infty$. Since

$$|R_{ZX}(q)|^2 \leq R_Z(0) R_X(0) \leq \frac{1}{b^2(n)} R_X^2(0), \quad (11.286)$$

then it is sufficient for $R_Z(p) = 0$ if $1/b(n)$ is bounded and $a_q(n)$ converges to zero as $q \to \infty$ (and $n \to \infty$). As explained in Section

11.9.4, this will be so if $X(i)$ is a regular process, because then $b^2(n) \geq b^2(\infty) > 0$ and a stable model exists, and $a_q \to 0$ as $q \to \infty$ for a stable model. [More generally, it can be shown that $Z(i)$ is white in the limit $n \to \infty$ even if $X(i)$ is not a regular process (cf. Section 13.6.1).]

34. To derive the formula (11.249) for the gain in entropy rate due to passage through a filter, proceed as follows:
   (a) Consider the *relative entropy* of the n-vector

   $$X_n \triangleq [X(0), X(1), X(2), \ldots, X(n-1)]^T,$$

   defined by

   $$H(X_n) \triangleq E\left\{\ln\left[\frac{1}{f_{X_n}(X_n)}\right]\right\}, \qquad (11.287)$$

   and show that the relative entropy of the linearly transformed vector

   $$Y_n \triangleq A X_n$$

   (where A is a nonrandom $n \times n$ matrix) is given by

   $$H(Y_n) = H(X_n) + \ln(|A|),$$

   for which $|A|$ is the absolute value of the determinant of A.
   Hint: Use (1.41).

   (b) Consider the lower triangular matrix

   $$A = \begin{bmatrix} g(0) & & & \\ g(1) & g(0) & & \\ g(2) & g(1) & g(0) & \\ \vdots & & & \\ g(n-1) & g(n-2) & \cdots & g(0) \end{bmatrix}$$

   corresponding to a causal time-invariant filter, with impulse response sequence $g(i)$, and show that

   $$\ln(|A|) = n \ln|g(0)|.$$

   (c) Use the results of (a) and (b) to show that the entropy rate (11.247) for the process defined by

   $$Y_n \triangleq [Y(0), Y(1), Y(2), \ldots, Y(n-1)]^T,$$

with $n \to \infty$, is given by

$$\overline{H}_Y = \overline{H}_X + \ln|g(0)|. \tag{11.288}$$

(d) It can be shown [Doob, 1953] that for a minimum-phase linear time-invariant transformation, with transfer function $G(f)$,

$$\ln|g(0)| = \int_{-1/2}^{1/2} \ln|G(f)|\, df. \tag{11.289}$$

Use (11.288) and (11.289) to prove that

$$\overline{H}_Y - \overline{H}_X = \frac{1}{2} \int_{-1/2}^{1/2} \ln|G(f)|^2\, df. \tag{11.290}$$

35. Derive the backward-prediction-error recursion (11.258e) by analogy with (11.258a) to (11.258d).

36. Show that the mean-squared linear-prediction error for an nth-order AR process cannot be made any smaller than that obtained with an nth-order linear predictor, regardless of how high the order of the predictor considered is.

37. (a) Let $R_X(j) = r^{|j|}$, where $|r| < 1$. Find an explicit formula for the nth-order linear predictor for this process. Comment on the difference in behavior of the process and the predictor for the two cases $r < 0$ and $r > 0$.

(b) Let $X(i)$ be the output of a second-order filter with unit-pulse response

$$h(i) = \begin{cases} a_1 r_1^i + a_2 r_2^i, & i \geq 1 \\ 0, & i \leq 0 \end{cases}$$

(where $|r_1| < 1$ and $|r_2| < 1$), and with white excitation. Find an explicit formula for the nth order linear predictor for $X(i)$.

Hint: $X(i)$ is a second-order Markov process.

12

Cyclostationary Processes

12.1 Introduction

Physical phenomena that involve periodicities give rise to random data for which appropriate probabilistic models exhibit periodically time-variant parameters. For example, in mechanical-vibration monitoring and diagnosis for machinery, periodicity arises from rotation, revolution, and reciprocation of gears, belts, chains, shafts, propellers, bearings, pistons, and so on; in atmospheric science (e.g., for weather forecasting), periodicity arises from seasons, caused primarily by rotation and revolution of the earth; in radio astronomy, periodicity arises from revolution of the moon, rotation and pulsation of the sun, rotation of Jupiter and revolution of its satellite Io, and so on, and can cause strong periodicities in time series (e.g., pulsar signals); in biology, periodicity in the form of biorhythms arises from both internal and external sources (e.g., circadian rhythms); in communications, telemetry, radar, and sonar, periodicity arises from sampling, scanning, modulating, multiplexing, and coding operations, and it can also be caused by rotating reflectors such as helicopter blades and air- and watercraft propellers. For these and many other examples, the periodicity can be an important characteristic that should be reflected in an appropriate probabilistic model. Therefore, stationary processes, with their time-invariant probabilistic parameters, are in general inadequate for the study of such phenomena, and cyclostationary processes are indicated. For example, various specific models of cyclostationary processes, including sampled-and-held noise, amplitude modulation, pulse-amplitude, -width, and -position modulation, and phase or frequency modulation, have been described in previous chapters.

A process, say $X(t)$, is said to be *cyclostationary in the wide sense** if its mean and autocorrelation are periodic with some period, say T:

$$m_X(t+T) = m_X(t), \tag{12.1}$$

$$R_X(t+T, u+T) = R_X(t, u), \tag{12.2}$$

for all t and u. We shall focus our attention on the autocorrelation function. Since (12.2) can be reexpressed as

$$R_X\left(t+T+\frac{\tau}{2}, t+T-\frac{\tau}{2}\right) = R_X\left(t+\frac{\tau}{2}, t-\frac{\tau}{2}\right), \tag{12.3}$$

then $R_X(t+\tau/2, t-\tau/2)$, which is a function of two independent variables, t and τ, is periodic in t with period T for each value of τ. It is assumed that the Fourier series representation for this periodic function converges, so that R_X can be expressed as

$$R_X\left(t+\frac{\tau}{2}, t-\frac{\tau}{2}\right) = \sum_\alpha R_X^\alpha(\tau) e^{i2\pi\alpha t}, \tag{12.4}$$

for which $\{R_X^\alpha\}$ are the Fourier coefficients,

$$R_X^\alpha(\tau) \triangleq \frac{1}{T} \int_{-T/2}^{T/2} R_X\left(t+\frac{\tau}{2}, t-\frac{\tau}{2}\right) e^{-i2\pi\alpha t} dt, \tag{12.5}$$

and α ranges over all integer multiples of the fundamental frequency $1/T$. This model for R_X is adequate for a phenomenon with a single periodicity. However, for a phenomenon with more than one periodicity it must be generalized. This is easily accomplished by letting α in (12.4) range over all integer multiples of all fundamental frequencies of interest, say $1/T_1, 1/T_2, 1/T_3, \ldots$, but then (12.5) must be modified as follows (exercise 1):

$$R_X^\alpha(\tau) \triangleq \lim_{Z \to \infty} \frac{1}{Z} \int_{-Z/2}^{Z/2} R_X\left(t+\frac{\tau}{2}, t-\frac{\tau}{2}\right) e^{-i2\pi\alpha t} dt. \tag{12.6}$$

Such a process is said to be *almost cyclostationary in the wide sense*, by analogy with the fact that the function $R_X(t+\tau/2, t-\tau/2)$ is said to be an *almost periodic function* (of t) [Gardner, 1978]. More generally, a nonstationary process $X(t)$ is said to *exhibit cyclostationarity* if there exists a periodicity frequency α for which the Fourier coefficient defined by (12.6) is not identically zero.

---

*For convenience, the modifier *wide-sense* is omitted in this chapter. This can be done without ambiguity, since only wide-sense properties are discussed, except when otherwise indicated.

This chapter provides an introduction to the theory of cyclostationary and almost cyclostationary processes, and more general nonstationary processes that exhibit cyclostationarity, by extending and generalizing some of the fundamental results for stationary processes presented in Chapters 8 to 10. As a matter of fact, the nonstationary processes studied in this chapter provide especially important examples of the very broad class of asymptotically mean stationary processes discussed in previous chapters. A primary objective of this chapter is to reveal the fundamental role that spectral correlation plays in the second-order theory of nonstationary processes that exhibit cyclostationarity.

12.2 Cyclic Autocorrelation and Cyclic Spectrum

The Fourier-coefficient function defined by (12.6) for $\alpha = 0$ is the time-averaged probabilistic autocorrelation defined for asymptotically mean stationary processes in Section 8.5,

$$\langle R_X \rangle(\tau) = R_X^\alpha(\tau) \qquad \text{for} \quad \alpha = 0. \tag{12.7}$$

This follows from the identity

$$\lim_{Z \to \infty} \frac{1}{Z} \int_{-Z/2}^{Z/2} R_X(t + \tau, t)\, dt = \lim_{Z \to \infty} \frac{1}{Z} \int_{-Z/2}^{Z/2} R_X\!\left(t + \frac{\tau}{2}, t - \frac{\tau}{2}\right) dt, \tag{12.8}$$

which reveals that $R_X^\alpha(\tau)$, for $\alpha = 0$, is invariant to time translation. That is, if $Y(t) \triangleq X(t + t_o)$, then

$$R_Y^\alpha(\tau) = R_X^\alpha(\tau) \qquad \text{for} \quad \alpha = 0. \tag{12.9}$$

Equation (12.8) is simply (12.9) with $t_o = -\tau/2$. In contrast to this translation invariance, the Fourier-coefficient functions defined by (12.6) for $\alpha \neq 0$ are not invariant to time translation. Rather they are cyclic (sinusoidal) (exercise 2):

$$R_Y^\alpha(\tau) = R_X^\alpha(\tau) e^{i2\pi \alpha t_o}. \tag{12.10}$$

Consequently, $R_X^\alpha(\tau)$ is referred to as the *cyclic autocorrelation function*, and α is called the *cycle frequency* parameter. It follows directly from (12.4) that the periodicity present in the autocorrelation function, $R_X(t + \tau/2, t - \tau/2)$, is completely characterized by the set of cyclic autocorrelations $\{R_X^\alpha(\tau)\}$ indexed by α. The set of values of α for which $R_X^\alpha(\tau) \neq 0$ is called the *cycle spectrum*.

The cyclic autocorrelation can be characterized in a way that reveals the role that periodicity in autocorrelation plays in the frequency domain. Specifically, let $U(t)$ and $V(t)$ be the frequency-shifted versions of $X(t)$ defined by

$$U(t) \triangleq X(t)e^{-i\pi\alpha t}, \qquad (12.11a)$$

$$V(t) \triangleq X(t)e^{+i\pi\alpha t}. \qquad (12.11b)$$

These processes have time-averaged autocorrelations given by (exercise 3)

$$\langle R_U \rangle(\tau) \triangleq \lim_{Z \to \infty} \frac{1}{Z} \int_{-Z/2}^{Z/2} E\left\{ U\left(t + \frac{\tau}{2}\right) U\left(t - \frac{\tau}{2}\right)^* \right\} dt$$

$$\equiv \langle R_X \rangle(\tau) e^{-i\pi\alpha\tau} \qquad (12.12a)$$

and

$$\langle R_V \rangle(\tau) \equiv \langle R_X \rangle(\tau) e^{+i\pi\alpha\tau}. \qquad (12.12b)$$

Furthermore, their time-averaged cross-correlation is given by (exercise 3)

$$\langle R_{UV} \rangle(\tau) \triangleq \lim_{Z \to \infty} \frac{1}{Z} \int_{-Z/2}^{Z/2} E\left\{ U\left(t + \frac{\tau}{2}\right) V\left(t - \frac{\tau}{2}\right)^* \right\} dt \equiv R_X^\alpha(\tau).$$

$$(12.13)$$

Therefore, the cyclic autocorrelation of the process X is simply the time-averaged cross-correlation between frequency-shifted versions of X. This reveals that a process exhibits cyclostationarity in the wide sense only if there exists correlation between some frequency-shifted versions of the process. It also reveals that a process can be stationary only if there does not exist any correlation between any frequency-shifted versions of the process, because only then can $R_X^\alpha \equiv 0$ for all $\alpha \neq 0$.

The time-averaged spectra (defined in Section 10.6) for the two frequency-shifted processes (12.11) follow from (12.12), and are given by (exercise 3)

$$\langle S_U \rangle(f) = \langle S_X \rangle(f + \alpha/2), \qquad (12.14a)$$

$$\langle S_V \rangle(f) = \langle S_X \rangle(f - \alpha/2), \qquad (12.14b)$$

and the time-averaged cross-spectrum follows from (12.13), and is given by

$$\langle S_{UV} \rangle(f) = \int_{-\infty}^{\infty} R_X^\alpha(\tau) e^{-i2\pi f \tau} d\tau \triangleq S_X^\alpha(f), \qquad (12.15)$$

which is called the *cyclic spectrum* or *cyclic spectral density*. It follows from (12.14) and (12.15) and the definition (10.53) that the time-averaged coherence function for $U(t)$ and $V(t)$ is

$$\rho_{UV}(f) \triangleq \frac{\langle S_{UV} \rangle(f)}{[\langle S_U \rangle(f) \langle S_V \rangle(f)]^{1/2}} \tag{12.16}$$

$$= \frac{S_X^\alpha(f)}{[\langle S_X \rangle(f + \alpha/2) \langle S_X \rangle(f - \alpha/2)]^{1/2}} \triangleq \rho_X^\alpha(f). \tag{12.17}$$

It can be concluded from the discussion of coherence in Section 10.3 that $\rho_X^\alpha(f)$ is the time-averaged correlation coefficient for the two frequency components of $X(t)$ at frequencies $f + \alpha/2$ and $f - \alpha/2$ [which are the frequency components of $U(t)$ and $V(t)$ at frequency f]. The function $\rho_X^\alpha(f)$ is therefore called the *spectral autocoherence function* for $X(t)$. To distinguish this from the coherence function for two arbitrary processes, the latter will be renamed the *cross-coherence function*. The preceding reveals that a process exhibits cyclostationarity in the wide sense only if there exists correlation between distinct frequency components of the process. Moreover, the separation between any two such frequencies, $(f + \alpha/2) - (f - \alpha/2) = \alpha$, must be a cycle frequency of the process.

A process $X(t)$ is said to be *completely coherent at spectral frequency f and cycle frequency α* if and only if the spectral autocoherence magnitude is unity, $|\rho_X^\alpha(f)| = 1$. Also a process $X(t)$ is said to be *completely incoherent* at spectral frequency f and cycle frequency α if and only if the spectral autocoherence is zero, $\rho_X^\alpha(f) = 0$.

As defined in Section 10.6, the instantaneous probabilistic spectral density, denoted by $S_X(t, f)$, is the Fourier transform of the instantaneous probabilistic autocorrelation,

$$S_X(t, f) \triangleq \int_{-\infty}^{\infty} R_X\left(t + \frac{\tau}{2}, t - \frac{\tau}{2}\right) e^{-i2\pi f \tau} d\tau. \tag{12.18}$$

It follows from (12.4), (12.15), and (12.18) that the instantaneous probabilistic spectral density for a cyclostationary or almost cyclostationary process is completely characterized by the cyclic spectra

$$S_X(t, f) = \sum_\alpha S_X^\alpha(f) e^{i2\pi \alpha t}. \tag{12.19}$$

In conclusion, the essence of the difference between stationary and cyclostationary or almost cyclostationary processes is that the latter exhibit spectral correlation. Furthermore, this spectral correlation is completely and conveniently characterized by the cyclic spectra $\{S_X^\alpha\}$ or equivalently by the cyclic autocorrelations $\{R_X^\alpha\}$. Another approach to characterizing cyclostationarity is presented in Section 12.8.4, where it is shown that the existence of a nonzero cyclic autocorrelation for $X(t)$ is necessary and sufficient for the existence of a quadratic time-invariant transformation of $X(t)$ that generates an additive sine-wave component in the mean of the transformed process. Thus, although a cyclostationary process does not necessarily contain additive periodic components (which give rise to spectral lines), it can be quadratically transformed into a process that does contain such components.

Example 1: Consider the amplitude-modulated sine wave

$$X(t) = Y(t)\cos(2\pi f_o t)$$
$$= \tfrac{1}{2}Y(t)e^{i2\pi f_o t} + \tfrac{1}{2}Y(t)e^{-i2\pi f_o t},$$

for which $Y(t)$ is a zero-mean stationary process. Since the sine-wave factor shifts the frequency of each spectral component of $Y(t)$ at frequency (say) f up to $f + f_o$ and down to $f - f_o$, then there is obviously spectral correlation for all pairs of frequency components centered at f and separated by $\alpha = 2f_o$. In fact the cyclic spectral density is given by

$$S_X^\alpha(f) = \tfrac{1}{4}S_Y(f), \qquad \alpha = 2f_o.$$

This can be obtained as follows:

$$R_X\!\left(t + \frac{\tau}{2}, t - \frac{\tau}{2}\right) = R_Y(\tau)\cos\!\left(2\pi f_o\!\left[t + \frac{\tau}{2}\right]\right)\cos\!\left(2\pi f_o\!\left[t - \frac{\tau}{2}\right]\right)$$
$$= \tfrac{1}{2}R_Y(\tau)[\cos(2\pi f_o \tau) + \cos(4\pi f_o t)].$$

Substitution into the definition (12.6), with $\alpha = 2f_o$, yields

$$R_X^\alpha(\tau) = \lim_{Z\to\infty}\frac{1}{Z}\int_{-Z/2}^{Z/2}\tfrac{1}{2}R_Y(\tau)\left[\cos(2\pi f_o \tau)\,e^{-i4\pi f_o t} + \tfrac{1}{2}e^{i8\pi f_o t} + \tfrac{1}{2}\right]dt$$
$$= \tfrac{1}{4}R_Y(\tau), \qquad \alpha = 2f_o,$$

where the limit

$$\lim_{Z\to\infty}\frac{1}{Z}\int_{-Z/2}^{Z/2}e^{i\omega t}\,dt = \lim_{Z\to\infty}\frac{\sin(\omega Z/2)}{\omega Z/2} = 0, \qquad \omega \neq 0$$

has been used. The desired result follows by Fourier transformation. The fact that additive periodic components in the mean can be generated with a quadratic time-invariant transformation of $X(t)$ is easy to see, since the squaring transformation yields the mean

$$E\{X^2(t)\} = E\{\tfrac{1}{2}Y^2(t)[1 + \cos(4\pi f_o t)]\},$$
$$= \tfrac{1}{2}R_Y(0)[1 + \cos(4\pi f_o t)],$$

which is periodic, whereas the mean before squaring is the constant

$$E\{X(t)\} = E\{Y(t)\cos(2\pi f_o t)\}$$
$$= m_Y \cos(2\pi f_o t) = 0.$$ ∎

Spectral Correlation. Although the cyclic spectrum has been defined to be the Fourier transform of the cyclic autocorrelation, it can be given an alternative equivalent definition that derives from its interpretation in terms of spectral correlation. Then the Fourier-transform relation (12.15) follows as a *result* that is a generalization of the Wiener-Khinchin relation (10.21), which is (12.15) with $\alpha = 0$. To accomplish this, consider the Fourier transform of the finite segment of $X(t)$ on the time interval $[t - W/2, t + W/2]$,

$$\tilde{X}_W(t,\nu) \triangleq \int_{t-W/2}^{t+W/2} X(u) e^{-i2\pi u \nu} \, du. \qquad (12.20)$$

The time-averaged correlation of the two spectral components with frequencies $\nu = f + \alpha/2$ and $\nu = f - \alpha/2$, normalized by W, is

$$\lim_{Z \to \infty} \frac{1}{Z} \int_{-Z/2}^{Z/2} \frac{1}{W} E\left\{ \tilde{X}_W\left(t, f + \frac{\alpha}{2}\right) \tilde{X}_W\left(t, f - \frac{\alpha}{2}\right)^* \right\} dt. \qquad (12.21)$$

Let us define the cyclic spectrum to be the limit of this spectral correlation as the spectral resolution $\Delta f = 1/W$ becomes infinitesimal:

$$S_X^\alpha(f) \triangleq \lim_{W \to \infty} \lim_{Z \to \infty} \frac{1}{Z} \int_{-Z/2}^{Z/2} \frac{1}{W} E\left\{ \tilde{X}_W\left(t, f + \frac{\alpha}{2}\right) \tilde{X}_W\left(t, f - \frac{\alpha}{2}\right)^* \right\} dt. \qquad (12.22)$$

By use of the convolution theorem it can be shown (exercise 4) that the product of Fourier transforms in (12.22) is the Fourier transform

$$P_X^\alpha(t,f)_W = \int_{-\infty}^\infty R_X^\alpha(t,\tau)_W e^{-i2\pi f \tau} \, d\tau = \int_{-W}^W R_X^\alpha(t,\tau)_W e^{-i2\pi f \tau} \, d\tau, \qquad (12.23)$$

where

$$P_X^\alpha(t,f)_W \triangleq \frac{1}{W} \tilde{X}_W\left(t, f + \frac{\alpha}{2}\right) \tilde{X}_W\left(t, f - \frac{\alpha}{2}\right)^* \qquad (12.24)$$

330 • **Random Processes**

and

$$R_X^\alpha(t,\tau)_W \triangleq \frac{1}{W} \int_{t-(W-|\tau|)/2}^{t+(W-|\tau|)/2} X\left(u+\frac{\tau}{2}\right) X\left(u-\frac{\tau}{2}\right) e^{-i2\pi\alpha u} \, du$$

$$= \frac{1}{W} \int_{-(W-|\tau|)/2}^{(W-|\tau|)/2} X\left(t+u+\frac{\tau}{2}\right) X\left(t+u-\frac{\tau}{2}\right) e^{-i2\pi\alpha(t+u)} \, du.$$

(12.25)

Therefore, (12.22) can be reexpressed as

$$S_X^\alpha(f) = \lim_{W\to\infty} \int_{-\infty}^{\infty} \frac{1}{W} \int_{-(W-|\tau|)/2}^{(W-|\tau|)/2} \lim_{Z\to\infty} \frac{1}{Z} \int_{-Z/2}^{Z/2} R_X\left(t+u+\frac{\tau}{2}, t+u-\frac{\tau}{2}\right)$$

$$\times e^{-i2\pi\alpha(t+u)} \, dt \, du \, e^{-i2\pi f\tau} \, d\tau$$

(12.26)

by substitution of (12.23) to (12.25) into (12.22), and interchange of the operations of Fourier transformation, expectation, and time averaging. Use of the definition (12.6) in (12.26) yields

$$S_X^\alpha(f) = \lim_{W\to\infty} \int_{-\infty}^{\infty} \frac{1}{W} \int_{-(W-|\tau|)/2}^{(W-|\tau|)/2} du \, R_X^\alpha(\tau) e^{-i2\pi f\tau} \, d\tau \quad (12.27)$$

$$= \lim_{W\to\infty} \int_{-\infty}^{\infty} \left(1 - \frac{|\tau|}{W}\right) R_X^\alpha(\tau) e^{-i2\pi f\tau} \, d\tau. \quad (12.28)$$

Application of the convolution theorem to (12.28) yields

$$S_X^\alpha(f) = \lim_{W\to\infty} V_{1/W}(f) \otimes \int_{-\infty}^{\infty} R_X^\alpha(\tau) e^{-i2\pi f\tau} \, d\tau, \quad (12.29)$$

where

$$V_{1/W}(f) \triangleq W\left[\frac{\sin(\pi W f)}{\pi W f}\right]^2. \quad (12.30)$$

Now, since

$$\lim_{W\to\infty} V_{1/W}(f) = \delta(f), \quad (12.31)$$

then the limit of the spectral correlation as the spectral resolution $\Delta f = 1/W$ becomes infinitesimal, (12.22), is given by

$$S_X^\alpha(f) = \int_{-\infty}^{\infty} R_X^\alpha(\tau) e^{-i2\pi f\tau} \, d\tau. \quad (12.32)$$

A dual interpretation of the cyclic spectrum can be obtained from the definition (12.22), by interchanging the operations of expectation and time

is nonzero and is a valid autocorrelation function. That is, there exists a stationary process, say $Y(t)$, such that

$$R_Y(\tau) = \langle R_X \rangle(t - u), \qquad \tau = t - u. \tag{12.38}$$

Similarly it can be shown that the *synchronized time-averaged autocorrelation* defined by

$$\langle R_X \rangle_T(t, u) \triangleq \lim_{N \to \infty} \frac{1}{2N + 1} \sum_{n=-N}^{N} R_X(t + nT, u + nT), \tag{12.39}$$

where $T = 1/\alpha_o$, is a valid autocorrelation function. That is, there exists a cyclostationary process, say $Z(t)$, such that

$$R_Z(t, u) = \langle R_X \rangle_T(t, u). \tag{12.40}$$

However, the processes Y and Z that satisfy (12.38) and (12.40) cannot in general be derived from the process X. That is, Y and Z are in general not stationary and cyclostationary components of the process X, in any sense. Nevertheless, $\langle R_X \rangle(t - u)$ and $\langle R_X \rangle_T(t, u)$ will be referred to as the *stationary* and *cyclostationary components*, respectively, of the autocorrelation $R_X(t, u)$. The stationary component is simply the $\alpha = 0$ cyclic autocorrelation of X, (12.7). Similarly, the cyclostationary component is the sum of cyclic autocorrelations corresponding to the period T (exercise 5),

$$\langle R_X \rangle_T(t, u) = \sum_{m=-\infty}^{\infty} R_X^{m/T}(t - u) e^{i\pi m(t+u)/T}. \tag{12.41}$$

The identity between (12.39) and (12.41), which can be reexpressed, using (12.6), as

$$\lim_{N \to \infty} \frac{1}{2N + 1} \sum_{n=-N}^{N} R_X(t + nT, u + nT)$$

$$\equiv \sum_{m=-\infty}^{\infty} \lim_{Z \to \infty} \frac{1}{Z} \int_{-Z/2}^{Z/2} R_X(t + v, u + v) e^{-i2\pi mv/T} dv, \tag{12.42}$$

is called the *synchronized averaging identity*. It describes the equivalence between two alternative ways to obtain a periodic component of a function, in this case the autocorrelation function. For an almost cyclostationary process, the autocorrelation can be represented by its cyclostationary and stationary components as follows:

$$R_X(t, u) = \langle R_X \rangle(t - u) + \sum_T \left[\langle R_X \rangle_T(t, u) - \langle R_X \rangle(t - u) \right], \tag{12.43}$$

in which the sum is over all fundamental periods of the almost periodic function R_X (exercise 5).

Example 2: Consider the sum of two amplitude-modulated sine waves,

$$X(t) = Y(t)\cos(2\pi f_1 t + \phi_1) + Z(t)\cos(2\pi f_2 t + \phi_2),$$

in which $Y(t)$ and $Z(t)$ are uncorrelated, zero-mean, stationary processes, and f_1/f_2 is irrational. The autocorrelation function is easily shown to be

$$R_X(t,u) = \tfrac{1}{2}R_Y(t-u)\big[\cos(2\pi f_1[t-u]) + \cos(2\pi f_1[t+u] + 2\phi_1)\big]$$
$$+ \tfrac{1}{2}R_Z(t-u)\big[\cos(2\pi f_2[t-u]) + \cos(2\pi f_2[t+u] + 2\phi_2)\big],$$

from which the definition (12.39) yields (using $T_1 = 1/f_1$ and $T_2 = 1/f_2$)

$$\langle R_X\rangle_{T_1}(t,u) = \tfrac{1}{2}R_Y(t-u)\big[\cos(2\pi f_1[t-u])$$
$$+ \cos(2\pi f_1[t+u] + 2\phi_1)\big]$$
$$+ \tfrac{1}{2}R_Z(t-u)\cos(2\pi f_2[t-u])$$

and

$$\langle R_X\rangle_{T_2}(t,u) = \tfrac{1}{2}R_Z(t-u)\big[\cos(2\pi f_2[t-u])$$
$$+ \cos(2\pi f_2[t+u] + 2\phi_2)\big]$$
$$+ \tfrac{1}{2}R_Y(t-u)\cos(2\pi f_1[t-u]),$$

and the definition (12.37) yields

$$\langle R_X\rangle(t-u) = \tfrac{1}{2}R_Y(t-u)\cos(2\pi f_1[t-u])$$
$$+ \tfrac{1}{2}R_Z(t-u)\cos(2\pi f_2[t-u]).$$

It is easily verified that (12.43) is satisfied with two terms in the sum, corresponding to $T = T_1$ and $T = T_2$. ∎

12.4 Linear Periodically Time-Variant Transformations

12.4.1 General Input-Output Relations

A particularly common situation in which cyclostationarity arises is that for which a process, say $X(t)$, with an autocorrelation possessing a nonzero stationary component (viz., an asymptotically mean stationary process), is subjected to a linear periodically time-variant (LPTV) transformation. For example, many modulation systems can be modeled as the scalar response of a multi-input LPTV transformation with stationary excitation. This includes amplitude modulation (double sideband, single sideband, vestigial sideband, and with or without suppressed carrier), phase and frequency modulation, quadrature amplitude modulation, pulse-amplitude modulation, all synchronous digital modulations such as phase-shift keying, frequency-shift keying, and so on (several of these examples are described in Section 12.5). Consequently, the study of cyclostationarity is facilitated by general formulas that describe cyclic spectra (spectral correlation functions) in terms of the parameters of LPTV transformations. This includes cyclic spectra that are *generated* by LPTV transformations of processes that exhibit no cyclostationarity, as well as cyclic spectra that are transformed by LPTV transformations of processes that do exhibit cyclostationarity.

Let us consider the LPTV transformation

$$Y(t) = \int_{-\infty}^{\infty} \boldsymbol{h}(t, u) \boldsymbol{X}(u) \, du, \qquad (12.44)$$

for which $\boldsymbol{X}(t)$ is a (column) vector excitation, $Y(t)$ is a scalar response, and $\boldsymbol{h}(t, u) = \boldsymbol{h}(t + T, u + T)$ is the (row) vector of impulse-response functions that specify the transformation. The function $\boldsymbol{h}(t + \tau, t)$ is periodic in t for each τ, and can therefore be represented by the Fourier series (assumed to converge)

$$\boldsymbol{h}(t + \tau, t) = \sum_{n=-\infty}^{\infty} \boldsymbol{g}_n(\tau) e^{i 2\pi n t / T}, \qquad (12.45)$$

where

$$\boldsymbol{g}_n(\tau) \triangleq \frac{1}{T} \int_{-T/2}^{T/2} \boldsymbol{h}(t + \tau, t) e^{-i 2\pi n t / T} \, dt. \qquad (12.46)$$

The *system function*, which is defined to be the Fourier transform

$$\boldsymbol{G}(t, f) \triangleq \int_{-\infty}^{\infty} \boldsymbol{h}(t, t - \tau) e^{-i 2\pi f \tau} \, d\tau, \qquad (12.47)$$

can therefore also be represented by a Fourier series; that is, substitution of (12.45) into (12.47) yields

$$\boldsymbol{G}(t, f) = \int_{-\infty}^{\infty} \left[\sum_{n=-\infty}^{\infty} \boldsymbol{g}_n(\tau) e^{-i 2\pi n \tau / T} e^{i 2\pi n t / T} \right] e^{-i 2\pi f \tau} \, d\tau$$

$$= \sum_{n=-\infty}^{\infty} \boldsymbol{G}_n\!\left(f + \frac{n}{T}\right) e^{i 2\pi n t / T}, \qquad (12.48)$$

where

$$\boldsymbol{G}_n(f) \triangleq \int_{-\infty}^{\infty} \boldsymbol{g}_n(\tau) e^{-i 2\pi f \tau} \, d\tau. \qquad (12.49)$$

By substitution of (12.45) into (12.44) into the definition of the cyclic autocorrelation, (12.6), it can be shown (exercise 6) that*

$$R_Y^\alpha(\tau) = \sum_{n,m=-\infty}^{\infty} \text{tr}\left\{ \left[R_X^{\alpha-(n-m)/T}(\tau) e^{-i\pi(n+m)\tau/T} \right] \otimes r_{nm}^\alpha(-\tau) \right\},$$

(12.50)

where R_X^β is the matrix of *cyclic cross-correlations* of the elements of $X(t)$,

$$R_X^\beta(\tau) \triangleq \lim_{Z \to \infty} \frac{1}{Z} \int_{-Z/2}^{Z/2} R_X\left(t + \frac{\tau}{2}, t - \frac{\tau}{2}\right) e^{-i2\pi\beta t} dt, \quad (12.51)$$

and r_{nm}^α is the matrix of *cyclic finite cross-correlations*,

$$r_{nm}^\alpha(\tau) \triangleq \int_{-\infty}^{\infty} g_n^T\left(t + \frac{\tau}{2}\right) g_m^*\left(t - \frac{\tau}{2}\right) e^{-i2\pi\alpha t} dt. \quad (12.52)$$

Fourier transformation of (12.50) and application of the convolution theorem yields (exercise 6)

$$S_Y^\alpha(f) = \sum_{n,m=-\infty}^{\infty} G_n\left(f + \frac{\alpha}{2}\right) S_X^{\alpha-(n-m)/T}\left(f - \frac{n+m}{2T}\right) G_m^T\left(f - \frac{\alpha}{2}\right)^*.$$

(12.53)

Equations (12.50) to (12.53) reveal that the set of cyclic autocorrelations and the set of cyclic spectra are each self-determinant characteristics under an LPTV transformation, in the sense that the only features of the excitation that determine the cyclic autocorrelations (cyclic spectra) of the response are the cyclic autocorrelations (cyclic spectra) of the excitation.

In the special case of a linear time-invariant transformation,

$$h(t, u) = h(t - u), \quad (12.54)$$

(12.50) to (12.53) reduce to

$$R_Y^\alpha(\tau) = \text{tr}\{ R_X^\alpha(\tau) \otimes r_h^\alpha(-\tau) \} \quad (12.55)$$

and

$$S_Y^\alpha(f) = H\left(f + \frac{\alpha}{2}\right) S_X^\alpha(f) H^T\left(f - \frac{\alpha}{2}\right)^*, \quad (12.56)$$

for which $r_h^\alpha(\tau)$ is the matrix of cyclic finite cross-correlations,

$$r_h^\alpha(\tau) \triangleq \int_{-\infty}^{\infty} h^T\left(t + \frac{\tau}{2}\right) h\left(t - \frac{\tau}{2}\right) e^{-i2\pi\alpha t} dt, \quad (12.57)$$

and $H(f)$ is the Fourier transform of $h(\tau)$. Also, in the special case for

---

*In (12.50), tr{·} is the trace operation, and in (12.52), the superscript T denotes matrix transposition, which should not be confused with the period T.

which the excitation $X(t)$ exhibits no cyclostationarity, (12.50) to (12.53) reduce to (exercise 6)

$$R_Y^\alpha(\tau) = \begin{cases} \sum_{n=-\infty}^{\infty} \text{tr}\{[\langle R_X \rangle(\tau)e^{-i\pi(2n-p)\tau/T}] \otimes r_{n(n-p)}^\alpha(-\tau)\}, & \alpha = p/T, \\ 0, & \alpha \neq p/T, \end{cases}$$
(12.58)

and

$$S_Y^\alpha(f) = \begin{cases} \sum_{m=-\infty}^{\infty} G_{m+p}\left(f + \frac{\alpha}{2}\right)\langle S_X \rangle\left(f - \frac{\alpha}{2} - \frac{m}{T}\right)G_m^T\left(f - \frac{\alpha}{2}\right)^*, \\ \alpha = p/T, \\ 0, \quad \alpha \neq p/T, \end{cases}$$
(12.59)

for all integers p. In this case, $Y(t)$ is cyclostationary with period T if $X(t)$ is stationary.

By substitution of (12.45) into (12.44) into the definition of the cyclic cross-correlation,

$$R_{XY}^\alpha(\tau) \triangleq \lim_{Z \to \infty} \frac{1}{Z} \int_{-Z/2}^{Z/2} R_{XY}\left(t + \frac{\tau}{2}, t - \frac{\tau}{2}\right) e^{-i2\pi\alpha t} dt, \quad (12.60)$$

it can be shown (exercise 7) that

$$R_{XY}^\alpha(\tau) = \sum_{m=-\infty}^{\infty} \left[R_X^{\alpha+m/T}(\tau) e^{i\pi m\tau/T}\right] \otimes \left[g_m^T(-\tau)^* e^{i\pi\alpha\tau}\right]. \quad (12.61)$$

Fourier transformation of (12.61) and application of the convolution theorem yields

$$S_{XY}^\alpha(f) = \sum_{m=-\infty}^{\infty} S_X^{\alpha+m/T}\left(f - \frac{m}{2T}\right) G_m^T\left(f - \frac{\alpha}{2}\right)^*, \quad (12.62)$$

where $S_{XY}^\alpha(f)$ is the correlation of the spectral components at frequencies $f + \alpha/2$ and $f - \alpha/2$ in X and Y, respectively, and

$$S_{XY}^\alpha(f) = \int_{-\infty}^{\infty} R_{XY}^\alpha(\tau) e^{-i2\pi f\tau} d\tau \quad (12.63)$$

In the special case of a linear time-invariant transformation, (12.61) and (12.62) reduce to (exercise 7)

$$R_{XY}^\alpha(\tau) = R_X^\alpha(\tau) \otimes \left[h^T(-\tau) e^{i\pi\alpha\tau}\right] \quad (12.64)$$

and

$$S_{XY}^\alpha(f) = S_X^\alpha(f) H^T\left(f - \frac{\alpha}{2}\right)^*. \quad (12.65)$$

Also, in the special case for which the excitation $X(t)$ exhibits no cyclostationarity, (12.61) and (12.62) reduce to (exercise 7)

$$R^\alpha_{XY}(\tau) = \begin{cases} [\langle R_X \rangle(\tau) e^{-i\pi\alpha\tau}] \otimes [g^T_p(-\tau) e^{i\pi\alpha\tau}], & \alpha = p/T, \\ 0, & \alpha \neq p/T, \end{cases}$$

(12.66)

and

$$S^\alpha_{XY}(f) = \begin{cases} \langle S_X \rangle(f + \alpha/2) G^T_p(-f + \alpha/2), & \alpha = p/T, \\ 0, & \alpha \neq p/T, \end{cases}$$

(12.67)

for all integers p.

It follows from (12.67) (exercise 7) that the Fourier coefficients of the system function (12.48) for an LPTV transformation can be determined from the time-averaged spectrum of the excitation and the cross-cyclic spectrum of the excitation and response,

$$G_n\left(f + \frac{n}{T}\right) = S^\alpha_{YX}\left(f + \frac{\alpha}{2}\right)^T \langle S_X \rangle^{-1}(f), \qquad \alpha = n/T. \quad (12.68)$$

This provides a means for identification of LPTV systems, which is discussed further in Section 12.8.5.

The general input-output formulas, (12.50) and (12.53), for periodic transformations are easily generalized for almost periodic transformations, for which the sum over the harmonically related frequencies $\alpha = n/T$ in (12.45) is generalized to include all values of α for which the Fourier coefficient functions,

$$g_n(\tau) \triangleq \lim_{Z \to \infty} \frac{1}{Z} \int_{-Z/2}^{Z/2} h(t + \tau, t) e^{-i2\pi\alpha_n t} dt,$$

are not identically zero. The generalized versions of (12.50) and (12.53) simply sum over all corresponding values of frequencies, rather than just the values n/T and m/T.

12.4.2 Rice's Representation

As explained in Section 11.2, any random process $X(t)$ can be expressed in the *quadrature-amplitude-modulation* (QAM) form

$$X(t) = U(t)\cos(2\pi f_o t) - V(t)\sin(2\pi f_o t), \qquad (12.69)$$

for any value of f_o, provided that $U(t)$ and $V(t)$ are given by

$$U(t) = X(t)\cos(2\pi f_o t) + Y(t)\sin(2\pi f_o t), \qquad (12.70a)$$
$$V(t) = Y(t)\cos(2\pi f_o t) - X(t)\sin(2\pi f_o t), \qquad (12.70b)$$

for any auxiliary process $Y(t)$. This is easily verified by substitution of (12.70) into (12.69) and use of a standard trigonometric identity. The QAM representation is particularly useful when $X(t)$ is a bandpass process with spectrum concentrated near $f = f_o$, because then $U(t)$ and $V(t)$ can both be made low-pass processes, with spectra concentrated around $f = 0$, by appropriate choice of $Y(t)$. An especially appropriate choice of $Y(t)$ is the Hilbert transform of $X(t)$,

$$Y(t) = h(t) \otimes X(t), \qquad (12.71a)$$

for which

$$h(t) = 1/\pi t, \qquad (12.71b)$$

$$H(f) = \begin{cases} -i, & f > 0, \\ +i, & f < 0. \end{cases} \qquad (12.71c)$$

In this case, it is shown in Section 11.2 that if $X(t)$ is bandlimited to $f \in (f_o - B, f_o + B)$ [and the image band $(-f_o - B, -f_o + B)$], then $U(t)$ and $V(t)$ are bandlimited to $f \in (-B, B)$. Furthermore, given any process in the form (12.69), with $U(t)$ and $V(t)$ bandlimited to $f \in (-B, B)$ for $B < f_o$, it is shown that $U(t)$ and $V(t)$ are *uniquely* determined by $X(t)$ and are given by (12.70), with $Y(t)$ defined by (12.71). Moreover, it can be shown that for any process $X(t)$, (12.69) to (12.71) yield a unique definition of envelope magnitude. Specifically, (12.69) can be reexpressed as

$$X(t) = A(t)\cos[2\pi f_o t + \Phi(t)], \qquad (12.72)$$

for which

$$A(t) = [U^2(t) + V^2(t)]^{1/2} \qquad (12.73a)$$

$$\Phi(t) = \tan^{-1}\left[\frac{V(t)}{U(t)}\right]. \qquad (12.73b)$$

Using the characterizations from (12.70),

$$U(t) = \text{Re}\{[X(t) + iY(t)]e^{-i2\pi f_o t}\}, \qquad (12.74a)$$

$$V(t) = \text{Im}\{[X(t) + iY(t)]e^{-i2\pi f_o t}\}, \qquad (12.74b)$$

it easily follows that

$$A(t) = |X(t) + iY(t)| \qquad (12.75a)$$

$$\Phi(t) = \arg\{X(t) + iY(t)\} - 2\pi f_o t. \qquad (12.75b)$$

Therefore, the envelope magnitude $A(t)$ in (12.73) is independent of the arbitrary choice for f_o; it is determined solely by $X(t)$ (and the Hilbert transform of $X(t)$).

This QAM representation (12.69)–(12.72), called *Rice's representation*, is valid regardless of the probabilistic model for $X(t)$. That is, $X(t)$ can be stationary, cyclostationary, almost cyclostationary, or more generally nonstationary. A complete study of the correlation and spectral properties, including the cyclic correlations and cyclic spectra, for $X(t)$ and its in-phase and quadrature components $U(t)$ and $V(t)$ can be based on one general formula for QAM processes. Specifically, let us consider a process, say $Q(t)$, in the QAM form

$$Q(t) = Z(t)\cos(2\pi f_o t) + W(t)\sin(2\pi f_o t). \quad (12.76)$$

This is a particular LPTV transformation of the two-dimensional vector of processes $\{Z(t), W(t)\}$, for which the vector of impulse-response functions is

$$\mathbf{h}(t, u) = \{\cos(2\pi f_o t)\delta(t - u), \sin(2\pi f_o t)\delta(t - u)\}, \quad (12.77)$$

and the vector of corresponding system functions is

$$\mathbf{G}(t, f) = \{\cos(2\pi f_o t), \sin(2\pi f_o t)\}. \quad (12.78)$$

Application of (12.50) yields (exercise 8)

$$R_Q^\alpha(\tau) = \frac{1}{2}[R_Z^\alpha(\tau) + R_W^\alpha(\tau)]\cos(2\pi f_o \tau)$$
$$+ \frac{1}{2}[R_{WZ}^\alpha(\tau) - R_{ZW}^\alpha(\tau)]\sin(2\pi f_o \tau)$$
$$+ \frac{1}{4}\sum_{n=-1,1}\{[R_Z^{\alpha+2nf_o}(\tau) - R_W^{\alpha+2nf_o}(\tau)]$$
$$+ ni[R_{WZ}^{\alpha+2nf_o}(\tau) + R_{ZW}^{\alpha+2nf_o}(\tau)]\}, \quad (12.79)$$

and application of (12.53) yields (exercise 8)

$$S_Q^\alpha(f) = \frac{1}{4}\sum_{n=-1,1}\{[S_W^\alpha(f + nf_o) + S_Z^\alpha(f + nf_o)]$$
$$+ ni[S_{WZ}^\alpha(f + nf_o) - S_{ZW}^\alpha(f + nf_o)]\}$$
$$+ \frac{1}{4}\sum_{n=-1,1}\{[S_Z^{\alpha+2nf_o}(f) - S_W^{\alpha+2nf_o}(f)]$$
$$+ ni[S_{WZ}^{\alpha+2nf_o}(f) + S_{ZW}^{\alpha+2nf_o}(f)]\}. \quad (12.80)$$

From (12.79) and its Fourier transform (12.80), we can determine all cyclic correlations and cyclic spectra for $X(t)$, $U(t)$, and $V(t)$, since each of the three representations (12.69), (12.70a), and (12.70b) is of the form (12.76). That is, the spectral correlation in $U(t)$ and $V(t)$ due to the spectral correlation in $X(t)$ can be determined, and similarly the spectral correlation in $X(t)$ due to that in $U(t)$ and $V(t)$ can be determined. For example, with the use of $Q = X$, $Z = U$, and $W = -V$, and selection of $\alpha = 0$, (12.79) yields (exercise 8)

$$\langle R_X \rangle(\tau) = \frac{1}{2}[\langle R_U \rangle(\tau) + \langle R_V \rangle(\tau)]\cos(2\pi f_o \tau)$$
$$+ \frac{1}{2}[\langle R_{UV} \rangle(\tau) - \langle R_{VU} \rangle(\tau)]\sin(2\pi f_o \tau)$$
$$+ \frac{1}{4} \sum_{n=-1,1} \{[R_U^{2nf_o}(\tau) - R_V^{2nf_o}(\tau)]$$
$$- ni[R_{UV}^{2nf_o}(\tau) + R_{VU}^{2nf_o}(\tau)]\}. \quad (12.81)$$

This result reveals that the conventional formula (e.g., [Papoulis, 1984]), which omits the terms in the sum over $n = -1, 1$, is correct only if $U(t)$ and $V(t)$ exhibit no cyclostationarity with frequency $\alpha = \pm 2f_o$ [e.g., if $U(t)$ and $V(t)$ are bandlimited to $f \in (-f_o, f_o)$], or the cyclostationarity is *balanced* in the sense that

$$R_U^{\pm 2f_o}(\tau) \equiv R_V^{\pm 2f_o}(\tau), \quad (12.82a)$$
$$R_{UV}^{\pm 2f_o}(\tau) \equiv -R_{UV}^{\pm 2f_o}(-\tau). \quad (12.82b)$$

As another example, with the use of $Q = X$, $Z = U$, and $W = -V$, selection of $\alpha = \pm 2f_o$, and the assumption that $U(t)$ and $V(t)$ exhibit no cyclostationarity, (12.79) yields (exercise 8)

$$R_X^{\pm 2f_o}(\tau) = \frac{1}{4}[\langle R_U \rangle(\tau) - \langle R_V \rangle(\tau)] \pm \frac{i}{4}[\langle R_{UV} \rangle(\tau) + \langle R_{VU} \rangle(\tau)],$$
$$(12.83a)$$

and also for other values of α

$$R_X^\alpha(\tau) \equiv 0, \quad \alpha \neq \pm 2f_o, 0. \quad (12.83b)$$

This result reveals that $X(t)$ exhibits no cyclostationarity if and only if the correlations of $U(t)$ and $V(t)$ are *balanced* in the sense that

$$\langle R_U \rangle(\tau) \equiv \langle R_V \rangle(\tau), \quad (12.84a)$$
$$\langle R_{UV} \rangle(\tau) \equiv -\langle R_{UV} \rangle(-\tau). \quad (12.84b)$$

Otherwise, $X(t)$ exhibits cyclostationarity, with period $1/2f_o$. Similarly, it can be shown through use of $Q = U$, $Z = X$, and $W = Y$, and also $Q = V$, $Z = Y$, and $W = -X$, in (12.80), that if $X(t)$ exhibits cyclostationarity with only the period $1/2f_o$, then $U(t)$ and $V(t)$ exhibit no cyclostationarity if and only if the cyclic spectrum of $X(t)$ is bandlimited:

$$S_X^{\pm 2f_o}(f) = 0, \qquad |f| \geq f_o. \tag{12.85}$$

This necessary and sufficient condition is satisfied if and only if either $X(t)$ is bandlimited so that

$$\langle S_x \rangle(f) = 0, \qquad |f| \geq 2f_o, \tag{12.86}$$

or $U(t)$ and $V(t)$ are *balanced out of band* in the sense that

$$\langle S_U \rangle(f) = \langle S_V \rangle(f), \qquad |f| > f_o, \tag{12.87a}$$
$$\langle S_{VU} \rangle(f) = -\langle S_{VU} \rangle(-f), \qquad |f| > f_o. \tag{12.87b}$$

Moreover, it can be shown that the cyclostationarity of $X(t)$ at cycle frequency α depends on the cyclostationarity of $U(t)$ and $V(t)$ at *only* the cycle frequency α if and only if the cyclic correlations are balanced in the sense that

$$R_U^{\alpha \pm 2f_o}(\tau) \equiv R_V^{\alpha \pm 2f_o}(\tau), \tag{12.88a}$$
$$R_{UV}^{\alpha \pm 2f_o}(\tau) \equiv -R_{UV}^{\alpha \pm 2f_o}(-\tau). \tag{12.88b}$$

Otherwise, there is dependence on the cyclostationarity of $U(t)$ and $V(t)$ at the cycle frequencies $\alpha \pm 2f_o$, as well as at α.

The only relations needed, in addition to (12.80), to completely determine the cyclic spectra of $U(t)$ and $V(t)$ in terms of the cyclic spectra of $X(t)$, are the following cyclic spectra for Hilbert transforms (which follow from (12.56), (12.65), and (12.71)):

$$S_Y^\alpha(f) = \begin{cases} -S_X^\alpha(f), & |f| < |\alpha|/2, \\ +S_X^\alpha(f), & |f| > |\alpha|/2, \end{cases} \tag{12.89}$$

$$S_{XY}^\alpha(f) = S_{YX}^\alpha(-f) = \begin{cases} -iS_X^\alpha(f), & f < \alpha/2, \\ +iS_X^\alpha(f), & f > \alpha/2. \end{cases} \tag{12.90}$$

Use of $Q = U$, $Z = X$, and $W = Y$ in (12.80), together with (12.89) and (12.90), yields (exercise 8)

$$S_U^\alpha(f) = S_X^\alpha(f+f_o)u\!\left(f+f_o-\frac{|\alpha|}{2}\right) + S_X^\alpha(f-f_o)u\!\left(-f+f_o-\frac{|\alpha|}{2}\right)$$
$$+ S_X^{\alpha+2f_o}(f)u\!\left(\frac{\alpha}{2}+f_o-|f|\right) + S_X^{\alpha-2f_o}(f)u\!\left(-\frac{\alpha}{2}+f_o-|f|\right). \tag{12.91a}$$

Similarly, use of $Q = V$, $Z = Y$, and $W = -X$ in (12.80), together with

(12.89) and (12.90), yields

$$S_V^\alpha(f) = S_X^\alpha(f+f_o)u\left(f+f_o - \frac{|\alpha|}{2}\right) + S_X^\alpha(f-f_o)u\left(-f+f_o - \frac{|\alpha|}{2}\right)$$
$$- S_X^{\alpha+2f_o}(f)u\left(\frac{\alpha}{2}+f_o - |f|\right) - S_X^{\alpha-2f_o}(f)u\left(-\frac{\alpha}{2}+f_o - |f|\right).$$
(12.91b)

Also, the cross-cyclic spectra of U and V are given by (exercise 8)

$$S_{UV}^\alpha(f) = i\left[S_X^\alpha(f+f_o)u\left(f+f_o - \frac{|\alpha|}{2}\right) - S_X^\alpha(f-f_o)u\left(-f+f_o - \frac{|\alpha|}{2}\right)\right.$$
$$\left. - S_X^{\alpha+2f_o}(f)u\left(\frac{\alpha}{2}+f_o - |f|\right) + S_X^{\alpha-2f_o}(f)u\left(-\frac{\alpha}{2}+f_o - |f|\right)\right].$$
(12.92)

In (12.91) and (12.92), u is the unit step function. It follows from (12.91) and (12.92) that if $X(t)$ exhibits no cyclostationarity, then neither $U(t)$ nor $V(t)$ exhibits cyclostationarity.

12.4.3 Sampling and Aliasing

Consider the discrete-time process $\{X(nT): n = 0, \pm1, \pm2, \pm3, \dots\}$ that is obtained by periodically sampling a continuous-time process $X(t)$, and let us determine the relationship between the cyclic spectra of $X(t)$ and $\{X(nT)\}$. Since the definition of the symmetric version of the cyclic autocorrelation, (12.6), cannot be directly extended to discrete-time processes (because the data $\{X(kT/2)\}$ do not exist for odd integers k), the asymmetric version,

$$\lim_{Z \to \infty} \frac{1}{Z} \int_{-Z/2}^{Z/2} R_X(t+\tau, t) e^{-i2\pi\alpha t} \, dt = R_X^\alpha(\tau) e^{i\pi\alpha\tau}, \quad (12.93)$$

is extended, and the sinusoidal factor $e^{-i\pi\alpha kT}$ suggested by (12.93) is introduced to obtain a discrete-time counterpart (an indirect extension) to the symmetric version. Specifically, the *cyclic autocorrelation* for a discrete-time process $\{X(nT)\}$ is defined by*

$$\tilde{R}_X^\alpha(kT) \triangleq \lim_{N \to \infty} \frac{1}{2N+1} \sum_{n=-N}^{N} \tilde{R}_X(nT+kT, nT) e^{-i2\pi\alpha(n+k/2)T}.$$
(12.94)

---

*Whereas the same notation is used for autocorrelations and spectra of both discrete-time and continuous-time processes in earlier chapters, the tilde ~ is used here to distinguish between these two cases.

Motivated by the cyclic Wiener-Khinchin relation (12.32), the *cyclic spectrum* for $\{X(nT)\}$ is defined by

$$\tilde{S}_X^\alpha(f) \triangleq \sum_{k=-\infty}^{\infty} \tilde{R}_X^\alpha(kT) e^{-i2\pi kTf}. \tag{12.95}$$

In order to determine the relationship between the cyclic autocorrelations of $X(t)$ and $\{X(nT)\}$, a form of the synchronized averaging identity (12.42) is applied to the definition (12.94) to obtain (exercise 9)

$$\tilde{R}_X^\alpha(kT) = \sum_{m=-\infty}^{\infty} R_X^{\alpha+m/T}(kT) e^{i\pi mk}. \tag{12.96}$$

Substitution of the relation (12.96) into the definition (12.95) yields

$$\tilde{S}_X^\alpha(f) = \frac{1}{T} \sum_{n,m=-\infty}^{\infty} S_X^{\alpha+m/T}\left(f - \frac{m}{2T} - \frac{n}{T}\right). \tag{12.97}$$

Thus, $\tilde{S}_X^\alpha(f)$ exhibits the periodicity properties

$$\tilde{S}_X^\alpha(f + 1/T) = \tilde{S}_X^\alpha(f), \tag{12.98a}$$

$$\tilde{S}_X^{\alpha+2/T}(f) = \tilde{S}_X^\alpha(f), \tag{12.98b}$$

$$\tilde{S}_X^{\alpha+1/T}(f - 1/2T) = \tilde{S}_X^\alpha(f). \tag{12.98c}$$

We see from (12.97) that the cyclic spectrum of $X(t)$ is, in general, not obtainable from the cyclic spectrum of $\{X(nT)\}$, due to aliasing effects in both α and f. However, if $X(t)$ is bandlimited in mean square to the Nyquist bandwidth,

$$\langle S_X \rangle(f) = 0, \qquad |f| \geq B < 1/2T, \tag{12.99}$$

then the bandwidth property (12.36), applied to (12.97), reveals that the support, in the (f, α) plane, of each of the terms in (12.97), indexed by n and m, is disjoint from the support of all other terms, as shown in Figure 12.2; that is, aliasing does not prevent recovery of S_X^α from $\tilde{S}_X^\alpha$:

$$S_X^\alpha(f) = \tilde{S}_X^\alpha(f), \qquad |f| < \frac{1}{2}\left[\frac{1}{T} - |\alpha|\right], \tag{12.100a}$$

$$S_X^\alpha(f) = 0, \qquad |f| \geq \frac{1}{2}\left[\frac{1}{T} - |\alpha|\right]. \tag{12.100b}$$

It should be noted that even if $X(t)$ is bandlimited to the Nyquist bandwidth, (12.99), the cyclic autocorrelation of $X(t)$ cannot be obtained simply

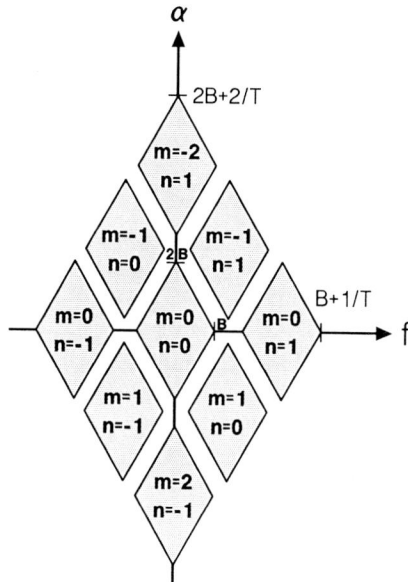

Figure 12.2 Support for the First Nine Terms in Equation (12.97) for the Cyclic Spectra of Mean-Square Bandlimited Time-Sampled Waveform for Which $S_X(f) = 0$ When $|f| \geq B$ (Vertical Scale Compressed)

by inverse Fourier transformation of the cyclic spectrum of $\{X(nT)\}$, because although (12.100a) holds, (12.100b) and (12.98) reveal that aliasing results in the inequality

$$\tilde{S}_X^\alpha(f) \neq S_X^\alpha(f), \qquad |f| \geq \frac{1}{2}\left[\frac{1}{T} - |\alpha|\right],$$

in general. Thus, $\tilde{S}_X^\alpha(f)$ does not equal $S_X^\alpha(f)$ for all $|f| < 1/2T$ and $|\alpha| < 1/T$. Hence, the inversion formula needed is

$$R_X^\alpha(\tau) = \int_{-B+|\alpha|/2}^{B-|\alpha|/2} \tilde{S}_X^\alpha(f) e^{i2\pi f\tau} df$$

for $B \leq 1/2T$. On the other hand, if $X(t)$ is bandlimited to half the Nyquist bandwidth ($B \leq 1/4T$), then R_X^α can be recovered from $\tilde{S}_X^\alpha$ by inverse Fourier transformation over the fixed band $[-B, B]$.

If $X(t)$ is not bandlimited, even the conventional spectrum is affected by aliasing in α as well as f:

$$\langle \tilde{S}_x \rangle(f) = \frac{1}{T} \sum_{n,m=-\infty}^{\infty} S_X^{m/T}\left(f - \frac{m}{2T} - \frac{n}{T}\right). \qquad (12.101)$$

This is generally unrecognized. Only in the case for which $X(t)$ exhibits no cyclostationarity do we obtain the relationship which is known for stationary processes,

$$\langle \tilde{S}_X \rangle(f) = \frac{1}{T} \sum_{n=-\infty}^{\infty} \langle S_X \rangle \left(f - \frac{n}{T} \right). \quad (12.102)$$

Examples of the difference between (12.101) and (12.102) are described in [Gardner, 1987b].

12.5 Examples of Cyclic Spectra for Modulated Signals

There are various ways that the spectral-correlation characteristics of modulated random signals can be exploited in practice. Specific problem areas where spectral correlation has been used or proposed include detection of signals masked by broadband noise and multiple interfering signals, sorting and identification of modulation type for multiple signals hidden in noise, synchronization to pulse timing and carrier phase, extraction of modulated and multiplexed received signals corrupted by noise, channel dispersion, and interference using periodically time-variant filters or arrays of sensors, and multiple source location using sensor arrays. In this section, the spectral-correlation characteristics of a number of the most commonly used analog and digital modulation systems are determined by calculating the cyclic autocorrelations and cyclic spectra, on the basis of specific probabilistic models.* As an extension of the well-known rule of thumb for calculating conventional spectra, it is often easier to calculate the cyclic autocorrelation first, and then employ the cyclic Wiener-Khinchin relation (12.32) to obtain the cyclic spectrum by Fourier transformation, rather than to calculate the cyclic spectrum directly via the cyclic periodogram and the definition (12.22).

12.5.1 Amplitude Modulation

Consider the amplitude-modulated (AM) sine wave

$$X(t) = A(t) \cos(2\pi f_o t + \phi_o). \quad (12.103)$$

This is an LPTV transformation of the process $A(t)$, and therefore (12.50) can be applied to obtain the following formula for the cyclic autocorrelation of $X(t)$ (exercise 10):

$$R_X^\alpha(\tau) = \tfrac{1}{2} R_A^\alpha(\tau) \cos(2\pi f_o \tau) + \tfrac{1}{4} R_A^{\alpha+2f_o}(\tau) e^{-i2\phi_o} + \tfrac{1}{4} R_A^{\alpha-2f_o}(\tau) e^{i2\phi_o}. \quad (12.104)$$

Fourier transformation of (12.104) yields the formula

$$\hat{S}_X^\alpha(f) = \tfrac{1}{4} \big[S_A^\alpha(f+f_o) + S_A^\alpha(f-f_o) \\ + S_A^{\alpha+2f_o}(f) e^{-i2\phi_o} + S_A^{\alpha-2f_o}(f) e^{i2\phi_o} \big] \quad (12.105)$$

for the cyclic spectrum of $X(t)$. This result can be used to obtain the cyclic

*A more comprehensive treatment is given in [Gardner, 1987a].

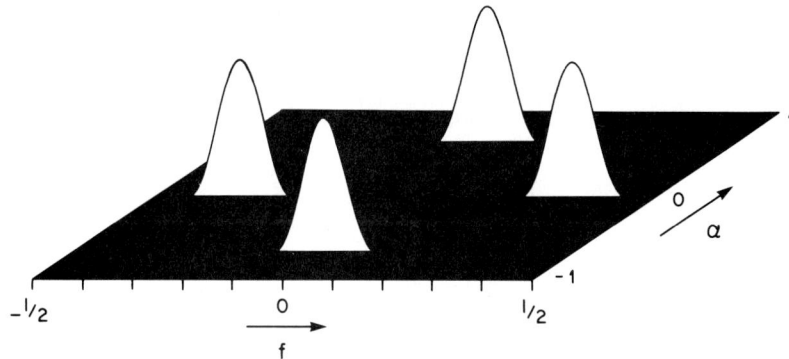

Figure 12.3 Spectral Correlation Surface for AM Signal with Stationary Amplitude

spectra for other types of modulation that involve an amplitude-modulated carrier. An example for which $A(t)$ is cyclostationary is binary phase-shift keying, which is treated in Section 12.5.5. The presence of the last two terms in (12.105) is often not recognized in studies involving frequency conversion. Examples of the error incurred by ignoring these terms are given in [Gardner, 1987b].

For the special case in which the amplitude $A(t)$ is stationary, $X(t)$ is cyclostationary with period $1/2f_o$, and (12.105) reduces to

$$S_X^\alpha(f) = \begin{cases} \frac{1}{4}S_A(f+f_o) + \frac{1}{4}S_A(f-f_o), & \alpha = 0, \\ \frac{1}{4}S_A(f)e^{\pm i2\phi_o}, & \alpha = \pm 2f_o, \\ 0, & \text{otherwise.} \end{cases} \quad (12.106)$$

A graph of a typical spectral-correlation surface, $|S_X^\alpha(f)|$, for an AM signal is shown in Figure 12.3.

It follows from the definition (12.17) and (12.106) that the magnitude of the spectral autocoherence of $X(t)$ is given by

$$|\rho_X^\alpha(f)| = \frac{S_A(f)}{\left[S_A(f)^2 + S_A(f+\alpha)S_A(f) + S_A(f-\alpha)S_A(f) + S_A(f+\alpha)S_A(f-\alpha)\right]^{1/2}}, \quad (12.107)$$

for $\alpha = \pm 2f_o$. Consequently, if $A(t)$ is bandlimited so that

$$S_A(f) \begin{cases} > 0, & |f| < B, \\ = 0, & |f| \geq B, \end{cases} \quad (12.108)$$

with $B < f_o$, then

$$|\rho_X^\alpha(f)| = \begin{cases} 1, & |f| < B \text{ and } |\alpha| = 2f_o, \; 0, \\ 0, & |f| \geq B \text{ or } |\alpha| \neq 2f_o, \end{cases} \quad (12.109)$$

and $X(t)$ is completely coherent for all frequencies f of interest and $\alpha = \pm 2f_o$. But if $B > f_o$, then for $\alpha = \pm 2f_o$

$$|\rho_X^\alpha(f)| < 1, \qquad |f| > 2f_o - B. \tag{12.110}$$

For example, if $A(t)$ is white, then

$$|\rho_X^\alpha(f)| = \tfrac{1}{2}, \qquad -\infty < f < \infty \text{ and } |\alpha| = 2f_o. \tag{12.111}$$

12.5.2 Quadrature Amplitude Modulation

Consider the QAM process

$$X(t) = C(t)\cos(2\pi f_o t) - S(t)\sin(2\pi f_o t). \tag{12.112}$$

The cyclic autocorrelation and cyclic spectrum for $X(t)$ are given by (12.79) and (12.80), respectively, with $Q = X$, $Z = C$, $W = -S$. The dependence of the cyclic spectrum of $X(t)$ on the cyclic spectra of $C(t)$ and $S(t)$ is described and discussed in Section 12.4.2. The results there can be used to obtain the cyclic spectrum for an amplitude-modulated and/or phase-modulated sine wave by use of the identity

$$X(t) = A(t)\cos[2\pi f_o t + \Phi(t)], \tag{12.113a}$$

for which

$$A(t) = \left[C(t)^2 + S(t)^2\right]^{1/2},$$

$$\Phi(t) = \tan^{-1}\left[\frac{S(t)}{C(t)}\right] \tag{12.113b}$$

and

$$C(t) = A(t)\cos\Phi(t),$$
$$S(t) = A(t)\sin\Phi(t). \tag{12.113c}$$

An example for which $C(t)$ and $S(t)$ are cyclostationary is quaternary phase-shift keying, which is treated in Section 12.5.5.

For the special case in which the in-phase and quadrature components, $C(t)$ and $S(t)$, are jointly stationary, $X(t)$ is cyclostationary with period $1/2f_o$, and we have (exercise 11)

$$\langle S_X \rangle(f) = \frac{1}{4}[S_C(f + f_o) + S_C(f - f_o) + S_S(f + f_o) + S_S(f - f_o)]$$
$$- \frac{1}{2}[S_{CS}(f + f_o)_i - S_{CS}(f - f_o)_i], \tag{12.114}$$

and

$$S_X^\alpha(f) = \frac{1}{4}[S_C(f) - S_S(f)] \pm \frac{i}{2}S_{CS}(f)_r, \qquad \alpha = \pm 2f_o, \tag{12.115}$$

with $S_X^\alpha = 0$ for $\alpha \neq 0$ and $\alpha \neq \pm 2f_o$. (In these formulas, the subscripts r and i denote the real and imaginary parts, respectively.) Thus, the cyclic spectrum is of the same general form as that shown in Figure 12.3 for AM, except that the full symmetry shown there is typically not exhibited by QAM. Only symmetry about the f axis and symmetry about the α axis is always exhibited by QAM. Moreover, it follows from the autocoherence inequality (12.34) that the heights of the features centered at $(f = 0, \alpha = \pm 2f_o)$ are lower than or equal to the heights centered at $(f = \pm f_o, \alpha = 0)$.

If $C(t)$ and $S(t)$ are bandlimited so that

$$S_C(f) = S_S(f) = 0, \qquad |f| \geq f_o, \qquad (12.116)$$

then the definition (12.17) and (12.114)–(12.115) yield (exercise 11)

$$|\rho_X^\alpha(f)|^2 = \frac{[S_C(f) - S_S(f)]^2 + 4[S_{CS}(f)_r]^2}{[S_C(f) + S_S(f)]^2 - 4[S_{CS}(f)_i]^2}, \qquad \alpha = \pm 2f_o \qquad (12.117)$$

for the autocoherence of $X(t)$. Moreover, it is established in Section 12.4.2 that any process $X(t)$ that is cyclostationary with period $1/2f_o$ and is bandlimited to $f \in (-2f_o, 2f_o)$ can be represented in the form (12.112) (Rice's representation) for which $C(t)$ and $S(t)$ are purely stationary and bandlimited to $f \in (-f_o, f_o)$. For practical purposes, this includes essentially all analog-modulated sine-wave carriers (with stationary modulating signals) used in conventional communications systems, and essentially all modulated periodic pulse trains (with stationary modulating sequences) with *excess bandwidth* (beyond the Nyquist bandwidth) of 100% or less. Thus, for all such signals, the autocoherence is given by (12.117). Furthermore, (12.117) can be reexpressed as

$$|\rho_X^\alpha(f)|^2 = 1 - \frac{4\left(1 - |\rho_{CS}(f)|^2\right) S_C(f) S_S(f)}{[S_C(f) + S_S(f)]^2 - 4[S_{CS}(f)_i]^2}, \qquad \alpha = \pm 2f_o, \qquad (12.118)$$

and the denominator can be reexpressed as

$$[S_C(f) + S_S(f)]^2 - 4[S_{CS}(f)_i]^2 = 16\left[\langle S_X \rangle (f - f_o)_e^2 - \langle S_X \rangle (f - f_o)_o^2\right],$$
$$0 \leq f < f_o, \qquad (12.119)$$

where the subscripts e and o denote the even and odd parts, respectively, of $S_X(f)$ about the point $f = f_o$ for $f > 0$. Consequently, for given spectrum S_X and spectral product $S_C S_S$, the autocoherence of $X(t)$ increases as the cross-coherence between $C(t)$ and $S(t)$ increases. Furthermore, for given cross-coherence $|\rho_{CS}|$ and spectral product $S_C S_S$, the autocoherence of $X(t)$ increases as the dominance of the even part (about f_o) of the spectrum S_X over the odd part increases [assuming $S_C(f)S_S(f) \neq 0$].

It follows from (12.117) that $X(t)$ is completely incoherent at $\alpha = \pm 2f_o$ and at any f,

$$\rho_X^\alpha(f) = 0, \qquad (12.120)$$

if and only if $C(t)$ and $S(t)$ are *balanced at f*, in the sense that

$$S_C(f) = S_S(f), \qquad (12.121\text{a})$$
$$S_{CS}(f)_r = 0. \qquad (12.121\text{b})$$

It also follows from (12.118) that $X(t)$ is completely coherent at $\alpha = \pm 2f_o$ and at any $|f| < f_o$,

$$|\rho_X^\alpha(f)| = 1, \qquad |f| < f_o, \qquad (12.122)$$

if and only if either (1),

$$S_C(f)S_S(f) = 0, \qquad (12.123\text{a})$$

or (2), $C(t)$ and $S(t)$ are completely cross-coherent at f,

$$|\rho_{CS}(f)| = 1, \qquad |f| < f_o, \qquad (12.123\text{b})$$

and either (12.121a) or (12.121b) (or both) is violated.

When (12.123b) holds, $C(t)$ and $S(t)$ are related (in the mean-square sense) by a linear time-invariant transformation (Section 10.3)

$$S(t) = h(t) \otimes C(t). \qquad (12.124)$$

But there do exist linear time-invariant transformations for which neither (12.121a) nor (12.121b) is violated, namely, those for which the transfer functions have unity magnitude and are purely imaginary with arbitrary signs at arbitrary values of f,

$$H(f) = \pm i. \qquad (12.125)$$

For example, the Hilbert transform (12.71), which yields a single-sideband signal $X(t)$, results in a time series that is completely incoherent for all f. In contrast, a transfer function that is a real constant yields a double-sideband signal $X(t)$ that is completely coherent for all $|f| < f_o$. Similarly, a vestigial-sideband signal, which is obtained by subjecting a double-sideband signal to a low-pass filtering operation with bandwidth, say, $B = f_o + b$, is completely coherent for $|f| < b$, and completely incoherent (for an ideal low-pass filter) for $|f| > b$. For the double-sideband signal, (12.113b) together with the fact that $H(f) = $ constant confirms that there is no phase modulation, $\Phi(t)$ = constant. However, if the double-sideband signal is filtered and the filter transfer function is asymmetric about the point $f = f_o$ (for $f > 0$), then $\Phi(t)$ is no longer constant, but $X(t)$ is still completely coherent for all $|f| < f_o$ (assuming $H(f) \neq 0$ for $|f| < 2f_o$).

In order to determine what type of stationary phase modulation annihilates coherence, one can use the fact that the necessary and sufficient condition, (12.121a) and (12.121b), for complete incoherence for all f [Equation (12.120)] is equivalent to the condition (exercise 12)

$$E\left\{A\left(t + \frac{\tau}{2}\right)A\left(t - \frac{\tau}{2}\right)\exp\left(i\left[\Phi\left(t + \frac{\tau}{2}\right) + \Phi\left(t - \frac{\tau}{2}\right)\right]\right)\right\} \equiv 0. \tag{12.126}$$

For example, if $A(t)$ is statistically independent of $\Phi(t)$ [e.g., $A(t)$ = constant], then (12.126) reduces to

$$\Psi_\tau(1,1) = 0, \quad -\infty < \tau < \infty, \tag{12.127}$$

in which Ψ is the joint characteristic function (which is denoted by Φ in Chapter 2) for the random variables $\Phi(t + \tau/2)$ and $\Phi(t - \tau/2)$. The condition (12.127) is a special condition that is not satisfied by many stationary processes $\Phi(t)$. For example, no stationary Gaussian process can satisfy (12.127), since then

$$\Psi_\tau(1,1) = \exp\{-[R_\Phi(\tau) + R_\Phi(0)]\}. \tag{12.128}$$

In summary, phase modulation of a sine wave by a stationary process yields a cyclostationary process, except in the case for which (12.127) holds, in which case cyclostationarity degenerates into stationarity.

12.5.3 Phase and Frequency Modulation

Consider the phase-modulated (PM) sine wave

$$X(t) = \cos[2\pi f_o t + \Phi(t)], \tag{12.129}$$

for which $\Phi(t)$ is generally nonstationary but exhibits no cyclostationarity (of any order). It can be shown that the cyclic autocorrelation for $X(t)$ is given by (exercise 13)

$$R_X^\alpha(\tau) = \begin{cases} \frac{1}{4}\langle\Psi_\tau\rangle(1,1) & \alpha = 2f_o, \\ \frac{1}{4}\langle\Psi_\tau\rangle(1,1)^*, & \alpha = -2f_o, \\ 0, & |\alpha| \neq 2f_o, 0, \end{cases} \tag{12.130}$$

in which $\langle\Psi_\tau\rangle$ is the time-averaged joint characteristic function for $\Phi(t + \tau/2)$ and $\Phi(t - \tau/2)$. Thus, $X(t)$ exhibits cyclostationarity with period $1/2f_o$, except when

$$\langle \Psi_\tau \rangle(1,1) \equiv 0. \tag{12.131}$$

Although (12.131) cannot hold for a stationary Gaussian process $\Phi(t)$, there are some processes $\Phi(t)$ that do satisfy it. An example is the balanced quaternary-valued process

$$\Phi(t) = \tan^{-1}\left[\frac{S(t)}{C(t)}\right],$$

for which $S(t)$ and $C(t)$ are statistically identical, uncorrelated, binary-valued (± 1) processes, with stationary transition times (e.g., a Poisson point process). [Also, the balanced quaternary-valued PAM process $\Phi(t)$, which yields the QPSK signal discussed in Section 12.5.5, satisfies (12.131). However, since this $\Phi(t)$ exhibits cyclostationarity because the transition times are periodic (with period, say, T_c), then $X(t)$ is cyclostationary with period T_c, even though it is not cyclostationary with period $1/2f_o$.] Furthermore, (12.131) is satisfied by some nonstationary processes, such as those that arise from frequency modulation. Specifically, let $\Phi(t)$ be given by

$$\Phi(t) = \int_0^t Z(u)\, du, \tag{12.132a}$$

for which $Z(t)$ is a stationary process. Then

$$Z(t) = \frac{d\Phi(t)}{dt}, \qquad t > 0, \tag{12.132b}$$

and $f_o + (1/2\pi)Z(t)$ is the instantaneous frequency of the modulated sine wave (12.129) (cf. Section 11.3). Now, if the spectrum $S_Z(f)$ is not high-pass (or bandpass), in the sense that it does not approach zero faster than linearly in f as $f \to 0$, then $\Phi(t)$ can satisfy (12.131). For example, if $Z(t)$ is WGN, then $\Phi(t)$ is the independent-increment Wiener process (a diffusion), for which it is well known that $X(t)$ is asymptotically stationary. Thus, frequency-modulated (FM) sine waves with low-pass (or all-pass) modulation do not exhibit cyclostationarity with period $1/2f_o$, whereas those with high-pass (or bandpass) modulation can exhibit cyclostationarity. For example, if $Z(t)$ is defined by (12.132b) for a stationary $\Phi(t)$, then $X(t)$ can be cyclostationary. Thus, we see that some FM sine waves exhibit no cyclostationarity, but most PM sine waves of practical interest do exhibit cyclostationarity. (It is noted that the marginal stability of the transformation (12.132a) can result in unusual behavior of the probabilistic model for FM [Gardner, 1987a].) Hence, there is a fundamental distinction to be made between the statistical properties of FM and PM sine waves. This is often not recognized, since it is common practice to introduce a random time-invariant phase variable to render all modulated sine waves stationary (cf. Section 5.4.5) and to adopt fairly arbitrary conventions (e.g., [Papoulis,

1984]) for distinguishing between phase modulation and frequency modulation. It is also common practice to deal with only the complex version of (12.129),

$$X'(t) \triangleq \exp\{i[2\pi f_o t + \Phi(t)]\}, \tag{12.133}$$

which is *apparently* stationary* for all stationary $\Phi(t)$, and to again adopt arbitrary conventions for distinguishing between the two types of modulation, FM and PM.

12.5.4 Pulse-Amplitude Modulation

Consider the pulse-amplitude-modulated (PAM) pulse train

$$X(t) = \sum_{m=-\infty}^{\infty} A(mT)q(t - t_o - mT), \tag{12.134}$$

for which $q(t)$ is a finite-energy pulse. Equation (12.134) is an LPTV transformation of the process $A(t)$, for which the impulse-response function is (exercise 14)

$$h(t, u) = \sum_{m=-\infty}^{\infty} q(t - t_o - mT)\delta(u - mT), \tag{12.135}$$

and the Fourier-series coefficient functions, defined by (12.46), are (exercise 14)

$$g_n(\tau) = \frac{1}{T}q(\tau - t_o). \tag{12.136}$$

Therefore, (12.50) can be applied, together with the identity

$$\sum_{n=-\infty}^{\infty} e^{i2\pi n\tau/T} \equiv T \sum_{j=-\infty}^{\infty} \delta(\tau - jT), \tag{12.137}$$

to obtain the cyclic autocorrelation (exercise 14)

$$R_X^\alpha(\tau) = \frac{1}{T} \sum_{j,k=-\infty}^{\infty} R_A^{\alpha-k/T}(jT)r_q^\alpha(\tau + jT)e^{-i2\pi(\alpha t_o + jk/2)}, \tag{12.138}$$

for which r_q^α is the cyclic finite autocorrelation for $q(t)$. It follows from the aliasing formula for sampled data, (12.96), that (12.138) can be reexpressed

---

*This disturbing fact that complex modulated sine waves are apparently stationary, even though their real parts can be cyclostationary is seen to be not so disturbing at all when it is recognized that the proper definition of a stationary complex time series is that its real and imaginary parts must be jointly stationary (real) time series. Thus, the time series and its complex conjugate must be jointly stationary. Hence, $X'(t)$ is cyclostationary, not stationary, if $X(t)$ is cyclostationary.

as* (exercise 14)

$$R_X^\alpha(\tau) = \frac{1}{T} \sum_{j=-\infty}^{\infty} \tilde{R}_A^\alpha(jT) r_q^\alpha(\tau + jT) e^{-i2\pi\alpha t_o}, \qquad (12.139)$$

in which $\tilde{R}_A^\alpha$ is the cyclic autocorrelation for the sequence $\{A(nT)\}$.

Fourier transformation of (12.138), and application of the *generalized Poisson sum formula* [Franks, 1969]

$$K(f) = \frac{1}{T} G(f) \sum_{n=-\infty}^{\infty} H\left(f + \frac{n}{T}\right) \qquad (12.140)$$

for the Fourier transform of

$$k(t) = \sum_{j=-\infty}^{\infty} h(jT) g(t + jT), \qquad (12.141)$$

yields the cyclic spectrum (exercise 14)

$$S_X^\alpha(f) = \left(\frac{1}{T}\right)^2 Q\left(f + \frac{\alpha}{2}\right) Q^*\left(f - \frac{\alpha}{2}\right) e^{-i2\pi\alpha t_o}$$

$$\times \sum_{j,k=-\infty}^{\infty} S_A^{\alpha - k/T}\left(f + \frac{j}{T} + \frac{k}{2T}\right). \qquad (12.142)$$

It follows from the aliasing formula for sampled data, (12.97), that (12.142) can be reexpressed as (exercise 14)

$$S_X^\alpha(f) = \frac{1}{T} Q\left(f + \frac{\alpha}{2}\right) Q^*\left(f - \frac{\alpha}{2}\right) \tilde{S}_A^\alpha(f) e^{-i2\pi\alpha t_o}, \qquad (12.143)$$

in which $\tilde{S}_A^\alpha$ is the cyclic spectrum for the sequence $\{A(nT)\}$.

For the special case in which $A(t)$ is stationary, $X(t)$ is cyclostationary with period T, and (12.142) reduces to

$$S_X^\alpha(f) = \begin{cases} \left(\frac{1}{T}\right)^2 Q\left(f + \frac{\alpha}{2}\right) Q^*\left(f - \frac{\alpha}{2}\right) e^{-i2\pi\alpha t_o} \sum_{j=-\infty}^{\infty} S_A\left(f + \frac{\alpha}{2} + \frac{j}{T}\right), & \alpha = \frac{k}{T}, \\ 0, & \alpha \neq \frac{k}{T}, \end{cases}$$

(12.144)

*Equations (12.139) and (12.143) are more general than (12.138) and (12.142) [since the former do not require that the sequence of random variables $\{A(nT)\}$ be interpreted as time samples of the process $A(t)$] and can be derived independently of the latter.

for all integers k. Furthermore, it follows from (12.144) and the definition (12.17) that

$$|\rho_X^\alpha(f)|^2 = 1, \qquad \alpha = k/T, \qquad (12.145)$$

for all integers k and all f for which $Q(f + \alpha/2) \neq 0$, $Q(f - \alpha/2) \neq 0$, and $\tilde{S}_A(f \pm \alpha/2) \neq 0$. Therefore PAM is completely coherent for all f and α for which it is not completely incoherent.

For the more realistic model of PAM which incorporates clock-timing jitter $\{J_n\}$,

$$X(t) = \sum_{n=-\infty}^{\infty} A(nT) q(t - nT - J_n),$$

it can be shown that if $\{J_n\}$ is stationary (in the strict sense), then $X(t)$ is in general cyclostationary (analogously to PM). However, if $\{J_n\}$ is an independent-increment sequence, then $X(t)$ exhibits no cyclostationarity (analogously to FM) [Gardner, 1987a].

A typical graph of the spectral correlation surface $|S_X^\alpha(f)|$ specified by (12.144), for a relatively broad amplitude spectrum $\tilde{S}_A(f)$ and a rectangle pulse of width T, is shown in Figure 12.4.

12.5.5 Digital Pulse and Carrier Modulation

General Case. Consider the digitally modulated pulse train

$$X(t) = \sum_{n=-\infty}^{\infty} \sum_{m=1}^{M} \delta_m(n) q_m(t - nT), \qquad (12.146)$$

for which $[\delta_1(n), \delta_2(n), \ldots, \delta_M(n)]^T \triangleq \boldsymbol{\delta}(n)$ is a random indicator vector (for each n) having one element equal to unity and the rest equal to zero. It

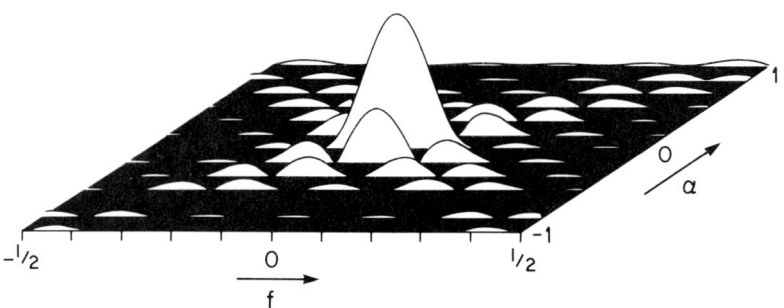

Figure 12.4 Typical Spectral Correlation Surface for PAM Signal with Relatively Broadband, Stationary Amplitude Sequence

can be shown (exercise 15) that the cyclic autocorrelation for $X(t)$ is

$$R_X^\alpha(\tau) = \frac{1}{T} \sum_{j=-\infty}^{\infty} \text{tr}\{r_q^\alpha(\tau + jT) \tilde{R}_\delta^\alpha(j)\}, \qquad (12.147)$$

in which $\tilde{R}_\delta^\alpha$ is the matrix of cyclic discrete-time cross-correlations

$$\tilde{R}_\delta^\alpha(j) \triangleq \lim_{N \to \infty} \frac{1}{2N+1} \sum_{n=-N}^{N} E\{\delta(n+j)\delta^T(n)\} e^{-i2\pi\alpha(n+j/2)T} \qquad (12.148)$$

and r_q^α is the matrix of cyclic finite cross-correlations

$$r_q^\alpha(\tau) \triangleq \int_{-\infty}^{\infty} q\left(t + \frac{\tau}{2}\right) q^T\left(t - \frac{\tau}{2}\right) e^{-i2\pi\alpha t} dt, \qquad (12.149)$$

where $q(t) \triangleq [q_1(t), q_2(t), \ldots, q_M(t)]^T$.

For the special case in which $\{\delta(n)\}$ exhibits no cyclostationarity, $X(t)$ in general exhibits cyclostationarity with only the period T, and (12.147) reduces to (exercise 15)

$$R_X^\alpha(\tau) = \begin{cases} \dfrac{1}{T} \sum\limits_{j=-\infty}^{\infty} \text{tr}\{r_q^\alpha(\tau + jT) \langle \tilde{R}_\delta \rangle(j)\} e^{i\pi\alpha jT}, & \alpha = \dfrac{k}{T}, \\ 0, & \alpha \neq \dfrac{k}{T}, \end{cases} \qquad (12.150)$$

for all integers k. For the special case in which the sequence $\{\delta(n)\}$ is stationary, uncorrelated, and uniformly distributed, (12.150) holds with

$$\langle \tilde{R}_\delta \rangle(j) = \begin{cases} \dfrac{1}{M} I, & j = 0, \\ \dfrac{1}{M^2} \mathbf{1}, & j \neq 0, \end{cases} \qquad (12.151)$$

for which I is the identity matrix and $\mathbf{1}$ is the matrix having all elements equal to unity.

Fourier transformation of (12.147) yields the cyclic spectrum for $X(t)$

$$S_X^\alpha(f) = \frac{1}{T} Q^T\left(f + \frac{\alpha}{2}\right) \tilde{S}_\delta^\alpha(f) Q^*\left(f - \frac{\alpha}{2}\right). \qquad (12.152)$$

For the special case in which $\{\delta(n)\}$ exhibits no cyclostationarity, (12.152) reduces to (exercise 15)

$$S_{\tilde{X}}^{\alpha}(f) = \begin{cases} \frac{1}{T} Q^T\left(f + \frac{\alpha}{2}\right) \langle \tilde{S}_{\delta} \rangle \left(f + \frac{\alpha}{2}\right) Q^*\left(f - \frac{\alpha}{2}\right), & \alpha = k/T, \\ 0, & \alpha \neq k/T, \end{cases}$$

(12.153)

for all integers k. For the special case in which the sequence $\{\delta(n)\}$ is stationary, uncorrelated, and uniformly distributed, (12.153) holds with

$$\langle \tilde{S}_{\delta} \rangle (f) = \left(\frac{1}{M} I - \frac{1}{M^2} \mathbf{1}\right) + \frac{1}{TM^2} \mathbf{1} \sum_{j=-\infty}^{\infty} \delta\left(f - \frac{j}{T}\right), \quad (12.154)$$

and therefore

$$S_{\tilde{X}}^{\alpha}(f) = \frac{1}{MT} \sum_{m=1}^{M} Q_m\left(f + \frac{\alpha}{2}\right) Q_m\left(f - \frac{\alpha}{2}\right)^*$$

$$+ \frac{1}{M^2 T} \left[\sum_{m=1}^{M} Q_m\left(f + \frac{\alpha}{2}\right)\right] \left[\sum_{n=1}^{M} Q_n\left(f - \frac{\alpha}{2}\right)^*\right]$$

$$\times \left[-1 + \frac{1}{T} \sum_{j=-\infty}^{\infty} \delta\left(f + \frac{\alpha}{2} - \frac{j}{T}\right)\right] \qquad (12.155)$$

for $\alpha = k/T$.

The cyclic spectra for a variety of digital pulse-modulated signals such as pulse-position modulation and pulse-width modulation, and digital carrier-modulated signals such as frequency-shift keying and phase-shift keying, can be obtained simply by substitution of the appropriate pulse-transforms into either (12.152) or, for stationary, uniformly distributed, uncorrelated data, (12.155) [Gardner, 1987a]. However, since frequency-shift keying and phase-shift keying cannot always be modeled by (12.146), these two modulation types are studied in more detail.

Frequency-Shift Keying. A frequency-shift-keyed (FSK) signal is said to be *carrier-phase-coherent* if it is generated from M continuously operating oscillators that are keyed off and on to produce the signal

$$X(t) = \cos\left\{\sum_{n=-\infty}^{\infty} q(t - nT_c) \left[\sum_{m=1}^{M} \delta_m(n)(2\pi f_m t + \theta_m)\right]\right\},$$

(12.156)

where $q(t)$ is the unity-height rectangle pulse of width T_c and $1/T_c$ is the keying rate (also called the *chip rate*). An FSK signal is said to be *phase-incoherent* if the phases of the T_c-second carrier bursts fluctuate randomly with the burst-time index n, in which case the M constants θ_m in (12.156) should be replaced with the single random time sequence $\{\Theta(n)\}$. Another type of FSK signal, which is referred to as *clock-phase coherent*, is that for which $\theta_m(n) + 2\pi f_m n T_c$, modulo 2π, is independent of n. This will occur, for example, if the carrier bursts are generated by exciting M zero-state narrowband filters with impulses, once every T_c seconds. In this case, each of the M carrier bursts starts off with the same phase every T_c seconds. If the carrier frequencies $\{f_m\}$ are integer multiples of the keying rate $1/T_c$, then clock-phase coherence yields carrier-phase coherence, and if all M phases $\{\theta_m\}$ are equal, the FSK signal has continuous phase from burst to burst. (Continuous phase is also possible without either carrier-phase coherence or clock-phase coherence.)

Phase-Coherent FSK. A clock-phase coherent FSK signal can be expressed in the form (12.146), with

$$q_m(t) = \cos(2\pi f_m t + \theta_m) q(t). \tag{12.157}$$

Therefore, the cyclic spectrum for this FSK signal [assuming the data sequence $\{\delta(n)\}$ is stationary] is given by (12.153) with $T = T_c$ and with

$$Q_m(f) = \frac{\sin[\pi(f - f_m)T_c] e^{i\theta_m}}{2\pi(f - f_m)} + \frac{\sin[\pi(f + f_m)T_c] e^{-i\theta_m}}{2\pi(f + f_m)}. \tag{12.158}$$

If the data sequence $\{\delta(n)\}$ is uncorrelated and has equiprobable values, then the cyclic spectrum is given by (12.155) with $T = T_c$. It follows that $|\hat{S}_x^\alpha(f)|$ has its maximum values at $\alpha = 0$ and $f = \pm f_m$, and if $\{f_n T_c\}$ are integers, then there are additional maxima at $\alpha = \pm 2f_m$ and $f = 0$. There are also secondary maxima (down by the factor $M - 1$ from the primary maxima) at $\pm \alpha = f_m \pm f_n$ and $\pm f = (f_m \mp f_n)/2$.

Phase-Incoherent FSK. A phase-incoherent FSK signal can be modeled by

$$X(t) = \sum_{n=-\infty}^{\infty} A_n(t) q(t - nT_c), \tag{12.159a}$$

for which $q(t)$ is a unity-height rectangle pulse of width T_c, and

$$A_n(t) = \cos(2\pi F_n t + \Theta_n), \tag{12.159b}$$

where $\{F_n\}$ and $\{\Theta_n\}$ are each independent identically distributed sequences that are independent of each other, F_n has an equiprobable M-ary discrete distribution, and Θ_n has a uniform distribution on $[-\pi, \pi)$. It can be shown (exercise 16) that

$$R_X\left(t + \frac{\tau}{2}, t - \frac{\tau}{2}\right) = \frac{1}{2} \sum_{n=-\infty}^{\infty} E\{\cos(2\pi F_n \tau)\}$$
$$\times q\left(t + nT_c + \frac{\tau}{2}\right) q\left(t + nT_c - \frac{\tau}{2}\right), \tag{12.160}$$

from which it follows that

$$R_X^\alpha(\tau) = \frac{1}{2} E\{\cos(2\pi F_n \tau)\} R_Y^\alpha(\tau), \tag{12.161}$$

where $Y(t)$ is a PAM process with a white unity-variance amplitude sequence, for which

$$S_Y^\alpha(f) = \frac{1}{T_c} Q\left(f + \frac{\alpha}{2}\right) Q^*\left(f - \frac{\alpha}{2}\right), \quad \alpha = \frac{k}{T_c}, \tag{12.162a}$$

where

$$Q(f) = \frac{\sin(\pi f T_c)}{\pi f}. \tag{12.162b}$$

It follows from (12.161) and (12.162) that

$$S_X^\alpha(f) = \frac{1}{4M} \sum_{m=1}^{M} S_Y^\alpha(f + f_m) + S_Y^\alpha(f - f_m), \quad \alpha = \frac{k}{T_c}. \tag{12.163}$$

Comparison of (12.163) [with (12.162a) substituted] and (12.155) [with (12.158) substituted] reveals that some terms from (12.155) are absent in (12.163), and as a result there are no impulses in $S_X^\alpha(f)$ for phase-incoherent FSK, and there are no peaks at $\alpha = \pm 2 f_m$. In fact, if $2 f_m T_c$ is not an integer, there is no contribution at all at $\alpha = \pm 2 f_m$. However, if the products $\{f_m T_c\}$ are sufficiently large to render negligible the overlap, in the (f, α) plane, of the 2M terms in (12.163), then the incoherent-phase FSK signal (i.e., incoherent at the frequencies of the bursts) is completely coherent (i.e., coherent at the burst rate and its harmonics) for all f and α for which one of these $2M$ terms is nonnegligible. (This includes only values of α that are integer multiples of the chip rate.) However, $f_m T_c$ is usually not very large in practice.

Phase-Shift Keying. A phase-shift-keyed (PSK) signal can be expressed by

$$X(t) = \sin\left[2\pi f_o t + \sum_{n=-\infty}^{\infty} q(t - nT_c) \sum_{m=1}^{M} \delta_m(n)\theta_m\right], \quad (12.164)$$

in which $1/T_c$ is the keying rate, and $q(t)$ is the unity-height rectangle pulse of width T_c. For the case in which the carrier frequency f_o is an integer multiple of the keying rate, (12.164) can be reexpressed in the form (12.146), with $T = T_c$ and

$$q_m(t) = \sin(2\pi f_o t + \theta_m) q(t). \quad (12.165)$$

Therefore, the cyclic spectrum for this PSK signal is given by (12.153) [assuming the data sequence $\{\delta(n)\}$ is stationary] with

$$Q_m(f) = \frac{\sin[\pi(f - f_o)T_c] e^{i\theta_m}}{2\pi i(f - f_o)} - \frac{\sin[\pi(f + f_o)T_c] e^{-i\theta_m}}{2\pi i(f + f_o)}.$$

$$(12.166)$$

For the case in which $f_o T_c$ is not an integer, another approach that exploits results on cyclic spectra for AM and QAM can be taken, as explained next.

BPSK. For binary PSK ($M = 2$), (12.164) (generalized to incorporate an arbitrary epoch parameter t_o and an arbitrary phase parameter ϕ_o) can be reexpressed as

$$X(t) = A(t)\cos(2\pi f_o t + \phi_o), \quad (12.167a)$$

in which

$$A(t) = \sum_{n=-\infty}^{\infty} A(nT_c) q(t - t_o - nT_c), \quad (12.167b)$$

where $\{A(nT_c)\}$ is a binary sequence with values

$$A(nT_c) = \frac{2}{\pi} \sum_{m=1}^{M} \delta_m(n)\theta_m = \pm 1. \quad (12.168)$$

Consequently, this BPSK signal is a composition of AM and PAM, which are treated in Sections 12.5.1 and 12.5.4. It follows that the cyclic spectrum for this BPSK signal can be obtained from (12.143) [with $X(t)$ and T

replaced by $A(t)$ and T_c] and (12.105), and is given by

$$S_X^\alpha(f) = \frac{1}{4T_c}\left\{\left[Q\left(f+f_o+\frac{\alpha}{2}\right)Q\left(f+f_o-\frac{\alpha}{2}\right)\tilde{S}_A^\alpha(f+f_o)\right.\right.$$
$$\left.+Q\left(f-f_o+\frac{\alpha}{2}\right)Q\left(f-f_o-\frac{\alpha}{2}\right)\tilde{S}_A^\alpha(f-f_o)\right]e^{-i2\pi\alpha t_o}$$
$$+Q\left(f+\frac{\alpha}{2}+f_o\right)Q\left(f-\frac{\alpha}{2}-f_o\right)\tilde{S}_A^{\alpha+2f_o}(f)e^{-i(2\pi[\alpha+2f_o]t_o+2\phi_o)}$$
$$\left.+Q\left(f+\frac{\alpha}{2}-f_o\right)Q\left(f-\frac{\alpha}{2}+f_o\right)\tilde{S}_A^{\alpha-2f_o}(f)e^{-i(2\pi[\alpha-2f_o]t_o-2\phi_o)}\right\},$$

(12.169a)

in which

$$Q(f) = \frac{\sin(\pi f T_c)}{\pi f}. \qquad (12.169b)$$

If the data sequence $\{A(nT_c)\}$ is stationary, and $f_o T_c \gg 1$, then only the first two terms in (12.169) are nonnegligible for $\alpha = k/T_c$, with $|k| \not\gg 1$ and $|f+f_o|T_c \not\gg 1$ or $|f-f_o|T_c \not\gg 1$ (k is an integer); and only the last two terms are nonnegligible for $\alpha = \pm 2f_o + k/T_c$ with $|k| \not\gg 1$ and $|f|T_c \not\gg 1$; otherwise all four terms are negligible. For uncorrelated data, $S_X^\alpha(f)$ peaks at $\alpha = 0$ and $\pm 2f_o$. A typical graph of the spectral correlation surface $|S_X^\alpha(f)|$ specified by (12.169) is shown in Figure 12.5 for a relatively broad data spectrum $\tilde{S}_A(f)$. If $f_o T_c$ is sufficiently large to render negligible the overlap, in the (f,α) plane, of the four terms in (12.169), then the BPSK signal is, to a close approximation, completely coherent for all f and α for which one of these four terms is nonnegligible.

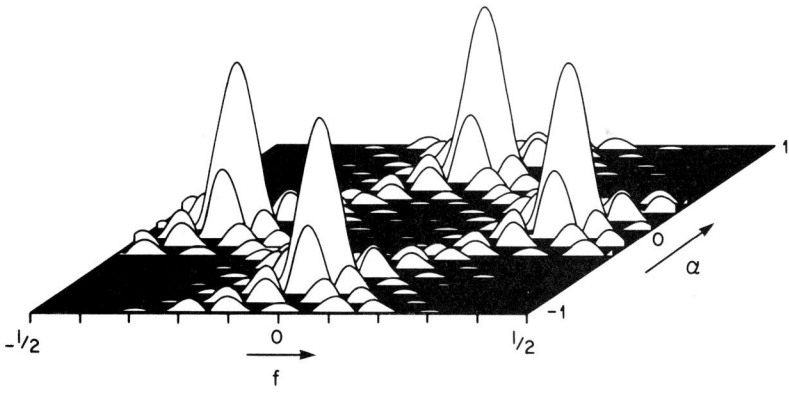

Figure 12.5 Typical Spectral Correlation Surface for BPSK Signal with Relatively Broadband Stationary Phase Sequence

QPSK. For quaternary PSK ($M = 4$), (12.164) (generalized to incorporate t_o and ϕ_o) can be reexpressed as

$$X(t) = C(t)\cos(2\pi f_o t + \phi_o) - S(t)\sin(2\pi f_o t + \phi_o), \quad (12.170a)$$

for which

$$C(t) = \sum_{n=-\infty}^{\infty} C(nT_c)q(t - t_o - nT_c),$$

$$S(t) = \sum_{n=-\infty}^{\infty} S(nT_c)q(t - t_o - nT_c), \quad (12.170b)$$

where $\{C(nT_c)\}$ and $\{S(nT_c)\}$ are binary sequences with values

$$C(nT_c) = \pm 1$$
$$S(nT_c) = \pm 1. \quad (12.171)$$

Consequently, this QPSK signal is a composition of QAM and PAM, which are treated in Sections 12.5.2 and 12.5.4. It follows that the cyclic spectrum for this QPSK signal can be obtained from (12.143) [with $X(t)$ and T replaced by $C(t)$ and T_c, and also by $S(t)$ and T_c] and (12.80) [with $Z(t)$ and $W(t)$ replaced by $C(t)$ and $-S(t)$]. For the common case in which $\{C(nT_c)\}$ and $\{S(nT_c)\}$ are uncorrelated and balanced in the sense that

$$\tilde{S}_C^\alpha - \tilde{S}_S^\alpha \equiv [\tilde{S}_{CS}^\alpha]_r \equiv 0, \quad (12.172)$$

(12.80) simplifies to yield

$$S_X^\alpha(f) = \frac{1}{2T_c}\left[Q\left(f + \frac{\alpha}{2} + f_o\right)Q\left(f - \frac{\alpha}{2} + f_o\right)\tilde{S}_C^\alpha(f + f_o)\right.$$
$$\left. + Q\left(f + \frac{\alpha}{2} - f_o\right)Q\left(f - \frac{\alpha}{2} - f_o\right)\tilde{S}_C^\alpha(f - f_o)\right]e^{-i2\pi\alpha t_o},$$
$$(12.173)$$

where $Q(f)$ is given by (12.169b). Thus, $S_X^\alpha(f)$ is nonnegligible only for $|f + f_o|T_c \not\gg 1$ or $|f - f_o|T_c \not\gg 1$. If the data sequences $\{C(nT_c)\}$ and $\{S(nT_c)\}$ are stationary, then $S_X^\alpha(f)$ is nonzero only for $\alpha = k/T_c$ (k is an integer). For uncorrelated data, $S_X^\alpha(f)$ peaks at $\alpha = 0$ and decreases as $|\alpha|$ increases. Whereas for BPSK the spectral-correlation surface (Figure 12.5) can be obtained by placing replicas of the surface for PAM (Figure 12.4) at the locations of the four peaks in the surface for AM (Figure 12.3), for QPSK the spectral-correlation surface can be obtained by placing replicas of the surface for PAM at the locations of the two peaks in the surface for balanced QAM (the two peaks on the f axis in Figure 12.3). If $f_o T_c$ is sufficiently large to render negligible the overlap, in the (f, α) plane, of the two terms in (12.173), the QPSK signal is, to a close approximation, completely coherent for all f and α for which one of these two terms is nonnegligible.

The inverse Fourier transform of (12.173) yields the limit cyclic autocorrelation for QPSK,

$$R_X^\alpha(\tau) = \frac{1}{T_c} \sum_{j=-\infty}^{\infty} \tilde{R}_C^\alpha(jT_c) r_q^\alpha(\tau + jT_c) \cos(2\pi f_o \tau) e^{-i2\pi\alpha t_o}.$$

(12.174)

12.6 Stationary Representations

A cyclostationary process $X(t)$ can be represented by a set of jointly stationary processes $\{A_p(t): p = 0, \pm 1, \pm 2, \ldots\}$ as follows:

$$X(t) = \sum_{p=-\infty}^{\infty} A_p(t) e^{i2\pi pt/T}, \quad (12.175a)$$

$$A_p(t) \triangleq [X(t) e^{-i2\pi pt/T}] \otimes w_T(t), \quad (12.175b)$$

for which

$$w_T(t) \triangleq \frac{\sin(\pi t/T)}{\pi t}, \quad (12.175c)$$

where T is the period of cyclostationarity. As an alternative, (12.175b) can be equivalently expressed as

$$A_p(t) = \{X(t) \otimes [w_T(t) e^{i2\pi pt/T}]\} e^{-i2\pi pt/T}. \quad (12.176)$$

This representation is most easily understood in the frequency domain. It partitions the frequency support of $X(t)$ into disjoint bands of width $1/T$ centered at integer multiples $\{p/T\}$ of $1/T$, so that the pth component in (12.175a) is simply the response of an ideal bandpass filter, with transfer function

$$H_p(f) = W_{1/T}(f - p/T) \triangleq \begin{cases} 1, & -\frac{1}{2T} < f - \frac{p}{T} \le \frac{1}{2T}, \\ 0, & \text{otherwise,} \end{cases}$$

(12.177)

to the excitation $X(t)$, as revealed by (12.176). The pth representor, $A_p(t)$, is the frequency-centered (low-pass) version of the pth component, as also revealed by (12.176). Thus, $A_p(t)$ has frequency support $(-1/2T, 1/2T]$

for each and every p. It follows that there are only a finite number of nonzero representors for a process with finite bandwidth. Although (12.175) is a valid representation for any process, it is especially appropriate for a cyclostationary process with period T because then the representors are jointly stationary (exercise 17). This representation is referred to as the *harmonic-series representation* (HSR) [Gardner and Franks, 1975].

It follows from (12.175) that the autocorrelation function for $X(t)$ is represented by

$$R_X(t,u) = \sum_{p,q=-\infty}^{\infty} R_{pq}(t-u) e^{i2\pi(pt-qu)/T}, \qquad (12.178a)$$

for which

$$R_{pq}(t-u) \triangleq E\{A_p(t) A_q^*(u)\}. \qquad (12.178b)$$

By comparing this with the Fourier-series representation (12.4), which can be reexpressed as

$$R_X(t,u) = \sum_{n=-\infty}^{\infty} R_X^{n/T}(t-u) e^{i\pi n(t+u)/T} \qquad (12.179a)$$

in which

$$R_X^{n/T}(t-u) \triangleq \frac{1}{T} \int_{-T/2}^{T/2} R_X(t+v, u+v) e^{-i2\pi nv/T} dv \, e^{-i\pi n(t+u)/T}, \qquad (12.179b)$$

it can be shown (exercise 18) that

$$R_X^{n/T}(\tau) e^{i\pi n\tau/T} \equiv \sum_{p=-\infty}^{\infty} R_{p(p-n)}(\tau) e^{i2\pi p\tau/T}. \qquad (12.180)$$

Fourier transformation of both sides of (12.180) yields (exercise 18)

$$S_X^{n/T}\left(f - \frac{n}{2T}\right) \equiv \sum_{p=-\infty}^{\infty} S_{p(p-n)}\left(f - \frac{p}{T}\right), \qquad (12.181)$$

where

$$S_{pq}(f) \triangleq \int_{-\infty}^{\infty} R_{pq}(\tau) e^{-i2\pi f\tau} d\tau. \qquad (12.182)$$

Furthermore, since $\{A_p(t)\}$ are bandlimited to $(-1/2T, 1/2T]$, then so too are $\{R_{pq}(\tau)\}$, that is

$$S_{pq}(f) = 0, \quad f \le -\frac{1}{2T} \text{ or } f > \frac{1}{2T}, \quad (12.183)$$

and it follows (exercise 18) from (12.181) and (12.183) that

$$S_{pq}(f) \equiv S_X^{(p-q)/T}\left(f + \frac{p+q}{2T}\right), \quad -\frac{1}{2T} < f \le \frac{1}{2T}. \quad (12.184)$$

Thus, the spectral density matrix $\{S_{pq}(f)\}$ and the cyclic spectra $\{S_X^{n/T}(f)\}$ can be obtained from each other quite simply with the aid of (12.181) and (12.184). It follows from (12.181) that the (potentially) cyclostationary process $X(t)$ is stationary if the spectral density matrix is diagonal.

A particularly useful form of the HSR spectral matrix is obtained if it is modified by multiplying each element, say the (p,q)th, by the normalizing factor

$$\left[S_{pp}(f)S_{qq}(f)\right]^{-1/2},$$

because then each and every element of the modified matrix is a cross-coherence function. Furthermore, it follows from (12.184) that these cross-coherences are given by

$$\frac{S_{pq}(f)}{\left[S_{pp}(f)S_{qq}(f)\right]^{1/2}} \equiv \frac{S_X^\alpha(f+\nu)}{\left[S_X(f+\nu+\alpha/2)S_X(f+\nu-\alpha/2)\right]^{1/2}},$$

$$-\frac{1}{2T} < f \le \frac{1}{2T}, \quad (12.185)$$

for which $\nu = (p+q)/2T$ and $\alpha = (p-q)/T$. Consequently, the cross-coherences are simply frequency-centered bandlimited (low-pass) versions of autocoherences:

$$\frac{S_{pq}(f)}{\left[S_{pp}(f)S_{qq}(f)\right]^{1/2}} \equiv \rho_X^{(p-q)/T}\left(f + \frac{p+q}{2T}\right), \quad -\frac{1}{2T} < f \le \frac{1}{2T}.$$

$$(12.186)$$

The HSR can be exploited in various applications. For example, for optimum and adaptive periodically time-variant filtering of cyclostationary signals, the HSR yields a decomposition into a set of optimum or adaptive time-invariant filters [Gardner and Franks, 1975]. Several other methods for representing a cyclostationary process in terms of jointly stationary processes can be similarly exploited [Gardner and Franks, 1975]. Perhaps the simplest of all such representations is the following *translation series representation* or *time series representation* (TSR) for a discrete-time cyclostationary process with period P:

$$X(n) = \sum_{p=0}^{P-1} Z_p([n]_P/P)\delta_{n-[n]_P-p}, \qquad (12.187a)$$

$$Z_p(m) \triangleq X(mP + p), \qquad m = 0, \pm 1, \pm 2, \ldots, \qquad (12.187b)$$

where $[n]_P$ denotes the greatest integer multiple of P that is less than or equal to n, and δ_k is the Kronecker delta. Thus, each representor $Z_p(m)$, is simply a subsequence of $X(n)$ obtained by sampling $X(n)$ once per period. It can be shown (exercise 20) that $\{Z_p(m)\}$ are jointly stationary if $X(n)$ is cyclostationary with period P. This decomposition of $X(n)$ into its TSR representors can be interpreted as a time-division demultiplexing operation, as depicted in Figure 12.6. Similarly the decomposition of $X(t)$ into the HSR representors $\{A_p(t)\}$ can be interpreted as a frequency-division demultiplexing operation, as depicted in Figure 12.7.

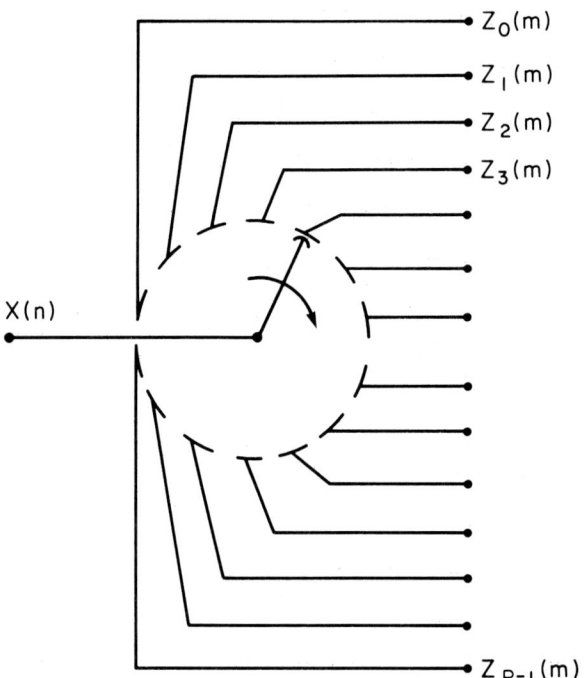

Figure 12.6 Decomposition of Cyclostationary Process into Stationary Components by Time-Division Demultiplexing Operation

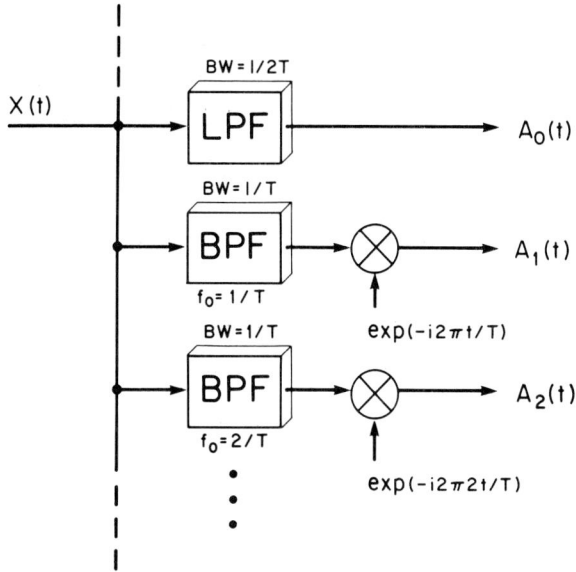

Figure 12.7 Decomposition of Cyclostationary Process into Stationary Components by Frequency-Division Demultiplexing Operation

12.7 Cycloergodicity and Duality

Consider the problem of estimating the *cyclic mean*, defined by

$$m_X^\alpha \triangleq \lim_{Z \to \infty} \frac{1}{Z} \int_{-Z/2}^{Z/2} m_X(t) e^{-i2\pi\alpha t} \, dt, \qquad (12.188)$$

of a nonstationary process, by using the estimate

$$\hat{m}_X^\alpha(W) \triangleq \frac{1}{W} \int_{-W/2}^{W/2} X(t) e^{-i2\pi\alpha t} \, dt. \qquad (12.189)$$

The expected value of this estimate is

$$E\{\hat{m}_X^\alpha(W)\} = \frac{1}{W} \int_{-W/2}^{W/2} m_X(t) e^{-i2\pi\alpha t} \, dt. \qquad (12.190)$$

In the limit $W \to \infty$ the estimate becomes the *empirical cyclic mean*

$$\lim_{W \to \infty} \hat{m}_X^\alpha(W) = \hat{m}_X^\alpha \triangleq \lim_{Z \to \infty} \frac{1}{Z} \int_{-Z/2}^{Z/2} X(t) e^{-i2\pi\alpha t} \, dt, \qquad (12.191)$$

and its expected value becomes the *probabilistic cyclic mean* defined by

(12.188),

$$\lim_{W \to \infty} E\{\hat{m}_X^\alpha(W)\} = m_X^\alpha. \qquad (12.192)$$

The variance of the estimate is

$$\text{Var}\{\hat{m}_X^\alpha(W)\} \triangleq E\{|\hat{m}_X^\alpha(W) - E\{\hat{m}_X^\alpha(W)\}|^2\}$$

$$= \frac{1}{W^2} \int_{-W/2}^{W/2} \int_{-W/2}^{W/2} K_X(t,u) e^{-i2\pi\alpha(t-u)} \, dt \, du.$$

$$(12.193)$$

If the variance approaches zero as $W \to \infty$, then $X(t)$ is said to exhibit *mean-square cycloergodicity of the mean at cycle frequency* α [Boyles and Gardner, 1983] (cf. Section 8.3).

By replacement of $X(t)$ with $Y_\tau(t) \triangleq X(t + \tau/2)X(t - \tau/2)$ in the preceding discussion (cf. Section 8.4), we obtain the definition of a process that exhibits *mean-square cycloergodicity of the autocorrelation*, in which case the *empirical cyclic autocorrelation* defined by

$$\hat{R}_X^\alpha(\tau) \triangleq \lim_{Z \to \infty} \frac{1}{Z} \int_{-Z/2}^{Z/2} X\left(t + \frac{\tau}{2}\right) X\left(t - \frac{\tau}{2}\right) e^{-i2\pi\alpha t} \, dt \quad (12.194)$$

equals its expected value, the probabilistic cyclic autocorrelation (12.6).

It follows from the cyclic Wiener-Khinchin relation that a process that has mean-square cycloergodicity of the autocorrelation has mean-square cycloergodicity of the time-variant spectral density $S_X(t, f)$, in the sense that its Fourier coefficients $S_X^\alpha(f)$ can be obtained from empirical cyclic spectra:

$$\hat{S}_X^\alpha(f) = \int_{-\infty}^{\infty} \hat{R}_X^\alpha(\tau) e^{-i2\pi f\tau} \, d\tau. \qquad (12.195)$$

Methods for measurement of cyclic spectra can be derived from known methods for measurement of cross-spectra of complex-valued processes by use of the characterization (12.15) of the cyclic spectrum as a cross-spectrum [Gardner, 1986a]. These methods are extensions and generalizations of the PSD measurement methods described in Section 11.4. The resolution and reliability properties of cyclic spectrum measurements are derived in [Gardner, 1987a].

It should be emphasized that the process $X(t)$ need not be cyclostationary or almost cyclostationary to exhibit cycloergodicity. This is analogous to the fact that a process need not be stationary to exhibit ergodicity. Cycloergodicity simply ensures that a cyclic component, say m_X^α or $R_X^\alpha(\tau)$, of a generally time-variant probabilistic parameter $m_X(t)$ or $R_X(t + \tau/2, t - \tau/2)$, can be estimated using sinusoidally weighted time averages of an individual sample path. Ergodicity then is just the special case $\alpha = 0$ of cycloergodicity. Similarly, the duality between the probabilistic theory of stationary processes and the deterministic theory based on limit time averages of an individual waveform (Section 8.6) is simply the special case $\alpha = 0$ of a more general duality between the probabilistic theory of cyclostationary processes and the deterministic theory based on limit sinusoidally weighted time averages of an individual waveform. The dual deterministic theory of cyclostationary processes is introduced in [Gardner, 1986c] and completely developed in [Gardner, 1987a]. Furthermore, these time averages can be reinterpreted as expected values on the basis of the fraction-of-time distribution. For example, the *cyclostationary fraction-of-time amplitude distribution with period T* is defined for a waveform $x(t)$ by

$$\hat{F}_{X(t)}(x) \triangleq \lim_{N \to \infty} \frac{1}{2N+1} \sum_{n=-N}^{N} u[x - x(t - nT)]$$

$$= \hat{F}_{X(t+T)}(x), \qquad (12.196)$$

and the corresponding *cyclostationary amplitude density* is defined by

$$\hat{f}_{X(t)}(x) \triangleq \frac{d\hat{F}_{X(t)}(x)}{dx}. \qquad (12.197)$$

It follows (cf. Section 8.6) that the expected value of $X(t)$ is given by (exercise 21)

$$\hat{E}\{X(t)\} \triangleq \int_{-\infty}^{\infty} x \hat{f}_{X(t)}(x) \, dx \qquad (12.198)$$

$$= \lim_{N \to \infty} \frac{1}{2N+1} \sum_{n=-N}^{N} x(t + nT) \triangleq \hat{m}_X(t). \quad (12.199)$$

Furthermore, the probabilistic and empirical cyclic means are simply the Fourier coefficients of this periodic mean (exercise 22),

$$\hat{m}_X^\alpha = \frac{1}{T} \int_{-T/2}^{T/2} \hat{m}_X(t) e^{-i2\pi\alpha t} \, dt, \qquad \alpha = \frac{p}{T}, \qquad (12.200)$$

for all integers p. Moreover, we can heuristically identify ensemble averages and time averages as one and the same if we define ensemble members by

$$X(t, \sigma) \triangleq x(t - \sigma T) \qquad (12.201)$$

for all integer values of σ, which denotes random samples from a sample space. This is an extension of Wold's isomorphism from stationary to cyclostationary processes (cf. Section 8.6). It should be emphasized that the probabilistic model that can be associated with an individual waveform via fraction-of-time distributions is not, in general, unique. For example, if

$X(t)$ exhibits cyclostationarity with more than one period, then for each period there is a different cyclostationary fraction-of-time distribution, and there is also a stationary fraction-of-time distribution. In order to obtain a unique composite probabilistic model, the fraction-of-time distributions for all periods must be added together in a form analogous to (12.43) for autocorrelations. For example, if the stationary amplitude distribution is denoted by $\hat{F}^0_{X(t)}$, and the cyclostationary amplitude distributions are denoted by $\hat{F}_{X(t);T}$ for each period T, then the composite distribution is given by [Gardner, 1987a]

$$\hat{F}_{X(t)} \triangleq \hat{F}^0_{X(t)} + \sum_T \left[\hat{F}_{X(t);T} - \hat{F}^0_{X(t)} \right], \qquad (12.202)$$

provided that no two periods are integrally related (e.g., $3T_1 = 4T_2$ for the two periods, T_1 and T_2). Moreover, if $X(t)$ is an almost cyclostationary random process with amplitude distribution $F_{X(t)}$, it can be shown that

$$E\left\{ \hat{F}_{X(t)} \right\} = F_{X(t)}, \qquad (12.203)$$

and if $X(t)$ exhibits appropriate cycloergodic properties, it is proposed that

$$\hat{F}_{X(t)} = E\left\{ \hat{F}_{X(t)} \right\}. \qquad (12.204)$$

Phase Randomization. A commonly employed technique for modifying a cyclostationary or almost cyclostationary process to obtain a stationary process is to introduce a random phase Θ with an appropriate distribution

$$Y(t) = X(t + \Theta), \qquad (12.205)$$

where Θ is independent of $X(t)$. This is explained in Section 5.4.5 for cyclostationary processes, and in [Gardner, 1978] for almost cyclostationary (and other nonstationary) processes. However, it should be emphasized that phase randomization destroys cycloergodicity regardless of the phase distribution.* For example, it follows from (12.10) that the empirical cyclic autocorrelation for the phase-randomized process (12.205) is

$$\hat{R}^\alpha_Y(\tau) = \hat{R}^\alpha_X(\tau) e^{i2\pi\alpha\Theta}, \qquad (12.206)$$

and its expected value is therefore

$$R^\alpha_Y(\tau) = E\{\hat{R}^\alpha_Y(\tau)\} = E\{\hat{R}^\alpha_X(\tau)\} E\{e^{i2\pi\alpha\Theta}\} = R^\alpha_X(\tau)\Phi_\Theta(2\pi\alpha), \qquad (12.207)$$

---

*The only exception to this is the distribution that has all of its mass at integer multiples of the period, T, of a cyclostationary process.

where Φ_Θ is the characteristic function for Θ, which satisfies

$$|\Phi_\Theta(\omega)| \le 1, \qquad (12.208)$$

with equality only for $\omega = 0$ (or ω = integer multiples of $2\pi/T$ for a Θ with all its mass at integer multiples of T). Thus, in general

$$R_Y^\alpha(\tau) \ne \hat{R}_Y^\alpha(\tau) \qquad (12.209)$$

even if $R_X^\alpha(\tau) \equiv \hat{R}_X^\alpha(\tau)$; that is, $Y(t)$ is not cycloergodic even if $X(t)$ is.

It can also be shown that if $X(t)$ is a Gaussian process, then $Y(t)$ cannot be Gaussian unless $X(t)$ is stationary. This follows from the fact that the probability distributions of $Y(t)$ are mixtures of nonidentical [for nonstationary $X(t)$] Gaussian distributions, and therefore cannot be Gaussian (cf. Chapter 5, exercise 7). Furthermore, it can be shown that if $X(t)$ is cyclostationary and $Y(t)$ is stationary, then $Y(t)$ cannot be Gaussian and ergodic, regardless of the particular probabilistic model for $X(t)$ [Gardner, 1987a]. Finally, it should be mentioned that if $X(t)$ exhibits several commensurate periodicities (such as the carrier frequency and symbol rate of a digitally modulated carrier) and if these periodicities are individually phase-randomized, then the resultant stationary process $Y(t)$ can be nonergodic even though the non-phase-randomized process $X(t)$ is ergodic [Gardner, 1987b].

12.8 Applications

12.8.1 General Discussion

The theory of cyclostationarity has the potential for useful application in any field in which random data from periodically time-variant phenomena are analyzed, processed, or utilized. For applications in which the periodicity itself is an object of study or is to be used for some purpose, the theory can be essential.* In the companion treatment [Gardner, 1987a], the potential for useful application of the theory of cyclostationarity is illustrated by further development of the theory for specific types of statistical inference and decision problems. These include problems of detecting the presence of random signals buried in noise and further masked

*The growing awareness of the usefulness of the concept of cyclostationarity is illustrated by recent work in the field of communication systems, including synchronization [Franks, 1980; Moeneclaey, 1982; Gardner, 1986b], cross-talk interference and modulation transfer noise [Campbell et al., 1983; Albuquerque et al., 1984], and transmitter and receiver filter design [Gardner and Franks, 1975; Mesiya et al., 1978; Ericson, 1981; Graef, 1983], and by work in the fields of digital signal processing [Pelkowitz, 1981; Ferrara, 1985], queueing [Kaplan, 1983; Ackroyd, 1984], noise in circuits [Strom and Signell, 1977], economics [Parzen and Pagano, 1979], and physical sciences such as hydrology, meteorology, and climatology [Jones and Brelsford, 1967; Hasselman and Barnett, 1981; Ortiz and Ruiz de Elvira, 1985; Vecchia, 1985]. The earliest engineering application appears to be [Bennett, 1958] on synchronization.

by interference; classifying such corrupted signals according to modulation type; estimating parameters of corrupted signals, such as time-difference-of-arrival, pulse rate and phase, and carrier frequency and phase; extracting signals from corrupted measurements; and identifying systems involving periodicity from corrupted input-output measurements.

Many of the methods proposed in [Gardner, 1987a] are derived as the solutions to specific optimization problems; others are practical adaptations of optimum but sometimes impractical solutions; and a few methods can be considered to be ad hoc. A theme that unifies these otherwise diverse methods is that spectral correlation is exploited to obtain discrimination against noise and interference or to otherwise obtain performance improvements relative to more conventional methods that exploit artificially imposed stationarity rather than cyclostationarity. For example, modulated signals that are severely masked by other interfering signals and noise can be more effectively detected, in some applications, by detection of spectral correlation rather than detection of energy. This is so, for instance, when the energy level of the background noise or interference fluctuates unpredictably, thereby complicating the problem of setting energy threshold levels [Gardner, 1988]. The utility of spectral correlation for detection is illustrated in Figure 12.8a where it is difficult to detect the contribution from the signal in the measured spectrum at $\alpha = 0$, but it is easy to detect the contribution in the measured spectral correlation function at $\alpha = \pm 2f_o$, $\alpha = \pm 1/T_c$, and $\alpha = \pm 2f_o \pm 1/T_c$. The signal here is BPSK (cf. Figure 12.5), the noise is white, and the interference consists of five AM signals (cf. Figure 12.3). The noise and each of the interfering signals have a power level equal to that of the BPSK signal. Figure 12.8a shows the measured spectral correlation magnitude surface for the corrupted BPSK signal, Figure 12.8b shows the measured surface for the uncorrupted BPSK signal, and Figure 12.8c shows the measured surface for the corruption alone. This example also illustrates that the unique patterns of spectral correlation that are exhibited by different types of modulated signals (cf. Section 12.5) can be effectively exploited for recognition of modulation type for severely corrupted signals. However, the production of reliable spectral correlation surfaces is a computationally intensive task. A survey of methods of measurement of spectral correlation is given in [Gardner, 1986a, 1987a], and computationally efficient digital algorithms are presented in [Brown, 1987; Brown and Loomis, 1988; Roberts, 1989].

Similarly, parameters such as the time-difference-of-arrival of a signal arriving at two sensors, which can be used for location of the source of a radiating signal, can be more effectively estimated for modulated signals in some applications by using measurements of cyclic cross-correlations and cyclic cross-spectra, rather than conventional cross-correlations and cross-spectra. This is so, for example, when there is a multiplicity of spectrally and temporally overlapping interfering modulated signals that cannot be

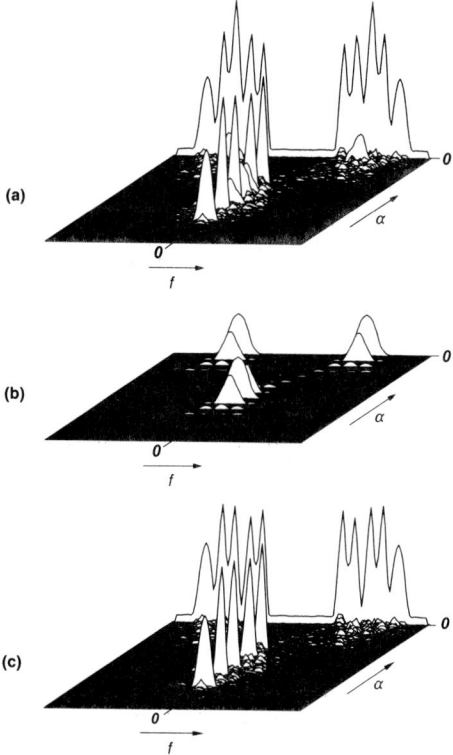

Figure 12.8 (a) Magnitude of Spectral Correlation Surface Computed from a Simulated BPSK Signal with Multiple AM Interference and Noise. (b) Same as (a) Except Signal Only. (c) Same as (a) Except Interference and Noise Only.

separated, or when the signal is weak relative to the noise [Gardner and Chen, 1988a, 1988b; Chen, 1989].

Likewise, linear combiners used with sensor arrays for beam- and null-steering (spatial filtering) for modulated signal extraction and interference rejection can be more effectively adapted in some applications when spectral correlation is exploited. This is a result of the fact that the need for transmission and storage of training signals or knowledge of the direction-of-arrival of the signal to be extracted can be dispensed with when knowledge of a cycle frequency, such as a doubled carrier frequency, or a baud rate, chip rate, or hop rate, is properly exploited. This exploitation involves generation of a reference (training) signal by frequency shifting a primary (received) signal by an amount equal to a cycle frequency of the signal to be extracted in order to decorrelate (between the reference and the primary) the noise and interference while maintaining high correlation for the signal components with that particular cycle frequency [Agee et al., 1987, 1988;

Agee, 1989]. Also, the inherent signal selectivity associated with measurements of the cyclic correlation matrix for a vector of signals received by an array of sensors can be used to substantially simplify and improve various methods for estimation of the directions of arrival of modulated-signal wavefronts impinging on the array (such as MUSIC, ESPRIT, and maximum likelihood) [Schell et al., 1989].

As another example of the utility of spectral correlation, modulated signals that are corrupted by noise and other interfering modulated signals can be more effectively extracted in some applications where only a single receiving antenna element is available by exploiting spectral correlation through the use of periodically (or almost periodically) time-variant filters, instead of more conventional time-invariant filters. This is so because such time-variant filters perform frequency-shifting operations as well as the frequency-dependent magnitude-weighting and phase-shifting operations performed by time-invariant filters. The value of this is easily seen for the simple example in which interference in some portions of the signal frequency band is so strong that it overpowers the signal in those partial bands. In this case, a time-invariant filter can only reject both the signal and the interference in those highly corrupted bands, whereas a time-variant filter can replace the rejected spectral components with spectral components from other uncorrupted (or at least less corrupted) bands that are highly correlated with the rejected components from the signal [Gardner, 1989].

Another example that is not as easy to explain involves two spectrally overlapping linearly modulated signals such as AM or ASK. It can be shown that in an ideal noise-free environment, regardless of the degree of spectral and temporal overlap, each of the two interfering signals can be perfectly extracted using frequency shifting and weighting, provided only that they do not share both the same carrier frequency and phase. Also, when broadband noise is present, extraction of each of the signals can be accomplished without substantial noise amplification [Brown, 1987].

A final example involves reduction of signal distortion due to frequency-selective fading caused by multipath propagation. Straightforward amplification in faded portions of the spectrum using a time-invariant filter suffers from the resultant amplification of noise. In contrast to this, a periodically time-variant filter can replace the faded spectral components with stronger highly correlated components from other bands. If these correlated spectral components are weaker than the original components before fading, there will be some noise enhancement when they are amplified. But the amount of noise enhancement can be much less than that which would result from the time-invariant filter, which can only amplify the very weak faded components [Gardner, 1989].

In this section, a brief sampling of applications is presented to spur the reader's interest in this developing area of statistical signal processing. The scope of this book, however, does not allow for detailed treatments of these applications. The discussions are therefore relatively terse, and the reader is referred to pertinent literature for more complete treatments and additional references.

In essentially all techniques described in the following subsections, the knowledge of the values of certain cycle frequencies is required. In some applications this is quite reasonable since such knowledge is required in standard practice (e.g., knowledge of carrier frequencies and baud rates required by communications receivers). However, in other applications such knowledge must be gained by measurements based on the same data that is to be processed. This can be accomplished by using spectral line regeneration techniques developed for detection and synchronization [Gardner, 1986b, 1987a, 1988] (cf. Subsection 12.8.4) and related techniques for spectral correlation measurement [Gardner 1986a, 1987a; Brown, 1987; Brown and Loomis, 1988; Roberts, 1989].

12.8.2 High-Resolution Direction Finding

CYCCOR Method 1. We consider the problem of using the received data from an array of n closely spaced sensors, such as radio or radar antenna elements, to estimate the directions of arrival (DOAs) of wavefronts impinging on the array from a multiplicity of m signal sources. For each one of the signal sources, we can model the n-vector of analytic signals (cf. Subsection 11.2.1) of the received data as

$$\boldsymbol{X}(t) = \boldsymbol{p}(\theta)S(t) + \boldsymbol{N}(t), \qquad (12.210)$$

where $S(t)$ is the signal from any one of the sources of interest and $\boldsymbol{p}(\theta)$ is the associated direction vector, which is determined by the direction of arrival θ and the directional characteristics of the array as reflected in the calibration function $\boldsymbol{p}(\cdot)$. The DOA can be a single angle in a plane, such as azimuth (assuming a planar array of sensors and a planar distribution of signal sources), or a pair of angles, such as azimuth and elevation. $\boldsymbol{N}(t)$ is the n-vector of complex envelopes consisting of the sum of a noise vector and $m - 1$ signal vectors from the other $m - 1$ signal sources. The model $\boldsymbol{p}(\theta)S(t)$ for the signal component of interest is an accurate approximation of the complex envelope when $S(t)$ has a sufficiently narrow relative bandwidth that the directional characteristics of the sensors and the impinging signal are nearly independent of frequency throughout the signal band.

The CYClic CORrelation (CYCCOR) matrix for this model is given by (exercise 27)

$$\boldsymbol{R}_X^\alpha(\tau) = R_S^\alpha(\tau)\boldsymbol{p}(\theta)\boldsymbol{p}(\theta)^\dagger + \boldsymbol{R}_N^\alpha(\tau), \qquad (12.211)$$

where $\boldsymbol{p}(\theta)^\dagger$ is the transpose conjugate of $\boldsymbol{p}(\theta)$. In obtaining (12.211), it is assumed that there is no correlation between $S(t)$ and $\boldsymbol{N}(t)$. If the cycle frequency parameter α is chosen to be a cycle frequency of the signal $S(t)$

but not of any of the other signals in $N(t)$, then (12.211) reduces to

$$\boldsymbol{R}_X^\alpha(\tau) = R_S^\alpha(\tau)\boldsymbol{p}(\theta)\boldsymbol{p}(\theta)^\dagger, \qquad (12.212a)$$

which is a rank-one matrix. Thus, as long as there are $n \geq 2$ sensors (so that the $n \times n$ matrix has a dimension exceeding unity), the direction vector $\boldsymbol{p}(\theta)$ is equal to the right eigenvector of $\boldsymbol{R}_X^\alpha(\tau)$ associated with the one nonzero eigenvalue.

Since $\boldsymbol{p}(\theta)$ is orthogonal to the $(n - 1)$-dimensional right null space of $\boldsymbol{R}_X^\alpha(\tau)$, then—if the calibration function $\boldsymbol{p}(\cdot)$ is known—the DOA θ can be determined by searching over ψ for the vector $\boldsymbol{p}(\psi)$ that is orthogonal to this null space, that is, orthogonal to the $n - 1$ right eigenvectors $\{\boldsymbol{e}_i : i = 2, 3, \ldots, n\}$ corresponding to the $n - 1$ zero-valued eigenvalues of $\boldsymbol{R}_X^\alpha(\tau)$. This can be accomplished by maximizing the function

$$Q(\psi) \triangleq \frac{\boldsymbol{p}^\dagger(\psi)\boldsymbol{p}(\psi)}{\left|\sum_{i=2}^n \boldsymbol{e}_i^\dagger \boldsymbol{p}(\psi)\right|^2}, \qquad (12.213)$$

which, ideally, approaches infinity as ψ approaches the correct DOA θ.

In practice, the matrix of cyclic correlations must be estimated using a finite-time average such as

$$\boldsymbol{R}_X^\alpha(\tau)_W \triangleq \frac{1}{W}\int_{-W/2}^{W/2} \boldsymbol{X}\!\left(t + \frac{\tau}{2}\right)\boldsymbol{X}^\dagger\!\left(t - \frac{\tau}{2}\right) e^{-i2\pi\alpha t}\, dt,$$

or a discrete-time counterpart. Such an estimate will typically have full rank (except in the limit as $W \to \infty$). However, for sufficiently long averaging time W, it will have an approximate rank of unity in the sense that one eigenvalue will be much larger than the $n - 1$ others. Thus, by using singular value decomposition (SVD) methods, the right eigenvectors associated with the $n - 1$ smallest eigenvalues of an estimated CYCCOR matrix can be computed and used to estimate the objective function $Q(\cdot)$, which can then be maximized. This same method can be applied repeatedly (sequentially or simultaneously) to the same received data set, but using different cycle frequencies, one for each cyclostationary-signal source of interest, in order to estimate DOA selectively.

This approach to DOA estimation differs from the more conventional SVD method called MUSIC, which is based on stationary process models for the signals, since that method is based on (12.211) evaluated at $\alpha = 0$ (and $\tau = 0$), in which case $\boldsymbol{R}_X^0(0)$ is in general a full-rank matrix. Consequently, it is required that the number n of sensors exceed the total number m of sources, and the contribution to $\boldsymbol{R}_X^0(0)$, namely $\boldsymbol{R}_N^0(0)$, from the noise must be estimated and subtracted from the estimate $\boldsymbol{R}_X^0(0)_W$ by using generalized SVD methods [Schmidt, 1979, 1981].

If there are $m_\alpha > 1$ statistically independent signals that exhibit cyclostationarity with the same cycle frequency $\alpha \neq 0$, then (12.212a) generalizes to

$$\boldsymbol{R}_X^\alpha(\tau) = \sum_{q=1}^{m_\alpha} R_{S_q}^\alpha(\tau) \boldsymbol{p}(\theta_q) \boldsymbol{p}(\theta_q)^\dagger, \qquad (12.212\text{b})$$

where $\{\theta_q\}$ are the DOAs of the m_α signals. It follows from (12.212b) that if $m_\alpha \leq n$ and $\{\boldsymbol{p}(\theta_k)\}$ are linearly independent, then the rank of $\boldsymbol{R}_X^\alpha(\tau)$ is m_α and its range space is spanned by the m_α direction vectors $\{\boldsymbol{p}(\theta_k)\}$. Thus, these direction vectors are orthogonal to the null space of $\boldsymbol{R}_X^\alpha(\tau)$. Hence, if $m_\alpha < n$, then the m_α DOAs can be found by searching over ψ for the vectors $\boldsymbol{p}(\psi)$ that are orthogonal to the null space of $\boldsymbol{R}_X^\alpha(\tau)$ by using the objective function $Q(\cdot)$ with the lower index of summation replaced by $m_\alpha + 1$ [Schell, et al., 1989], whereas the MUSIC method uses the null space of $R_X^0(0) - R_N^0(0)$ and replaces the lower index of summation with m [Schmidt, 1979, 1981].

Since $\boldsymbol{e}_i^\dagger \boldsymbol{p}(\theta) = 0$, then we see from (12.210) that

$$\boldsymbol{e}_i^\dagger \boldsymbol{X}(t) = \boldsymbol{e}_i^\dagger \boldsymbol{N}(t).$$

That is, using any of the $n - 1$ vectors $\boldsymbol{e}_i^\dagger$ of n complex weights to linearly combine the n received signals nulls out the signal of interest $S(t)$. Because of the high resolution associated with nulls in the spatial reception pattern of a sensor array, in contrast to the relatively low resolution of the beams, this method of DOA estimation yields high resolution relative to beamforming methods that search for the beam direction ψ at which the received signal strength is maximum.

The cyclic correlation in (12.211) is defined by

$$\boldsymbol{R}_X^\alpha(\tau) \triangleq \left\langle E\left\{ \boldsymbol{X}\left(t + \frac{\tau}{2}\right) \boldsymbol{X}^\dagger\left(t - \frac{\tau}{2}\right) \right\} e^{-i2\pi\alpha t} \right\rangle \qquad (12.214\text{a})$$

and is appropriate if the cycle frequency α to be used is a baud rate, chip rate, or hop rate. However, if a doubled (down-converted) carrier frequency is to be used for α, then the appropriate quantity is the *cyclic conjugate correlation* [Brown, 1987] defined by

$$\boldsymbol{R}_{XX^*}^\alpha(\tau) \triangleq \left\langle E\left\{ \boldsymbol{X}\left(t + \frac{\tau}{2}\right) \boldsymbol{X}^T\left(t - \frac{\tau}{2}\right) \right\} e^{-i2\pi\alpha t} \right\rangle, \qquad (12.214\text{b})$$

in which the second vector is transposed but not conjugated (cf. exercise 28). In this case, $\boldsymbol{p}(\theta)\boldsymbol{p}(\theta)^\dagger$ in (12.211) and (12.212) gets replaced with $\boldsymbol{p}(\theta)\boldsymbol{p}(\theta)^T$. These same considerations apply to the second CYCCOR method for DOA estimation, which is discussed in the next subsection.

A number of simulations of the CYCCOR DOA-estimation technique, applied to various signal, noise, and interference environments, have been conducted. These simulations demonstrate that, with adequate averaging time, this technique has several advantages, relative to the MUSIC method, stemming from its selectivity for signals exhibiting cyclostationarity. For the purpose of illustration, the results of two simulations are shown in Figure 12.9. In these simulations, the antenna array consists of four isotropic sensors spaced uniformly on a circle having diameter equal to half the wavelength corresponding to the center frequency of the narrow reception band. The noise vector $N(t)$ consists of independent, identically distributed, white Gaussian sensor/receiver noise. The averaging time used to measure $R_X^\alpha(\tau)$ is $800T_c$ where T_c is the keying interval of the BPSK signal(s) of interest.

In the first simulation the BPSK signal arrives with a DOA of 60° (relative to a reference angle) and a closely spaced, equal-power APK interference (with a keying interval of $4T_c/3$) arrives with a DOA of 63°. Each signal is 10 dB above the noise level. For the MUSIC method ($\alpha = 0$), the parameter value $m = 2$ is used, whereas $m_\alpha = 1$ is used for the CYCCOR method ($\alpha = 1/T_c$). Because of the similarity of the CYCCOR method to the MUSIC method, it is denoted by Cyclic MUSIC in Figure 12.9. Graphs of the DOA-estimation function $Q(\psi)$ in (12.213) for the MUSIC and CYCCOR methods are shown in Figure 12.9a. It can be seen that CYCCOR (which uses knowledge of α) accurately resolves the DOA

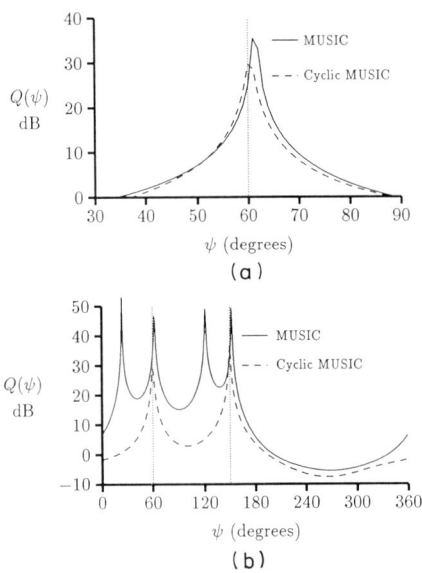

Figure 12.9 (a) Graphs of the DOA Estimation Function $Q(\psi)$ in (12.213) for the MUSIC and Cyclic MUSIC Methods Applied to a Four-Element Circular Array with One Signal of Interest and One Interference. (b) Graphs of the DOA Estimation Function $Q(\psi)$ in (12.213) for the MUSIC and Cyclic MUSIC Methods Applied to a Four-Element Circular Array with Two Signals of Interest and Three Interferences.

of the signal of interest, whereas MUSIC (which uses knowledge of R_N) is unable to resolve the signal and interference. If α is changed to the keying rate of the APK signal instead of the BPSK signal, then CYCCOR accurately resolves the DOA of the APK signal (simulation result not shown).

In the second simulation, there are two equal-power BPSK signals of interest with the same keying rate T_c, and with DOAs of 60° and 150°; and there are three interferences (TV, FM, and pulsed radar) with DOAs of 10°, 20°, and 70°. Each of the five signals is 10 dB above the noise level. Since $m = 5$ exceeds $n = 4$, there is no null space for the MUSIC method. Thus, the largest allowable parameter value, $m = 3$, is used in the MUSIC method. For the CYCCOR method, we have $m_\alpha = 2$. Graphs of the DOA estimation function $Q(\psi)$ are shown in Figure 12.9b. It can be seen that CYCCOR (Cyclic MUSIC) accurately determines the DOAs of the two signals of interest, whereas MUSIC does not resolve the pair at 10° and 20° or the pair at 60° and 70°, and it produces a spurious peak at 120°.

CYCCOR Method 2. We consider essentially the same problem as the preceding one, but we now assume that we have two arrays of sensors in the same plane that are identical to each other in every respect except that one array is linearly displaced from the other by a distance Δ. In this case, the received data from one array is still modeled as (12.210), but that from the displaced array is modeled as

$$Y(t) = p(\theta)\phi S(t) + M(t), \qquad (12.215)$$

where $M(t)$ plays a role analogous to that of $N(t)$ in (12.210), and the scalar ϕ is given by

$$\phi = \exp\left[i2\pi f_o \frac{\Delta}{c} \sin(\theta)\right], \qquad (12.216)$$

in which f_o is the center frequency of the signal, c is the speed of propagation, and θ is the angle of arrival of the signal $S(t)$, relative to the array-displacement direction. By the same argument leading up to (12.212), we arrive at (exercise 27)

$$R_{XY}^\alpha(\tau) = R_S^\alpha(\tau)\phi^* p(\theta) p(\theta)^\dagger. \qquad (12.217)$$

With the use of (12.212) and (12.217), we obtain the equation

$$R_X^\alpha(\tau) - \lambda R_{XY}^\alpha(\tau) = R_S^\alpha(\tau)[1 - \lambda\phi^*]\, p(\theta) p(\theta)^\dagger. \qquad (12.218a)$$

Consequently, $\lambda = 1/\phi^*$ is the generalized eigenvalue of the rank-one pair of $n \times n$ CYCCOR matrices $R_X^\alpha(\tau)$ and $R_{XY}^\alpha(\tau)$. Thus, by using generalized SVD methods, we can estimate ϕ from estimates [cf. (12.213)] of these two CYCCOR matrices. From an estimate of ϕ and knowledge of f_o we can obtain an estimate of the DOA θ using (12.216), without requiring array calibration [knowledge of $p(\cdot)$].

This approach to DOA estimation differs from the more conventional SVD method called ESPRIT, which is based on stationary process models for the signals, since that method is based on (12.211) and its counterpart for $R_{XY}^\alpha(\tau)$, both evaluated at $\alpha = 0$ (and $\tau = 0$), in which case both $R_X^0(0)$ and $R_{XY}^0(0)$ are in general full-rank matrices. Thus, the conventional method again requires $n > m$ and the contributions to these two matrices from the noise, namely $R_N^0(0)$ and (if it is not zero) $R_{NM}^0(0)$, must again be estimated and subtracted from the estimates $R_X^0(0)_W$ and $R_{XY}^0(0)_W$, respectively, by using generalized SVD methods [Paulraj et al., 1986; Roy et al., 1986].

If there are $m_\alpha > 1$ statistically independent signals that exhibit cyclostationarity with the same cycle frequency $\alpha \neq 0$, then (12.218a) generalizes to

$$R_X^\alpha(\tau) - \lambda R_{XY}^\alpha(\tau) = \sum_{q=1}^{m_\alpha} R_{S_q}^\alpha(\tau)[1 - \lambda \phi_q^*] p(\theta_q) p(\theta_q)^\dagger, \quad (12.218b)$$

which is a rank-m_α matrix [assuming $m_\alpha \leq n$ and $\{p(\theta_q)\}$ are linearly independent]. The m_α generalized eigenvalues of the pair of CYCCOR matrices $R_X^\alpha(\tau)$ and $R_{XY}^\alpha(\tau)$ are $\{\lambda_q\} = \{1/\phi_q^*\}$, from which estimates of the m_α DOAs $\{\theta_q\}$ can be obtained using (12.216), as is done in the ESPRIT method using $[R_X^0(0) - R_N^0(0)] - \lambda[R_{XY}^0(0) - R_{NM}^0(0)]$.

In contrast to the conventional methods, since only $n \geq m_\alpha + 1$ sensors are required in the array for CYCCOR method 1 and in each of the two arrays for CYCCOR method 2, both the arrays and the SVD computations required to estimate the DOAs can be considerably simplified when $m_\alpha \ll m$ (exercise 29). However, the amount of time averaging used to estimate the CYCCOR matrices will typically have to be much greater than that needed for the conventional correlation matrices, since the signals not of interest and the noise must both decorrelate sufficiently well to yield negligible contributions $R_N^\alpha(\tau)_W$ and $R_{NM}^\alpha(\tau)_W$ to the estimated CYCCOR matrices $R_X^\alpha(\tau)_W$ and $R_{XY}^\alpha(\tau)_W$. The value of the lag parameter τ can be important here since it determines the strength of the factor $R_S^\alpha(\tau)_W$ in the desired component of the estimated CYCCOR matrices (exercise 30).

There are several ways to possibly enhance the CYCCOR methods so that shorter averaging times can be used. One way is to integrate a weighted version of the CYCCOR matrix over a range of lag values τ. An appropriate weighting function for model (12.210) is $R_S^\alpha(\tau)^*$, which yields the CYCCOR matrix replacement

$$\int R_X^\alpha(\tau)_W R_S^\alpha(\tau)^* \, d\tau \simeq \int |R_S^\alpha(\tau)|^2 \, d\tau \, p(\theta) p(\theta)^\dagger. \quad (12.219a)$$

If $R_S^\alpha(\tau)^*$ is unknown, it can be replaced with

$$\sum_k R_{X_k}^\alpha(\tau)_W^*,$$

where X_k is the kth element of X. An alternative to (12.219a) is

$$\int S_X^\alpha(f)_W S_S^\alpha(f)^* \, df \simeq \int |S_S^\alpha(f)|^2 \, df \, p(\theta)p(\theta)^\dagger, \qquad (12.219b)$$

and if $S_S^\alpha(f)$ is unknown, it can be replaced with

$$\sum_k S_{X_k}^\alpha(f)_W^*.$$

In this last expression, $S_{X_k}^\alpha(f)_W$ is an estimate of the cyclic spectrum $S_{X_k}^\alpha(f)$. For example, $S_{X_k}^\alpha(f)_W$ could be a frequency-smoothed version of the Fourier transform of $R_{X_k}^\alpha(\tau)_W$.

12.8.3 Time-Difference-of-Arrival Estimation

SPECCORR Method. Another approach to direction finding that exploits cyclostationarity and requires only a single pair of widely separated sensors for each planar direction to be determined can be arrived at by the system-identification approach that leads to the conventional generalized cross-correlation methods for time-difference-of-arrival (TDOA) estimation [Gardner and Chen, 1988a]. Considering for the moment a single wavefront impinging on a pair of sensors, and interpreting the (real or complex) data $X(t)$ received by one sensor as the input to a system and the data $Y(t)$ received by the other sensor as the system output, the TDOA, from which the DOA can be approximated, can be interpreted as the delay of the system. The delay τ_0 can be determined from the location of the peak in the impulsive impulse-response function $h(\tau) = \delta(\tau - \tau_0)$ of the pure-delay system. The impulse-response function can in turn be obtained by inverse Fourier transformation of the transfer function $H(f)$ of the system. The transfer function can be determined from the ratio of spectra

$$H(f) = \frac{S_{YX}(f)}{S_X(f)}, \qquad (12.220)$$

which follows from the input-output relation (10.29) for the spectra associated with the input and output of a linear time-invariant system. Since inverse Fourier transformation of this ratio of spectra,

$$h(\tau) = \int_{-\infty}^{\infty} \frac{S_{YX}(f)}{S_X(f)} e^{i2\pi f \tau} \, df, \qquad (12.221)$$

yields the impulse-response function, then this function can be estimated using formula (12.221) with the ideal spectra replaced by their estimates obtained from the data $X(t)$ and $Y(t)$ (cf. Section 11.4). This is one of the generalized cross-correlation methods for TDOA estimation. Others can be arrived at by similar arguments which lead to different denominators in

(12.221). When the denominator is replaced by unity, this produces the classic cross-correlation method (cf. exercise 13 in Chapter 10).

The generalized cross-correlation method based on (12.221) is actually optimum in the sense that the system model with transfer function given by (12.220) minimizes the MSE between the output $Y(t)$ of the actual system (which might not be either linear or time invariant) and the system model output, when both the actual system and its model have the same input $X(t)$ (cf. Section 13.4). Nevertheless, this criterion of optimality is not particularly appropriate when the system input-output measurements $X(t)$ and $Y(t)$ are contaminated with noise or interference that is correlated from input to output. Unfortunately, this is precisely what happens when interfering signal wavefronts impinge on the pair of sensors. The alternative method described next provides tolerance to such contamination by using frequency shifting to decorrelate the contamination from input to output.

Analogous to the input-output relation for conventional spectra which leads to (12.220), we have an input-output spectral correlation relation [cf. (12.65)] that leads to the SPECtral CORrelation Ratio (SPECCORR) formula

$$H(f) = \frac{S_{YX}^{\alpha}(f - \alpha/2)}{S_{X}^{\alpha}(f - \alpha/2)} \qquad (12.222)$$

for the transfer function of any linear time-invariant system with input $X(t)$ and output $Y(t)$, both of which are cyclostationary with cycle frequency α. By choosing α to be a cycle frequency of only one particular signal wavefront impinging on the pair of sensors, we can reject contributions to the spectral correlation functions used in (12.222) from noise and other impinging signal wavefronts that do not exhibit cyclostationarity with this same cycle frequency. This can be seen as follows. We express the two received data sets as

$$X(t) = S(t) + N(t) \qquad (12.223a)$$

$$Y(t) = S(t - \tau_o) + M(t), \qquad (12.223b)$$

where $S(t)$ is the signal of interest, τ_o is the TDOA of $S(t)$ at the two sensors, and $N(t)$ and $M(t)$ each represent the sum of noise and all other signals impinging on the sensors. If we assume there is no correlation between $S(t)$ and $N(t)$ and between $S(t)$ and $M(t)$, then we obtain the spectral correlation functions (exercise 31)

$$S_{YX}^{\alpha}(f) = S_{S}^{\alpha}(f)\, e^{-i2\pi(f+\alpha/2)\tau_o} + S_{MN}^{\alpha}(f) \qquad (12.224a)$$

$$S_{X}^{\alpha}(f) = S_{S}^{\alpha}(f) + S_{N}^{\alpha}(f). \qquad (12.224b)$$

With the appropriate choice for α as just described, we have

$$S_S^\alpha(f) \neq 0 \tag{12.225a}$$

$$S_{MN}^\alpha(f) \equiv 0 \tag{12.225b}$$

$$S_N^\alpha(f) \equiv 0, \tag{12.225c}$$

and (12.222) therefore becomes

$$H(f) = e^{-i2\pi f \tau_o}, \tag{12.226a}$$

from which the TDOA can be obtained by inverse Fourier transformation:

$$h(\tau) = \delta(\tau - \tau_o). \tag{12.226b}$$

However, for the conventional method that is based on (12.222) with $\alpha = 0$, as in (12.221), the contributions $S_{MN}^0(f)$ and $S_N^0(f)$ due to noise and all other impinging signals will not in general vanish as in (12.225). Thus, the conventional transfer-function-estimation formula (12.220) becomes

$$H(f) = \frac{e^{-i2\pi f \tau_o} + S_{MN}(f)/S_S(f)}{1 + S_N(f)/S_S(f)}. \tag{12.227}$$

The desired peak at τ_o in the inverse transform of (12.227) can be severely masked by contributions from the terms $S_{MN}(f)/S_S(f)$ and $S_N(f)/S_S(f)$. In fact, the former term can give rise to other peaks corresponding to the TDOAs of the interfering signals.

The substantial superiority of the SPECCORR method based on (12.222) over the conventional method based on (12.220) is demonstrated in simulations in [Gardner and Chen, 1988a, 1988b; Chen, 1989], where various cases of strong single and multiple narrowband and broadband interfering signals are studied. For the purpose of illustration, the results of one simulation are shown in Figure 12.10. Part (a) of this figure shows a graph of $h(\tau)$ obtained from (12.221) with the ratio of spectra replaced with the ratio of estimated spectra. Part (b) shows a graph of $h(\tau)$ obtained from (12.221) with the ratio of spectra replaced with the ratio of estimated cyclic spectra, as in (12.222). The signal here is (real) BPSK, its TDOA is $\tau_o = 48$ (which represents three keying intervals), the cycle frequency used is the keying rate $\alpha = 1/T_c$, and the data segment used for measuring the spectral correlation functions has length $2048 T_c$. The signal contamination consists of white noise and five interfering AM signals with various TDOAs. The estimated spectra and cyclic spectra (magnitudes) for this signal and contamination are shown in Figure 12.8. It can be seen from Figure 12.10b that the correct TDOA is unambiguously determined by the location of the strongest peak for the SPECCORR method. However, the conventional method in Figure 12.10a yields many peaks, the strongest of which is not located at the correct TDOA.

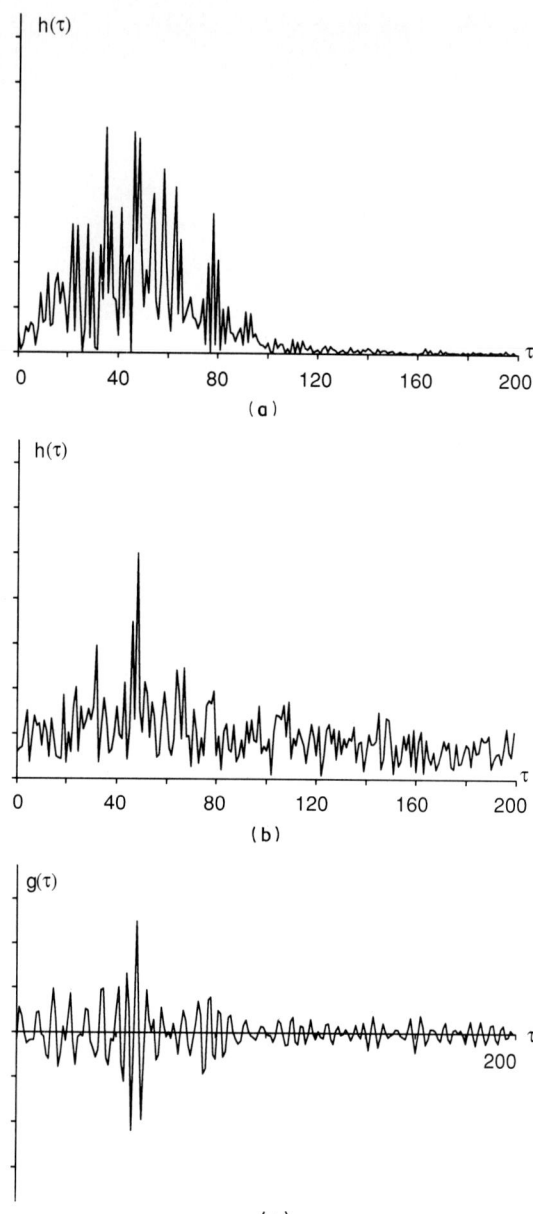

Figure 12.10 (a) Graph of the Conventional Generalized Cross-Correlation Function for a Segment of Length $2048T_c$ of the Corrupted Signal with the Spectrum and Cyclic Spectrum Shown in Figure 12.8a. (b) Graph of the Generalized Cross-Correlation Function from the SPECCORR Method for the Same Corrupted Signal Segment as in Part (a). (c) Graph of the Function (12.229) From the SPECCORP Method for the Same Corrupted Signal Segment as in Part (a).

SPECCORP Method. A second approach to TDOA estimation that exploits spectral correlation to obtain tolerance to noise and interference is the SPECtral CORrelation Product (SPECCORP) method introduced in [Gardner and Chen, 1988b]. The SPECCORP method seeks a peak in the real part of the inverse Fourier transform of the frequency-shifted conjugate product (rather than the ratio, as in SPECCORR) of the spectral correlation functions:

$$g(\tau) = \text{Re}\left\{\int_{-\infty}^{\infty} S_{YX}^{\alpha}\left(f - \frac{\alpha}{2}\right) S_{X}^{\alpha}\left(f - \frac{\alpha}{2}\right)^{*} e^{i2\pi f\tau}\, df\right\}. \quad (12.228)$$

Maximizing (12.228) is equivalent to minimizing the integrated magnitude-squared difference between $S_{YX}^{\alpha}(f)$ and $S_{X}^{\alpha}(f)\, e^{-i2\pi(f+\alpha/2)\tau}$, which is motivated by the fact that this difference is zero when $X(t)$ and $Y(t)$ are given by (12.223) and $\tau = \tau_0$, as can be seen from (12.224) and (12.225).* Substitution of (12.224) and (12.225) into (12.228) yields

$$g(\tau) = \text{Re}\left\{\left[R_S^{\alpha}(u) \otimes R_S^{\alpha}(-u)^{*}\right] e^{i\pi\alpha u}\bigg|_{u=\tau-\tau_0}\right\}, \quad (12.229)$$

which does indeed reach its maximum value at $\tau = \tau_0$. For signal types whose cyclic autocorrelation $R_S^{\alpha}(\tau)$, when correlated with itself as in (12.229), produces a sufficiently distinct peak, the SPECCORP method can provide TDOA estimates that are comparable in accuracy to those provided by the SPECCORR method, while using substantially shorter data segments [Gardner and Chen, 1988b; Chen, 1989]. This is illustrated in Figure 12.10c, which shows a graph of $g(\tau)$ for the same signal, noise, and interference environment used to illustrate the SPECCORR method in Figure 12.10b. The length of the data segment used in the SPECCORP method is only $256 T_c$, whereas that used in the SPECCORR method is $2048 T_c$. The peak occurs at the correct TDOA, $\tau = \tau_0 = 48$.

12.8.4 Spatial Filtering

We consider the problem of adaptively adjusting the magnitudes and phases of a set of waveforms with narrow relative bandwidth received by an array of n sensors such that their addition enhances a desired signal impinging on the array while suppressing impinging noise and interfering signals. This technique results in an effective spatial reception pattern with a beam in the direction of arrival of the signal of interest and up to $n - 1$ nulls in the directions of arrival of the interfering signals. Thus, it accomplishes spatial filtering. For this pupose, we consider the same model (12.210) as that used in the previous subsection (except the signals are

---

*In [Gardner and Chen, 1988b], the SPECCORP method is referred to as SPECCOA, which stands for SPEctral COherence Alignment.

downconverted from analytic signals to complex envelopes). The magnitudes and phases of the complex envelopes $\{X_j(t) : j = 1, 2, \ldots, n\}$ of the n received data sets are to be adjusted and the resultant compensated complex signals added together to form an estimate

$$\hat{S}(t) \stackrel{\Delta}{=} \mathbf{g}\mathbf{X}(t) = \sum_{j=1}^{n} g_j X_j(t) \qquad (12.230)$$

of the signal of interest $S(t)$. The complex n-vector (row) of weights $\mathbf{g}$ in this estimate models the magnitude and phase adjustments and is called the *steering vector*. The problem of interest here is to find the set of weights that minimizes the time-averaged mean-squared error (TAMSE) between the signal and its estimate,

$$\text{TAMSE} = \left\langle E\{|S(t) - \hat{S}(t)|^2\} \right\rangle. \qquad (12.231)$$

By substituting (12.230) into (12.231) and equating to zero the derivatives of TAMSE with respect to the real and imaginary parts of the n elements of $\mathbf{g}$, we obtain the optimum weight vector (exercise 32)

$$\mathbf{g}_o = \langle \mathbf{R}_{SX}^T \rangle \langle \mathbf{R}_X \rangle^{-1}, \qquad (12.232)$$

where

$$\langle \mathbf{R}_X \rangle \stackrel{\Delta}{=} \langle E\{\mathbf{X}(t)\mathbf{X}^\dagger(t)\} \rangle \qquad (12.233)$$

$$\langle \mathbf{R}_{SX} \rangle \stackrel{\Delta}{=} \langle E\{S(t)\mathbf{X}^\dagger(t)\} \rangle. \qquad (12.234)$$

Use of the model (12.210) (with the explicit denotation of θ suppressed) in (12.234) and substitution of the result into (12.232) yields the more explicit result (exercise 32)

$$\mathbf{g}_o = \mathbf{p}^\dagger \langle \mathbf{R}_X \rangle^{-1} \langle R_S \rangle(0), \qquad (12.235)$$

which is the well-known optimum weight vector for forming a beam in the direction of arrival of $S(t)$ and up to $n - 1$ nulls in the directions of arrival of interfering signals [Monzingo and Miller, 1980]. However, the well-known solution is generalized here from stationary data to cyclostationary data by virtue of the fact that time-averaged cyclostationary correlations here replace stationary correlations in the well-known solution.

Unfortunately, this optimum weight vector cannot in general be determined in practice without knowledge of $S(t)$ throughout a training period during which the correlation $\langle \mathbf{R}_{SX} \rangle$ can be measured either directly or indirectly through the use of an iterative algorithm that seeks the minimum of the time-averaged squared error (which is the same as the minimum of the TAMSE for ergodic processes). This problem can be circumvented when the direction of arrival of $S(t)$ is known because then the problem is to nonadaptively constrain a beam to form in this known direction while adaptively forming nulls in the unknown directions of arrival of the interfering signals by simply minimizing the time-averaged squared output $\hat{S}(t)$ of the constrained array. But, there are important applications where this direction is not known. Nevertheless, if the signal of interest $S(t)$ exhibits cyclostationarity with known (or measured) cycle frequency α, and the noise and interfering signals $N(t)$ do not exhibit cyclostationarity with this same cycle frequency, then the optimum weight vector (12.235) can, in principle, be determined to within a scale factor (which has no effect on the spatial reception pattern of the array), even when $S(t)$ and its direction of arrival are unknown. To see this, we replace the training signal $S(t)$ in (12.231) with the time- and frequency-shifted waveform

$$\tilde{S}(t) \triangleq X_j(t + \tau) e^{-i2\pi\alpha t}, \qquad (12.236)$$

where $X_j(t)$ is any one of the n received waveforms. Then we have in place of $\langle R_{SX} \rangle$

$$\langle R_{\tilde{S}X} \rangle = R^\alpha_{X_jX}(\tau) e^{i\pi\alpha\tau}, \qquad (12.237)$$

which upon substitution of the model (12.210) becomes (exercise 32)

$$\langle R_{\tilde{S}X} \rangle = \left[R^\alpha_S(\tau) p_j p^* + R^\alpha_{N_jN}(\tau) \right] e^{i\pi\alpha\tau}. \qquad (12.238)$$

But, by the appropriate choice of α just described, we have

$$R^\alpha_S(\tau) \neq 0 \qquad (12.239a)$$

$$R^\alpha_{N_jN}(\tau) \equiv 0, \qquad (12.239b)$$

in which case (12.232), with $S(t)$ replaced by $\tilde{S}(t)$, becomes

$$g_* = a g_o, \qquad (12.240)$$

where g_o is given by (12.235) and the scale factor a is given by

$$a \triangleq \frac{R^\alpha_S(\tau)}{\langle R_S \rangle(0)} p_j e^{i\pi\alpha\tau}. \qquad (12.241)$$

Although this scale factor has no effect on the spatial reception pattern, in practice it must not be too small, which means that the jth element's attenuation p_j [for the direction of arrival of $S(t)$] and the cyclic correlation coefficient $|R^\alpha_S(\tau)|/\langle R_S \rangle(0)$ (for the value of τ used) must not be too small.

In practice, by discretizing time, the time-averaged squared error

$$\text{TASE} \triangleq \langle |\tilde{S}(t) - \hat{S}(t)|^2 \rangle \qquad (12.242)$$

can be minimized iteratively using the recursive least-squares algorithm described in Subsection 13.2.2.

The essence of the approach to array adaptation just described is to derive from the received data $X(t)$ a training signal replacement $\tilde{S}(t)$ that is correlated with the component $S(t)$ in the estimate $\hat{S}(t)$, but is uncorrelated with the corruptive component $N(t)$ in $\hat{S}(t)$. In fact, the minimum-TAMSE solution (12.240) is identical to that obtained by maximizing the magnitude of the correlation coefficient ρ for $\hat{S}(t)$ and $\tilde{S}(t)$,

$$|\rho| \triangleq \frac{|\langle R_{\hat{S}\tilde{S}} \rangle|}{[\langle R_{\hat{S}} \rangle \langle R_{\tilde{S}} \rangle]^{1/2}}, \quad (12.243)$$

subject to the constraint that the correlation coefficient for $\tilde{S}(t)$ and $S(t)$ be nonzero and the correlation coefficient for $\tilde{S}(t)$ and $N(t)$ be equal to zero (exercise 32). Because of this equivalence, the speed of convergence of the iterated least-squares algorithm can be increased [by enhancing the required empirical decorrelation of $\tilde{S}(t)$ with $N(t)$ relative to the correlation of $\tilde{S}(t)$ with $S(t)$] by designing the training signal replacement $\tilde{S}(t)$ to maximize jointly with $\hat{S}(t)$ the correlation coefficient (12.243). In order to satisfy the constraint that $\tilde{S}(t)$ and $S(t)$ be correlated and $\tilde{S}(t)$ and $N(t)$ not be correlated, we can constrain $\tilde{S}(t)$ to be of the general form

$$\tilde{S}(t) = k[X(t) \otimes h(t)]e^{-i2\pi\alpha t}, \quad (12.244)$$

where k is called the *control vector* (row) and $h(t)$ is the impulse-response function of a scalar filter. The correlation constraints can then be satisfied as before with (12.236) by choosing α to be a cycle frequency of $S(t)$ but not of $N(t)$.

The jointly optimum control vector ideally would act like a steering vector to form a beam in the direction of arrival of $S(t)$ and nulls in the directions of arrival of the interfering signals in $N(t)$. This vector can be obtained by first substituting the optimum steering vector g into (12.243). The result of this substitution is the Rayleigh quotient (exercise 33)

$$|\rho|^2 = \frac{k R_{YX}^\alpha \langle R_X \rangle^{-1} [R_{YX}^\alpha]^\dagger k^\dagger}{k \langle R_Y \rangle k^\dagger}, \quad (12.245)$$

where $Y(t) \triangleq h(t) \otimes X(t)$. This reveals that the optimum control vector k is given by the generalized eigenvector $k^\dagger = k_o^\dagger$ associated with the largest generalized eigenvalue λ of the generalized eigenequation

$$R_{YX}^\alpha \langle R_X \rangle^{-1} [R_{YX}^\alpha]^\dagger k_o^\dagger = \lambda \langle R_Y \rangle k_o^\dagger. \quad (12.246)$$

Alternatively, k_o is given by (exercise 33)

$$k_o = v_o^\dagger \langle R_Y \rangle^{-1/2}, \quad (12.247)$$

where v_o is the eigenvector associated with the largest eigenvalue of the matrix M defined by

$$M \triangleq \langle R_Y \rangle^{-1/2} R_{YX}^\alpha \langle R_X \rangle^{-1} [R_{YX}^\alpha]^\dagger \langle R_Y \rangle^{-1/2}. \quad (12.248)$$

Thus, SVD methods are required for optimization of the control vector k.

With the constraint (12.244), the performance coefficient (12.243) becomes (exercise 34)

$$|\rho| = \frac{|R_{\hat{S}U}^\alpha|}{[\langle R_{\hat{S}}\rangle\langle R_U\rangle]^{1/2}}, \quad (12.249)$$

which is a type of cross-coherence coefficient for the signal estimate $\hat{S}(t)$ and the *control signal*

$$U(t) \stackrel{\Delta}{=} k[X(t) \otimes h(t)]. \quad (12.250)$$

This control signal is appropriate if the cycle frequency α used in (12.244) is a baud, chip, or hop rate. However, if a doubled (down-converted) carrier frequency is to be used for α, then $Y(t) = h(t) \otimes X(t)$ in (12.250) should be conjugated or, equivalently, the cyclic cross-correlation $R_{\hat{S}U}^\alpha$ in (12.249) should be replaced with the cyclic conjugate cross-correlation $R_{\hat{S}U^*}^\alpha$ (cf. exercise 28).

Since both $\hat{S}(t)$ and $U(t)$ in the cross-coherence coefficient (12.249) are linear transformations of the received data $X(t)$, then maximization of this coefficient with respect to either g or k can be interpreted as optimally restoring the self-coherence in $X(t)$ that has been reduced by the additive signal corruption $N(t)$. Consequently, this approach to array adaptation is called Self COherence REstoral (SCORE). A computationally efficient algorithm for iteratively solving for the jointly optimum steering and control vectors g_o and k_o can be obtained by using the fact that joint maximization of the cross-coherence coefficient (12.249) with respect to g and k is equivalent to joint minimization of the TAMSE $\langle E\{|\hat{S}(t) - \tilde{S}(t)|^2\}\rangle$ subject to the constraint that either $\langle E\{|\hat{S}(t)|^2\}\rangle = g\langle R_X\rangle g^\dagger$ or $\langle E\{|\tilde{S}(t)|^2\}\rangle = k\langle R_V\rangle k^\dagger$ be held equal to a constant. Here, $V(t)$ is the reference signal vector

$$V(t) \stackrel{\Delta}{=} [X(t) \otimes h(t)]e^{-i2\pi\alpha t}, \quad (12.251)$$

in terms of which the training signal replacement $\tilde{S}(t)$ can be expressed as

$$\tilde{S}(t) = kV(t), \quad (12.252)$$

which is analogous to

$$\hat{S}(t) = gX(t). \quad (12.253)$$

The algorithm consists of alternately iterating two normalized recursive least-squares algorithms for minimizing the TASE (12.242) with respect to each of g and k. The normalization consists of scaling g and k every time they are updated to maintain $\langle|\hat{S}(t)|^2\rangle$ or $\langle|\tilde{S}(t)|^2\rangle$ equal to a constant.

A number of simulations of the SCORE technique applied to a four-element circular array in various signal, noise, and interference environments are reported in [Agee et al., 1988a, 1988b; Agee, 1989]. These simulations demonstrate that this technique, when implemented with a pair of alternating normalized recursive least squares algorithms, can indeed provide rapid convergence to a scaled version of the optimum steering vector (12.232) when the signal to be extracted is the only signal in the environment that exhibits cyclostationarity with the cycle frequency α used to form the reference signal (12.251). These simulations also show that the SCORE technique implemented directly in terms of the solution to the generalized eigenequation (12.246) can provide L near-optimum steering vectors corresponding to the L largest generalized eigenvalues when L uncorrelated signals all exhibiting cyclostationarity with the same cycle frequency α that is used in the algorithm are present in the environment.

For the purpose of illustration, the results of one simulation are shown in Figure 12.11. The simulated planar environment consists of four signals: a signal of interest, BPSK (DOA = 60°, SNR = 20 dB); and three interfering signals, APK (16-QAM) (DOA = −45°, SNR = 15 dB), FM (120-channel FDM-FM) (DOA = 30°, SNR = 30 dB), and TV (television) (DOA = −110°, SNR = 40 dB). The received signals are down-converted to complex baseband, and the resultant carrier offsets are 0 Hz for BPSK and APK, −500 kHz for FM, and 2 MHz for TV. The keying rates for BPSK and APK are 4 MHz and 3 MHz, respectively. The sensor array that is simulated is a four-element uniformly spaced circular array with half-wavelength diameter, isotropic sensors, white Gaussian sensor noise, and a bandwidth of 10 MHz.

The performance measure used in the simulation is the ratio (denoted by SINR) of the average power in the signal component of $\hat{s}(t)$ to the average power in the interference and noise component of $\hat{s}(t)$. This SINR is evaluated for various steering vectors obtained during the convergence process of the recursive least squares algorithms. The array processor in which the control vector is fixed at $k = [1, 0, 0, 0]$ is denoted by LS-SCORE, and the processor in which k and g are jointly optimized is denoted

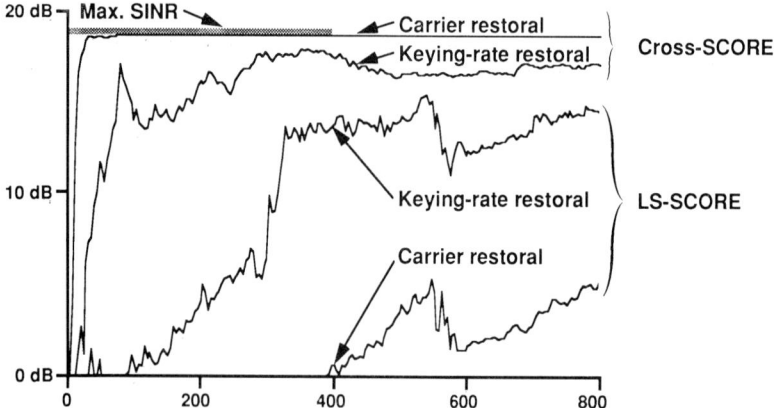

Figure 12.11 Graphs of the Output SNR Versus the Adaptation Time of the LS-SCORE and Cross-SCORE Algorithms.

by cross-SCORE. Both processors are operated using for α both the keying rate ($\alpha = 4$ MHz) and the carrier frequency offset ($\alpha = 0$ Hz). Thus, there are four graphs of SINR versus number of iterations shown in Figure 12.11. It can be seen from the performance graphs in Figure 12.11 that the cross-SCORE processor converges much faster than the LS-SCORE processor.

12.8.5 Detection of Cyclostationary Signals

Multicycle Detector. As explained in Section 11.8, the quadratic device that maximizes deflection (and is optimum in several other senses) for the detection of a random signal $S(t)$ in additive white Gaussian noise (with spectral density N_o) is given by

$$Y = \frac{1}{N_o^2 W} \int_{-W/2}^{W/2} \int_{-W/2}^{W/2} R_S(u, v) X(u) X(v) \, du \, dv, \qquad (12.254)$$

where $\{X(t) : -W/2 \leq t \leq W/2\}$ is the observed noise or the observed signal plus noise. The detection statistic Y is used in the threshold test

$$Y \begin{cases} > \gamma & \Rightarrow \text{ decide } S(t) \text{ is present} \\ < \gamma & \Rightarrow \text{ decide } S(t) \text{ is absent} \end{cases} \qquad (12.255)$$

for some appropriate threshold level γ. Furthermore, it is shown there that the value of the maximized deflection is given by

$$D_{\max} = \left[\frac{1}{2N_o^2} \int_{-W/2}^{W/2} \int_{-W/2}^{W/2} R_S^2(u, v) \, du \, dv \right]^{1/2} \qquad (12.256)$$

Now, if $S(t)$ is cyclostationary or almost cyclostationary, then the autocorrelation function is given in terms of cyclic autocorrelations by

$$R_S(u, v) = \sum_\alpha R_S^\alpha(u - v) e^{i\pi\alpha(u+v)}. \qquad (12.257)$$

Substitution of (12.257) into (12.254) and application of a change of variables (cf. Chapter 7, exercise 4) yields

$$Y = \frac{1}{N_o^2 W} \sum_\alpha \int_{-W}^{W} R_S^\alpha(\tau) \int_{-(W-|\tau|)/2}^{(W-|\tau|)/2} X\left(t + \frac{\tau}{2}\right) X\left(t - \frac{\tau}{2}\right) e^{i2\pi\alpha t} dt \, d\tau. \qquad (12.258)$$

Substitution of the definition (12.25) for the cyclic correlogram (at $t = 0$) for $X(t)$ into (12.258) yields

$$Y = \sum_\alpha \frac{1}{N_o^2} \int_{-W}^{W} R_S^\alpha(\tau)^* R_X^\alpha(\tau)_W \, d\tau. \qquad (12.259)$$

Thus, it can be seen that this optimum detector measures the cyclic correlograms of the observed waveform for all cycle frequencies α contained in the signal to be detected, and then correlates these with stored replicas of the cyclic autocorrelations of the signal. An alternative characterization of this detector follows from (12.259) with an application of Parseval's relation, together with (12.15) (with X replaced by S) and (12.23):

$$Y = \sum_\alpha \frac{1}{N_o^2} \int_{-\infty}^{\infty} S_S^\alpha(f)^* P_X^\alpha(f)_W \, df. \tag{12.260}$$

Thus, the optimum detector measures the cyclic periodograms of the observed waveform for all cycle frequencies α contained in the signal to be detected, and then correlates these with stored replicas of the cyclic spectral densities of the signal. In the special case for which the signal is stationary, only the $\alpha = 0$ term remains in (12.259) and (12.260), and in this case the detector is the optimum radiometer discussed in Section 11.8. Direct implementation of the optimum detector for a cyclostationary (or almost cyclostationary) signal requires knowledge of the phase of the signal to be detected, since the quantities R_S^α and S_S^α depend on this phase [cf. (12.10)]. It is not likely that a detector would possess knowledge of the phase of a random signal that might not even be present. Thus, the implementation of the optimum detector requires a search over the unknown phase parameter to maximize Y.

Let us now consider the formula for maximum deflection (12.256). Substitution of (12.257) into (12.256) and application of a change of variables yields

$$D_{max} \simeq \left[\frac{W}{2N_o^2} \sum_\alpha \int_{-W}^{W} |R_S^\alpha(\tau)|^2 \left(1 - \frac{|\tau|}{W}\right) d\tau \right]^{1/2}, \tag{12.261}$$

which is a close approximation when W is much larger than the largest period of cyclostationarity and also much larger than the widths of the cyclic autocorrelation functions (exercise 35). Since the observation time W is assumed to greatly exceed the widths of $\{R_S^\alpha\}$, then (12.261) yields the close approximations

$$D_{max} \simeq \left[\frac{W}{2N_o^2} \sum_\alpha \int_{-W}^{W} |R_S^\alpha(\tau)|^2 \, d\tau \right]^{1/2} \tag{12.262}$$

$$\simeq \left[\frac{W}{2N_o^2} \sum_\alpha \int_{-\infty}^{\infty} |S_S^\alpha(f)|^2 \, df \right]^{1/2}. \tag{12.263}$$

In the special case for which the signal is stationary, only the $\alpha = 0$ term remains in (12.261) to (12.263), and this is the deflection for the optimum radiometer discussed in Section 11.8. Thus the $\alpha \neq 0$ terms directly indicate how much larger the deflection can be if the cyclostationarity of the signal is not ignored by introduction of a random phase to obtain a stationary model. For example, it can be shown that deflection is increased by 4.8 dB for a BPSK signal by using the multicycle detector rather than the radiometer [Chen, 1989].

Single-Cycle Detector. In order to avoid the search over the unknown phase parameter required to implement the optimum *multicycle detector* specified by (12.260), we can consider deleting all but one of the terms in the sum over α and then comparing the magnitude of this term to a threshold as in (12.255). As already mentioned, if only the term corresponding to $\alpha = 0$ is retained, we obtain the optimum radiometer. However, if a term corresponding to $\alpha \neq 0$ (such as a baud rate or doubled carrier frequency) is retained, we obtain what is called a *single-cycle detector* in contrast to the multicycle detector (12.260). The single-cycle detector also is optimum in the sense that it is an optimum spectral-line regenerator.

To explain the optimality of the single-cycle detector, we consider the following quadratic time-invariant transformation of the observed process $X(t)$,

$$Y(t) = \int_{-\infty}^{\infty} \int_{-\infty}^{\infty} k(u, v) X(t - u) X(t - v) \, du \, dv. \qquad (12.264)$$

The mean of the transformed process $Y(t)$ is given by

$$m_Y(t) = \int_{-\infty}^{\infty} \int_{-\infty}^{\infty} k(u, v) R_X(t - u, t - v) \, du \, dv. \qquad (12.265)$$

Although it is assumed that $X(t)$ exhibits no spectral lines, it is possible for spectral lines to be regenerated from $X(t)$ by the quadratic transformation. Thus $Y(t)$ and, therefore, its mean $m_Y(t)$ can indeed exhibit spectral lines. The average power associated with a spectral line in $m_Y(t)$ can be determined from the amplitude of the associated sine wave, say, $\exp(i2\pi\alpha t)$, in $m_Y(t)$. That is, the average power is given by the limit

$$P_Y^{\alpha} \triangleq \left| \lim_{Z \to \infty} \frac{1}{Z} \int_{-Z/2}^{Z/2} m_Y(t) e^{-i2\pi\alpha t} \, dt \right|^2. \qquad (12.266)$$

Substitution of (12.265) into (12.266) and use of the convolution theorem yields the formula (exercise 36)

$$P_Y^{\alpha} = \left| \int_{-\infty}^{\infty} K\left(f + \frac{\alpha}{2}, f - \frac{\alpha}{2}\right) S_X^{\alpha}(f) \, df \right|^2, \qquad (12.267)$$

where $K(\mu, \nu)$ is the double Fourier transform of the kernel $k(u, v)$ in (12.264),

$$K(\mu, \nu) \triangleq \int_{-\infty}^{\infty} \int_{-\infty}^{\infty} k(u, v) e^{-i2\pi(\mu u - \nu v)} \, du \, dv. \qquad (12.268)$$

We see from this that a spectral line can be regenerated from $X(t)$ at frequency α using a quadratic transformation if and only if the spectral correlation function with frequency separation α is not identically zero, $S_X^{\alpha}(f) \neq 0$. If a signal is present in $X(t)$, then $X(t) = S(t) + N(t)$, where $N(t)$ is assumed to be stationary white Gaussian noise, and we therefore have the identity

$$S_X^\alpha(f) \equiv S_S^\alpha(f), \qquad \alpha \neq 0, \tag{12.269}$$

which can be used in (12.267).

We want to consider the approach of detecting the presence of the signal $S(t)$ in the corrupted observations $X(t)$ by detecting the presence of a regenerated spectral line in $Y(t)$. To optimize this type of detector, we choose the particular quadratic transformation, which is specified by the kernel $k(u, v)$ or its transform $K(\mu, v)$, that regenerates the strongest possible spectral line relative to the strength of noise (when the signal is absent) at its output.* For the purpose of detecting the presence of the spectral line at frequency α, it is appropriate to pass $Y(t)$ through a narrowband filter with bandwidth, say, B, center frequency α, and with unity gain at band center. In this case, we need only be concerned with the strength of the output noise spectral density at frequency α. When the signal is absent from $X(t)$, we have $X(t) = N(t)$, and in this case it can be shown that the spectral density of $Y(t)$ at frequency α is given by (exercise 36)

$$S_Y(\alpha) = 2N_0^2 \int_{-\infty}^{\infty} \left| K\left(f + \frac{\alpha}{2}, f - \frac{\alpha}{2}\right) \right|^2 df, \qquad \alpha \neq 0. \tag{12.270}$$

As a measure of SNR at the output of the quadratic transformation, we use the ratio of the average spectral-line-power from the signal to the average noise-power in the band passed by the narrowband filter:

$$\text{SNR} \triangleq \frac{P_Y^\alpha}{B S_Y(\alpha)}, \qquad \alpha \neq 0. \tag{12.271}$$

Application of the Cauchy-Schwarz inequality to (12.271), with (12.267), (12.269), and (12.270) substituted in, shows that the SNR-maximizing kernel transform must satisfy the necessary and sufficient condition (exercise 36)

$$K\left(f + \frac{\alpha}{2}, f - \frac{\alpha}{2}\right) = c[S_S^\alpha(f)]^* \tag{12.272}$$

for any nonzero constant c. Use of this condition in (12.271) yields the maximum value of the SNR (exercise 36)

$$\text{SNR}_{\max} = \frac{1}{2BN_0^2} \int_{-\infty}^{\infty} |S_S^\alpha(f)|^2 df, \tag{12.273}$$

which is determined by the strength of spectral correlation in the signal.

It is shown in exercise 37 that when the quadratic transformation with kernel transform $K(\mu, v)$ is followed by an ideal bandpass filter with center

---

*The solution to this optimization problem has application to synchronization (e.g., for weak-signal demodulation) as well as detection [Gardner, 1986b, 1987a].

frequency α, bandwidth $B \to 0$, and unity gain at band center, the kernel transform for the combination of the quadratic transformation and the filter (this combination is another quadratic transformation) is given by

$$K'(\mu, \nu) = \begin{cases} K(\mu, \nu), & \mu - \nu = \alpha, \\ 0, & \mu - \nu \neq \alpha. \end{cases} \qquad (12.274a)$$

Consequently, the values of the kernel transform $K(\mu, \nu)$ for $\mu - \nu \neq \alpha$ are irrelevant. Therefore the constraint (12.272) is essentially a full specification.

When an ideal bandpass filter with very narrow bandwidth B is applied to a finite segment of data with segment length W ($B \ll 1/W$), the amplitude at the output of the filter is proportional to WB and, therefore, approaches zero as $B \to 0$. To avoid this problem for detection based on a finite segment of observations $X(t)$, we can simply make the gain of the filter proportional to $1/B$. In the limit as $B \to 0$, (12.274a) then becomes (exercise 37)

$$K'(\mu, \nu) = K(\mu, \nu)\delta(\mu - \nu - \alpha). \qquad (12.274b)$$

Let us now substitute the idealized kernel, specified by (12.272) and (12.274b), into (12.264), but with the time parameters u and v limited to the interval $[-W/2, W/2]$, to obtain a practical implementation of the maximum-SNR spectral-line-regenerating detector that uses a finite segment of observations $\{X(u): |t - u| < W/2\}$. The result is (exercise 37)

$$Y(t) = \frac{1}{N_0^2} \int_{-\infty}^{\infty} S_S^\alpha(f)^* P_X^\alpha(t, f)_W \, df \, e^{i2\pi\alpha t}, \qquad (12.275)$$

where $P_X^\alpha(t, f)_W$ is the time-variant cyclic periodogram defined by (12.20) and (12.24). This detection waveform $Y(t)$ can be used at each value of t for detection in the interval $[t - W/2, t + W/2]$, and for $t = 0$ it is identical to the single-cycle detector obtained from (12.260).

In conclusion, we see that the single-cycle detector is actually a maximum-SNR spectral-line regenerator. Also, we see that with $B = 1/W$, the maximized SNR (12.273) is equal to the single term in the maximized squared deflection (12.263) corresponding to the single cycle frequency α used by the spectral-line regenerator.

The preceding derivation of the maximum-SNR spectral-line regenerator can be generalized to the case of nonwhite Gaussian noise and/or interference $N(t)$. As long as the observation time W greatly exceeds the reciprocal of the narrowest peaks or valleys in the spectrum $S_N(f)$ and cyclic spectrum $S_S^\alpha(f)$, then (12.270), (12.273), and (12.275) can be accurately generalized simply by moving the factor N_0^2 inside the integral and replacing it with $S_N(f + \alpha/2)S_N(f - \alpha/2)$; also, (12.272) can be accurately generalized by dividing the right-hand side by $S_N(f + \alpha/2)S_N(f - \alpha/2)$ (cf. [Gardner, 1987a]).

396 • **Random Processes**

Comparison with the Radiometer. When the observation interval for the optimum radiometer derived in Section 11.8 is allowed to slide along with time, as done with the spectral-line regenerator, it can be shown that the radiometer output contains a spectral line at frequency $\alpha = 0$ regardless of whether or not the signal is present in the noise, and that the output is bandlimited to the narrow band of width $1/W$ centered at frequency zero. In contrast to this, the single cycle detector output contains a spectral line at frequency $\alpha \neq 0$ only if the signal is present (assuming the noise and interference do not exhibit cyclostationarity with cycle frequency α). These facts about spectral lines follow directly from formula (12.267) for the spectral-line power at the output of any quadratic time-invariant detector, and the fact that $S_X^\alpha(f)$ satisfies

$$S_X^\alpha(f) = \begin{cases} S_S^\alpha(f), & \alpha \neq 0, \text{ signal present,} \\ S_S(f) + S_N(f), & \alpha = 0, \text{ signal present,} \\ 0, & \alpha \neq 0, \text{ signal absent,} \\ S_N(f), & \alpha = 0, \text{ signal absent.} \end{cases} \quad (12.276)$$

This is illustrated with the simulation shown in Figure 12.8 for a BPSK signal (cf. Figure 12.5) corrupted by white noise and five interfering AM signals (cf. Figure 12.3). Thus, the radiometer must distinguish between the strength of the spectral line at $\alpha = 0$ due to signal-plus-noise/interference and the spectral line at frequency $\alpha = 0$ due only to noise/interference, whereas the single-cycle detector need only distinguish between the presence and absence of a spectral line at $\alpha \neq 0$. This can greatly complicate the problem of setting the energy-threshold level to be used with the radiometer and renders the radiometric approach to detection (which ignores cyclostationarity) inherently more susceptible to unknown and changing noise or interference, especially for weak signals. This can be important for applications such as spread-spectrum radio signal interception [Gardner, 1988] and detection of sonic noise produced by submarine propeller cavitation.

The maximum SNR and maximum deflection are explicitly evaluated for the single-cycle and multicycle detectors for various types of PSK signals in [Chen, 1989]. Evaluations of probability of detection versus probability of false alarm obtained from simulations for various noise and interference environments have confirmed the superiority of single-cycle detectors over radiometers in environments with fluctuating backgrounds. For the purpose of illustration, the performance graphs from one simulation are shown in Figure 12.12. The signal here is BPSK with carrier frequency f_o, and the corruption consists of white noise with average power (averaged over all observation intervals) in the signal band equal to that of the signal, and spectral intensity that fluctuates randomly (from one observation interval to another) with a coefficient of variation (variance normalized by squared mean) of 1/10. The observation time W contains 128 keying intervals of

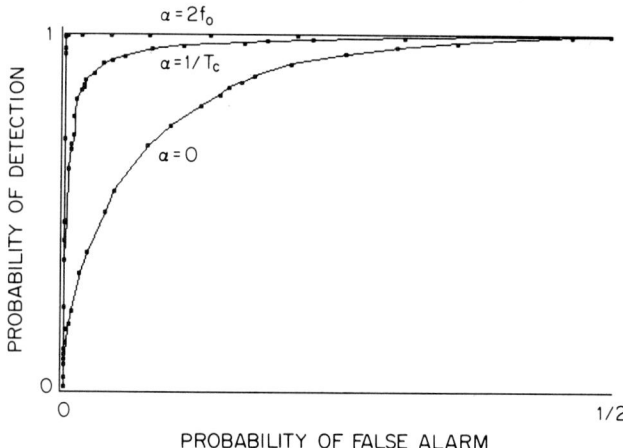

Figure 12.12 Graphs of Probability of Detection Versus Probability of False Alarm Computed From Simulations for Two Single-Cycle Detectors and the Radiometer. The Signal is BPSK and the Noise is White with Random Spectral Intensity Having Coefficient of Variation of 1/10. The Average SNR is 0 dB and the Time-Bandwidth Product of the Signal is 256.

length T_c, each of which contains 16 time samples. It can be seen that for a given probability of false alarm (obtained by selection of an appropriate value of the threshold for each detector), the probabilities of detection for the single-cycle detectors are considerably higher than those for the radiometer.

12.8.6 Estimation of Cyclostationary Signals

Signal Extraction. Let us consider the problem of extracting a random signal $S(t)$ from corrupted measurements $X(t)$ by filtering. This problem is considered in Section 11.7, but there it is assumed that $S(t)$ is WSS and that the corruption consists of linear time-invariant distortion, with impulse-response function $g(t)$, and additive WSS noise $N(t)$,

$$X(t) = S(t) \otimes g(t) + N(t). \tag{12.277}$$

Let us assume here that either or both of $S(t)$ and $N(t)$ exhibit cyclostationarity. This accounts for modulated signals $S(t)$ and for additive corruption $N(t)$ that consists of modulated interfering signals as well as noise. In fact, the derivation presented here allows for any model for $X(t)$ as long as $X(t)$ and $S(t)$ are jointly almost cyclostationary.

In Section 11.7, it was appropriate to consider a time-invariant filter since it is shown in Section 13.4 that such a filter is optimum when $S(t)$ and $X(t)$ are jointly WSS. However, when $S(t)$ and $X(t)$ are jointly almost cyclostationary, the optimum filter is almost periodically time-variant. Let the impulse-response function for this filter be denoted by $h(t, u)$. Then the signal estimate provided by this filter is given by

$$\hat{S}(t) = \int_{-\infty}^{\infty} h(t, u)X(u)\, du. \tag{12.278}$$

In Section 11.7, the MSE between $\hat{S}(t)$ and $S(t)$ is adopted as a measure of signal extraction performance. This is appropriate because the joint stationarity of $S(t)$ and $X(t)$ and the time invariance of the filter considered there result in a MSE that is independent of time. However, here MSE is an almost periodic function of time. In order to obtain a single number as a performance measure, we adopt the time-averaged value of the MSE,*

$$\text{TAMSE} \triangleq \langle E\{[\hat{S}(t) - S(t)]^2\}\rangle. \tag{12.279}$$

In Section 13.3, it is shown that the signal estimate that minimizes the TAMSE must satisfy the same necessary and sufficient orthogonality condition as that described in Section 11.7, namely

$$E\{[\hat{S}(t) - S(t)]X(v)\} = 0 \tag{12.280}$$

for all values of t and v. Substitution of (12.278) into (12.280) yields the following necessary and sufficient condition on the optimum signal-estimation filter $h_o(t, u)$ (exercise 38)

$$\int_{-\infty}^{\infty} h_o(t, u)R_X(v, u)\, du = R_{SX}(t, v) \tag{12.281}$$

for all values of t and v. Also, the minimum value of TAMSE can be obtained by using (12.278) and (12.280) in (12.279) (exercise 38). The result is given by

$$\text{TAMSE} = \left\langle R_S(t, t) - \int_{-\infty}^{\infty} h_o(t, u)R_{SX}(t, u)\, du \right\rangle. \tag{12.282}$$

In order to obtain more explicit results, we can use the Fourier series representation

$$h_o(t, u) = \sum_{\beta} g_\beta(t - u)e^{i2\pi\beta u}, \tag{12.283a}$$

where

$$g_\beta(\tau) \triangleq \langle h_o(t + \tau, t)e^{-i2\pi\beta t}\rangle, \tag{12.283b}$$

as well as the Fourier series representation (12.4) for $R_X(v, u)$ and similarly for $R_{SX}(t, v)$. Substitution of these representations into (12.281) and (12.282)

---

*As a matter of fact, only the time average is needed; the expectation can be deleted, and results that are completely equivalent to those obtained herein can be obtained [Gardner, 1987a].

yields (exercise 39) the design equation

$$\sum_{\beta} g_{\beta}(\tau) \otimes [R_X^{\alpha-\beta}(\tau)e^{i\pi(\alpha+\beta)\tau}] = R_{SX}^{\alpha}(\tau)e^{i\pi\alpha\tau}, \qquad (12.284)$$

which must hold for all τ, and the performance formula

$$\text{TAMSE}_{\min} = R_S^0(0) - \sum_{\beta} \int_{-\infty}^{\infty} g_{\beta}(\tau)R_{SX}^{\beta}(\tau)^{*}e^{-i\pi\beta\tau} \, d\tau. \qquad (12.285)$$

Equation (12.284) must hold for all values of α and β for which $R_X^{\alpha-\beta}(\tau)$ and $R_{SX}^{\alpha}(\tau)$ are not identically zero. The same set of values of β must be used in (12.285) as well as in (12.283). The equations (12.284) and (12.285) are the generalizations of (11.197) and (11.205) from the case of jointly WSS $S(t)$ and $X(t)$ to jointly almost cyclostationary $S(t)$ and $X(t)$. By analogy with the simplifications obtained in Section 11.7, we can simplify these generalized equations by reexpressing them in the frequency domain. Specifically, application of the convolution theorem to (12.284) yields (exercise 39) the design equation

$$\sum_{\beta} G_{\beta}(f) S_X^{\alpha-\beta}\left(f - \frac{\alpha+\beta}{2}\right) = S_{SX}^{\alpha}\left(f - \frac{\alpha}{2}\right), \qquad (12.286)$$

which must hold for all f, where

$$G_{\beta}(f) \triangleq \int_{-\infty}^{\infty} g_{\beta}(\tau) e^{-i2\pi f \tau} d\tau. \qquad (12.287)$$

Similarly, the performance formula (12.285) can be reexpressed as (exercise 39)

$$\text{TAMSE}_{\min} = \int_{-\infty}^{\infty} \left[S_S^0(f) - \sum_{\beta} G_{\beta}(f) S_{SX}^{\beta}\left(f - \frac{\beta}{2}\right)^{*} \right] df. \qquad (12.288)$$

Substitution of the representation (12.283) into the estimation formula (12.278) yields the more explicit expression

$$\hat{S}(t) = \sum_{\beta} g_{\beta}(t) \otimes [X(t)e^{i2\pi\beta t}] \qquad (12.289)$$

for the optimum signal estimate. This reveals that this estimate can be obtained by simply frequency shifting the observations $X(t)$ by the amounts β and then filtering each of these frequency-shifted processes using time-invariant filters with transfer functions $G_{\beta}(f)$, which are the solutions to the design equation (12.286).

When the signal $S(t)$ and observations $X(t)$ are bandlimited processes (as they always are in practical models), the number of values of β required in (12.289) and the number of equations in (12.286) will be finite [assuming there exists a minimum distance between all cycle frequencies exhibited by the processes $S(t)$ and $X(t)$, as will always be the case for practical models].

Also, suboptimum (or optimum constrained) performance can be obtained by constraining at the outset the number and values of the parameter β used in (12.289). For this case, it is shown in [Gardner, 1987a] that the same design equation (12.286) and performance formula (12.288) result, except that the set of values of β (which is identical to the set of values of α) in these equations is restricted according to the constraint imposed at the outset.

As briefly described in the introductory Section 12.8.1, the optimum FREquency-SHift (FRESH) filter solved for here can provide substantial improvements in signal extraction performance, including both superior distortion-reduction (e.g., for multipath distortion) and superior interference- and noise-suppression. The improvements in performance are obtained through the use of frequency shifting in order to exploit spectral correlation. Specific calculations of degree of performance improvement for the case where corruption consists only of additive white noise are reported in [Gardner and Franks, 1975]. More impressive theoretical results for interference suppression by FRESH filtering of amplitude-modulated signals such as AM and ASK are reported in [Brown, 1987]. Also, a more in-depth discussion including simulations of interference suppression and multipath-distortion reduction for bauded signals, such as BPSK, QPSK, and APK, obtained through exploitation of spectral correlation, by use of adaptive fractionally spaced (Nyquist rate) equalizers and their FRESH counterparts in place of more common baud-rate equalizers, is given in [Gardner, 1989], and the results of simulations using modulus restoral algorithms are given in [Agee, 1989]. In addition, techniques for adaptive implementation of FRESH filters are developed in [Ferrara, 1985; Gardner, 1987a].

The results presented here are for real-valued processes only. The required generalizations for complex-valued processes are derived in [Brown, 1987]. This generalization to complex-valued, almost cyclostationary processes is not as straightforward as it is for WSS processes and time-invariant filters. For example, the real-valued, optimum, time-variant filtering of a real-valued, almost cyclostationary process is not equivalent to complex-valued filtering of the complex representation of the process; the conjugate of this representation also must be filtered by a different complex-valued filter, and the two filtered processes must be added together.

For the purpose of illustration, a graph of $\text{TAMSE}_{\min}$ versus the number of frequency shifts used [values of β in (12.283) and its conjugate counterpart] is shown in Figure 12.13. The desired signal $S(t)$ here is the complex envelope for an AM signal with triangular spectrum, (down-converted) carrier frequency $f_s = 0$ and the bandwidth B, and the corruption $N(t)$ consists of the complex envelope of the sum of white noise, with average power in the signa band 20 dB below that of the signal (0 dB), and an AM interfering signal with triangular spectrum, bandwidth B, power 0 dB, and carrier frequency $f_i = \delta B$, where $\delta = 0.225$ in case 1 and $\delta = 0.1$ in case

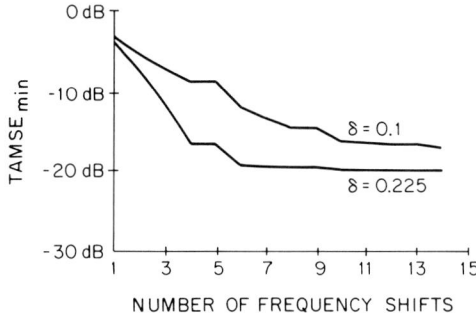

Figure 12.13 Theoretical MSE Performance of Optimum FRESH Filter Versus the Number of Frequency Shifts Used for Two Interfering AM Signals in Noise. The Parameter δ is the Bandwidth-normalized Carrier Offset between the Two Signals.

2. It can be seen that with six frequency shifts, the interference is essentially completely removed in case 1 and, since the TAMSE$_{min}$ is on the order of the noise power, the signal distortion introduced is negligible. For case 2, a larger number of frequency shifts is required.

The approach taken here of optimizing only the receiving filter can be generalized to that of jointly optimizing both transmitting and receiving filters. This more general problem is studied in [Ericson, 1981; Graef, 1983].

System Identification. We consider the problem of identifying a periodically or almost periodically time-variant linear system using measurements of its WSS excitation $X(t)$ and its (almost) cyclostationary response $Y(t)$, and also knowledge of the periods of its periodicities. To accomplish this task of system identification, we can try to fit a model to the system. Let the time-variant impulse-response function for the model be denoted by $h(t, u)$ and let $\hat{Y}(t)$ denote its response to the excitation $X(t)$. We want to solve for the model impulse-response function $h(t, u)$ that minimizes the TAMSE between the measurements $Y(t)$ of the actual system's response and the response $\hat{Y}(t)$ of the model. By interpreting $S(t)$ and $\hat{S}(t)$ in the previous subsection as $Y(t)$ and $\hat{Y}(t)$, all results derived there apply to the system identification problem here. The values of β used in (12.283) and, therefore, in (12.286) and (12.288) are obtained from our assumed knowledge of the periods of the periodicities exhibited by the otherwise unknown system, although these too can be determined from a spectral correlation analysis of the system response $Y(t)$.

Since the excitation $X(t)$ is WSS, then only the terms in (12.286) for which $\beta = \alpha$ are nonzero. Consequently, (12.286) reduces to

$$G_\beta(f)S_X(f - \beta) = S_{YX}^\beta\left(f - \frac{\beta}{2}\right), \tag{12.290}$$

which admits the explicit solution

$$G_\beta(f) = \frac{S_{YX}^\beta(f - \beta/2)}{S_X(f - \beta)}. \tag{12.291}$$

Thus, the best-fitting model has the impulse-response function

$$h(t, u) = \sum_\beta g_\beta(t - u)e^{i2\pi\beta u} \tag{12.292}$$

and the corresponding system function [cf. (12.47)]

$$G(t, f) = \sum_\beta G_\beta(f + \beta)e^{i2\pi\beta t}, \tag{12.293}$$

where g_β and G_β are a Fourier transform pair. The resultant minimized modeling error can be determined by substitution of (12.291) into (12.288). The result is

$$\text{TAMSE}_{\min} = \int_{-\infty}^{\infty} S_Y^0(f)\left[1 - \sum_\beta \left|\rho_{YX}^\beta\left(f - \frac{\beta}{2}\right)\right|^2\right] df, \tag{12.294}$$

where

$$|\rho_{YX}^\beta(f)|^2 \triangleq \frac{|S_{YX}^\beta(f)|^2}{S_Y^0(f + \beta/2)S_X^0(f - \beta/2)}. \tag{12.295}$$

If the unknown transformation from $X(t)$ to $Y(t)$ is exactly linear and (almost) periodically time-variant, and if β is exactly one of the periodicity frequencies of this transformation, then it follows from (12.68) that formula (12.291) gives the exact values for the component $G_\beta(f)$ in the actual system function. Consequently, an exact model for the system can be obtained from (12.291) and (12.293). Unlike conventional methods for identification of generally time-variant systems [Gardner, 1987a], the method proposed here is not subject to any limitations on simultaneous resolvability in time t and frequency f of the system function. Several approaches to adaptive implementation for minimum-TAMSE identifiers for periodic systems are described in [Gardner, 1985].

12.9 Summary

A process is said to *exhibit cyclostationarity* in the wide sense if its time-variant autocorrelation function $R_X(t + \tau/2, t - \tau/2)$ contains finite additive sine-wave components (12.6). The Fourier coefficient $R_X^\alpha(\tau)$ for each such component is called the *cyclic autocorrelation*. The frequency α of the sine-wave component is called the *cycle frequency*. The Fourier transform $S_X^\alpha(f)$ of the cyclic autocorrelation is called the *cyclic spectrum* (12.15), and it is a *spectral correlation function*, that is, its value at f and α is the time-averaged correlation of the spectral components of $X(t)$ at frequencies $f + \alpha/2$ and $f - \alpha/2$ (12.22). The corresponding correlation coefficient, as a function of f and α, is called the *autocoherence function* (12.17). Thus, a process exhibits cyclostationarity with cycle frequency α if and only if it exhibits correlation between spectral components at frequencies separated by α. A process is said to be *completely coherent* at cycle frequency α and spectral frequency f if and only if its autocoherence magnitude is unity at these frequencies. Similarly it is said to be *completely incoherent* if the autocoherence is zero. The cyclic spectrum also has the interpretation of being the time-averaged cross-spectrum (12.15) for the two frequency-shifted versions, $X(t)e^{i\pi\alpha t}$ and $X(t)e^{-i\pi\alpha t}$, of the process $X(t)$, and similarly the cyclic autocorrelation has the interpretation of being the time-averaged cross-correlation (12.13).

If a process exhibits cyclostationarity in the wide sense, then its time-variant autocorrelation function contains additive periodic components, which themselves are valid autocorrelation functions (12.39), and the coefficients in the Fourier series representation of such a periodic component $\langle R_X \rangle_T(t, u)$ are the cyclic autocorrelations with cycle frequencies equal to the harmonics (integer multiples) of the reciprocal of the period $\alpha = m/T$, $m = 1, 2, 3, \ldots,$ (12.41).

The cyclic spectra at the output of a linear periodically time-variant transformation are given by the simple formula (12.53), which is merely a sum of weighted translated input cyclic spectra, in which the weighting functions $\mathbf{G}_n(f)$ are the Fourier coefficient functions for the system function $G(t, f)$ (12.47).

For a process that exhibits cyclostationarity, the conventional formulas relating power spectra of a bandpass process and its low-pass components in Rice's representation can be generalized to provide relations among cyclic spectra. These formulas, (12.80) and (12.302), reveal the need for correction terms for the conventional formulas for power spectra when the bandpass process exhibits cyclostationarity at twice the center frequency f_o in Rice's representation. Similarly, for a process that exhibits cyclostationarity, the conventional power-aliasing formula for the power spectrum of a time-sampled signal can be generalized to a spectral-correlation aliasing formula (12.97), which reveals the need for correction terms for the conventional formula when the process being sampled exhibits cyclostationarity at some harmonic of the sampling rate.

The cyclic autocorrelation functions, cyclic spectra, and autocoherence functions are calculated for a wide variety of types of modulated signals in Section 12.5, and the magnitudes of the cyclic spectra are graphed as the heights of surfaces above the bifrequency (f, α) plane for a few of these modulation types. This provides a means for visually characterizing and classifying modulation types.

There are several ways to represent a wide-sense cyclostationary process (i.e., a process with a periodically time-variant autocorrelation) in terms of jointly wide-sense stationary processes (i.e., processes with time-invariant auto- and cross-correlations). The *harmonic series representation* (HSR) (12.175) is one method, and the auto- and cross-spectra of the representors are simply cyclic spectra defined over certain spectral bands (12.184). The *time-series representation* (TSR) (12.187) is another method which is, in a sense, the time-frequency dual of the HSR.

Processes that exhibit cyclostationarity can also exhibit the property of *cycloergodicity*, whereby cyclic autocorrelations and cyclic spectra can be determined from a single sample path of the process. For example, a process exhibits mean-square cycloergodicity of the autocorrelation at cycle frequency α if the empirical cyclic autocorrelation (12.194) is mean-square equivalent to the probabilistic cyclic autocorrelation (12.6).

The duality between the determininistic (time-average) and probabilistic (ensemble-average) theories of stationary processes that is developed in previous chapters can be extended to cyclostationary processes, and also to almost cyclostationary processes, and this can be explained in terms of cyclostationary fraction-of-time distributions (12.196).

Every cyclostationary or almost cyclostationary process can be made stationary by phase randomization. Although this is a commonly used technique, it is important to recognize that phase-randomization destroys cycloergodicity.

The theory of cyclostationary processes has important applications in many problem areas, including modulated-signal source location, synchronization, detection of weak modulated signals masked by noise and interference, extraction of modulated signals from corrupted observations including interference and noise suppression and distortion reduction by frequency-shift filtering, adaptive spatial filtering for modulated-signal wavefront extraction in interference environments, periodic-system identification, and modeling and forecasting for seasonal time-series.

A more comprehensive and nonprobabilistic development of theory and method for random data from periodic phenomena can be found in [Gardner, 1987a].

EXERCISES

★1. Verify the Fourier-coefficient formula (12.6) by substitution of (12.4) with α replaced by β.

2. (a) Verify the cyclic property (12.10), where $Y(t) = X(t + t_o)$.
(b) Verify the symmetry relations for a real vector process $\boldsymbol{X}(t)$:

$$\boldsymbol{R}_X^\alpha(-\tau) = \boldsymbol{R}_X^\alpha(\tau)^T, \qquad (12.296\text{a})$$

$$\boldsymbol{R}_X^{-\alpha}(\tau) = \boldsymbol{R}_X^\alpha(\tau)^*. \qquad (12.296\text{b})$$

Cyclostationary Processes • 405

3. (a) Verify the formulas (12.12) to (12.14) for $U(t)$ and $V(t)$ defined by (12.11).
   (b) Verify the symmetry relations for a real vector process $X(t)$:
   $$S_X^\alpha(-f) = S_X^\alpha(f)^T, \qquad (12.297a)$$
   $$S_X^{-\alpha}(f)^T = S_X^\alpha(f)^*. \qquad (12.297b)$$

4. Verify the *cyclic-periodogram–cyclic-correlogram relation* (12.23).

★5. (a) Verify (12.41), for the special case of an almost cyclostationary process, by substitution of (12.4) into (12.39). [Do not assume $\alpha = m/T$ in (12.4).]
   (b) Verify (12.43) by substitution of (12.41) and comparison with (12.4).

6. (a) Derive the input-output cyclic-autocorrelation relation (12.50) from (12.44) to (12.46) and (12.6).

   Hint: Direct substitution using $h(t + \tau/2, u)$ and $h(t - \tau/2, v)$, and interchange of expectation with integration and summation, followed by evaluation of the limit [using the definition (12.51)] yields

   $$R_Y^\alpha(\tau) = \int_{-\infty}^{\infty}\int_{-\infty}^{\infty} \sum_{n,m=-\infty}^{\infty} g_n\left(t' + \frac{\tau + \tau'}{2}\right) R_X^{\alpha-(n-m)/T}(-\tau')$$
   $$\times e^{-i\pi(n+m)\tau'/T} g_m^T\left(t' - \frac{\tau + \tau'}{2}\right)^* e^{-i2\pi\alpha t'} dt'\, d\tau',$$
   (12.298)

   where the change of variables $t' = t - (u + v)/2$ and $\tau' = v - u$ ($u = t - t' - \tau'/2$, $v = t - t' + \tau'/2$) has been introduced. Now, use of the matrix identities

   $$g_n R_X g_m^{T*} = \mathrm{tr}\{R_X^T g_n^T g_m^*\},$$
   $$R_X^T(-\tau) = R_X(\tau)$$

   and evaluation of the integral with respect to t', using the definition (12.52), yields

   $$R_Y^\alpha(\tau) = \int_{-\infty}^{\infty} \sum_{n,m=-\infty}^{\infty} \mathrm{tr}\{R_X^{\alpha-(n-m)/T}(\tau')e^{-i\pi(n+m)\tau'/T}$$
   $$\times r_{nm}^\alpha(\tau' + \tau)\}\, d\tau', \qquad (12.299)$$

   from which (12.50) follows.

(b) Derive (12.53) from (12.50) by use of the convolution theorem.
(c) Verify that the general formulas (12.50) to (12.53) reduce to the special cases (12.55) to (12.57) for a linear time-invariant transformation.
(d) Verify that the general formulas (12.50) to (12.53) reduce to the special cases (12.58) to (12.59) for a process $X(t)$ that exhibits no cyclostationarity.

7. (a) Derive (12.61) from (12.44) to (12.46) and (12.60).

   Hint: Follow a method analogous to the hint in exercise 6.

   (b) Derive (12.62) from (12.61) by use of the convolution theorem.
   (c) Verify that the general formulas (12.61) and (12.62) reduce to the special cases (12.64) and (12.65) for a linear time-invariant transformation.
   (d) Verify that the general formulas (12.61) and (12.62) reduce to the special cases (12.66) and (12.67) for a process $X(t)$ that exhibits no cyclostationarity. Then derive (12.68) from (12.67).

8. (a) Derive the QAM form (12.76) from (12.44) and (12.77), with $X(t) = [Z(t)W(t)]^T$.
   (b) Use (12.50) and (12.77) to derive (12.79).
   (c) Use (12.53) and (12.78) to derive (12.80).
   (d) Derive (12.81) from (12.79).
   (e) Derive (12.83) from (12.79).
   (f) Derive (12.91) and (12.92) as follows: Consider the complex envelope

$$\Gamma(t) \triangleq [X(t) + iY(t)]e^{-i2\pi f_o t} \quad (12.300a)$$
$$= U(t) + iV(t). \quad (12.300b)$$

Solve this equation to obtain

$$X(t) = \tfrac{1}{2}\Gamma(t)e^{i2\pi f_o t} + \tfrac{1}{2}\Gamma^*(t)e^{-i2\pi f_o t}, \quad (12.301a)$$
$$U(t) = \tfrac{1}{2}\Gamma(t) + \tfrac{1}{2}\Gamma^*(t), \quad (12.301b)$$
$$V(t) = \frac{1}{2i}\Gamma(t) - \frac{1}{2i}\Gamma^*(t). \quad (12.301c)$$

Then show that*

$$S_X^\alpha(f) = \tfrac{1}{4}\big[S_\Gamma^\alpha(f - f_o) + S_{\Gamma*}^\alpha(f + f_o)$$
$$+ S_{\Gamma\Gamma*}^{\alpha - 2f_o}(f) + S_{\Gamma*\Gamma}^{\alpha + 2f_o}(f)\big], \quad (12.302a)$$

---

*According to the notational convention in this book, $S_\Gamma = F\{R_\Gamma\}$ and $S_{\Gamma\Gamma*} = F\{R_{\Gamma\Gamma*}\}$, where $R_\Gamma(\tau) \triangleq E\{\Gamma(t + \tau/2)\Gamma^*(t - \tau/2)\}$ and $R_{\Gamma\Gamma*}(\tau) \triangleq E\{\Gamma(t + \tau/2)\Gamma(t - \tau/2)\}$, and similarly for S_Γ^α and $S_{\Gamma\Gamma*}^\alpha$.

$$S_\Gamma^\alpha(f) = 4S_X^\alpha(f + f_o)u\left(f + f_o - \frac{|\alpha|}{2}\right), \quad (12.302b)$$

$$S_{\Gamma\Gamma^*}^\alpha(f) = 4S_X^{\alpha+2f_o}(f)u\left(\frac{\alpha}{2} + f_o - |f|\right), \quad (12.302c)$$

$$S_U^\alpha(f) = \tfrac{1}{4}[S_\Gamma^\alpha(f) + S_{\Gamma^*}^\alpha(f)$$
$$+ S_{\Gamma\Gamma^*}^\alpha(f) + S_{\Gamma^*\Gamma}^\alpha(f)], \quad (12.302d)$$

$$S_V^\alpha(f) = \tfrac{1}{4}[S_\Gamma^\alpha(f) + S_{\Gamma^*}^\alpha(f)$$
$$- S_{\Gamma\Gamma^*}^\alpha(f) - S_{\Gamma^*\Gamma}^\alpha(f)], \quad (12.302e)$$

$$S_{UV}^\alpha(f) = \frac{i}{4}[S_\Gamma^\alpha(f) - S_{\Gamma^*}^\alpha(f)$$
$$- S_{\Gamma\Gamma^*}^\alpha(f) + S_{\Gamma^*\Gamma}^\alpha(f)]. \quad (12.302f)$$

Substitute (12.302b) and (12.302c) into (12.302d) to (12.302f) to obtain (12.91) and (12.92). It should be noted that it follows from (12.302b) and (12.302c) that the supports in the (f, α) plane of the four terms in (12.302a) are disjoint, as shown in Figure 12.14.

★9. (a) Use the synchronized averaging identity (12.42) in the definition (12.94) to derive (12.96).

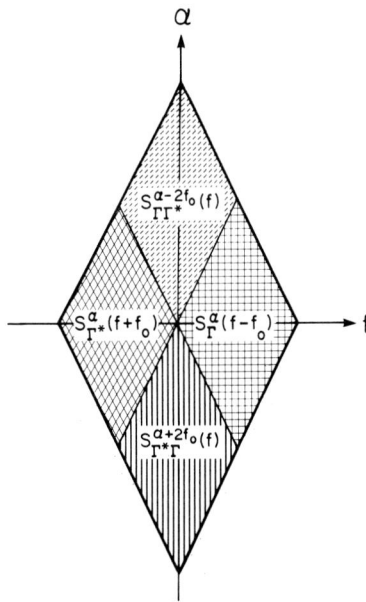

Figure 12.14 Support Regions for the Spectral Correlation Functions Associated with the Complex Envelope of a Bandlimited Signal (Exercise 8).

408 • **Random Processes**

(b) Derive the aliasing formula (12.97) from (12.96) and (12.95), and then verify (12.98) (see Chapter 11, exercise 1).

★**10.** (a) For the AM signal (12.103), verify (12.104) for the cyclic autocorrelation, using (12.50).
(b) Derive (12.105), for the cyclic spectrum, from (12.104).

11. (a) For the QAM signal (12.112), derive (12.114) and (12.115) for the spectrum and cyclic spectrum from (12.80).
(b) Derive the formula (12.117) for the autocoherence magnitude of a QAM signal, using (12.114) to (12.116).
(c) For the autocoherence magnitude, verify (12.118) using (12.117).
(d) Verify (12.119) using (12.114).

12. Use (12.113c) to show that (12.126) is equivalent to (12.121).

13. Derive (12.130) from (12.129) for PM-FM using the definition

$$\langle \Psi_\tau \rangle(\omega_1, \omega_2) \triangleq \lim_{Z \to \infty} \frac{1}{Z} \int_{-Z/2}^{Z/2} E\left\{ \exp\left(i\left[\Phi\left(t + \frac{\tau}{2}\right)\omega_1 \right.\right.\right.$$
$$\left.\left.\left. + \Phi\left(t - \frac{\tau}{2}\right)\omega_2 \right]\right)\right\} dt. \qquad (12.303)$$

14. (a) Show that (12.44) and (12.135) yield the PAM signal (12.134).
(b) Verify (12.136) using (12.135) and (12.46).
(c) Use (12.50), (12.136), and (12.137) to derive the cyclic autocorrelation (12.138).
(d) Use (12.96) to reexpress (12.138) as (12.139).
(e) Use the generalized Poisson sum formula (12.140)–(12.141) to derive the cyclic spectrum (12.142) from (12.138).
(f) Use (12.97) to reexpress (12.142) as (12.143).

15. (a) Derive the formula (12.147) for the cyclic autocorrelation for digital modulation from (12.50).
(b) Show that if $\delta(n)$ exhibits no cyclostationarity, then (12.147) reduces to (12.150).
(c) Derive (12.153), for the cyclic spectrum, from (12.150).

16. Derive (12.160) from (12.159), for the phase-incoherent FSK signal.

17. (a) *Staggered QPSK* (SQPSK), which is sometimes called *offset QPSK* (OQPSK), is a modified form of QPSK in which either $C(t)$ or $S(t)$ (not both) in (12.170) is delayed by half the keying interval, $T_c/2$. Show that the cyclic spectrum for SQPSK (with stationary uncorrelated data) is identical to that of BPSK for $\alpha = \pm 2f_o + k/T_c$ with k any odd integer and for $\alpha = k/T_c$ with k any even integer, and is identically zero for all other values of α.
(b) *Minimum-shift keying* (MSK) is a modification of SQPSK in which the rectangle pulse q of width T_c is replaced with the positive half

cycle of a sine wave with period $2T_c$. For what values of α is the spectral correlation nonzero (assuming stationary uncorrelated data)?

18. Verify that if $X(t)$ is cyclostationary with period T, the representors $\{A_p(t)\}$ defined by (12.175) are jointly stationary.

19. (a) Use (12.178a) and (12.179b) to verify the relation (12.180).
    (b) Derive (12.181) from (12.180).
    (c) Use a sketch to explain how (12.184) follows from (12.181) and (12.183).

20. Use a sketch to verify that (12.187a) follows from (12.187b).

21. Verify that if $X(n)$ is cyclostationary with period P, then $\{Z_p(m)\}$ in (12.187) are jointly stationary.

22. Show that (12.196) to (12.198) yield (12.199) for the expected value of $X(t)$.

23. Use (12.199) to verify (12.200) for the cyclic mean.

24. For a cyclostationary process $X(t)$ with cycle frequency α, define the two processes
$$C(t) \triangleq X(t)\cos(\pi\alpha t),$$
$$S(t) \triangleq X(t)\sin(\pi\alpha t).$$
Show that
$$\operatorname{Re}\{R_X^\alpha(\tau)\} = \langle R_C \rangle(\tau) - \langle R_S \rangle(\tau), \qquad (12.304\mathrm{a})$$
$$\operatorname{Im}\{R_X^\alpha(\tau)\} = -[\langle R_{CS} \rangle(\tau) + \langle R_{SC} \rangle(\tau)]. \qquad (12.304\mathrm{b})$$

25. Let $X(t)$ be cyclostationary with period T, and let $Y(t)$ be the phase-randomized process $Y(t) = X(t + \Theta)$, where Θ is independent of $X(t)$. Show that a sufficient condition for $Y(t)$ to be WSS is that its probability density function satisfy the condition
$$\sum_{n=-\infty}^{\infty} f_\Theta(\theta - nT) = \frac{1}{T}, \qquad -\infty < \theta < \infty. \qquad (12.305)$$

Hint: Use (12.207) to conclude that
$$\Phi_\Theta(2\pi m/T) = 0 \qquad \text{for all integers } m \neq 0$$
is a sufficient condition; then Fourier-transform the function
$$\sum_{m=-\infty}^{\infty} \Phi_\Theta(2\pi m/T)\delta(t - m/T) \equiv \delta(\tau) \qquad -\infty < t < \infty.$$

Observe that (12.305) is equivalent to *Nyquist's zero-intersymbol-inter-*

ference criterion for an interpolating pulse Φ_Θ whose (conjugate) Fourier transform is f_Θ [Nyquist, 1928b].

26. (a) Prove that, for a zero-mean process, the ratio

$$\rho \triangleq \frac{R_X^\alpha(\tau)}{\langle R_X \rangle(0)} \qquad (12.306)$$

is a correlation coefficient for each and every value of α and τ.

Hint: Use (12.11).

(b) Prove that the inequality

$$\int_{-\infty}^{\infty} |R_X^\alpha(\tau)|^2 \, d\tau \le \int_{-\infty}^{\infty} [\langle R_X \rangle(\tau)]^2 \, d\tau \qquad (12.307)$$

is valid.

Hint: Use (12.34).

27. (a) Show that the model (12.210), in which $S(t)$ and $N(t)$ are not correlated, leads to the equation (12.211) for the cyclic correlation matrix.

(b) Show that the models (12.210) and (12.215), in which $S(t)$ and $N(t)$ are not correlated and $S(t)$ and $M(t)$ are not correlated and neither $N(t)$ nor $M(t)$ exhibit cyclostationarity with cycle frequency α, leads to the equation (12.217) for the cyclic cross-correlation matrix.

(c) Show that when (12.212a) is valid, we obtain

$$\int R_X^\alpha(\tau) \sum_{k=1}^{n} R_{X_k}^\alpha(\tau)^* \, d\tau = \int |R_S^\alpha(\tau)|^2 \, d\tau [p^\dagger(0)p(0)]p(0)p(0)^\dagger.$$

28. Let $S(t)$ be the complex envelope for a BPSK process $X(t)$ with carrier frequency f_o and bandlimited spectrum $S_X(f) = 0$ for $|f| > b$. Assume that the phase-modulating sequence is stationary and that f_oT_c is not an integer. Show that $R_S^\alpha = 0$ and $R_{SS^*}^\alpha \ne 0$ for $\alpha = 2f_*$, and that $R_S^\alpha \ne 0$ and $R_{SS^*}^\alpha = 0$ for $\alpha = 1/T_c$, where $f_* = f_o - f_o'$ and f_o' is the center frequency in Rice's representation from which the complex envelope is obtained.

Hint: Use (12.302b) and (12.302c).

29. Consider the generalized eigenequation

$$Rv - \lambda Sv = 0, \qquad (12.308)$$

where R and S are 2×2 matrices, S is nonsingular, and

$$S^{-1}R = \begin{bmatrix} a & b \\ c & d \end{bmatrix}. \qquad (12.309)$$

Show that the two generalized eigenvalues are given by

$$\lambda = \frac{a+d}{2} \pm \left[\frac{a^2+d^2}{4} - \frac{1}{2}ad + bc\right]^{1/2} \quad (12.310)$$

and the corresponding generalized eigenvectors are given by

$$v = \frac{b}{[b^2 + (\lambda - a)^2]^{1/2}} \begin{bmatrix} 1 \\ \frac{(\lambda - a)}{b} \end{bmatrix}. \quad (12.311)$$

30. Consider a PAM signal $S(t)$ with white amplitude sequence and full-duty-cycle rectangular pulses of width T. Show that $|R_S^\alpha(\tau)|$ peaks at $\tau = \pm T/2$ for $\alpha = 1/T$. Then consider a QPSK signal $S(t)$ with a white phase sequence and a full-duty-cycle rectangular pulse envelope $q(t)$ of width $T_c = T$, and determine the location of the highest peak in $|R_S^\alpha(\tau)|$ for $\alpha = 1/T$.

    Hint: Use (12.174).

31. Derive equations (12.224a) and (12.224b) for the auto- and cross-spectral correlation functions for the TDOA model (12.223).

32. (a) Show that the weight vector given by (12.232) minimizes the TAMSE (12.231). Do this by expanding the square and then equating to zero the derivatives with respect to the real and imaginary parts of each of the elements of the weight vector to obtain the necessary and sufficient orthogonality condition

$$\langle E\{[gX(t) - S(t)]X^\dagger(t)\}\rangle = 0. \quad (12.312)$$

   (b) Use the model (12.210) to derive the simplification (12.235) of (12.232).

   (c) Use the sensor-array model (12.210) and the result of part (a) with $S(t)$ replaced by $\tilde{S}(t)$ defined by (12.236) to verify (12.238) and its consequence (12.240)–(12.241).

   (d) Let $\hat{S}(t)$ be given by (12.230), $X(t)$ be given by (12.210), and assume that $\langle E\{\tilde{S}(t)S(t)^*\}\rangle \neq 0$ and $\langle E\{\tilde{S}(t)N(t)^\dagger\}\rangle = \mathbf{0}$. Prove that maximization of the correlation coefficient (12.243) with respect to g yields a scaled version of the vector g_o that minimizes the TAMSE (12.231).

   Hint: The ratio $|gv|^2/gMg^\dagger$, where M is Hermitian and positive definite, is maximized by $g = av^\dagger M^{-1}$, and this can be shown using the decomposition $M = M^{1/2}M^{1/2}$, where $M^{1/2}$ is Hermitian, and the Cauchy-Schwarz inequality for n dimensional space: $|gh^\dagger| \leq \|g\| \|h\|$ with equality if and only if $g = ah$.

33. (a) Show that the correlation coefficient magnitude (12.243), with $\hat{S}$

given by (12.230) and (12.232) with S replaced by $\tilde{S}$, where $\tilde{S}$ is defined by (12.244), reduces to (12.245).

Hint: Evaluate $|\langle R_{\tilde{S}\hat{S}}\rangle|$ directly and then use the fact that $\langle R_{\tilde{S}\hat{S}}\rangle = \langle R_{\hat{S}}\rangle$ when $\hat{S}$ is the minimum-TAMSE estimate of $\tilde{S}$.

(b) Show that (12.245) is indeed a Rayleigh quotient of the form

$$Q \triangleq \frac{v^\dagger M v}{v^\dagger v}, \qquad (12.313)$$

where v is the column vector defined by (12.247) and M is the Hermitian symmetric matrix defined by (12.248).

(c) To verify the fact that Q in part (b) is maximum when v is the eigenvector corresponding to the largest eigenvalue of M, proceed as follows. The SVD of M is given by

$$M = U \Lambda U^\dagger, \qquad (12.314)$$

where U is the $n \times n$ orthonormal matrix

$$U = [v_1, v_2, \ldots, v_n]$$

and Λ is the diagonal matrix

$$\Lambda = \mathrm{diag}\{\lambda_1, \lambda_2, \ldots, \lambda_n\},$$

where $\{\lambda_i\}$ and $\{v_i\}$ are the n eigenvalues in descending order and the corresponding n orthonormal eigenvectors of M. Since these n eigenvectors span n-dimensional space, then any vector v can be expressed as a linear combination of the form

$$v = \sum_{i=1}^{n} a_i v_i. \qquad (12.315)$$

Substitute (12.315) and (12.314) into (12.313) to show that Q can be expressed as

$$Q = \frac{\sum_{i=1}^{n} \lambda_i |a_i|^2}{\sum_{i=1}^{n} |a_i|^2}. \qquad (12.316)$$

Then prove that among all sets $\{a_i\}$ of scalars, the particular set $\{1, 0, 0, 0, \ldots, 0\}$ maximizes Q.

34. Derive the performance coefficient (12.249) from (12.243) using definitions (12.244) and (12.250).

35. Derive the approximation (12.261) from the maximum-deflection formula (12.256).

Hint: Substitute (12.257) into (12.256), and make a change of variables of integration (cf. Chapter 7, exercise 4) to obtain

$$D^2_{max} = \sum_\alpha \sum_\beta \int_{-W}^{W} \int_{-(W-|\tau|/2)}^{(W-|\tau|/2)} R^\alpha_S(\tau)R^\beta_S(\tau)e^{i2\pi(\alpha+\beta)t}\,dt\,d\tau \qquad (12.317)$$

$$= \sum_\alpha \sum_\beta \int_{-W}^{W} R^\alpha_S(\tau)R^\beta_S(\tau)\frac{\sin[\pi(\alpha+\beta)(W-|\tau|)]}{\pi(\alpha+\beta)}\,d\tau.$$

Then use the assumption that W greatly exceeds the widths of $\{R^\alpha_S(\tau)\}$ and the reciprocal of the minimum nonzero value of $|\alpha+\beta|$ (viz., the period T), to show that only the $\alpha+\beta=0$ terms are nonnegligible, which yields (12.261).

36. (a) Show that the spectral-line power at frequency α at the output of a quadratic time-invariant transformation is given by (12.267).

Hint: First substitute (12.265) into (12.266) to show that

$$P^\alpha_Y = \left| \int_{-\infty}^{\infty} \int_{-\infty}^{\infty} k(u,v)R^\alpha_X(u-v)e^{-i\pi\alpha(u+v)}\,du\,dv \right|^2. \qquad (12.318)$$

(b) Show that the spectral density at frequency α at the output of a quadratic time-invariant transformation with symmetrical kernel $k(u,v) = k(v,u)$ and with WGN at the input is given by (12.270).

Hint: Determine the autocorrelation for $Y(t)$ given by (12.264) using Isserlis's formula (5.59). Then evaluate the Fourier transform of this autocorrelation at frequency $\alpha \neq 0$.

(c) Use the Cauchy-Schwarz inequality (11.186)–(11.187) to verify that (12.272) maximizes the SNR (12.271) for a spectral-line regenerator.

(d) Use the result of (c) to verify formula (12.273) for the maximum value of SNR.

37. (a) Consider the convolution $Z(t) = Y(t) \otimes h(t)$ with $Y(t)$ given by (12.264). Use a change of variables of integration to show that this yields another quadratic time-invariant transformation of the form (12.264) for $Z(t)$, but with kernel

$$k'(u,v) = \int_{-\infty}^{\infty} k(u-s, v-s)h(s)\,ds. \qquad (12.319a)$$

Use this result to show that the kernel transform is given by

$$K'(\mu, \nu) = K(\mu, \nu)H(\mu - \nu). \qquad (12.319b)$$

Finally, show that if $H(f) = 0$ for $f \neq \alpha$, then (12.274a) results.

(b) Let the transfer function $H(f)$ in part (a) be of the form

$$H(f) = \begin{cases} 1/B, & |f - \alpha| \leq B/2, \\ 0, & |f - \alpha| > B/2, \end{cases}$$

and show that in the limit $B \to 0$ (12.319b) becomes (12.274b).

(c) Show that the inverse double Fourier transform of (12.274b), with (12.272) substituted in, yields

$$k(u, v) = cR_S^a(u - v)^* e^{i\pi\alpha(u+v)}, \qquad (12.320)$$

and that this kernel when substituted into (12.264), with limits of integration reduced to $-W/2$ and $+W/2$ and with $c = 1/N_o^2 W$, yields

$$Y(t) = \frac{1}{N_o^2} \int_{-\infty}^{\infty} R_S^a(\tau)^* R_X^a(t, \tau)_W \, d\tau \, e^{i2\pi\alpha t}, \qquad (12.321)$$

where the cyclic correlogram $R_X^a(t, \tau)_W$ is zero for $|\tau| > W$. Then use Parseval's relation to obtain the final result (12.275).

38. (a) Use the orthogonality condition (12.280) to obtain the necessary and sufficient condition (12.281) for the optimum filter.
    (b) Use the orthogonality condition (12.280) to show that the minimum TAMSE can be expressed by

$$\text{TAMSE}_{\min} = \langle E\{S^2(t)\}\rangle - \langle E\{S(t)\hat{S}(t)\}\rangle. \qquad (12.322)$$

Then use (12.278) in this expression to obtain the formula (12.282).

39. (a) Use the Fourier series representations (12.283a), (12.4), and a similar one for $R_{SX}(t, u)$ to derive the design equation (12.284) from (12.281) and to derive the performance formula (12.285) from (12.282).
    (b) Show that (12.286) is equivalent to the design equation (12.284), and that (12.288) is equivalent to the performance formula (12.285).

40. It follows from (12.65) that for a single-input–single-output linear time-invariant system, the transfer function can be expressed as

$$H(f) = \frac{S_{XY}^a(f + \alpha/2)}{S_X^a(f + \alpha/2)}.$$

If the input $X(t)$ and output $Y(t)$ of an unknown system are accessible, they can be used to estimate the numerator and denominator in this formula and thereby to estimate the unknown transfer function. Consider the situation where the input and output are corrupted by additive measurement noise, $N(t)$ and $M(t)$, and only the corrupted measurements,

$$X'(t) = X(t) + N(t)$$
$$Y'(t) = Y(t) + M(t)$$

are available. Assuming that $N(t)$ and $M(t)$ are statistically independent of the actual system input and output, $X(t)$ and $Y(t)$, express the transfer function approximation

$$\hat{H}(f) \triangleq \frac{S^{\alpha}_{X'Y'}(f + \alpha/2)}{S^{\alpha}_{X'}(f + \alpha/2)}.$$

in terms of only $H(f)$ and the cyclic spectra for $X(t)$, $N(t)$, and $M(t)$. Compare the two cases corresponding to $\alpha = 0$ and $\alpha = \alpha_o \neq 0$, where $X(t)$ exhibits cyclostationarity with cycle frequency α_o, but $N(t)$ and $M(t)$ do not. Assume that $N(t)$ and $M(t)$ are statistically dependent, $S_{NM}(f) \neq 0$,(cf. [Gardner, 1990]).

41. Let $X(t)$ and $Y(t)$ be statistically independent, almost cyclostationary processes, and consider the product $Z(t) \triangleq X(t)Y(t)$. Show that the cyclic spectra for $Z(t)$ are given by the double (discrete and continuous) convolution

$$S^{\alpha}_Z(f) = \sum_{\beta} \int_{-\infty}^{\infty} S^{\alpha-\beta}_X(f - v)S^{\beta}_Y(v)\, dv. \qquad (12.323)$$

Hint: First find the autocorrelation function for $Z(t)$. Then use the Fourier series representations for R_X and R_Y to determine the cyclic autocorrelations for $Z(t)$.

13

Minimum-Mean-Squared-Error Estimation

IN THIS CHAPTER random variables are given a geometrical interpretation as vectors in an inner-product space, analogous to Euclidean space. Based on this interpretation, the geometrical foundation for problems of minimum-mean-squared-error (MMSE) estimation is developed in terms of orthogonal projection. Then brief introductory studies of three classes of MMSE problems are presented. These are the problems commonly referred to as non-causal Wiener filtering, causal Wiener filtering, and Kalman filtering. In addition, brief introductions to the problems of recursive least-squares estimation and of optimum periodically time-variant filtering are included.

13.1 *The Notion of Minimum-Mean-Squared-Error Estimation*

In order to introduce the concept of minimum-mean-squared-error estimation, we consider the problem of determining the amplitude, frequency, and phase of a weak sinusoidally time-variant magnetic field by measuring the current that it induces in a conducting wire. The current $y(t)$ measured in the conductor will be the sum of the unknown sine-wave current $x(t)$ due to the magnetic field, and a current $n(t)$ with highly irregular fluctuations due to thermal noise:

$$y(t) = x(t) + n(t), \qquad T_i \leq t \leq T_f, \qquad (13.1)$$

where T_i and T_f are the initial and final times of measurement. Since the magnetic field is weak, the desired current $x(t)$ is masked by the noise current $n(t)$.

We could try an ad hoc approach, such as bandpass-filtering the noisy current, in the hope that the signal would be passed and most of the noise would be filtered away. Then we could use standard methods for measuring amplitude, phase, and frequency of a nearly sinusoidal wave. But how can we determine an appropriate bandwidth and center frequency for the bandpass filter? Can this be done optimally?

We shall take a general approach to this signal-extraction problem. In particular, we shall imbed our problem in a class of problems with similar attributes, by proposing an abstract mathematical model for our measurements $y(t)$ and the desired signal $x(t)$. We shall then propose a measure of performance by which we can mathematically predict the accuracy of proposed estimates of $x(t)$. Finally, we shall obtain a mathematical solution for the *optimum* (according to our performance criterion) estimate $\hat{x}_o(t)$, that is, the estimate that maximizes performance. We shall then be in a better position to evaluate some of the merits of ad hoc approaches to estimation.

The mathematical model that we shall adopt is a probabilistic model in which the measurements $y(t)$ and desired signal $x(t)$ are interpreted as sample paths from random processes $Y(t)$ and $X(t)$. The performance measure that we shall adopt is the instantaneous mean squared error (MSE),

$$\text{MSE} = E\{[X(t) - \hat{X}(t)]^2\}, \tag{13.2}$$

in which we consider t to be fixed, but arbitrary. The final ingredient in the mathematical definition of our problem is the specification of an *admissible* set A of estimation functions $g(t, \cdot)$, where

$$\hat{x}(t) = g\big(t, \{y(u) : T_i \le u \le T_f\}\big), \tag{13.3}$$

which we abbreviate $\hat{x}(t) = g(t, \{y\})$. An example is the set of all linear functions, that is, functions of the form

$$g(t, \{y\}) = \int_{T_i}^{T_f} h(t, u) y(u) \, du, \tag{13.4}$$

of which a bandpass filter is one example.

In conclusion, for each fixed but arbitrary time t, we want to solve for the particular estimation function $g_o(t, \cdot)$ in the admissible set A that minimizes MSE,

$$\min_{g(t, \cdot) \in A} E\{[X(t) - \hat{X}(t)]^2\}, \tag{13.5}$$

where

$$\hat{X}(t) = g(t, \{Y\}),$$

and we want to evaluate the resultant minimum value of MSE,

$$\text{MSE}_o \triangleq E\{[X(t) - \hat{X}_o(t)]^2\}, \qquad (13.6)$$

$$\hat{X}_o(t) \triangleq g_o(t, \{Y\}).$$

In order to enhance intuition for this MSE approach to estimation, we shall begin by developing a geometrical interpretation of minimum-MSE estimation. Then we shall use this interpretation to obtain a general solution to the minimization problem (13.5) and to obtain a general formula for the resultant optimum performance (13.6). As one application, this general approach to estimation is used in Section 13.4.2 to determine an optimum bandpass filter for the signal-extraction problem considered in this section.

13.2 Geometric Foundation*

13.2.1 Euclidean Space

Three-dimensional Euclidean space is an example of an inner-product space; that is, it is a linear vector space on which an inner product is defined. The vectors are 3-tuples of real numbers, $X = \{\alpha_1, \alpha_2, \alpha_3\}$, and the set L of all such 3-tuples is a linear vector space because L contains all linear combinations of pairs of vectors from L; that is, for every pair of vectors $X = \{\alpha_1, \alpha_2, \alpha_3\}$ and $Y = \{\beta_1, \beta_2, \beta_3\}$ in L and every pair of scalars (real numbers) σ and γ, the vector

$$Z = \sigma X + \gamma Y = \{\sigma\alpha_1 + \gamma\beta_1, \sigma\alpha_2 + \gamma\beta_2, \sigma\alpha_3 + \gamma\beta_3\} \qquad (13.7)$$

is in L. The *inner product* of two vectors X and Y in L is a real number defined by

$$(X, Y) \triangleq \alpha_1\beta_1 + \alpha_2\beta_2 + \alpha_3\beta_3. \qquad (13.8a)$$

The *norm* (length) of a vector X is a real number defined by

$$\|X\| = (X, X)^{1/2} = [\alpha_1^2 + \alpha_2^2 + \alpha_3^2]^{1/2}. \qquad (13.9a)$$

The *distance* between two vectors X and Y is given by the norm of their difference, $\|X - Y\|$. The *angle* between two vectors X and Y is given by

$$\theta_{XY} = \cos^{-1}\left[\frac{(X, Y)}{\|X\|\|Y\|}\right]. \qquad (13.10)$$

---

*In order to have the flexibility of notation needed in this chapter the convention of denoting vectors in Euclidean space by boldface is not followed in this section. Instead lightface capitals are used to denote vectors.

Thus, X and Y are orthogonal (perpendicular) if and only if their inner product is zero:

$$X \perp Y \Leftrightarrow (X, Y) = 0. \tag{13.11}$$

The *Pythagorean theorem* establishes that two vectors X and Y are orthogonal if and only if the squared norm of their sum equals the sum of their squared norms:

$$(X, Y) = 0 \Leftrightarrow \|X \pm Y\|^2 = \|X\|^2 + \|Y\|^2. \tag{13.12}$$

If the lines connecting the points (vectors) X and Y to the origin (zero vector) are interpreted as two sides of a right triangle as illustrated in Figure 13.1, then (13.12) can be interpreted as "the square of the hypotenuse equals the sum of the squares of the other two sides."

The *parallelogram law* states that the sum of the squares of the two diagonals of a parallelogram equals the sum of the squares of the four sides. If the lines connecting the points (vectors) 0 and X, 0 and Y, X and $X + Y$, and Y and $X + Y$ are interpreted as the sides of a parallelogram (i.e., $0, X, Y, X + Y$ are the vertices), then the parallelogram equality can be expressed as

$$\|X + Y\|^2 + \|X - Y\|^2 = 2\|X\|^2 + 2\|Y\|^2. \tag{13.13}$$

The *law of cosines* states that the square of the side of a triangle opposite one of its angles, θ, is equal to the sum of the squares of the other two sides

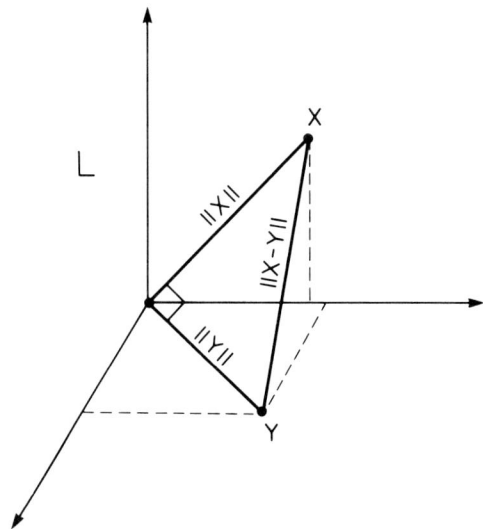

Figure 13.1 Graphical Illustration of Pythagorean Theorem for Vectors

minus twice their product times $\cos\theta$. If the lines connecting the vectors 0 and X, 0 and Y, and X and Y are interpreted as the sides of a triangle, then the law of cosines can be expressed as

$$\|X - Y\|^2 = \|X\|^2 + \|Y\|^2 - 2\|X\|\,\|Y\|\cos\theta, \qquad (13.14a)$$

which, using (13.10), reduces to

$$\|X - Y\|^2 = \|X\|^2 + \|Y\|^2 - 2(X, Y). \qquad (13.14b)$$

Observe that an obvious consequence of the fact that $|\cos\theta| \leq 1$ for all angles θ is the *Cauchy-Schwarz inequality*,

$$|(X, Y)| \leq \|X\|\,\|Y\|, \qquad (13.15)$$

which holds for every pair of vectors X, Y. Furthermore, it is geometrically obvious that equality holds in (13.15) if and only if X and Y are colinear ($X = \alpha Y$).

As we shall see in Sections 13.2.3 and 13.2.4, the geometrical relationships (13.12) to (13.15) apply to inner-product spaces other than Euclidean space, as do the following geometrical constructs.

Orthogonal Projection. Any subset M of vectors in L is a *sub-linear-vector-space* (*subspace*) if and only if M contains all *linear combinations* of its own vectors. For example, the set M of vectors with zero third element ($X = \{\alpha_1, \alpha_2, 0\}$) is a subspace of L, since if $X \in M$ and $Y \in M$, then for every pair of scalars σ and γ, we have

$$Z = \sigma X + \gamma Y \triangleq \{\sigma\alpha_1 + \gamma\beta_1, \sigma\alpha_2 + \gamma\beta_2, 0\},$$

and therefore $Z \in M$. Similarly, the set of all vectors of the form $X = \{\alpha_1, \alpha_3 - \alpha_1, \alpha_3\}$ is another example of a subspace of L.

It is easy to see geometrically that the closest vector $\hat{X}_o$ in a subspace M to a given vector X in L can be obtained by an orthogonal projection of X onto M as shown in Figure 13.2. Furthermore, it is easy to see geometrically that the difference between X and the closest vector $\hat{X}_o$ is orthogonal to every vector $\hat{X}$ in M; that is,

$$(X - \hat{X}_o, \hat{X}) = 0 \qquad \forall \hat{X} \in M. \qquad (13.16)$$

Thus, the unique solution to the minimum-norm-error approximation problem

$$\min_{\hat{X} \in M} \|X - \hat{X}\| \qquad (13.17)$$

is given by the orthogonal projection $\hat{X}_o$ of X onto the space M, and a necessary and sufficient condition for a vector $\hat{X}_o$ in M to be this closest vector to X is that it satisfies the orthogonality condition (13.16).

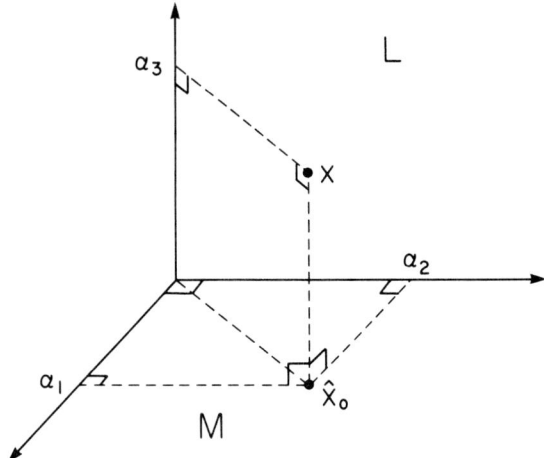

Figure 13.2 Graphical Illustration of Orthogonal Protection $\hat{X}_o$ of $X \in L$ onto M

Equation (13.16) can be solved for $\hat{X}_o$ as follows. Let the two-dimensional subspace M be spanned by the pair of basis vectors Y_1 and Y_2; that is, every vector $\hat{X}$ in M can be expressed as a linear combination of Y_1 and Y_2,

$$\hat{X} = h_1 Y_1 + h_2 Y_2, \qquad (13.18)$$

so that for a given basis Y_1, Y_2, the vector $\hat{X}_o$ is determined by the two-tuple of scalars $H_o = [h_1^o, h_2^o]^T$. To solve for these two scalars, the representations for $\hat{X}$ and $\hat{X}_o$ in terms of Y_1 and Y_2 are substituted into (13.16). This yields

$$(X - h_1^o Y_1 - h_2^o Y_2, h_1 Y_1 + h_2 Y_2) = 0 \qquad \forall h_1, h_2,$$

which can be reexpressed (by use of the linearity property of inner products) as

$$h_1[(X, Y_1) - h_1^o(Y_1, Y_1) - h_2^o(Y_2, Y_1)]$$
$$+ h_2[(X, Y_2) - h_1^o(Y_1, Y_2) - h_2^o(Y_2, Y_2)] = 0 \qquad \forall h_1, h_2.$$

But this equation holds for all h_1 and h_2 if and only if each of the two bracketed terms is zero. Writing these two equations in matrix form yields

$$\begin{bmatrix} (Y_1, Y_1) & (Y_2, Y_1) \\ (Y_1, Y_2) & (Y_2, Y_2) \end{bmatrix} \begin{bmatrix} h_1^o \\ h_2^o \end{bmatrix} = \begin{bmatrix} (X, Y_1) \\ (X, Y_2) \end{bmatrix}. \qquad (13.19a)$$

Hence, the two scalars h_1^o and h_2^o are given by the solution to the two simultaneous linear equations (13.19a). For example, if M is the space of all vectors with zero third element, then a basis for M is $Y_1 = \{1, 0, 0\}$ and $Y_2 = \{0, 1, 0\}$, and the unique solution from (13.19a) is $\hat{X}_o = \{\alpha_1, \alpha_2, 0\}$, for $X = \{\alpha_1, \alpha_2, \alpha_3\}$.

Although these general results for three-dimensional Euclidean space are equally valid for higher-dimensional Euclidean spaces, they are not as easy to visualize. The inner product and norm for the l-dimensional Euclidean space L, for which the vectors are l-tuples of real numbers, are defined by

$$(X, Y) \triangleq \sum_{i=1}^{l} \alpha_i \beta_i \qquad (13.8\text{b})$$

$$\|X\| = (X, X)^{1/2} = \left[\sum_{i=1}^{l} \alpha_i^2\right]^{1/2}, \qquad (13.9\text{b})$$

of which (13.8a) and (13.9a) are the special cases corresponding to $l = 3$. The geometrical relationships (13.12) to (13.15) hold for (13.8b) and (13.9b). Similarly, given an m-dimensional ($m < l$) subspace M of L (which is a hyperplane through the origin of L), the closest vector (l-tuple) $\hat{X}_o$ in M to a given vector X in L is the orthogonal projection of X onto M, and is given by the unique solution to the orthogonality condition (13.16). The condition (13.16) can be expressed more explicitly in terms of a set of basis vectors, say $\{Y_1, Y_2, \ldots, Y_m\}$, for the subspace M. Specifically, by substituting the representation for $\hat{X}$,

$$\hat{X} = \sum_{i=1}^{m} h_i Y_i, \qquad (13.20)$$

in terms of the m basis vectors and m corresponding scalars $\{h_i\}$, into (13.16), the following equivalent condition on $\hat{X}_o$ is obtained:

$$(X - \hat{X}_o, Y_i) = 0, \qquad i = 1, 2, \ldots, m. \qquad (13.21)$$

Furthermore, by substituting the representation

$$\hat{X}_o = \sum_{j=1}^{m} h_j^o Y_j \qquad (13.22)$$

into (13.21), the following equivalent condition on the scalars $\{h_j^o\}$ that specify $\hat{X}_o$ is obtained:

$$\sum_{j=1}^{m} h_j^o (Y_j, Y_i) = (X, Y_i), \qquad i = 1, 2, \ldots, m. \qquad (13.23)$$

Let the m-vectors with ith elements (X, Y_i) and h_i^o be denoted by $\tilde{R}_{xY}$ and H_o, respectively, and let the $m \times m$ matrix with (i, j)th element (Y_j, Y_i) be denoted by $\tilde{R}_Y$. Then the m simultaneous linear equations (13.23) can be reexpressed as

$$\tilde{R}_Y H_o = \tilde{R}_{xY}, \qquad (13.19\text{b})$$

of which (13.19a) is the special case corresponding to $m = 2$. The set of m simultaneous linear equations (13.19b) are commonly called the *normal equations* because they embody the orthogonality condition (and *orthogonal* and *normal* are synonyms).

13.2.2 Least-Squares Estimation

As an application of orthogonal projection, consider the problem of estimating l values of a time series X,

$$X = [x_1, x_2, \ldots, x_l]^T, \qquad (13.24)$$

by filtering (convolving) another time series Y (e.g., corrupted measurements of X) with a sequence of m weights

$$H = [h_0, h_1, \ldots, h_{m-1}]^T. \qquad (13.25)$$

The estimate is then of the form

$$\hat{x}_n = \sum_{i=0}^{m-1} h_i y_{n-i}, \qquad n = 1, 2, \ldots, l. \qquad (13.26)$$

This estimate can be reexpressed in vector notation as

$$\hat{X} = \sum_{i=0}^{m-1} h_i Y_i, \qquad (13.27)$$

where

$$Y_i \triangleq [y_{1-i}, y_{2-i}, \ldots, y_{l-i}]^T. \qquad (13.28)$$

The sum of squared errors is adopted as a measure of the estimation error:

$$\sum_{n=1}^{l} [x_n - \hat{x}_n]^2 = \|X - \hat{X}\|^2,$$

and the solution of the optimization problem

$$\min_{H} \|X - \hat{X}\| \qquad (13.29)$$

is sought. The weight vector H_o that minimizes this error norm is given by the orthogonality condition (13.19b) for which the m-vector $\tilde{R}_{xY}$ has i-th element

$$[\tilde{R}_{xY}]_i = \sum_{n=1}^{l} x_n y_{n-i}, \qquad (13.30)$$

and the $m \times m$ matrix $\tilde{R}_Y$ has (ij)-th element

$$[\tilde{R}_Y]_{ij} = \sum_{n=1}^{l} y_{n-j} y_{n-i}. \tag{13.31}$$

Thus, the optimum weight vector H_o is specified by $\tilde{R}_{xY}$, which is a finite cross-correlation of the two time series X and Y, and $\tilde{R}_Y$, which is a finite autocorrelation of the time series Y.

Recursive Least Squares. The optimum weight vector H_o can be reexpressed as

$$H_o(l) = \bar{R}_Y^{-1}(l) \bar{R}_{xY}(l), \tag{13.32}$$

in which

$$\bar{R}_{xY}(l) \triangleq \frac{1}{l} \sum_{n=1}^{l} x_n Y(n) \tag{13.33}$$

and

$$\bar{R}_Y(l) \triangleq \frac{1}{l} \sum_{n=1}^{l} Y(n) Y^T(n), \tag{13.34}$$

where

$$Y(n) \triangleq [y_n, y_{n-1}, \ldots, y_{n-m+1}]^T. \tag{13.35}$$

If the data from the time series Y are arranged in the form of an $l \times m$ matrix,

$$M \triangleq \begin{bmatrix} y_1 & y_0 & \cdots & y_{2-m} \\ y_2 & y_1 & \cdots & y_{3-m} \\ \vdots & \vdots & & \vdots \\ y_l & y_{l-1} & \cdots & y_{l+1-m} \end{bmatrix}, \tag{13.36}$$

then $Y(n)$ is the transpose of the nth row vector of M, and Y_i is the ith column vector of M. Now, if the parameter l is interpreted as an update time index, then the solution (13.32) to (13.34) can be manipulated into a recursion from which the solution $H_o(l)$ can be obtained by updating the solution $H_o(l-1)$. Specifically, it follows directly from (13.33) and (13.34) that $\bar{R}_{xY}$ and $\bar{R}_Y$ obey the recursions

$$\bar{R}_{xY}(l) = \frac{l-1}{l} \bar{R}_{xY}(l-1) + \frac{1}{l} x(l) Y(l), \tag{13.37}$$

$$\bar{R}_Y(l) = \frac{l-1}{l} \bar{R}_Y(l-1) + \frac{1}{l} Y(l) Y^T(l). \tag{13.38}$$

Application of Woodbury's identity (exercise 5) to (13.38) yields the recursion

$$\overline{R}_Y^{-1}(l) = \frac{l}{l-1}\left[\overline{R}_Y^{-1}(l-1) - \frac{\overline{R}_Y^{-1}(l-1)Y(l)Y^T(l)\overline{R}_Y^{-1}(l-1)}{(l-1) + Y^T(l)\overline{R}_Y^{-1}(l-1)Y(l)}\right],$$

$$l \geq m.$$

(13.39)

Substitution of

$$\overline{R}_Y(l)H_o(l) = \overline{R}_{xY}(l) \tag{13.40}$$

into (13.37) yields

$$\overline{R}_Y(l)H_o(l) = \frac{l-1}{l}\overline{R}_Y(l-1)H_o(l-1) + \frac{1}{l}x(l)Y(l), \quad (13.41)$$

which upon substitution of (13.38) becomes

$$\overline{R}_Y(l)H_o(l) = \left[\overline{R}_Y(l) - \frac{1}{l}Y(l)Y^T(l)\right]H_o(l-1) + \frac{1}{l}x(l)Y(l).$$

(13.42)

Multiplication of both sides of (13.42) by $\overline{R}_Y^{-1}(l)$ yields the recursion

$$H_o(l) = H_o(l-1) + K(l)e(l), \quad l \geq m, \tag{13.43}$$

where $e(l)$ is the estimation error, $x(l) - \hat{x}(l)$, prior to the update of $H_o(l-1)$,

$$e(l) \triangleq x(l) - H_o^T(l-1)Y(l), \tag{13.44}$$

and $K(l)$ is the *gain vector*,

$$K(l) \triangleq \frac{1}{l}\overline{R}_Y^{-1}(l)Y(l). \tag{13.45a}$$

Thus, $H_o(l)$ can be updated using (13.43) to (13.45), for which (13.39) can be used to update $K(l)$. An alternative form for the gain vector (13.45a) can be obtained by use of the identity

$$\frac{1}{l}\overline{R}_Y^{-1}(l)Y(l) = \frac{\overline{R}_Y^{-1}(l-1)Y(l)}{(l-1) + Y^T(l)\overline{R}_Y^{-1}(l-1)Y(l)}, \tag{13.45b}$$

which can be derived (exercise 6) from (13.39).

13.2.3 Hilbert Space of Functions of a Real Variable

If a set L of functions of a real variable t on some interval T [e.g., $X = \{\alpha(t) : t \in T\}$] contains all linear combinations of its own elements (functions), then it is a linear vector space, and the functions are called *vectors*. If the functions are square-integrable, then parallel to (13.8b) and (13.9b), an inner product and norm can be defined on L by

$$(X, Y) \triangleq \int_T \alpha(t)\beta(t)\, dt, \tag{13.8c}$$

$$\|X\| \triangleq \left[\int_T \alpha^2(t)\, dt\right]^{1/2}. \tag{13.9c}$$

The discrete coordinate index i in (13.8b) and (13.9b) is replaced with the continuous coordinate index t in (13.8c) and (13.9c), and the discrete sums are replaced with continuous sums (integrals). An infinite-dimensional linear vector space of this type is called a *Hilbert space* (cf. [Epstein, 1970]). It can be shown that definitions (13.8c) and (13.9c) satisfy the geometrical relationships (13.12) to (13.15). Thus, functions of a real variable can be given geometrical interpretations. For example, two functions X and Y are said to be orthogonal in Hilbert space if their inner product (13.8c) is zero; for example, $X = \cos 2\pi t$ and $Y = \sin 2\pi t$ are orthogonal for $T = [0, 1]$ (exercise 3).

13.2.4 Hilbert Space of Random Variables

Let S be the sample space for a probability space on which two random variables X and Y are defined, and let P be the probability measure on S. Then X and Y are functions of event points $s \in S$; that is,

$$X = \{\alpha(s) : s \in S\},$$
$$Y = \{\beta(s) : s \in S\}.$$

[The alternative notation $X(s) \equiv \alpha(s)$ is used in previous chapters.] The *correlation* of X and Y is the expected value of their product,

$$E\{XY\} = \sum_{s \in S} \alpha(s)\beta(s)P(s) \tag{13.46}$$

or

$$E\{XY\} = \int_{s \in S} \alpha(s)\beta(s)\, dP(s). \tag{13.47}$$

The first formula applies for a finite (or denumerable) sample space S, and the second formula applies more generally. Of course $E\{XY\}$ can be reexpressed in terms of the joint probability distribution function F_{XY} as

$$E\{XY\} = \int\int xy\, dF_{XY}(x, y), \qquad (13.48)$$

or in terms of the joint probability density function f_{XY} as

$$E\{XY\} = \int\int xy f_{XY}(x, y)\, dx\, dy. \qquad (13.49)$$

However, it is the formulation (13.46) to (13.47) that is reminiscent of the inner-product formulas (13.8c) and (13.8b). In fact, (13.46) and (13.47) are valid inner products [with weighting function $P(\cdot)$] on the linear vector space L of all *finite-mean-square* random variables defined on a given probability space, that is, all random variables for which

$$E\{X^2\} = \int_{s\in S} \alpha^2(s)\, dP(s) < \infty. \qquad (13.50)$$

Therefore, we can use the geometrical notation

$$(X, Y) \triangleq E\{XY\} \qquad (13.8d)$$

$$\|X\| \triangleq \left[E\{X^2\}\right]^{1/2}, \qquad (13.9d)$$

and we can give random variables a geometrical interpretation. The inner product and norm (13.8d) and (13.9d) satisfy the geometrical relationships (13.12) to (13.15) (exercise 8). For example, two random variables X and Y are *orthogonal* if and only if their correlation is zero:

$$(X, Y) \equiv E\{XY\} = 0. \qquad (13.51)$$

13.3 Minimum-Mean-Squared-Error Estimation

13.3.1 Nonlinear Estimation

Let M denote the subspace of L consisting of all finite-mean-square functions $g(\cdot)$ of a specific random variable Y, that is, all $g(Y)$ for which $E\{g^2(Y)\} < \infty$. Consider the following minimum-mean-squared-error estimation problem:

$$\min_{\hat{X}=g(Y)} E\{(X - \hat{X})^2\} \qquad (13.52)$$

for some X in L. This problem can be reexpressed as

$$\min_{\hat{X}\in M} \|X - \hat{X}\|, \qquad X \in L. \qquad (13.53)$$

Therefore, the best estimator $\hat{X}_o$ is the orthogonal projection of X onto M. Furthermore, this best estimator is the unique solution to the orthogonality condition (13.16), which can be reexpressed as

$$E\{(X - \hat{X}_o)\hat{X}\} = 0 \quad \forall \hat{X} \in M, \quad (13.54)$$

and also as

$$E\{[X - g_o(Y)]g(Y)\} = 0 \quad \forall g(\cdot), \quad (13.55)$$

where $g_o(\cdot)$ is the optimum estimation function being sought, and $g(\cdot)$ is any admissible estimation function $[E\{g^2(Y)\} < \infty]$. Let us try the following candidate for solution:

$$\hat{X}_o = E\{X|Y\}, \quad (13.56)$$

that is,

$$g_o(\cdot) = E\{X|(\cdot)\}. \quad (13.57)$$

Substitution of (13.57) into the necessary and sufficient condition (13.55) yields

$$E\{Xg(Y)\} = E\{E\{X|Y\}g(Y)\} \quad \forall g(\cdot). \quad (13.58)$$

Furthermore, since

$$E\{X|Y\}g(Y) = E\{Xg(Y)|Y\} \quad (13.59)$$

for all X, Y, and $g(\cdot)$, then (13.58) becomes

$$E\{Xg(Y)\} = E\{E\{Xg(Y)|Y\}\}. \quad (13.60)$$

But (13.60) is indeed true for every X, Y, and $g(\cdot)$. It simply states that the expected value of a conditional mean is the unconditional mean (cf. Section 2.4). Hence (13.56) is indeed the optimum estimator: the minimum-mean-squared-error estimator for X in terms of Y is the mean of X conditioned on Y; that is, it is simply what one *expects* X to be, having observed Y.

Not only have we discovered that the random variable $E\{X|Y\}$ is a minimum-mean-squared-error estimator, but we have also discovered that it is an orthogonal projection. Recall that, as explained in Section 2.4, although $E\{X|y\}$ is a number, $E\{X|Y\}$ is a random variable. By convention the random variable $\hat{X}$ is called an *estimator* and a sample $\hat{x}$ is called an *estimate*. However, the estimation function $g(\cdot)$ also is sometimes called an estimator.

13.3.2 Linear Estimation

One-Dimensional Case. Let M be the subspace consisting of all linear functions of the random variable Y; that is, M consists of all $\hat{X}$ of the form

$$\hat{X} = g(Y) = hY, \tag{13.61}$$

where h is a number. Consider the problem of finding the linear function $g_o(Y) = h_o Y$ that minimizes the mean squared error in estimating a random variable X:

$$\min_{\hat{X}=hY} E\{(X - \hat{X})^2\}. \tag{13.62}$$

The optimum estimate is the solution to the orthogonality condition

$$E\{(X - h_o Y)hY\} = 0 \quad \forall h. \tag{13.63}$$

Use of the linearity property of expectation yields

$$h[E\{XY\} - h_o E\{Y^2\}] = 0 \quad \forall h, \tag{13.64}$$

which admits the unique solution

$$h_o = \frac{E\{XY\}}{E\{Y^2\}}. \tag{13.65}$$

Furthermore, it follows from the orthogonality condition (13.63) and the solution (13.65) that the resultant minimum value of the mean squared error is (exercise 10)

$$\begin{aligned}
\text{MSE}_o &\triangleq E\{(X - \hat{X}_o)^2\} \\
&= E\{X^2\} - E\{\hat{X}_o^2\} \\
&= E\{X^2\} - E\{X\hat{X}_o\},
\end{aligned} \tag{13.66}$$

and for this one-dimensional case we have

$$\text{MSE}_o = E\{X^2\} - \frac{(E\{XY\})^2}{E\{Y^2\}}.$$

Thus,

$$\text{MSE}_o = E\{X^2\}(1 - \rho^2), \tag{13.67}$$

where ρ is defined by

$$\rho \triangleq \frac{E\{XY\}}{\sqrt{E\{X^2\}E\{Y^2\}}}, \tag{13.68}$$

and is the correlation coefficient (if X and Y have zero means). It can be concluded from (13.67) that $|\rho|$ is a measure of the degree to which two random variables X and Y are linearly related.

n-Dimensional Case. Let M be the subspace consisting of all linear functions of the n random variables*

$$\mathbf{Y} = [Y_1, Y_2, \ldots, Y_n]^T; \qquad (13.69)$$

that is, M consists of all random variables of the form†

$$\hat{X} = g(\mathbf{Y}) = \sum_{i=1}^{n} h(i) Y_i = \mathbf{h}^T \mathbf{Y} = \mathbf{Y}^T \mathbf{h}. \qquad (13.70)$$

The solution to the minimization problem

$$\min_{\hat{X} \in M} E\{(X - \hat{X})^2\} \qquad (13.71)$$

is given by the solution $\hat{X}_o = \mathbf{h}_o^T \mathbf{Y}$ to the orthogonality condition

$$E\{(X - \mathbf{h}_o^T \mathbf{Y}) \mathbf{Y}^T \mathbf{h}\} = 0 \quad \forall \mathbf{h}. \qquad (13.72)$$

But (13.72) can be reexpressed (by use of the linearity property of $E\{\cdot\}$) as

$$E\{(X - \mathbf{h}_o^T \mathbf{Y}) \mathbf{Y}^T\} \mathbf{h} = 0 \quad \forall \mathbf{h}, \qquad (13.73)$$

and (13.73) is satisfied if and only if

$$E\{(X - \mathbf{h}_o^T \mathbf{Y}) \mathbf{Y}^T\} = \mathbf{0}, \qquad (13.74)$$

which is equivalent (because of the linearity of $E\{\cdot\}$) to

$$E\{X \mathbf{Y}^T\} = \mathbf{h}_o^T E\{\mathbf{Y} \mathbf{Y}^T\}. \qquad (13.75)$$

If the n-vector of correlations is denoted by

$$\mathbf{R}_{XY} \triangleq E\{X\mathbf{Y}\}, \qquad (13.76)$$

and the $n \times n$ matrix of correlations is denoted by

$$\mathbf{R}_Y \triangleq E\{\mathbf{Y}\mathbf{Y}^T\} \equiv \mathbf{R}_Y^T, \qquad (13.77)$$

then (13.75) (transposed) becomes

$$\mathbf{R}_Y \mathbf{h}_o = \mathbf{R}_{XY}, \qquad (13.78)$$

---

*In Section 13.2, lightface capitals were used to denote n-tuples; but since capitals are used also to denote random variables, now boldface is used to denote n-tuples, and boldface capitals are used to denote n-tuples of random variables, as in previous chapters.

†In Section 13.2, the notation h_i is used for the elements of a real non-random n-tuple; however $h(i)$ will be used from this point forward in order to be consistent with the notation used for filters in Chapter 9.

which can be written out as

$$\sum_{j=1}^{n} R_Y(i,j) h_o(j) = R_{XY}(i), \qquad i = 1, 2, 3, \ldots, n. \tag{13.79}$$

[Notice the analogy between (13.78) and (13.19b).] When the inverse matrix R_Y^{-1} exists, the solution to this set of n simultaneous linear equations can be expressed as

$$\mathbf{h}_o = \mathbf{R}_Y^{-1} \mathbf{R}_{XY}, \tag{13.80}$$

which obviously includes (13.65) as the special case $n = 1$.

A simplification that is developed in the preceding solution method is particularly useful. Specifically, the two orthogonality conditions

$$E\{(X - \hat{X}_o)\hat{X}\} = 0 \qquad \forall X \in M \tag{13.81}$$

and

$$E\{(X - \hat{X}_o) Y_i\} = 0, \qquad i = 1, 2, 3, \ldots, n, \tag{13.82}$$

are equivalent. The reason is that the n random variables $\{Y_i : i = 1, \ldots, n\}$ span the space M, in the sense that every $\hat{X} \in M$ is a linear combination of these n Y_i. [Notice the analogy to (13.21)]. Observe that (13.82) requires that the estimation error $X - \hat{X}_o$ be orthogonal to all of the observed random variables Y_i.

Parallel to the formulas (13.66) and (13.67) for MSE_o for $n = 1$, the general expression (13.66) for MSE_o for $n \geq 1$ can be reduced to (exercise 10)

$$\text{MSE}_o = R_X(1 - \rho^2), \tag{13.83}$$

where

$$\rho^2 \triangleq \mathbf{R}_{XY}^T \mathbf{R}_Y^{-1} \mathbf{R}_{XY} / R_X \tag{13.84}$$

and

$$R_X = E\{X^2\}. \tag{13.85}$$

Analogous to the interpretation of ρ in (13.68), ρ in (13.84) can be interpreted as a measure of the degree to which one random variable X is linearly related to n random variables $\{Y_i\}_1^n$.

Infinite-Dimensional Case. Let M be the subspace consisting of all linear combinations of the random variables $\{Y(t) : t \in V\}$ from a random process Y; that is, M consists of all random variables of the form

$$\hat{X} = \int_V h(t) Y(t) \, dt. \tag{13.86}$$

It is easily shown, parallel to the preceding argument for the n-dimensional case, that the solution to the minimization problem

$$\min_{\hat{X} \in M} E\{(X - \hat{X})^2\} \qquad (13.87)$$

is given by the solution to the orthogonality condition

$$E\left\{\left[X - \int_V h_o(u)Y(u)\,du\right]Y(t)\right\} = 0 \qquad \forall t \in V, \qquad (13.88)$$

This condition can be reexpressed (by use of the linearity of $E\{\cdot\}$) as

$$\int_V R_Y(t,u)h_o(u)\,du = R_{XY}(t) \qquad \forall t \in V, \qquad (13.89)$$

which is analogous to (13.79). Furthermore, the general expression (13.66) for MSE_o can be reduced to

$$\text{MSE}_o = R_X(1 - \rho^2), \qquad (13.90)$$

where

$$\rho^2 \triangleq \frac{1}{R_X} \int_V R_{XY}(t)h_o(t)\,dt, \qquad (13.91)$$

which is analogous to (13.84) with (13.80) substituted in.

13.3.3 Waveform Estimation

Given a sample $\{y(u): u \in V\}$ of a segment of a random process $Y(u)$, it is desired to estimate the value of a related, but unobservable, random variable X. It is desired to use a linear estimation rule,

$$\hat{X} = \int_V h(u)Y(u)\,du, \qquad (13.92)$$

and to determine the particular linear rule that minimizes the mean squared error:

$$\min_{h(\cdot)} E\{(X - \hat{X})^2\}. \qquad (13.93)$$

It has been determined in Section 13.3.2 that the optimum (MMSE) estimation function $h_o(\cdot)$ is implicitly specified by the orthogonality condition

$$E\{(X - \hat{X}_o)Y(v)\} = 0 \qquad \forall v \in V, \qquad (13.94)$$

where

$$\hat{X}_o \triangleq \int_V h_o(u)Y(u)\,du. \qquad (13.95)$$

Substitution of (13.95) into (13.94) yields, after interchange of expectation and integration,

$$\int_V h_o(u) E\{Y(v)Y(u)\} \, du = E\{XY(v)\} \qquad \forall v \in V. \quad (13.96)$$

This desired result will be called the *design equation*; its solution $h_o(\cdot)$ is the desired optimum estimation function. It is noted that $h_o(\cdot)$ is completely specified (implicitly) by the autocorrelation R_Y of the observations together with the cross-correlation R_{XY} of the observations and the unobservable variable.

With the use of (13.94), it can be shown that the minimum MSE reduces to

$$\text{MSE}_o = E\{X^2\} - E\{\hat{X}_o^2\} = E\{X^2\} - E\{X\hat{X}_o\}. \quad (13.97)$$

Substitution of (13.95) into (13.97) yields

$$\text{MSE}_o = E\{X^2\} - \int_V h_o(u) E\{XY(u)\} \, du. \quad (13.98)$$

This desired result will be called the *performance formula*. Substitution of the solution $h_o(\cdot)$ to (13.96) into (13.98) yields the value of the minimum MSE.

Consider now the slightly more general problem of estimating the value of a random process $X(t)$ at each and every value of time t in some desired set T. Since the optimum estimation function can be different for different values of t, its dependence on t is denoted by $h_o(t, \cdot)$. Thus, it is desired to determine the set $\{h_o(t, \cdot) : t \in T\}$ of linear rules that minimize the set of MSEs:

$$\min_{h(t, \cdot)} E\{[X(t) - \hat{X}(t)]^2\} \qquad \forall t \in T, \quad (13.99)$$

$$\hat{X}(t) = \int_V h(t, u) Y(u) \, du \qquad \forall t \in T. \quad (13.100)$$

It follows directly from (13.95) and (13.98) that the design equation is

$$\int_V h_o(t, u) E\{Y(v)Y(u)\} \, du = E\{X(t)Y(v)\} \qquad \forall v \in V, t \in T,$$

$$(13.101)$$

and the performance formula is

$$\text{MSE}_o(t) = E\{X^2(t)\} - \int_V h_o(t, u) E\{X(t)Y(u)\} \, du \qquad \forall t \in T.$$

$$(13.102)$$

It follows from (13.100) that the estimation function $h(t, u)$ is the impulse-response function of the linear transformation from $\{Y(u): u \in V\}$ to $\{\hat{X}(t): t \in T\}$.

A variety of different types of estimation problems can be obtained by prescribing different sets T and V. For example, if

$$V \triangleq \{u : u \le t + \alpha\}, \qquad (13.103)$$

then the estimate is given by

$$\hat{X}(t) = \int_{-\infty}^{t+\alpha} h(t, u) Y(u) \, du. \qquad (13.104)$$

Thus, if $\alpha = 0$, $\hat{X}(t)$ is a *causal estimate*; if $\alpha < 0$, $\hat{X}(t)$ is a *causal extrapolated* or *predicted estimate*; and if $\alpha > 0$, $\hat{X}(t)$ is a *noncausal interpolated* or *smoothed estimate*. In Section 13.4 the extreme case of noncausal estimation with unlimited data,

$$V = T = (-\infty, \infty), \qquad (13.105)$$

is focused on, and it is assumed that $X(t)$ and $Y(t)$ are jointly WSS. This yields a particularly tractable problem from which considerable insight into estimation in general can be gained. Then, in Section 13.5, an introduction to the less tractable case of causal estimation is presented. Treatments with broader scope are available in books on optimum filtering (e.g., [Van Trees, 1968], [Anderson and Moore, 1979], [Kailath, 1981]). Also, the case of noncausal estimation for jointly almost cyclostationary $X(t)$ and $Y(t)$ is discussed in some detail in Subsection 12.8.5 and also in [Gardner, 1987a].

13.4 Noncausal Wiener Filtering

13.4.1 General Formulas

If $X(t)$ and $Y(t)$ are jointly WSS, then

$$R_Y(v, u) = R_Y(v - u),$$
$$R_{XY}(t, v) = R_{XY}(t - v), \qquad (13.106)$$

and as a result, the solution to the design equation (13.101) [with (13.105)] is time-invariant:

$$h_o(t, u) = h_o(t - u). \qquad (13.107)$$

Thus, (13.100) [with (13.105)] becomes

$$\hat{X}_o(t) = \int_{-\infty}^{\infty} h_o(t - u) Y(u) \, du$$
$$= h_o(t) \otimes Y(t), \qquad (13.108)$$

which is a convolution, and the estimate $\hat{X}_o(t)$ can be obtained by passing the observations $Y(t)$ through a *linear time-invariant filter* with impulse-response function $h_o(\cdot)$. Using the change of variables $t - u = w$ and $t - v = \tau$ in (13.101) [with (13.105) to (13.107)] yields (exercise 15)

$$\int_{-\infty}^{\infty} h_o(w) R_Y(\tau - w) \, dw = R_{XY}(\tau) \qquad \forall \tau, \qquad (13.109)$$

which is a convolution

$$h_o(\tau) \otimes R_Y(\tau) = R_{XY}(\tau). \qquad (13.110)$$

Thus, the optimum filter can be solved for by application of the convolution theorem for Fourier transforms to (13.110) to obtain

$$H_o(f) S_Y(f) = S_{XY}(f), \qquad (13.111)$$

which yields the solution

$$\boxed{H_o(f) = \frac{S_{XY}(f)}{S_Y(f)}.} \qquad (13.112)$$

This is the transfer function of what is called the *noncausal Wiener filter*, in honor of Norbert Wiener's pioneering work [Wiener, 1949].

A similar change of variables in the performance formula (13.102) [with (13.105) to (13.107)] yields (exercise 15)

$$\mathrm{MSE}_o = R_X(0) - \int_{-\infty}^{\infty} h_o(\tau) R_{XY}(\tau) \, d\tau, \qquad (13.113)$$

which can be reexpressed (by use of Parseval's relation for Fourier transforms) as

$$\mathrm{MSE}_o = \int_{-\infty}^{\infty} S_X(f) \, df - \int_{-\infty}^{\infty} H_o(f) S_{XY}^*(f) \, df. \qquad (13.114)$$

Substitution of (13.112) into (13.114) yields

$$\boxed{\mathrm{MSE}_o = \int_{-\infty}^{\infty} S_X(f) \left[1 - |\rho(f)|^2\right] df,} \qquad (13.115)$$

where

$$\boxed{|\rho(f)| \triangleq \frac{|S_{XY}(f)|}{[S_X(f) S_Y(f)]^{1/2}},} \qquad (13.116)$$

which is the magnitude of the coherence between $X(t)$ and $Y(t)$ at frequency f. Recall that coherence is a measure of linear relationship; for example if $\rho(f) \equiv 0$, then $H_o(f) \equiv 0$ and $\text{MSE}_o = E\{X^2(t)\}$; that is, if $X(t)$ and $Y(t)$ have no linear relationship (in the mean-square sense), then $X(t)$ cannot be estimated from $Y(t)$ using a linear filter. At the other extreme, if $|\rho(f)| \equiv 1$, then there exists a filter with transfer function denoted by $G(f)$, such that

$$Y(t) = g(t) \otimes X(t). \tag{13.117}$$

Furthermore, the optimum filter is, from (13.112),

$$H_o(f) = \frac{1}{G(f)}, \tag{13.118}$$

and $\text{MSE}_o = 0$. Thus, if $X(t)$ and $Y(t)$ are completely linearly related (in the mean-square sense), then $X(t)$ can be *perfectly* estimated from $Y(t)$ with a linear filter.

It should be noted that if $S_Y(f) = 0$ at any point or in any interval, then $S_{XY}(f) = 0$ at the same point or in the same interval, and therefore a solution H_o to (13.111) always exists, but can be nonunique (at frequencies where Y contains no power density). Nevertheless, if Y contains an arbitrarily small amount of white noise, then $S_Y(f) > 0$, and the solution H_o is unique.

13.4.2 Joint Equalizing and Noise Filtering

Let $Y(t)$ be the sum of a distorted signal $g(t) \otimes S(t)$ and zero-mean noise $N(t)$ that is uncorrelated with the signal,

$$Y(t) = g(t) \otimes S(t) + N(t), \tag{13.119}$$

and let $X(t)$ be the signal

$$X(t) = S(t). \tag{13.120}$$

It follows from (13.112) that the optimum filter for $S(t)$ is

$$H_o(f) = \frac{G^*(f)S_S(f)}{|G(f)|^2 S_S(f) + S_N(f)}, \tag{13.121}$$

and the coherence magnitude [which determines MSE_o from (13.115)] is

$$|\rho(f)|^2 = \frac{r(f)}{1 + r(f)}, \tag{13.122}$$

where

$$r(f) \triangleq \frac{|G(f)|^2 S_S(f)}{S_N(f)}, \tag{13.123}$$

Minimum-Mean-Squared-Error Estimation • 437

which is a ratio of the PSD of the signal component of $Y(t)$ to the PSD of the noise component of $Y(t)$; i.e., $r(f)$ is a *PSD signal-to-noise ratio* (PSD SNR). It follows from the reexpression of (13.121) as

$$H_o(f) = \left[\frac{1}{G(f)}\right]|\rho(f)|^2 \tag{13.124}$$

that at frequencies where the coherence is small, because the PSD SNR is small, the optimum filter acts as a strong attenuator; and at frequencies where the coherence is near its maximum of 1, because the PSD SNR is large, the optimum filter acts as an equalizer that removes the distortion.

The overall measure of performance, MSE, can be decomposed into two component measures—one for signal distortion and the other for noise—as explained in the following. The estimated signal is given by

$$\hat{S}(t) = h_o(t) \otimes [g(t) \otimes S(t) + N(t)]. \tag{13.125}$$

Thus, the estimation error is

$$\hat{S}(t) - S(t) = d(t) \otimes S(t) + a(t) \otimes N(t), \tag{13.126}$$

where

$$d(t) \triangleq h_o(t) \otimes g(t) - \delta(t),$$
$$a(t) \triangleq h_o(t), \tag{13.127}$$

and $\delta(t)$ is the Dirac delta. Hence, $d(t)$ is the resultant signal distortion function, and $a(t)$ is the noise attenuation function. The transfer functions for these are

$$D(f) = H_o(f)G(f) - 1 = \frac{-1}{1 + r(f)}, \tag{13.128}$$

$$A(f) = \frac{1}{G(f)}\left[\frac{r(f)}{1 + r(f)}\right]. \tag{13.129}$$

It follows from (13.126) that (exercise 18)

$$\text{MSE}_o = \int_{-\infty}^{\infty} |D(f)|^2 S_S(f)\, df + \int_{-\infty}^{\infty} |A(f)|^2 S_N(f)\, df$$
$$\triangleq \text{MSE}_S + \text{MSE}_N. \tag{13.130}$$

The frequency-dependent contribution to MSE due to distortion is completely characterized by the signal PSD and the distortion factor $D(f)$, and the contribution to MSE due to noise is completely characterized by the noise PSD and the noise attenuation factor $A(f)$.

The *output signal-to-noise ratio* (SNR_o) is defined to be the ratio of the power in the signal-dependent part ($h_o \otimes g \otimes S$) of the output, $\hat{S}(t)$, to the power in the noise-dependent part ($h_o \otimes N$). Therefore (exercise 18),

$$\text{SNR}_o = \frac{E\{[h_o(t) \otimes g(t) \otimes S(t)]^2\}}{E\{[h_o(t) \otimes N(t)]^2\}}$$

$$= \frac{\int_{-\infty}^{\infty} |1 + D(f)|^2 S_S(f)\, df}{\int_{-\infty}^{\infty} |A(f)|^2 S_N(f)\, df}. \qquad (13.131)$$

Hence, SNR_o is completely characterized by the two factors $D(f)$ and $A(f)$, and the signal and noise PSDs.

Equivalent Linear Model. When no explicit model like (13.119) is available to describe the relationship between observations $Y(t)$ and a desired signal $X(t)$, a model of the form (13.119) can be constructed to fit the statistical description of $X(t)$ and $Y(t)$, namely, the spectral densities S_X, S_{XY}, S_Y. Specifically, if a distortion transfer function is *defined* by

$$G(f) \triangleq \frac{S_{XY}^*(f)}{S_X(f)}, \qquad (13.132a)$$

then a noise process $N(t)$ can be defined in terms of $Y(t)$ and $X(t)$ by (13.119) and (13.120):

$$N(t) \triangleq Y(t) - g(t) \otimes X(t). \qquad (13.132b)$$

It can be shown that $N(t)$ is uncorrelated with $X(t)$ and has PSD (exercise 19)

$$S_N(f) = S_Y(f) - \frac{|S_{XY}(f)|^2}{S_X(f)} = S_Y(f)\left[1 - |\rho(f)|^2\right], \qquad (13.133)$$

where $\rho(f)$ is the coherence function for $X(t)$ and $Y(t)$. Thus, if $X(t)$ is interpreted as a signal $S(t)$, then (13.132b) yields the model (13.119).

Example 1: Consider the signal-extraction problem described in Section 13.1, and assume that the noisy sine-wave current can be modeled by

$$Y(t) = A\cos(2\pi Kt + \Phi) + N(t), \qquad (13.134)$$

where $N(t)$ is white noise with spectral density N_o; Φ is uniformly distributed on $[-\pi, \pi)$; K has probability density $f_K(\cdot)$; and $N(t)$, Φ, and K are mutually independent. It follows from example 1 in Chapter 10 that the signal

$$S(t) = A\cos(2\pi Kt + \Phi) \qquad (13.135)$$

has spectral density

$$S_S(f) = \frac{E\{A^2\}}{4} f_K(f) + \frac{E\{A^2\}}{4} f_K(-f). \qquad (13.136)$$

The optimum filter is, from (13.121),

$$H_o(f) = \frac{S_S(f)}{S_S(f) + N_o}. \qquad (13.137)$$

For a weak signal, for which

$$S_S(f) \ll N_o, \qquad (13.138)$$

(13.136) to (13.138) yield

$$H_o(f) \simeq \frac{E\{A^2\}}{4N_o} f_K(f) + \frac{E\{A^2\}}{4N_o} f_K(-f). \qquad (13.139)$$

Thus, if the random frequency K has mean $m_K = f_o$ and standard deviation $\sigma_K = B/2$, then $H_o(f)$ is the transfer function of a bandpass filter with center frequency f_o and bandwidth on the order of B. The precise shape of the passband characteristic of the filter is given by the probability density f_K. ∎

Example 2: Consider the signal-in-noise model

$$Y(t) = S(t) + N(t), \qquad (13.140)$$

for which $N(t)$ is white noise with spectral density N_o, and $S(t)$ is uncorrelated with $N(t)$ and has autocorrelation function

$$R_S(\tau) = ae^{-\alpha|\tau|}. \qquad (13.141)$$

It is desired to estimate a delayed version of the signal,

$$X(t) = S(t - t_o), \qquad (13.142)$$

based on observations of the noisy signal $Y(t)$. The cross-correlation of $X(t)$ and $Y(t)$ is

$$R_{XY}(\tau) = R_S(\tau - t_o), \qquad (13.143)$$

and the autocorrelation of $Y(t)$ is

$$R_Y(\tau) = R_S(\tau) + N_o \delta(\tau). \qquad (13.144)$$

Thus, the cross-spectrum of $X(t)$ and $Y(t)$ is

$$S_{XY}(f) = \frac{2\alpha a e^{-i2\pi f t_o}}{(2\pi f)^2 + \alpha^2}, \qquad (13.145)$$

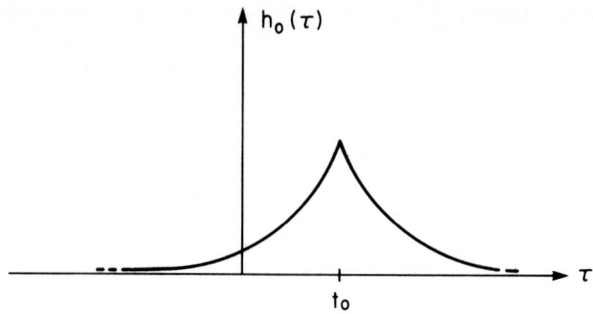

Figure 13.3 Optimum Impulse-Response Function for Example 2

and the spectrum of $Y(t)$ is

$$S_Y(f) = \frac{2\alpha a}{(2\pi f)^2 + \alpha^2} + N_o. \tag{13.146}$$

Hence, the optimum filter, which is specified by (13.112), is

$$H_o(f) = \frac{b e^{-i2\pi f t_o}}{(2\pi f)^2 + \beta^2}, \tag{13.147}$$

where

$$\beta^2 \triangleq \alpha^2 + 2a\alpha/N_o, \tag{13.148}$$

$$b \triangleq 2a\alpha/N_o. \tag{13.149}$$

It follows that the optimum impulse-response function is

$$h_o(\tau) = \frac{a}{N_o\sqrt{1 + 2a/\alpha N_o}} e^{-\beta|\tau - t_o|}, \tag{13.150}$$

which is shown in Figure 13.3. It can be seen that for a sufficiently long delay t_o, an arbitrarily accurate causal approximation to the optimum noncausal filter can be obtained simply by truncation of $h_o(\tau)$ for $\tau < 0$, that is, by retaining only the causal part of $h_o(\tau)$. This approximation will be accurate if $t_o \gg 1/\beta$. ∎

In applications where $H_o(f)$ is a rational function, its causal part, denoted by $\hat{H}_o(f)$, can be obtained without first inverse-transforming as in example 2. It can be shown that

$$\hat{H}_o(f) = [\overline{H}_o(i2\pi f)]_+,$$

where

$$[\overline{H}_o(s)]_+ \triangleq \sum_{\text{LHP}_*} \{\text{partial-fraction expansion of } \overline{H}_o(s)\}, \quad (13.151)$$

and $\overline{H}_o(s)$ is the *bilateral Laplace transform* of $h_o(\tau)$ [i.e., the Fourier transform is $H_o(f) = \overline{H}_o(i2\pi f)$], and LHP$_*$ denotes all terms in the partial-fraction expansion that correspond to poles in the left half s plane, plus any constant term or positive-powers-of-s terms. As an illustration, consider example 2 with $t_o = 0$. Then H_o is rational and

$$\overline{H}_o(s) = \frac{b}{(\beta + s)(\beta - s)}. \quad (13.152)$$

The truncated (causal part) can be obtained by expanding (13.152) into the sum of partial fractions

$$\overline{H}_o(s) = \frac{c_+}{\beta + s} + \frac{c_-}{\beta - s} \quad (13.153)$$

to obtain

$$[\overline{H}_o(s)]_+ = \frac{c_+}{\beta + s}. \quad (13.154)$$

Thus

$$\hat{H}_o(f) = \frac{c_+}{\beta + i2\pi f}. \quad (13.155)$$

As explained in Section 13.4.1, an alternative approach to obtaining a causal filter is to include causality as a constraint in defining the interval of integration, V, in (13.101), as in (13.103). This leads to an optimum causal filter as explained in Section 13.5.

Example 3: Consider the model

$$Y(t) = G(t) \otimes S(t) + N(t) \quad (13.156)$$

for a noisy distorted signal, and assume that $G(t)$ is a random function, and that $S(t)$ and $N(t)$ are orthogonal and independent of $G(t)$. It follows that the cross-correlation function for $S(t)$ and $Y(t)$ is

$$R_{SY}(\tau) = R_S(\tau) \otimes E\{G(-\tau)\}, \quad (13.157)$$

and the autocorrelation function for $Y(t)$ is

$$R_Y(\tau) = R_S(\tau) \otimes E\{r_G(\tau)\} + R_N(\tau). \quad (13.158)$$

Therefore the cross-spectrum for $S(t)$ and $Y(t)$ is

$$S_{SY}(f) = S_S(f) E\{\tilde{G}^*(f)\}, \qquad (13.159)$$

$[\tilde{G}$ is the Fourier transform of $G]$ and the spectrum for $Y(t)$ is

$$S_Y(f) = S_S(f) E\{|\tilde{G}(f)|^2\} + N_o. \qquad (13.160)$$

Thus, the optimum filter, which is specified by (13.112) with $X(t) = S(t)$, is

$$H_o(f) = \frac{E\{\tilde{G}^*(f)\} S_S(f)}{E\{|\tilde{G}(f)|^2\} S_S(f) + S_N(f)}. \qquad (13.161)$$

The identity

$$E\{|\tilde{G}(f)|^2\} \equiv [E\{|\tilde{G}(f)|\}]^2 + \text{Var}\{|\tilde{G}(f)|\} \qquad (13.162)$$

reveals that the optimum filter for random distortion cannot be obtained from the optimum filter for deterministic distortion, (13.121), simply by replacement of the deterministic function $\tilde{G}(f)$ with its expected value. In addition to $E\{\tilde{G}(f)\}$, both the mean magnitude $E\{|\tilde{G}(f)|\}$ and the variance of $|\tilde{G}(f)|$ play a role in determining the optimum filter (cf. [Maurer and Franks, 1970]). ∎

13.5 Causal Wiener Filtering

Consider the observation interval

$$V = \{u : u \leq t\} \qquad (13.163)$$

and the estimation interval

$$T = \{t : -\infty < t < \infty\}. \qquad (13.164)$$

Then the optimum filter specified by the design equation (13.101) is constrained to be causal:

$$h_o(t, u) = 0, \quad u > t. \qquad (13.165)$$

If $X(t)$ and $Y(t)$ are jointly WSS, then (13.106) holds, and as a result the solution to (13.101) [with (13.163) and (13.164)] is time-invariant:

$$h_o(t, u) = h_o(t - u). \qquad (13.166)$$

Thus, (13.100) [with (13.163)] becomes

$$\boxed{\hat{X}(t) = \int_{-\infty}^{t} h(t - u) Y(u) \, du,} \qquad (13.167)$$

which is a convolution. Using a change of variables in (13.101) [with (13.163) and (13.164)] yields the design equation

$$\int_0^\infty h_o(w) R_Y(\tau - w) \, dw = R_{XY}(\tau), \qquad \tau \geq 0. \qquad (13.168)$$

Similarly, the performance formula (13.102) [with (13.163) and (13.164)] reduces to

$$\text{MSE}_o = R_X(0) - \int_0^\infty h_o(\tau) R_{XY}(\tau) \, d\tau. \qquad (13.169)$$

Observe that the necessary and sufficient design equation (13.168) need hold only for $\tau \geq 0$. In fact, it can be shown that when the solution h_o is determined and then used to evaluate the left-hand side of (13.168) for $\tau < 0$, it will not (in general) equal the right-hand side. Thus, it is sometimes helpful to multiply both sides of (13.168) by a unit step function $u(\tau)$ to obtain an equation that holds for *all* τ:

$$\int_0^\infty h_o(w) R_Y(\tau - w) \, dw \, u(\tau) = R_{XY}(\tau) u(\tau) \qquad \forall \tau. \qquad (13.170)$$

Although the design equation (13.170), which is called the *Wiener-Hopf equation* (after Norbert Wiener and Eberhard Hopf [Wiener and Hopf, 1931]), looks similar to the design equation (13.109) for the optimum noncausal filter, it is surprisingly more difficult to solve.* For example, if the Fourier transforms of both sides of (13.170) are equated, the equivalent design equation

$$\int_0^\infty R_{XY}(\tau) e^{-i2\pi f \tau} \, d\tau = \int_0^\infty h_o(w) e^{-i2\pi f w} \int_0^\infty R_Y(\tau - w) e^{-i2\pi f(\tau - w)} \, dw \, d\tau$$

$$= \int_0^\infty h_o(w) e^{-i2\pi f w} \int_{-w}^\infty R_Y(u) e^{-i2\pi f u} \, du \, dw$$

(13.171)

is obtained. The presence of the variable $-w$, rather than a constant (such as 0 or $-\infty$), in the lower limit of integration prevents us from factoring the right-hand side of (13.171) into the product of two transforms, as was done in solving for the optimum noncausal filter. Hence, this approach does not yield a solution. In order to develop an alternative approach to solution of (13.168), we digress in the next subsection to introduce the useful technique

---

*It is unfortunate that the term *Wiener-Hopf equation* is being used indiscriminately in recent literature to denote any linear equation for minimum-norm linear estimation, including (13.19b), (13.78), (13.101), and (13.109).

of invertible whitening (originally introduced in [Kolmogorov, 1941] for discrete time, and [Bode and Shannon, 1950] and [Zadeh and Ragazzini, 1950] for continuous time; cf. [Kailath, 1974]).

13.5.1 Invertible Whitening and Spectral Factorization

Given a *colored* (nonwhite) process $Y(t)$, it is desired to find a transfer function $G(f)$ such that both $G(f)$ and $1/G(f)$ correspond to stable causal filters, and such that

$$Z(t) = \int_{-\infty}^{t} g(t-\tau)Y(\tau)\,d\tau \qquad (13.172)$$

is a white process $[S_Z(f) \equiv 1]$. Since

$$S_Z(f) = |G(f)|^2 S_Y(f), \qquad (13.173)$$

then it is required that

$$\left|\frac{1}{G(f)}\right|^2 = S_Y(f). \qquad (13.174)$$

Thus, the problem is to *factor* $S_Y(f)$ into the product*

$$\left[\frac{1}{G(f)}\right]\left[\frac{1}{G^*(f)}\right] = S_Y(f) \qquad (13.175)$$

such that the inverse transforms,

$$g(t) = \int_{-\infty}^{\infty} G(f)e^{i2\pi ft}\,df \qquad (13.176)$$

and

$$g^{-1}(t) = \int_{-\infty}^{\infty} \frac{1}{G(f)}e^{i2\pi ft}\,df, \qquad (13.177)$$

satisfy *stability*,

$$g(t) \to 0 \text{ and } g^{-1}(t) \to 0 \quad \text{as} \quad t \to \infty, \qquad (13.178)$$

and *causality*,

$$g(t) = 0 \text{ and } g^{-1}(t) = 0 \quad \text{for} \quad t < 0. \qquad (13.179)$$

The reason it is desired to require $g^{-1}(t)$ to be stable and causal is that it is desired to be able to recover the colored process $Y(t)$ from the white process

*The factor $1/G(f)$ in this section is the same as the factor $G(f)$ in Section 11.9.4.

$Z(t)$ with a stable, causal filter

$$Y(t) = \int_{-\infty}^{t} g^{-1}(t - \tau) Z(\tau) \, d\tau. \tag{13.180}$$

Observe that since $G(f)[1/G(f)] \equiv 1$, then $g(t) \otimes g^{-1}(t) = \delta(t)$. Thus, (13.180) follows from (13.172), (13.176), and (13.177).

It can be shown that as long as $S_Y(f)$ can be decomposed into the sum of a constant, say N_o, and a positive remainder, say S'_Y, that is square-integrable and satisfies the *Paley-Wiener condition* [Wiener and Paley, 1934]

$$\int_{-\infty}^{\infty} \frac{|\log S'_Y(f)|}{1 + f^2} \, df < \infty, \tag{13.181}$$

then the desired factorization of $S_Y(f)$ exists. The condition (13.181) requires that $S'_Y(f)$ decay ($f \to \infty$) toward zero slowly enough, and $S'_Y(f) \neq 0$. In the following subsection, we shall discuss the special case where $S_Y(f)$ is a bounded rational function of f^2. Such spectra can always be interpreted as having arisen from a process $\tilde{Y}(t)$ generated by passing white noise $\tilde{Z}(t)$ through a stable, causal, time-invariant, linear filter with a rational transfer function, say $T(f)$ [i.e., $S_Y(f) \equiv S_{\tilde{Y}}(f)$]; however, $Y(t) \equiv \tilde{Y}(t)$ need not be true [e.g., $\tilde{Z}(t)$ could be either WGN or WPIN or some combination of WGN and WPIN; all that is required is that $S_{\tilde{Z}}(f) \equiv 1$]. The inverse, $1/T(f)$, of this system is not necessarily causal or stable (since the original system may have zeros in the right half s plane). Nevertheless, such spectra do satisfy (13.181), and therefore do admit the desired factorization. Interestingly, when $Y(t) \equiv \tilde{Y}(t)$, this means that although both $\tilde{Z}(t)$ and $Z(t)$ (see Figure 13.4) are white, they are not necessarily the same process, since, although $|T(f)| \equiv |G^{-1}(f)|$, the phases of $T(f)$ and $G^{-1}(f)$ are not necessarily identical (exercise 31).

Observe that the procedure described in the following paragraph is a frequency-domain technique for obtaining a *multiplicative* decomposition of a rational function of f^2 into causal and causally invertible factors, whereas the procedure described in Section 13.4.2, (13.151), is a frequency-domain technique for obtaining an *additive* decomposition of a rational function of f^2 into causal and anticausal terms.

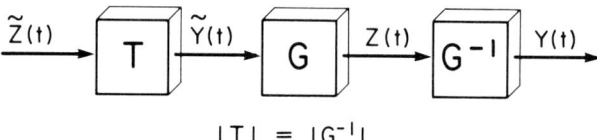

Figure 13.4 Relationship between a White Process $\tilde{Z}$ and the Innovations Process Z

Pole-Zero Assignment. Let $\bar{S}_Y$ be the bilateral Laplace transform of $R_Y(\tau)$,

$$\bar{S}_Y(s) \triangleq \int_{-\infty}^{\infty} R_Y(\tau) e^{-s\tau} d\tau.$$

This bilateral Laplace transform is related to the Fourier transform $S_Y(f)$ by $\bar{S}_Y(s) = S_Y(s/i2\pi)$ for s purely imaginary if and only if $\bar{S}_Y(s)$ has no poles on the imaginary axis of the complex s-plane. (Otherwise S_Y contains impulses not present in $\bar{S}_Y$). But since $S_Y(f)$ is assumed to be bounded, then $\bar{S}_Y(s)$ cannot have poles on the imaginary s axis. Furthermore, since $S_Y(f)$ is assumed to be rational in f^2, then $\bar{S}_Y(s)$ is rational in s^2. As a result $\bar{S}_Y(s) = \bar{S}_Y(-s)$. Furthermore, since $R_Y(\tau)$ is real, then $\bar{S}_Y(\sigma + i\omega) = \bar{S}_Y(\sigma - i\omega)$ where σ and ω are the real and imaginary parts of s. Hence, $\bar{S}_Y(s)$ has *quadrantal symmetry* in the s-plane. A typical pole-zero diagram for $\bar{S}_Y(s)$ is shown in Figure 13.5. As a result of this symmetry, $\bar{S}_Y(s)$ can be factored several different ways into the form

$$\bar{S}_Y(s) = F(s) F(-s). \tag{13.182}$$

However, if it is desired that both $1/F(s)$ and $F(s)$ be stable and causal, then $F(s)$ must have all of its poles and zeros in the left half plane (LHP), and $F(-s)$ must have all of its poles and zeros in the right half plane (RHP). Thus, there is only one factorization of the desired form, namely

$$F(s) = \sqrt{S_o} \, \frac{\prod_{\text{LHP}z_i} (s - z_i)}{\prod_{\text{LHP}p_i} (s - p_i)}, \tag{13.183}$$

where $\{z_i\}$ and $\{p_i\}$ are the locations of zeros and poles, respectively, of

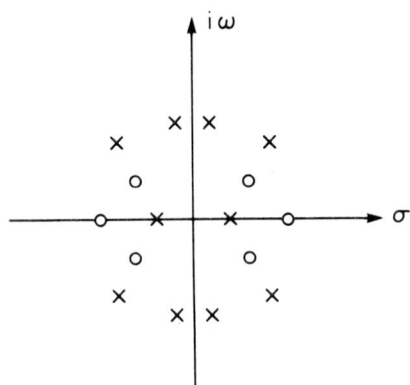

Figure 13.5 Typical Pole-Zero Diagram for Bilateral Laplace Transform $\bar{S}_Y(s)$ of an Autocorrelation Function, Illustrating Quadrantal Symmetry

$\bar{S}_Y(s)$. This is called the *canonical factorization* of $\bar{S}_Y(s)$, and is denoted by

$$\bar{S}_Y(s) = [\bar{S}_Y(s)]^+ [\bar{S}_Y(s)]^-. \tag{13.184}$$

Hence, the desired stable, causal whitener is

$$G(f) = \frac{1}{[\bar{S}_Y(i2\pi f)]^+}, \tag{13.185a}$$

and due to symmetry,

$$G^*(f) = \frac{1}{[\bar{S}_Y(i2\pi f)]^-}, \tag{13.185b}$$

and therefore

$$\frac{1}{|G(f)|^2} = [\bar{S}_Y(i2\pi f)]^+ [\bar{S}_Y(i2\pi f)]^- = \bar{S}_Y(i2\pi f). \tag{13.186}$$

(Regarding notation, recall that, from (13.151), $[\bar{S}_Y]_+$ is obtained from an additive decomposition of $\bar{S}_Y$, whereas $[\bar{S}_Y]^+$ is obtained from a multiplicative decomposition of $\bar{S}_Y$. However, in both cases, the inverse transform vanishes for $\tau < 0$.) Observe that the hypothetical transfer function $T(f)$ discussed earlier is *one* of the nonunique factors $F(i2\pi f)$, but not necessarily the canonical factor.

Innovations Representation. The unique unit-intensity white process $Z(t)$ obtained from a process $Y(t)$ by a stable, causal, invertibly stable, and invertibly causal filter, is called the *innovations representation* for $Y(t)$. The reason for this term is that since $Z(t)$ is uncorrelated with all of its past, the information that $Z(t)$ provides at each instant is entirely new information —an *innovation*. The concept of an innovations representation plays a fundamental role in the theory of causal filtering, as illustrated in the next subsection, and also in Section 13.6. (See Section 11.9 for a discussion of the innovations representation for a discrete-time process.)

13.5.2 Solution of the Wiener-Hopf Equation

Given an invertible, causal, stable whitener $G(f)$, the observations $Y(t)$ can be transformed into the innovations process $Z(t)$, and then the optimum causal filter for $X(t)$, denoted by h_*, that uses the transformed observations $Z(t)$ can be sought:

$$\begin{aligned}\hat{X}_o(t) &= h_*(t) \otimes Z(t) \\ &= h_*(t) \otimes g(t) \otimes Y(t).\end{aligned} \tag{13.187}$$

In order for this two-step procedure to yield the optimum estimate of $X(t)$ corresponding to observations $Y(t)$, the function h_* must turn out to be

$$h_*(t) = h_o(t) \otimes g^{-1}(t), \tag{13.188}$$

since only then will it be true that

$$\begin{aligned}\hat{X}_o(t) &= h_*(t) \otimes g(t) \otimes Y(t) \\ &= h_o(t) \otimes g^{-1}(t) \otimes g(t) \otimes Y(t) \\ &= h_o(t) \otimes Y(t). \end{aligned} \tag{13.189}$$

It should be emphasized that if g had not been chosen such that g^{-1} is causal, then (13.188) could not be satisfied by a causal h_*. In other words, we are allowed to replace the original observations Y with transformed observations Z only if the admissible filters (viz., causal filters) can undo (invert) this transformation; otherwise the procedure of transforming Y to Z might discard useful information that could not be recovered with an admissible transformation.

Now, the Wiener-Hopf equation for h_* is

$$\int_0^\infty h_*(\tau) R_Z(t-\tau)\, d\tau = R_{XZ}(t) \qquad \forall t \geq 0. \tag{13.190}$$

But

$$R_Z(t-\tau) = \delta(t-\tau) \tag{13.191a}$$

and

$$R_{XZ}(t) = R_{XY}(t) \otimes g(-t), \tag{13.191b}$$

and therefore (13.190) reduces to

$$h_*(t) = R_{XY}(t) \otimes g(-t) \qquad \forall t \geq 0. \tag{13.192}$$

That is, $h_*(t)$ is the causal part of $R_{XY}(t) \otimes g(-t)$, which is expressed in the frequency domain by

$$H_*(f) = \left[S_{XY}(f) G^*(f)\right]_+. \tag{13.193}$$

It follows from (13.187) to (13.189) that

$$H_o(f) = G(f) H_*(f). \tag{13.194}$$

Thus,

$$H_o(f) = G(f)\left[S_{XY}(f) G^*(f)\right]_+, \tag{13.195}$$

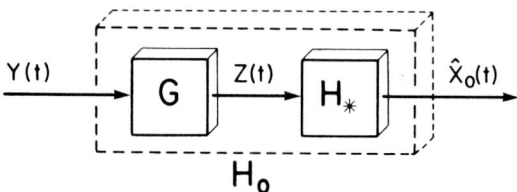

Figure 13.6 Causal Wiener Filter H_o, Decomposed into Whitening Filter G and Optimum Filter H_* for White Observations

from which (13.185) yields

$$H_o(f) = \frac{1}{[\bar{S}_Y(i2\pi f)]^+} \left[\frac{S_{XY}(f)}{[\bar{S}_Y(-i2\pi f)]^+} \right]_+ . \quad (13.196)$$

This is the transfer function of what is called the *causal Wiener filter*, in honor of Norbert Wiener's pioneering work [Wiener, 1949]. A block diagram of the causal Wiener filter, decomposed into the cascade connection of a whitening filter and the optimum filter for white observations, is shown in Figure 13.6.

Example 4: For the model in example 2 with $t_o = 0$,

$$S_{XY}(f) = S_S(f) = \frac{2\alpha a}{\alpha^2 + (2\pi f)^2} \quad (13.197)$$

and

$$S_Y(f) = S_S(f) + N_o = N_o \left[\frac{\gamma^2 + (2\pi f)^2}{\alpha^2 + (2\pi f)^2} \right], \quad (13.198)$$

where γ is a function of α, N_o, a. Therefore, it follows from (13.185) that

$$G(f) = N_o^{-1/2} \left[\frac{\alpha + i2\pi f}{\gamma + i2\pi f} \right], \quad (13.199)$$

which yields

$$\begin{aligned} S_{XY}(f)G^*(f) &= \left[\frac{2\alpha a}{\alpha^2 + (2\pi f)^2} \right] N_o^{-1/2} \left[\frac{\alpha - i2\pi f}{\gamma - i2\pi f} \right] \\ &= \frac{2\alpha a N_o^{-1/2}}{(\alpha + i2\pi f)(\gamma - i2\pi f)} \\ &= \frac{d_+}{\alpha + i2\pi f} + \frac{d_-}{\gamma - i2\pi f}, \quad (13.200) \end{aligned}$$

from which we obtain

$$[S_{XY}(f)G^*(f)]_+ = \frac{d_+}{\alpha + i2\pi f}. \tag{13.201}$$

Hence, (13.195), (13.199), and (13.201) yield

$$H_o(f) = \frac{N_o^{-1/2} d_+}{\gamma + i2\pi f}, \tag{13.202}$$

and therefore

$$h_o(t) = ce^{-\gamma t} u(t). \tag{13.203}$$

■

Signal in Additive White Noise. It can be shown (exercise 30) that, for the optimum causal filter for a signal in uncorrelated additive white noise,

$$\text{MSE}_o = N_o h_o(0^+). \tag{13.204}$$

Also, it can be shown that for a signal in uncorrelated additive white noise, (13.196) reduces to

$$H_o(f) = 1 - \frac{\sqrt{N_o}}{[N_o + \bar{S}_S(i2\pi f)]^+}. \tag{13.205}$$

Furthermore, (13.205) can be used to show that (13.204) can be reexpressed as

$$\text{MSE}_o = N_o \int_{-\infty}^{\infty} \ln\left[1 + \frac{S_S(f)}{N_o}\right] df. \tag{13.206}$$

Observe that the evaluation of (13.206) requires neither factorization nor additive decomposition of S_Y. Equations (13.205) and (13.206) are known as the *Yovits-Jackson formulas*, after the originators [Yovits and Jackson, 1955].

A theory of discrete-time Wiener filtering—both causal and noncausal—that is analogous to the continuous-time theory presented in Section 13.4 and this Section 13.5 can be developed by methods similar to those used here. The primary distinction is that the bilateral Z transform is used in place of the bilateral Laplace transform [Tretter, 1976; Anderson and Moore, 1979]. (See also Section 11.9.)

13.6 Kalman Filtering*

In this section an alternative approach to the formulation and solution of linear MMSE estimation problems is developed. The motivation is to obtain a characterization of the solution that is an alternative to the linear integral equation (13.101)—an alternative that is particularly useful for nonstationary processes. This is accomplished by restricting the class of estimation problems to those involving signals in additive noise, and by further restriction to signals exhibiting a specific dynamical structure. This particular approach, originally developed by Rudolf Emil Kalman [Kalman 1960; Kalman and Bucy, 1961], and others (see [Kailath, 1974]), has proven to be a significant complement to Wiener's approach (summarized in Section 13.5) for both continuous- and discrete-time problems, and especially for nonstationary processes. But even for stationary processes, Kalman's approach is a significant complement in that it provides recursive solutions, which are particularly amenable to computationally efficient implementation. For pedagogical reasons and for the sake of brevity, only the discrete-time case is considered here.

13.6.1 *Innovations Representation and Linear Prediction*

The innovations-representation (whitening) approach will be used to derive the Kalman filtering equations. This approach was introduced by Andrei Nikolaevich Kolmogorov [Kolmogorov, 1941], and exploited by Kalman [Kalman, 1960] for discrete time, and developed by Thomas Kailath [Kailath, 1968] for continuous time (cf. [Kailath, 1974]). As we shall see, the essence of the difference between the Kalman filtering approach and the Wiener filtering approach is that a time-domain method based on difference equations and correlation functions replaces the frequency-domain method based on rational spectral density functions for solving for the whitening filter. The various connections between these two approaches are discussed in [Kailath, 1974, 1981].

Given a sample of the past and present of the evolving process

$$Y(j) = X(j) + N(j), \qquad j = 0, 1, 2, \ldots, i, \qquad (13.207)$$

it is desired to estimate the present value $X(i)$ of the signal process X. The noise process N is independent of X, has zero mean, and is stationary and white:

$$R_N(i) = N_o \delta_i, \qquad (13.208)$$

---

*The convention of denoting Euclidean vectors with boldface italic that was abandoned in Section 3.2 is readopted in this section, as it was in Section 3.3.

where δ_i is the Kronecker delta

$$\delta_i = \begin{cases} 1, & i = 0, \\ 0, & i \neq 0. \end{cases}$$

The linear MMSE one-step noise-free predictor of the observations $Y(i)$ is denoted by $\hat{Y}(i|i-1)$, and the linear MMSE one-step noisy predictor of the signal $X(i)$ is denoted by $\hat{X}(i|i-1)$. Both predictors are based on observations $\{Y(j)\}_0^{i-1} = \{Y(0), Y(1), \ldots, Y(i-1)\}$. It can be shown that

$$\hat{Y}(i|i-1) = \hat{X}(i|i-1). \tag{13.209}$$

This follows (exercise 34) from the necessary and sufficient orthogonality conditions

$$E\{[\hat{Y}(i|i-1) - Y(i)]Y(i-j)\} = 0, \quad j = 1, 2, \ldots, i, \tag{13.210}$$

$$E\{[\hat{X}(i|i-1) - X(i)]Y(i-j)\} = 0, \quad j = 1, 2, \ldots, i. \tag{13.211}$$

[Note that (13.209) is equivalent to $\hat{N}(i|i-1) = 0$.] The noise-free prediction error process is denoted by

$$Z(i) = Y(i) - \hat{Y}(i|i-1). \tag{13.212}$$

It follows (exercise 35) from the orthogonality condition that $Z(i)$ is white (but, in general, nonstationary):

$$R_Z(i, j) = R_Z(i, i)\delta_{i-j}. \tag{13.213}$$

Let $f(\cdot, \cdot)$ denote the time-variant impulse-response function for the predictor

$$\hat{Y}(i|i-1) = \sum_{j=0}^{i-1} f(i, j) Y(j). \tag{13.214}$$

Then (13.212) and (13.214) yield

$$Z(i) = \sum_{j=0}^{i} g(i, j) Y(j), \tag{13.215}$$

$$g(i, j) = \delta_{i-j} - f(i, j) u_{i-j-1}, \tag{13.216}$$

where u_i is the unit-step sequence

$$u_i = \begin{cases} 1 & i \geq 0, \\ 0 & i < 0. \end{cases} \tag{13.217}$$

Thus, $g(i, j) = 0$ for $j > i$; hence, the white process Z is obtained from the observations Y with a linear *causal* transformation. Furthermore, since $\{g(k, j): j, k = 0, 1, 2, \ldots, i\}$, with $g(k, j) = 0$ for $j > k$, can be interpreted as the elements of an $(i + 1) \times (i + 1)$ lower triangular matrix, and since the inverse of such a matrix is another lower triangular matrix, then the transformation from Y to Z is invertibly causal. Therefore, Z is the innovations representation for Y (cf. Section 11.9). Consequently, the desired causal signal estimate, denoted by $\hat{X}(i|i)$, can be obtained from Z by a linear causal transformation,

$$\hat{X}(i|i) = \sum_{j=0}^{i} h_*(i, j) Z(j) \qquad (13.218)$$

[parallel to (13.187)]. Thus, $\hat{X}(i|i)$ can be obtained from Y, using the composition of the two transformations g and h_* in (13.215) and (13.218).

Since $\{Z(j)\}_0^i$ is an orthogonal set of random variables, then the orthogonal projection of $X(i)$ onto the $(i + 1)$-dimensional linear space spanned by $\{Z(j)\}_0^i$ is the sum of orthogonal projections of $X(i)$ onto each of the $i + 1$ one-dimensional linear spaces spanned by each of the $i + 1$ random variables $Z(j)$ (this can be visualized geometrically as the equivalence between the projection onto a plane and the sum of the projections onto the orthogonal coordinate axes of the plane). It follows that

$$\hat{X}(i|i) = \hat{X}(i|i - 1) + k(i) Z(i), \qquad (13.219)$$

where $k(i)$ is the scalar that orthogonally projects $X(i)$ onto the space spanned by the current innovation, namely (exercise 36)

$$k(i) \triangleq R_Z^{-1}(i, i) R_{XZ}(i, i), \qquad (13.220)$$

and $\hat{X}(i|i - 1)$ is the orthogonal projection of $X(i)$ onto the space spanned by the preceding $i - 1$ innovations. But (13.209) and (13.212) yield

$$Z(i) = Y(i) - \hat{X}(i|i - 1), \qquad (13.221)$$

with which (13.219) can be reexpressed as

$$\hat{X}(i|i) = [1 - k(i)] \hat{X}(i|i - 1) + k(i) Y(i). \qquad (13.222)$$

We can conclude that the filtered estimate $\hat{X}(i|i)$ can be obtained directly from the predicted estimate $\hat{X}(i|i - 1)$ and the current observation $Y(i)$. To obtain a recursive solution for the predicted estimate, we consider the class of signal processes X that are generated recursively.

13.6.2 State-Variable Model

It is assumed that the signal process X can be modeled as the response of a finite-order linear system, possible time-variant, characterized by an nth-order difference equation, driven by white noise V. Thus X can be characterized by the following recursive state equations as discussed in Section 9.4.1:

$$X(i) = c(i)W(i),$$
$$W(i+1) = A(i)W(i) + b(i)V(i), \quad (13.223)$$

where $W(i)$ is the n-dimensional state vector, $A(i)$ is the $n \times n$ state feedback matrix, $b(i)$ is the n-dimensional input column matrix, and $c(i)$ is the n-dimensional output row matrix. In terms of the state transition matrix $\Phi(i, j)$, we have

$$A(i) = \Phi(i+1, i). \quad (13.224)$$

This signal model (13.223) is said to be *recursive* because the next signal state, $W(i+1)$, is determined by the current signal state $W(i)$ and a perturbation $b(i)V(i)$. Thus, the signal state repeatedly recurs. A vector-signal-flow diagram of the signal-generating mechanism is shown in Figure 13.7. Since the process X is uncorrelated with the process N in (13.207), $W(0)$ and V in the model for X must be taken to be uncorrelated with N. In addition, $W(0)$ is taken to be uncorrelated with V. The white process V has zero mean and covariance function

$$R_V(i, j) = \delta_{i-j}. \quad (13.225)$$

This state-variable model for the signal process arises naturally in many practical problems, such as control-system problems for which Y represents noisy measurements on the output X of a control system described by the matrices A, b, c. From another point of view, this state-variable signal model can be justified by observing that (as we discovered in Section 13.3.3) the linear MMSE estimator is completely determined by only the auto- and cross-correlation functions for the signal being estimated, X, and the observations Y; consequently, the only features of a signal model that affect

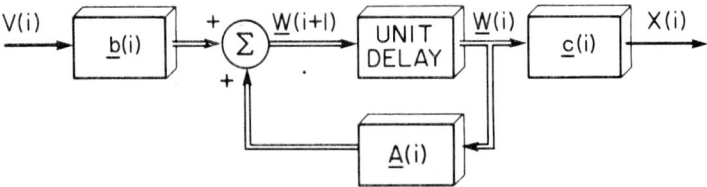

Figure 13.7 Signal-Flow Diagram for Signal-Generating Mechanism for Kalman Filtering

the estimator solution per se are these correlation functions—and it can be shown that many correlation functions of practical interest can be accurately approximated using state-variable signal models. For example, the problem of determining a linear system (e.g., filter) which, when driven by a white process, produces an output process with a specified autocorrelation (e.g., inverse transform of a specified spectral density for a stationary process) is a generalization of the problem of determining the inverse of a whitening filter. Consequently, for stationary processes with rational spectral densities, the spectral factorization method developed in Section 13.5.1 provides a solution to the problem of determining a state-variable model for a signal with a given rational spectral density. For nonstationary processes, there are generalizations of the spectral factorization method that are applicable [Kailath, 1974; Anderson and Moore, 1979].

Since the signal $X(i)$ is a linear transformation, (13.223), on the state $W(i)$, then (exercise 37) the linear MMSE predicted estimate of $X(i + 1)$ can be obtained from the linear MMSE predicted estimate of $W(i + 1)$ (based on the same observations) by the same linear transformation,

$$\hat{X}(i + 1|i) = c(i + 1)\hat{W}(i + 1|i). \tag{13.226}$$

Consequently, we focus on the underlying problem of predicting the state $W(i + 1)$. It follows (exercise 38) from the orthogonality condition that the predicted state estimate is given by

$$\hat{W}(i + 1|i) = \sum_{j=0}^{i} h_*(i, j) Z(j), \tag{13.227}$$

where h_* [which is not directly related to h_* in (13.218)] is specified by

$$\sum_{j=0}^{i} R_Z(k, j) h_*(i, j) = R_{WZ}(i + 1, k), \quad k = 0, 1, 2, \ldots, i.$$

$$\tag{13.228}$$

Equations (13.227), (13.228), and (13.213) yield

$$\hat{W}(i + 1|i) = \sum_{j=0}^{i} R_Z^{-1}(j, j) R_{WZ}(i + 1, j) Z(j). \tag{13.229}$$

13.6.3 Kalman Recursions

With the use of the state equations (13.223), (13.229) can be manipulated (exercise 39) into the following recursive form, called the *Kalman state-predictor recursion*:

$$\boxed{\hat{W}(i+1|i) = A(i)\hat{W}(i|i-1) + k(i+1,i)Z(i),} \quad (13.230)$$

where $k(\cdot,\cdot)$ is the *Kalman-prediction gain vector*

$$k(i+1,i) \triangleq R_Z^{-1}(i,i)R_{WZ}(i+1,i), \quad (13.231)$$

and where the innovations sequence $Z(i)$ is [from (13.207), (13.221), (13.223), (13.226)] given by

$$\boxed{Z(i) = Y(i) - c(i)\hat{W}(i|i-1)} \quad (13.232a)$$

or

$$Z(i) = c(i)\tilde{W}(i|i-1) + N(i), \quad (13.232b)$$

where $\tilde{W}(i|i-1)$ is the *state prediction error*

$$\tilde{W}(i|i-1) = W(i) - \hat{W}(i|i-1). \quad (13.233)$$

Using (13.232b), the Kalman gain vector can be expressed as (exercise 40)

$$\boxed{k(i+1,i) = A(i)k(i),} \quad (13.234)$$

where

$$\boxed{k(i) = \frac{R_{\tilde{W}}(i|i-1)c^T(i)}{c(i)R_{\tilde{W}}(i|i-1)c^T(i) + N_o}, \quad i \geq 1,} \quad (13.235a)$$

in which $R_{\tilde{W}}(i|i-1)$ is the correlation matrix of the state prediction error. The following recursion for this correlation matrix is derived in exercise 42:

$$\boxed{\begin{aligned} R_{\tilde{W}}(i|i-1) = &\, A(i-1)R_{\tilde{W}}(i-1|i-1)A^T(i-1) \\ &+ b(i-1)b^T(i-1), \quad i \geq 1, \end{aligned}}$$

$$(13.235b)$$

$$\boxed{\begin{aligned} R_{\tilde{W}}(i-1|i-1) = &\, R_{\tilde{W}}(i-1|i-2) \\ &- \frac{R_{\tilde{W}}(i-1|i-2)c^T(i-1)c(i-1)R_{\tilde{W}}(i-1|i-2)}{N_o + c(i-1)R_{\tilde{W}}(i-1|i-2)c^T(i-1)}, \\ & \quad i \geq 2. \end{aligned}}$$

$$(13.235c)$$

In these recursions, the initial value $R_{\tilde{W}}(0|0)$ must be given (cf. example 5).

Paralleling the argument leading to (13.219), the following relationship between the filtered and predicted state estimates can be obtained (directly):

$$\hat{W}(i|i) = \hat{W}(i|i-1) + k(i)Z(i), \qquad (13.236)$$

where $k(\cdot)$ is the *Kalman-filtering gain vector*

$$k(i) \triangleq R_Z^{-1}(i,i) R_{WZ}(i,i), \qquad (13.237)$$

which can be reexpressed by (13.235a). Substitution of (13.236) [solved for $\hat{W}(i|i-1)$] into (13.230) yields (exercise 41)

$$\hat{W}(i+1|i) = A(i)\hat{W}(i|i). \qquad (13.238)$$

Thus, a future prediction $\hat{W}(i+1|i)$ can be obtained from a current filtered estimate $\hat{W}(i|i)$ simply by linear transformation $A(i)$. However, a current filtered estimate $\hat{W}(i|i)$ cannot be obtained from a current prediction $\hat{W}(i|i-1)$ without using the current innovation $Z(i)$ in (13.236). Substitution of (13.238) (with i replaced by $i-1$) into (13.236) directly yields the following recursive form, called the *Kalman state-filter recursion*:

$$\boxed{\hat{W}(i|i) = A(i-1)\hat{W}(i-1|i-1) + k(i)Z(i).} \qquad (13.239)$$

Notice that the recursion (13.239) for the filtered estimate is analogous to the recursion (13.230) for the predicted estimate. The only difference is in the gain vectors and time indices.

Parallel to (13.226), the filtered estimate of the signal $X(i)$ can be obtained from the filtered estimate of the state simply by the linear transformation

$$\hat{X}(i|i) = c(i)\hat{W}(i|i). \qquad (13.240)$$

The filtering recursion (13.239), together with the relations (13.232), (13.238), and (13.240), is described by the vector-signal-flow diagram shown in Figure 13.8. Note that the filtering system is simply the signal-generating system shown in Figure 13.7, modified by state feedback through $c(i)$ to the

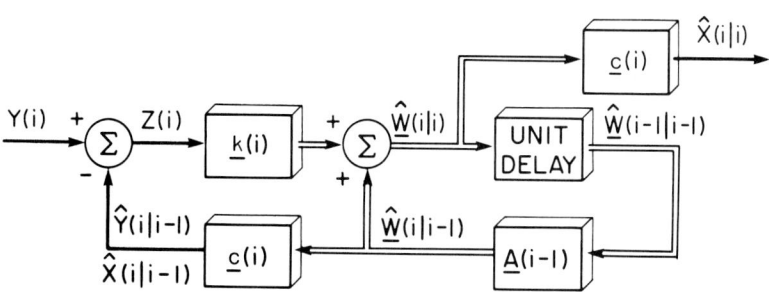

Figure 13.8 Signal-Flow Diagram for Kalman Filter

input, with the input matrix $b(i)$ replaced by the Kalman gain matrix $k(i)$, with the output taken through $c(i)$ before the delay (rather than after), and with the feedback matrix $A(i)$ delayed one time unit. Note also that this realization of the filtering system, with unit-pulse response h, is not a tandem connection of two separate systems—the whitening system with unit-pulse response g in (13.215), and the optimum filter (based on the whitened observations Z) with unit-pulse response h_* in (13.218). Since both systems, g and h_*, are simple modifications of the state-estimation system, they can be combined into a single system, namely that shown in Figure 13.8. (Of course this single system h is *equivalent* to a tandem connection of the two subsystems g and h_*.)

Generalizations. This state-variable approach to recursive estimation can be generalized in several ways (cf. [Kailath, 1974], [Anderson and Moore, 1979], [Kailath, 1981]). The additive uncorrelated white-noise model (13.207) for the observations can be generalized to account for signal distortion, nonwhite noise, and correlation between signal and noise. However, in order to retain the recursive nature of the estimator solution, the signal-distortion mechanism and noise-coloring mechanism must both be modeled using finite-order linear difference equations, and the correlation between signal and noise must be causal; that is, the current signal can depend on only past noise, not current or future noise. Such causal correlation occurs in feedback systems for which measurement noise at the system output is fed back to the system input.

The linear state-variable model (13.223) also can be generalized to a nonlinear state-variable model for which the linear state transformations $A(i)$ and $c(i)$ are replaced with nonlinear transformations. Then by linearizing the model about the current state estimate $\hat{W}(i|i)$, one can obtain an approximate state estimator that is specified by a nonlinear recursion, called the *extended Kalman filter* [Anderson and Moore, 1979].

Equivalence between Kalman and Wiener Filtering. Since the Kalman filtering recursion is a solution to a causal linear MMSE estimation problem, it must agree with the causal Wiener filter for the same problem. But the Wiener filtering problem was formulated as a *steady-state* estimation problem, that is, the desired signal X and observations Y were assumed to exist for all time and to be stationary; whereas the Kalman filtering problem is formulated with a finite starting time. Nevertheless, if the white process $V(i)$ that generates the signal is stationary and the state-variable signal-generating system is time-invariant (A, b, c are constant), then the Kalman filter will exhibit transient behavior [in the Kalman gain vector $k(i)$] that dies away, yielding a time-invariant steady-state filter that is identical to the Wiener filter (for the same desired-signal and observation correlation functions). Similar equivalences exist for other estimation problems, including prediction (extrapolation) and smoothing (interpolation).

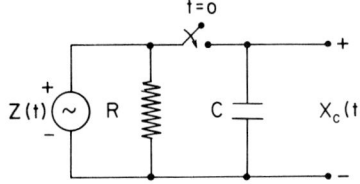

Figure 13.9 Circuit Diagram for Example 5

Applications. Applications of Kalman filtering are diverse. The advantages of recursive computation make this approach to estimation very attractive. A particularly important application is to problems of identification and control of dynamical systems. The dynamical system model inherent in the Kalman filtering approach is natural for applications involving the tracking of moving targets. This includes various problems in navigation using radar and sonar as well as video tracking and guidance and control of vehicles, ballistic missiles, and many other dynamical systems. The reader is referred to the following reference books for extensions, generalizations, and applications: [Leondes, 1970; Gelb, 1974; Schwartz and Shaw, 1975; Tretter, 1976; Srinath and Rajasekaran, 1979; Maybeck, 1979; Anderson and Moore, 1979; Kailath, 1981; Brown, 1983; and Ljung, 1987].

Example 5: Consider the resistive-capacitive circuit shown in Figure 13.9. The voltage $X_c(t)$ across the capacitor is governed by the differential equation

$$\frac{dX_c(t)}{dt} + \frac{1}{RC} X_c(t) = \frac{1}{C} Z_c(t), \qquad t \geq 0, \qquad (13.241a)$$

for which $Z_c(t)$ is a thermal-noise current (a white process) with spectral density N_*. It is desired to measure the voltage, but the measurement instrument introduces thermal noise $N_c(t)$ and thereby produces the noisy measurement

$$Y_c(t) = X_c(t) + N_c(t), \qquad (13.242a)$$

for which $N_c(t)$ has spectral density N_o. Let us consider the problem of obtaining the MMSE estimate of $X_c(t)$ based on the noisy measurement $Y_c(t)$. Continuous-time Kalman filtering theory could be applied directly to this problem. However, since it is desired to implement the filtering algorithm on a digital computer, the equations will be discretized, and the discrete-time Kalman filtering theory will be applied. It can be shown [Kailath, 1980] that an appropriate discrete-time model is given by

$$W(i+1) = aW(i) + bV(i), \qquad (13.241b)$$
$$Y(i) = X(i) + N(i), \qquad (13.242b)$$
$$X(i) = cW(i), \qquad (13.243)$$

for $i \geq 0$, where

$$a = e^{-\Delta/RC},$$
$$b = R(1 - e^{-\Delta/RC})\sqrt{N_*},$$
$$c = 1, \qquad (13.244)$$

and $N(i)$ and $V(i)$ are zero-mean stationary white-noise processes with variances of N_o and 1, respectively, and Δ is the sampling increment. The state $W(i) = X(i)$ of this system represents the capacitor voltage. Let us determine the Kalman filter for $\hat{W}(i|i)$. Equation (13.239) yields

$$\hat{W}(i|i) = a\hat{W}(i-1|i-1) + k(i)Z(i), \qquad (13.245)$$

where $Z(i)$ is, from (13.232a) and (13.238),

$$Z(i) = Y(i) - ac\hat{W}(i-1|i-1). \qquad (13.246)$$

Substitution of (13.246) into (13.245) yields

$$\hat{W}(i|i) = a[1 - k(i)c]\hat{W}(i-1|i-1) + k(i)Y(i). \qquad (13.247)$$

The gain sequence $k(i)$ is given by (13.235a), which reduces to

$$k(i) = \frac{cR_{\tilde{W}}(i|i-1)}{N_o + c^2 R_{\tilde{W}}(i|i-1)}, \qquad i \geq 1, \qquad (13.248)$$

and the state prediction error correlation is given by (13.235b) and (13.235c), which reduce to

$$R_{\tilde{W}}(i|i-1) = a^2 R_{\tilde{W}}(i-1|i-1) + b^2, \qquad i \geq 1, \qquad (13.249a)$$

where

$$R_{\tilde{W}}(i-1|i-1) = R_{\tilde{W}}(i-1|i-2) - \frac{c^2 R_{\tilde{W}}^2(i-1|i-2)}{N_o + c^2 R_{\tilde{W}}(i-1|i-2)},$$
$$i \geq 2. \qquad (13.249b)$$

Substitution of (13.249b) into (13.249a) yields

$$R_{\tilde{W}}(i|i-1) = \frac{a^2 N_o R_{\tilde{W}}(i-1|i-2)}{N_o + c^2 R_{\tilde{W}}(i-1|i-2)} + b^2, \quad i \geq 2. \tag{13.250}$$

Thus, (13.247), (13.248), and (13.250) provide a set of explicit update equations for $\hat{W}(i|i)$. However, to start off these recursions, the initial estimate $\hat{W}(0|0)$ must be specified. Then the initial prediction error is given by

$$R_{\tilde{W}}(1|0) = a^2 R_{\tilde{W}}(0|0) + b^2$$
$$= a^2 [R_W(0) - \hat{W}^2(0|0)] + b^2. \tag{13.251}$$

It is shown in exercise 43 that for this scalar problem the equations (13.248) and (13.250) can be solved to obtain the explicit recursion

$$k(i) = \frac{a^2 k(i-1) + b^2 c/N_o}{1 + (bc)^2/N_o + a^2 ck(i-1)}, \quad i \geq 2, \tag{13.252}$$

for the gain sequence, and it is also shown that, since $W(0) = 0$, then

$$k(1) = \frac{cb^2}{N_o + (cb)^2}. \tag{13.253}$$

The steady-state gain value k_∞ can be obtained from (13.252) by use of $k(i) = k(i-1) = k_\infty$. The result is easily shown to be the solution to a quadratic algebraic equation. The corresponding steady-state Kalman filter is given by

$$\hat{W}(i|i) = a(1 - k_\infty c)\hat{W}(i-1|i-1) + k_\infty Y(i). \tag{13.254}$$

This first-order, linear, time-invariant difference equation can be solved, and the steady-state solution is

$$\hat{W}(i|i) = h(i) \otimes Y(i), \tag{13.255}$$

where the Kalman-filter impulse response is given by

$$h(i) = k_\infty [a(1 - k_\infty c)]^i, \quad i \geq 0. \tag{13.256}$$

It can be shown that this is identical to the discrete-time causal Wiener filter obtained by the method of spectral factorization. ■

13.7 *Optimum Periodically Time-Variant Filtering*

The Kalman filtering approach to optimum filtering applies to nonstationary as well as stationary processes. For example, it applies to cyclostationary processes. For a cyclostationary signal process, the state-variable model exhibits periodically time-variant parameters A, b, and c. Thus, the Kalman filter also exhibits periodically time-variant parameters A, b, and c, and an asymptotically periodic gain, k (cf. [Rootenberg and Ghozati, 1978]). However, there are alternative frequency-domain approaches to optimum filtering that are analogous to the Wiener filtering approach for stationary processes, and for the problem of noncausal filtering these frequency-domain approaches can be especially tractable, and can therefore provide considerable insight into optimum periodically time-variant filtering. For a number of cyclostationary signal models of interest, these approaches yield explicit solutions that can be directly interpreted in terms of modulation-demodulation, multiplexing-demultiplexing, sampling-interpolation, and time-invariant Wiener filtering operations [Gardner and Franks, 1975] (see also [Ericson, 1981; Graef, 1983]). An obvious approach to the problem of optimum noncausal filtering of cyclostationary processes is to use one of the representations, described in Section 12.6, for a scalar-valued cyclostationary process in terms of a vector-valued (possibly infinite-dimensional for a continuous-time process) stationary process, and to then apply multivariate (vector) Wiener filtering theory. Several such representations have been exploited for this purpose, and the reader is referred to the literature [Gardner and Franks, 1975]. An alternative approach, which is described in Section 12.8.5, is based on the representation of the autocorrelation function in terms of cyclic autocorrelations, and yields a filter design equation for the optimum system function in terms of cyclic spectra.

13.8 *Summary*

The geometry of vectors in three-dimensional Euclidean space can be extended not only to vectors in higher-dimensional Euclidean space, but also to functions interpreted as vectors, including functions of real numbers, as well as functions of points in an abstract point set, such as random variables which are functions of sample points. This allows the correlation $E\{XY\}$ of two random variables, X and Y, to be interpreted as an inner product of two vectors. One of the most important geometrical constructs that can be extended from Euclidean space to the space of random variables is the *orthogonal projection theorem* and its associated *orthogonality condition* for minimum-norm-error estimation—both linear and nonlinear. This orthogonality condition is a powerful tool in the derivation and interpretation of estimator design equations, performance formulas, and recursive algorithms. This is demonstrated for nonprobabilistic linear least-squares estimation, as well as probabilistic nonlinear estimation, and probabilistic linear estimation including continuous-time Wiener filtering and discrete-time Kalman filtering.

For optimum causal filtering, including both steady-state Wiener filtering and transient-plus-steady-state Kalman filtering, the concept and associated methods of the *innovations representation* of a process play a fundamental role, as does the intimately related concept and techniques for *spectral factorization*. In addition to their part in Wiener and Kalman filtering theory and method, they play a fundamental role in autoregressive modeling and linear prediction, as explained in Section 11.9, as well as in autoregressive spectrum estimation, as explained in [Gardner, 1987a].

This chapter provides only an introduction to a highly developed and broad subject area with diverse applications. Topics not covered include continuous-time Kalman filtering theory, both continuous-time and discrete-time Kalman smoothing (noncausal filtering) theory, and discrete-time Wiener filtering, smoothing, and prediction theory. Also, the increasingly important topic of computationally efficient and/or VLSI-implementable algorithms for filtering, smoothing, and prediction is not covered (except for a brief introduction in Section 11.9). Nevertheless, this chapter does provide a sound treatment of the fundamentals that underly these various topics. Supplementary treatments can be found in the literature cited throughout this chapter.

EXERCISES

1. Show that each subspace of three-dimensional Euclidean space is either a plane through the origin or a line through the origin, and that there are infinitely many such subspaces.

2. Determine the solution $\hat{H}_o$ to (13.19a) for the particular subspace M containing all vectors of the form $\hat{X} = \{\alpha_1, \alpha_3 - \alpha_1, \alpha_3\}$.

3. Verify that $X = \cos 2\pi t$ and $Y = \sin 2\pi t$ are orthogonal for $T = [0, 1]$. Are they orthogonal for all finite intervals T?

4. Find the orthogonal projection of the function $X = \alpha(t)$, which is a rectangle of unity height and width π, onto the space M consisting of all functions of the form
$$\hat{X} = \hat{\alpha}(t) = \sigma \cos t + \gamma \sin t \qquad \forall t \in [0, \pi]$$
for all σ, γ.

5. **Woodbury's identity.** Let A be a nonsingular $n \times n$ matrix, and let b and c^T be n-dimensional column vectors. Verify *Woodbury's identity*

$$[A + bc]^{-1} = A^{-1} - \frac{A^{-1}bcA^{-1}}{1 + cA^{-1}b}. \qquad (13.257)$$

6. Use (13.39) to verify (13.45b).

7. **Pseudoinverse matrix.** Verify that the solution to the least-squares estimation problem (13.32) to (13.34) can be reexpressed as

$$H_o = M^{(-1)}X, \qquad (13.258)$$

where

$$M^{(-1)} \triangleq [M^T M]^{-1} M^T. \qquad (13.259)$$

$M^{(-1)}$ is referred to as the $m \times l$ *pseudoinverse* of the $l \times m$ matrix M defined by (13.36). Observe that if $m = l$ and M is invertible, the pseudoinverse becomes the true inverse and it can be shown that the estimation error is zero. Also, observe that the least-squares estimation problem (13.29), whose solution is (13.32) to (13.34), can be reinterpreted as a problem of finding the best approximate solution H to a set of l inconsistent simultaneous linear equations in m unknowns, expressed in vector form as

$$MH \neq X;$$

that is,

$$\min_H \|X - \hat{X}\|,$$

where

$$\hat{X} = MH.$$

This is the reason for calling $M^{(-1)}$ the *pseudoinverse* of M.

8. Prove that the definitions (13.8d) and (13.9d) satisfy (13.12) to (13.15).

9. Verify that (13.56) yields the desired minimum of (13.52). Do this as follows:
   (a) Add and subtract $\hat{X}_o = E\{X|Y\}$ in $E\{(X - \hat{X})^2\} = \text{MSE}$.
   (b) Use

$$E\{(X - E\{X|Y\})g(Y)\} = 0 \quad \forall g(\cdot)$$

[as proved by (13.55) to (13.60)] to obtain the expression

$$\text{MSE} = E\{(X - E\{X|Y\})^2\} + E\{(\hat{X} - E\{X|Y\})^2\}.$$

Then conclude that MSE is minimized by $\hat{X} = E\{X|Y\}$.

10. Verify (13.66). Then verify (13.83)-(13.84) and (13.90)-(13.91).

11. *Linear conditional mean.* Let X and Y be jointly Gaussian random variables. Show that

$$E\{X|y\} = h_o y + f_o \qquad (13.260)$$

for some constants h_o and f_o, and show that if X and Y have zero means, then $f_o = 0$. Hence, the (unconstrained) MMSE estimator for jointly Gaussian variables is a linear estimator.

12. *Wide-sense conditional mean.* Denote the random variable $f_{X|Y}(x|Y)$ by Z, and denote the generalized ($f_1 \neq 0$) linear MMSE estimate of Z, based on the observation Y, by $\hat{Z} \equiv \hat{f}_{X|Y}(x|Y) = h_1 Y + f_1$. Use the orthogonality condition

$$E\{(\hat{Z} - Z)(hY + f)\} = 0 \qquad \text{for all } h, f$$

to show that*

$$\hat{f}_{X|Y}(x|Y) = f_X(x)\left[1 + \frac{(E\{Y|x\} - m_Y)}{\sigma_Y^2}(Y - m_Y)\right].$$

(13.261)

Let $\hat{E}\{X|Y\}$ denote the integral

$$\hat{E}\{X|Y\} \triangleq \int_{-\infty}^{\infty} x \hat{f}_{X|Y}(x|Y)\,dx. \qquad (13.262)$$

Show that

$$\hat{E}\{X|Y\} = m_X + \frac{K_{XY}}{\sigma_Y^2}(Y - m_Y), \qquad (13.263)$$

and verify that this is the generalized linear MMSE estimate of X based on the observation Y,

$$\hat{E}\{X|Y\} = \hat{X} = h_o Y + f_o. \qquad (13.264)$$

Finally, use the result of exercise 11 to show that if X and Y are jointly Gaussian, then

$$\hat{E}\{X|Y\} = E\{X|Y\}. \qquad (13.265)$$

Note: The linear MMSE approximation $\hat{E}\{X|y\}$ to the conditional mean $E\{X|y\}$ is called the *wide-sense conditional mean* [Doob, 1953, Chapter II]. As explained in Section 13.3.1, $E\{X|Y\}$ is the orthogonal

---

*The notation $\hat{f}$ here denotes a MMSE estimate of a probability density, whereas in other chapters $\hat{f}$ denotes a fraction-of-time probability density.

projection of X onto the linear space of all finite-mean-square functions of Y. By comparison, $\hat{E}\{X|Y\}$ is the orthogonal projection of X onto the linear space of all *linear* functions of Y (cf. [Doob, 1953, Chapter IV, para. 3; Chapter VIII, para. 1]). Application of linear MMSE estimation of conditional densities $f_{X|Y}(x|Y)$ (and conditional discrete distributions) to the design of structurally constrained receivers for signal detection and estimation is treated in [Gardner, 1976b].

13. Let X and Y be jointly Gaussian with zero means. Show that the unconditional and conditional MSE are identical; that is, for all y, we have

$$E\{(X - \hat{X})^2 | y\} = E\{(X - \hat{X})^2\}, \qquad (13.266)$$

where

$$\hat{X} = E\{X | y\}.$$

This indicates that no matter what the measurement y is, the error, averaged over all x, is the same. Hence, there is no such thing as a *bad measurement* y (although there can be a *bad sample* x in the sense that $(x - \hat{x})^2 \gg E\{(X - \hat{X})^2\}$).

14. (a) Derive (13.96) from (13.94).
    (b) Verify that (13.95) and (13.96) do indeed minimize MSE by adding $\hat{X}_o$ to and subtracting it from $X - \hat{X}$ to obtain the form

    $$\text{MSE} = \text{MSE}_o + \Delta,$$

    where $\Delta \geq 0$, and $\Delta = 0$ if and only if $\hat{X} = \hat{X}_o$.

    Hint: Use (13.94).

15. Verify that (13.109) follows from (13.101). Similarly, verify that (13.113) follows from (13.102).

16. Prove that if $h_o(t, u)$ is the optimum noncausal filter for $X(t)$ specified by (13.101) and (13.105), then the optimum noncausal filter for $X'(t) \triangleq X(t) \otimes g(t)$ is

    $$h'_o(t, u) = h_o(t, u) \otimes g(t)$$
    $$= \int_{-\infty}^{\infty} h_o(t - v, u) g(v) \, dv$$

    for any function $g(t)$.

17. *Coherence measurement.* Show that the magnitude of the coherence function (13.116) can be factored as

    $$|\rho(f)|^2 = H_{X|Y}(f) H_{Y|X}(f), \qquad (13.267)$$

where $H_{X|Y}(f)$ [or $H_{Y|X}(f)$] is the noncausal Wiener filter for $X(t)$ [or $Y(t)$], given observations $Y(t)$ [or $X(t)$]. This characterization of $|\rho(f)|$ can be exploited in the development of techniques for measuring $|\rho(f)|$, whereby $H_{X|Y}$ and $H_{Y|X}$ are determined using adaptive filtering techniques (e.g., the recursive least-squares algorithm described in Section 13.2.2).

18. Verify (13.130) and (13.131).
19. *Equivalent linear model.*
    (a) Verify that for a given $X(t)$ and $Y(t)$, the function $g(\cdot)$ defined by (13.132a) minimizes the mean squared value of $N(t)$ defined by (13.132b). Consequently, among the infinite number of functions g and corresponding residuals N that render the model (13.119)-(13.120) valid for two given processes Y and X, the particular g defined by (13.132a) yields the closest match of $g \otimes X$ to Y.
    (b) Verify (13.133). Then verify that the definitions (13.132a) and (13.132b) are consistent with the model (13.119)-(13.120) in the sense that
    $$S_Y = |G|^2 S_X + S_N.$$
    (c) As an application of linear modeling to system identification, consider the following. Let $Y(t)$ be the response of an unknown system, which is possibly nonlinear and/or time-variant, to the excitation $X(t)$. Show that (13.132a) is the transfer function of the *best-fitting* linear time-invariant system, in the sense that the response of this system,
    $$\hat{Y}(t) = X(t) \otimes g(t),$$
    to the excitation $X(t)$ is closer, in the mean-square sense, to the response $Y(t)$ of the unknown system, than is the response of any other linear time-invariant system. This suggests a robust method for identification and modeling of unknown systems. (See the results of exercise 24 for means of extending the result of this exercise beyond the framework of WSS processes.)
20. *SNR gain.* Investigate the behavior of D, A, MSE_S, MSE_N, and SNR_o in (13.128) to (13.131) for $r \gg 1$ and $r \ll 1$. Also show that, for the model (13.119), the PSD SNR at the output of any linear time-invariant filter [with $H(f) \neq 0$] is the same as the PSD SNR at the input; contrast this (for $r \gg 1$) with the output SNR, SNR_o given by (13.131), relative to the input SNR, defined by

$$\text{SNR}_i \triangleq \frac{E\{[g(t) \otimes S(t)]^2\}}{E\{[N(t)]^2\}} = \frac{\int_{-\infty}^{\infty} |G(f)|^2 S_S(f)\, df}{\int_{-\infty}^{\infty} S_N(f)\, df}.$$

(13.268)

21. *Narrowbandpass filtering.* Let Y and S be related by

$$Y(t) = S(t) + N(t),$$

where

$$R_{SN}(\tau) \equiv 0,$$
$$R_N(\tau) = N_o \delta(\tau),$$
$$R_S(\tau) = ae^{-\alpha|\tau|}\cos(2\pi f_o \tau).$$

Show that if $S(t)$ is narrowband (i.e., $\alpha \ll 2\pi f_o$), then to a close approximation,

$$H_o(f) \simeq \frac{b}{\left[\beta^2 + (2\pi)^2(f - f_o)^2\right]}, \qquad f \geq 0,$$

$$H_o(f) = H_o(-f) \qquad \text{for all } f,$$

$$\text{MSE}_o \simeq \frac{a}{[1 + a/\alpha N_o]^{1/2}}$$

for appropriate values of b and β. Thus, the center frequency of the filter is equal to the carrier frequency of the signal; but since $\beta > \alpha$, the bandwidth of the filter exceeds that of the signal.

Hint: Use $\alpha \ll 2\pi f_o$ to justify the approximation

$$\left[\alpha^2 + (2\pi)^2(f - f_o)^2\right]^{-1} + \left[\alpha^2 + (2\pi)^2(f + f_o)^2\right]^{-1}$$
$$\simeq \left[\alpha^2 + (2\pi)^2(f - f_o)^2\right]^{-1}$$

for $f \geq 0$ (draw a picture).

22. *Filter performance for white noise.* Verify that, for the model

$$Y(t) = S(t) + N(t),$$

where $N(t)$ is uncorrelated white noise,

$$\text{MSE}_o = N_o h_o(0) = \int_{-\infty}^{\infty} \frac{N_o S_S(f)}{N_o + S_S(f)} \, df. \qquad (13.269)$$

Use this formula to check the result in exercise 21.

23. Verify (13.152) to (13.155) for example 2.

24. Duality.
  (a) Consider the dual to the minimum-MSE estimation problem for WSS processes,

$$\min_{h(\cdot)} \left\{ \lim_{T \to \infty} \frac{1}{T} \int_{-T/2}^{T/2} [X(t) - \hat{X}(t)]^2 \, dt \right\}, \qquad (13.270)$$

$$\hat{X}(t) = h(t) \otimes Y(t).$$

Show that the solution is given in terms of the empirical spectra (cf. Chapter 10),

$$H_o(f) = \frac{\hat{S}_{XY}(f)}{\hat{S}_Y(f)}. \qquad (13.271)$$

  (b) Consider the problem of minimizing the time-averaged MSE

$$\min_{h(\cdot)} \left\{ \lim_{T \to \infty} \frac{1}{T} \int_{-T/2}^{T/2} E\{[X(t) - \hat{X}(t)]^2\} \, dt \right\}, \qquad (13.272)$$

$$\hat{X}(t) = h(t) \otimes Y(t),$$

where $X(t)$ and $Y(t)$ are jointly regular nonstationary (asymptotically mean stationary) processes. Show that the solution is given in terms of the time-averaged probabilistic spectra (cf. Chapter 10),

$$H_o(f) = \frac{\langle S_{XY} \rangle(f)}{\langle S_Y \rangle(f)}. \qquad (13.273)$$

Hint:

$$(X, Y) \triangleq \lim_{T \to \infty} \frac{1}{T} \int_{-T/2}^{T/2} E\{X(t)Y(t)\} \, dt$$

is a valid inner product for jointly asymptotically mean stationary processes.

25. Consider two processes $X(t)$ and $Y(t)$ whose probabilistic model depends on some random parameters Φ, and let

$$S_{XY|\Phi}(f) \quad \text{and} \quad S_{Y|\Phi}(f)$$

denote the conditional spectra. Show that the optimum filter for $X(t)$, based on observations $Y(t)$, is

$$H_o(f) = \frac{E\{S_{XY|\Phi}(f)\}}{E\{S_{Y|\Phi}(f)\}}, \qquad (13.274)$$

where the expectation is over Φ. An example of this is example 3, for which Φ is the random distortion function. Observe the duality between (13.273) and (13.274).

26. Determine the noncausal Wiener filter for the Wiener-increment process $S(t) = W(t) - W(t - T)$ [for which $W(t)$ is the Wiener process] observed in independent white noise $N(t)$. Show that for a weak signal $[S_S(f) \ll N_o]$, the optimum filter for $S(t - t_o)$ is closely approximated by a causal filter for $t_o > T$.

27. *Noise-free prediction.* Consider the problem of predicting the value of a process X at time $t + \alpha$, using noise-free observations of $X(u)$ for $-\infty < u \le t$. Consider the process X with autocorrelation function,
$$R_X(\tau) = ae^{-\beta|\tau|},$$
and show that the causal predictor that minimizes the mean squared error,
$$E\{[X(t + \alpha) - \hat{X}(t + \alpha)]^2\},$$
$$\hat{X}(t + \alpha) \triangleq \int_{-\infty}^{t} h(t - u) X(u) \, du,$$
has transfer function
$$H(f) = e^{-\alpha\beta},$$
and therefore yields the estimate
$$\hat{X}(t + \alpha) = e^{-\alpha\beta} X(t).$$
The simplicity of this predictor is a result of the continuous-time counterpart of a general property of discrete-time linear predictors for processes with all-pole models (autoregressive processes) that is explained in exercise 11.36.

28. *Comparison of causal and noncausal filters.* Verify (13.199), and evaluate the parameters γ, d_+, c in terms of α, N_o, a in example 4. Compare the gain and bandwidth parameters c and γ for the optimum causal filter (13.203) with the gain and bandwidth parameters k and β for the optimum noncausal filter (13.150), for $t_o = 0$, which is denoted by
$$h'_o(t) = ke^{-\beta|t|}.$$
Verify that $\gamma = \beta$ and
$$c = \frac{2k}{1 + kN_o/a}.$$

Note: The great similarity between the optimum causal filter and the causal approximation to the optimum noncausal filter obtained by truncation, as in example 2, is an atypical coincidence.

29. *Performances of causal and noncausal filters.* Use (13.150), the result of exercise 22, (13.203), and (13.204) to compare the minimum MSE for the optimum causal and noncausal filters in terms of the parameters a, N_o, and α. Graph the two MSEs, normalized by $R_S(0)$, as functions of the SNR parameter $\lambda = a/\alpha N_o$.

   Note: For signals that are more sharply bandlimited than that in this example, the difference in performances of causal and noncausal filters can be substantially larger [Van Trees, 1968].

30. Use (13.168) and (13.169) to verify (13.204).

31. Propose a specific rational transfer function $T(f)$ for a causal stable filter that converts a unit-intensity white process $\tilde{Z}(t)$ into a colored process $Y(t)$, but for which $\tilde{Z}(t)$ differs from the innovations process $Z(t)$.

32. Derive (13.205) from (13.196).

    Hint: First reexpress (13.196) as

    $$H_o(f) = \frac{1}{[\bar{S}_Y(i2\pi f)]^+} \left\{ [\bar{S}_Y(i2\pi f)]^+ - \frac{N_o}{[\bar{S}_Y(i2\pi f)]^-} \right\}_+, \quad (13.275)$$

    and then manipulate this into (13.205), using the identities $[A + B]_+ \equiv A_+ + B_+$, $[A^+]_+ \equiv A^+$, and (provided that A is a proper rational function)$[1/(1 + A)^-]_+ \equiv 1$.

33. *Complementary filters.* For $Y(t) = S(t) + N(t)$, with $S(t)$ and $N(t)$ uncorrelated, let H_1 be the optimum causal filter for $S(t)$, and let H_2 be the optimum causal filter for $N(t)$. Use (13.196) to show that $H_1(f) + H_2(f) = 1$. Explain why.

34. Using only (13.207), (13.210), and (13.211), verify (13.209).

35. Use (13.210) and (13.212) to verify (13.213).

36. Verify that (13.220) yields the orthogonal projection $k(i)Z(i)$ of $X(i)$ onto $Z(i)$.

37. Verify that the orthogonal projection of $c_1 W_1 + c_2 W_2$ onto Y is $c_1 \hat{W}_1 + c_2 \hat{W}_2$, where $\hat{W}_1$ and $\hat{W}_2$ are the orthogonal projections of W_1 and W_2 onto Y. (Draw a geometrical picture to understand the result, but use the orthogonality condition to prove it.)

38. Verify that (13.228) is the correct design equation for the optimum predictor: then use (13.227) and (13.228) to verify (13.229).

39. Use (13.223) to verify that

    $$R_{WZ}(i+1, j) = A(i) R_{WZ}(i, j), \quad j \leq i. \quad (13.276)$$

    Hint: $R_{VZ}(i, j) = 0$ for $j \leq i$.

Then use this result to derive (13.230) from (13.229). [Note that the second term in (13.230) is simply the last term in the sum in (13.229).]

40. Use (13.285) and definition (13.237) to verify (13.234). Then use (13.231) and (13.232) to verify (13.235a).

41. Solve (13.236) for $\hat{W}(i|i-1)$, and substitute into (13.230) to verify (13.238).

42. *Kalman gain recursion.* Substitute (13.232b) into (13.236) and subtract $W(i)$ from both sides of the resulting equation to obtain

$$\tilde{W}(i|i) = [I - k(i)c(i)]\tilde{W}(i|i-1) - k(i)N(i), \quad (13.277)$$

where

$$\tilde{W}(i|i) \triangleq W(i) - \hat{W}(i|i). \quad (13.278)$$

Multiply each side of (13.277) by its own transpose, and evaluate the expected value to obtain

$$R_{\tilde{W}}(i|i) = [I - k(i)c(i)]R_{\tilde{W}}(i|i-1)[I - k(i)c(i)]^T$$
$$+ N_o k(i)k^T(i). \quad (13.279)$$

Substitute (13.235a) into (13.279) to obtain

$$R_{\tilde{W}}(i|i) = R_{\tilde{W}}(i|i-1) - \frac{R_{\tilde{W}}(i|i-1)c^T(i)c(i)R_{\tilde{W}}(i|i-1)}{N_o + c(i)R_{\tilde{W}}(i|i-1)c^T(i)}.$$
$$(13.280)$$

Furthermore, subtract $W(i)$ from both sides of (13.238) (with i replaced by $i-1$), and use (13.223) on the right side to obtain

$$\tilde{W}(i|i-1) = A(i-1)\tilde{W}(i-1|i-1) + b(i-1)V(i-1).$$
$$(13.281)$$

Then show that this yields

$$R_{\tilde{W}}(i|i-1) = A(i-1)R_{\tilde{W}}(i-1|i-1)A^T(i-1)$$
$$+ b(i-1)b^T(i-1). \quad (13.282)$$

Equations (13.280) and (13.282) are the desired recursions for the filtering- and prediction-error correlation matrices. Observe that these recursions require knowledge of an initial correlation matrix, $R_{\tilde{W}}(0, 0) = R_W(0) - \hat{W}(0|0)\hat{W}^T(0|0)$. Obtain an alternative form involving $k(i)$ for the recursion (13.280) by substitution of (13.235a) into (13.280).

43. (a) Show that for the first-order Kalman filter in example 5, the Kalman gain recursion is given by (13.252).

Hint: Solve (13.248) for $R_{\tilde{W}}(i|i-1)$, substitute this for both $R_{\tilde{W}}(i|i-1)$ and $R_{\tilde{W}}(i-1|i-2)$ into (13.250), and then solve for $k(i)$.

(b) Since it is known that $X_c(0) = 0$, then $W(0) = 0$, and therefore $R_W(0) = 0$ and the most appropriate initial estimate is $\hat{W}(0|0) = 0$. Use these initial values to show that $k(1)$ is given by (13.253).

(c) Show that the Kalman prediction recursion is given by

$$\hat{W}(i|i-1) = a\hat{W}(i-1|i-2) + ak(i-1)Z(i-1),$$

and that the impulse response $g(i)$ of the steady-state prediction filter is

$$g(i) = ah(i),$$

where $h(i)$ is the impulse response of the steady-state filter, (13.256).

(d) Let $a = \frac{3}{4}$, $b = 1$, $c = 1$, and sketch the gain sequence (13.252) for $N_o = 1$ and $N_o = \frac{1}{4}$. Also sketch the steady-state response (13.256) for $N_o = 1$ and $N_o = \frac{1}{4}$.

Solutions to Selected Exercises

Chapter 1

12. (a) (i) $\alpha = 1, \beta = 0$.

$$I^2 = \iint_{-\infty}^{\infty} \frac{1}{2\pi} \exp\left\{-\frac{1}{2}(x^2 + y^2)\right\} dx\, dy$$

$$= \int_0^{2\pi} \int_0^{\infty} \frac{1}{2\pi} \exp\left\{-\frac{1}{2}r^2\right\} r\, dr\, d\theta$$

$$= \int_0^{2\pi} \frac{1}{2\pi} d\theta \int_0^{\infty} \exp\{-t\}\, dt = 1.$$

Thus,

$$I = \int_{-\infty}^{\infty} \frac{1}{\sqrt{2\pi}} \exp\left\{-\frac{1}{2}x^2\right\} dx = 1.$$

(ii) $\alpha \neq 1, \beta \neq 0$. A change of variables in

$$J \stackrel{\Delta}{=} \int_{-\infty}^{\infty} \frac{1}{\sqrt{2\pi\alpha^2}} \exp\left\{-\frac{1}{2}\left(\frac{x-\beta}{\alpha}\right)^2\right\} dx$$

yields

$$J = \int_{-\infty}^{\infty} \frac{1}{\sqrt{2\pi}} \exp\left\{-\frac{1}{2}z^2\right\} dz.$$

Then it follows from (i) that $J = 1$.
(b) From the hint, we have $dw = (1/\alpha')(1/\sqrt{1 - \gamma^2})\, dy$. Therefore, using the hint, we have

$$f_X(x) = \int_{-\infty}^{\infty} f_{XY}(x, y)\, dy$$

$$= \int_{-\infty}^{\infty} \frac{1}{2\pi\alpha\alpha'\sqrt{1 - \gamma^2}} \exp\left\{-\frac{1}{2}\left(\frac{x - \beta}{\alpha}\right)^2\right\}$$

$$\times \exp\left\{-\frac{1}{2}w^2\right\} \alpha'\sqrt{1 - \gamma^2}\, dw$$

$$= \frac{1}{\alpha\sqrt{2\pi}} \exp\left\{-\frac{1}{2}\left(\frac{x - \beta}{\alpha}\right)^2\right\}$$

$$\times \int_{-\infty}^{\infty} \frac{1}{\sqrt{2\pi}} \exp\left\{-\frac{1}{2}w^2\right\} dw$$

$$= \frac{1}{\alpha\sqrt{2\pi}} \exp\left\{-\frac{1}{2}\left(\frac{x - \beta}{\alpha}\right)^2\right\}.$$

(c) To simplify the proof, let $Z = U + V$, where $U = aX$ and $V = bY$. Then,

$$E\{U\} = \beta_U = a\beta, \qquad \text{Var}\{U\} = \alpha_U^2 = (a\alpha)^2,$$

$$E\{V\} = \beta_V = b\beta', \qquad \text{Var}\{V\} = \alpha_V^2 = (b\alpha')^2,$$

$$\text{Cov}\{U, V\} = ab\, \text{Cov}\{X, Y\},$$

from which it follows that $\gamma_{UV} = \gamma$ and

$$f_{UV}(u, v) = \frac{1}{|a|\,|b|} f_{XY}(x, y).$$

It follows from (1.45) that

$$f_Z(z) = \int_{-\infty}^{\infty} f_{UV}(z - v, v)\, dv.$$

Therefore,

$$f_Z(z) = \frac{1}{2\pi\alpha_U\alpha_V\sqrt{1 - \gamma_{UV}^2}} \int_{-\infty}^{\infty}$$

$$\exp\left\{-\frac{\left(\frac{z - v - \beta_U}{\alpha_U}\right)^2 - 2\gamma_{UV}\left(\frac{z - v - \beta_U}{\alpha_U}\right)\left(\frac{v - \beta_V}{\alpha_V}\right) + \left(\frac{v - \beta_V}{\alpha_V}\right)^2}{2(1 - \gamma_{UV}^2)}\right\} dv.$$

Carrying out the above integration, we obtain

$$f_Z(z) = \frac{1}{\sqrt{2\pi}\sqrt{\alpha_U^2 + 2\gamma_{UV}\alpha_U\alpha_V + \alpha_V^2}}$$
$$\times \exp\left\{-\frac{1}{2}\frac{(z - \beta_U - \beta_V)^2}{\alpha_U^2 + 2\gamma_{UV}\alpha_U\alpha_V + \alpha_V^2}\right\}$$
$$= \frac{1}{\sigma\sqrt{2\pi}} \exp\left\{-\frac{1}{2}\left(\frac{z - m}{\sigma}\right)^2\right\},$$

which is a Gaussian probability density function [with $m = a\beta - b\beta'$ and $\sigma^2 = (a\alpha)^2 + 2\gamma ab\alpha\alpha' + (b\alpha')^2$].

(d) From the definition of conditional probability density,

$$f_{X|Y}(x|Y = y) = \frac{f_{XY}(x, y)}{f_Y(y)}$$

$$= \frac{1}{2\pi\alpha\alpha'\sqrt{1 - \gamma^2}}$$
$$\times \exp\left\{-\frac{\left(\frac{x-\beta}{\alpha}\right)^2 - 2\gamma\left(\frac{x-\beta}{\alpha}\right)\left(\frac{y-\beta'}{\alpha'}\right) + \left(\frac{y-\beta'}{\alpha'}\right)^2}{2(1 - \gamma^2)}\right\}$$

$$\times \alpha'\sqrt{2\pi}\exp\left\{\frac{1}{2}\left(\frac{y - \beta'}{\alpha'}\right)^2\right\}$$

$$= \frac{1}{\alpha\sqrt{2\pi}\sqrt{1 - \gamma^2}}$$
$$\times \exp\left\{-\frac{1}{2}\left(\frac{x - \left[\beta + \gamma\left(\frac{y-\beta'}{\alpha'}\right)\alpha\right]}{\alpha\sqrt{1 - \gamma^2}}\right)^2\right\}$$

$$= \frac{1}{\alpha''\sqrt{2\pi}} \exp\left\{-\frac{1}{2}\left(\frac{x - \beta''}{\alpha''}\right)^2\right\}.$$

14. Let
$$Y = Y_1 = X_1 + X_2 \quad \text{and} \quad Y_2 = X_2$$
or, equivalently,
$$Y = AX,$$
where
$$Y = [Y_1\ Y_2]^T, \quad X = [X_1\ X_2]^T, \quad A = \begin{bmatrix} 1 & 1 \\ 0 & 1 \end{bmatrix}.$$

Then, from (1.41), we have

$$f_Y(y) = \frac{f_X(A^{-1}y)}{|A|} = f_X\left(\begin{bmatrix} y_1 - y_2 \\ y_2 \end{bmatrix}\right)$$

$$= f_{X_1 X_2}(y_1 - y_2, y_2) = f_{X_1 X_2}(y - x_2, x_2).$$

Thus from (1.43), we have

$$f_Y(y) = \int_{-\infty}^{\infty} f_Y(y)\, dy_2 = \int_{-\infty}^{\infty} f_{X_1 X_2}(y - x_2, x_2)\, dx_2.$$

15. Let

$$Y = [Y_1\ Y_2]^T, \qquad X = [X_1\ X_2]^T$$

and

$$Y = g(X) = \begin{bmatrix} \sqrt{X_1^2 + X_2^2} \\ \tan^{-1}(X_2/X_1) \end{bmatrix}$$

or, equivalently,

$$X = g^{-1}(Y) = \begin{bmatrix} Y_1 \cos Y_2 \\ Y_1 \sin Y_2 \end{bmatrix}.$$

Then from (1.39), we have

$$f_Y(y) = f_X[g^{-1}(y)] \left|\frac{\partial g^{-1}(y)}{\partial y}\right| = f_{X_1}(x_1) f_{X_2}(x_2) \begin{vmatrix} \cos y_2 & -y_1 \sin y_2 \\ \sin y_2 & y_1 \cos y_2 \end{vmatrix}$$

$$= \frac{y_1}{2\pi} \exp\left\{-\frac{1}{2} y_1^2\right\}, \qquad y_1 \geq 0,\ 0 \leq y_2 < 2\pi.$$

Marginal densities:

$$f_{Y_1}(y_1) = \int_{-\infty}^{\infty} f_Y(y)\, dy_2 = \int_0^{2\pi} \frac{y_1}{2\pi} \exp\left\{-\frac{1}{2} y_1^2\right\} dy_2$$

$$= y_1 \exp\left\{-\frac{1}{2} y_1^2\right\}, \qquad y_1 \geq 0$$

$$f_{Y_2}(y_2) = \int_{-\infty}^{\infty} f_Y(y)\, dy_1 = \int_0^{\infty} \frac{y_1}{2\pi} \exp\left\{-\frac{1}{2} y_1^2\right\} dy_1$$

$$= \frac{1}{2\pi}, \qquad 0 \leq y_2 < 2\pi.$$

Independence: It follows directly from above that

$$f_{Y_1 Y_2}(y_1, y_2) = f_{Y_1}(y_1) f_{Y_2}(y_2).$$

19. (a) Since $0 \leq Y \leq 1$ and $dg(x)/dx = f_X(x)$, then it follows from (1.36)

that

$$f_Y(y) = \frac{f_X(x)}{|dg(x)/dx|} = 1, \quad 0 \le y \le 1.$$

(b) From (1.36), we have

$$f_Z(z) = f_Y(h^{-1}(z)) \left|\frac{dh^{-1}(z)}{dz}\right| = 1 \cdot \frac{dF_W(z)}{dz} = f_W(z).$$

(c) Let $k(\cdot)$ be the composition of $F_W^{-1}(\cdot)$ and $F_X(\cdot)$. Then $Z = F_W^{-1}(F_X(X))$, and it follows from parts (a) and (b) that $F_Z(\cdot) = F_W(\cdot)$.

22.

$$\begin{aligned}F_Y(y) &= P[\{Y < y\} \cap \{(X < 0) \cup (X \ge 0)\}] \\ &= P[(Y < y) \cap (X < 0)] + P[(Y < y) \cap (X \ge 0)] \\ &= P[Y < y \mid X < 0]P[X < 0] + P[Y < y \mid X \ge 0]P[X \ge 0] \\ &= F_{Y|X<0}(y)P + F_{Y|X\ge 0}(y)(1 - P),\end{aligned}$$

where

$$P \triangleq P[X < 0].$$

Thus,

$$f_Y(y) = f_{Y|X<0}(y)P + f_{Y|X\ge 0}(y)(1 - P).$$

But, if $X < 0$, then $Y = 0$; therefore,

$$f_{Y|X<0}(y) = \delta(y).$$

Also, if $X \ge 0$, then $Y = X$; therefore (cf. exercise 8 in Chapter 1),

$$f_{Y|X\ge 0}(y) = \frac{f_X(y)u(y)}{1 - P}$$

where

$$u(y) = \begin{cases} 1, & y > 0, \\ 0, & y \le 0. \end{cases}$$

Hence,

$$f_Y(y) = P\delta(y) + f_X(y)u(y),$$

where

$$P \triangleq \int_{-\infty}^{0} f_X(x)\, dx.$$
∎

Chapter 2

4. Use of $Y = g(X)$ and (1.36) yields

$$E\{Y\} = \int_{-\infty}^{\infty} y f_Y(y) \, dy = \int_{y=-\infty}^{\infty} g(x) \frac{f_X(x)}{|dg(x)/dx|} \frac{dg(x)}{dx} dx$$

$$= \begin{cases} \int_{x=-\infty}^{\infty} g(x) f_X(x) \, dx, & \frac{dg}{dx} > 0 \\ -\int_{x=\infty}^{-\infty} g(x) f_X(x) \, dx, & \frac{dg}{dx} < 0 \end{cases}$$

$$= \int_{-\infty}^{\infty} g(x) f_X(x) \, dx \triangleq E\{g(X)\}.$$

5. (a)

$$\sigma_X^2 = E\{(X - m_X)^2\} = E\{X^2 - 2m_X X + m_X^2\}$$
$$= E\{X^2\} - 2m_X E\{X\} + m_X^2 = E\{X^2\} - m_X^2.$$

(b)

$$K_{XY} = E\{(X - m_X)(Y - m_Y)\}$$
$$= E\{XY\} - m_X E\{Y\} - m_Y E\{X\} + m_X m_Y$$
$$= R_{XY} - m_X m_Y.$$

6. (a) The mean is given by

$$m_X = E\{X\} = \int_{-\infty}^{\infty} \frac{x}{\alpha\sqrt{2\pi}} \exp\left\{-\frac{1}{2}\left(\frac{x-\beta}{\alpha}\right)^2\right\} dx$$

$$= \int_{-\infty}^{\infty} \frac{\alpha y + \beta}{\alpha\sqrt{2\pi}} \exp\left\{-\frac{y^2}{2}\right\} \alpha \, dy$$

$$= \beta \int_{-\infty}^{\infty} \frac{1}{\sqrt{2\pi}} \exp\left\{-\frac{y^2}{2}\right\} dy + \alpha \int_{-\infty}^{\infty} \frac{y}{\sqrt{2\pi}} \exp\left\{-\frac{y^2}{2}\right\} dy.$$

The first integral is unity as shown in exercise 12a in Chapter 1. The second integral is zero because its integrand is the product of an even function and an odd function and is, therefore, an odd function; thus,

$$m_X = \beta 1 + \alpha 0 = \beta.$$

The variance is given by

$$\sigma_X^2 = E\{X^2\} - m_X^2$$

$$= \int_{-\infty}^{\infty} \frac{x^2}{\alpha\sqrt{2\pi}} \exp\left\{-\frac{1}{2}\left(\frac{x-\beta}{\alpha}\right)^2\right\} dx - \beta^2$$

$$= \int_{-\infty}^{\infty} \frac{(\alpha y + \beta)^2}{\alpha\sqrt{2\pi}} \exp\left\{-\frac{y^2}{2}\right\} \alpha\, dy - \beta^2$$

$$= \frac{1}{\sqrt{2\pi}} \int_{-\infty}^{\infty} (\alpha^2 y^2 + 2\alpha\beta y + \beta^2) \exp\left\{-\frac{y^2}{2}\right\} dy - \beta^2$$

$$= \alpha^2 1 + 2\alpha\beta 0 + \beta^2 1 - \beta^2 = \alpha^2.$$

The coefficients 0 and 1 in the second and third terms are obtained as in the first part. The coefficient 1 in the first term is obtained from integration by parts:

$$\int_{-\infty}^{\infty} y^2 \exp\left\{-\frac{y^2}{2}\right\} dy = \sqrt{2\pi}.$$

(b) We have

$$\rho \triangleq \frac{K_{XY}}{\sigma_X \sigma_Y} = \frac{R_{XY} - m_X m_Y}{\sigma_X \sigma_Y},$$

where

$$R_{XY} = E\{XY\} = E\{E\{X \mid Y\}Y\}.$$

From exercise 12d in Chapter 1, we also have

$$f_{X|Y}(x \mid Y = y) = \frac{1}{\alpha''\sqrt{2\pi}} \exp\left\{-\frac{1}{2}\left(\frac{x-\beta''}{\alpha''}\right)^2\right\},$$

where

$$\beta'' = m_X + \frac{\gamma(y - m_Y)\sigma_X}{\sigma_Y} \quad \text{and} \quad \alpha'' = \sigma_X\sqrt{1-\gamma^2}.$$

Therefore, $E\{X \mid Y\} = \beta''$ and

$$R_{XY} = E\{E\{X \mid Y\}Y\}$$

$$= E\left\{\left(m_X + \frac{\gamma(Y - m_Y)\sigma_X}{\sigma_Y}\right)Y\right\}$$

$$= \left(m_X - \gamma m_Y \frac{\sigma_X}{\sigma_Y}\right) m_Y + \gamma \frac{\sigma_X}{\sigma_Y}(\sigma_Y^2 + m_Y^2)$$

$$= \gamma \sigma_X \sigma_Y + m_X m_Y.$$

Consequently, $\rho = \gamma$.

11. From (1.32), we have

$$f_Z(z) = \int_{-\infty}^{\infty} f_X(z - u) f_Y(u)\, du.$$

From (2.13), we have

$$\Phi_Z(\omega) = \int_{-\infty}^{\infty} e^{i\omega z} f_Z(z)\, dz = \iint_{-\infty}^{\infty} e^{i\omega z} f_X(z - u)\, dz\, f_Y(u)\, du$$

$$= \int_{-\infty}^{\infty} \Phi_X(\omega) e^{i\omega u} f_Y(u)\, du = \Phi_X(\omega)\Phi_Y(\omega).$$

12. We have

$$Y = \frac{1}{n}\sum_{k=1}^{n} X_k^2 - \frac{2}{n^2}\sum_{k,j=1}^{n} X_k X_j + \frac{1}{n^2}\sum_{i,j=1}^{n} X_i X_j$$

$$= \frac{1}{n}\sum_{k=1}^{n} X_k^2 - \frac{1}{n^2}\sum_{i,j=1}^{n} X_i X_j.$$

Therefore,

$$E\{Y\} = \frac{1}{n}\sum_{k=1}^{n} \sigma_X^2 - \frac{1}{n^2}\sum_{i=j=1}^{n} \sigma_X^2$$

$$= \sigma_X^2 - \frac{\sigma_X^2}{n} = \frac{n-1}{n}\sigma_X^2 \to \sigma_X^2, \quad \text{as } n \to \infty.$$

Also, we have

$$Y^2 = \frac{1}{n^2}\sum_{k,l=1}^{n} X_k^2 X_l^2 + \frac{1}{n^4}\sum_{k,l,i,j=1}^{n} X_k X_l X_i X_j - \frac{2}{n^3}\sum_{k,i,j=1}^{n} X_k^2 X_i X_j.$$

Using the hint yields

$$E\{Y^2\} = \frac{1}{n^2}\sum_{k,l=1}^{n} [E\{X_k^2\}E\{X_l^2\} + 2E^2\{X_k X_l\}]$$

$$+ \frac{1}{n^4}\sum_{k,l,i,j=1}^{n} [E\{X_k X_l\}E\{X_i X_j\}$$

$$+ E\{X_k X_i\}E\{X_l X_j\} + E\{X_k X_j\}E\{X_l X_i\}]$$

$$- \frac{2}{n^3}\sum_{k,i,j=1}^{n} [E\{X_k^2\}E\{X_i X_j\} + 2E\{X_k X_i\}E\{X_k X_j\}]$$

$$= \left(\sigma_X^4 + \frac{2}{n}\sigma_X^4\right) + \frac{1}{n^4}(3n^2\sigma_X^4) - \frac{2}{n^3}(n^2\sigma_X^4 + 2n\sigma_X^4)$$

$$= \frac{(n^2 - 1)}{n^2}\sigma_X^4.$$

Thus,
$$\text{Var}\{Y\} = E\{Y^2\} - E^2\{Y\} = \frac{2(n-1)}{n^2}\sigma_X^4 \to \frac{2\sigma_X^4}{n} \to 0 \quad \text{as } n \to \infty.$$

14. Let $w(x, y) = f_{XY}(x, y)$, $h(x, y) = x$, and $g(x, y) = y$. Then the Cauchy-Schwarz inequality yields
$$|E\{XY\}|^2 \le E\{X^2\}E\{Y^2\}.$$
This must also be true for $X' = X - m_X$ and $Y' = Y - m_Y$. Therefore,
$$|\text{Cov}\{X, Y\}|^2 \le \text{Var}\{X\}\text{Var}\{Y\},$$
from which we obtain $|\rho| \le 1$.

15. (a) Let $w(x, y) = f_{XY}(x, y)$, $g(x, y) = x - m_X$, and $h(x, y) = y - m_Y$. Then the triangle inequality yields
$$E\{[X + Y - (m_X + m_Y)]^2\}^{1/2}$$
$$\le E\{(X - m_X)^2\}^{1/2} + E\{(Y - m_Y)^2\}^{1/2},$$
from which we obtain
$$\sigma_{X+Y} \le \sigma_X + \sigma_Y.$$
For $W = X + Y + Z$, this yields
$$\sigma_W \le \sigma_{X+Y} + \sigma_Z \le \sigma_X + \sigma_Y + \sigma_Z,$$
and similarly for the sum of any finite number of random variables.

(b) We have
$$E\{(X+Y)^2\} - E^2\{X+Y\} = E\{X^2\} + E\{Y^2\} + 2E\{XY\}$$
$$- E^2\{X\} - E^2\{Y\} - 2E\{X\}E\{Y\}.$$
Since $E\{XY\} = E\{X\}E\{Y\}$, this reduces to
$$\sigma_{X+Y}^2 = \sigma_X^2 + \sigma_Y^2.$$
For $W = X + Y + Z$, this yields
$$\sigma_W^2 = \sigma_{X+Y}^2 + \sigma_Z^2 = \sigma_X^2 + \sigma_Y^2 + \sigma_Z^2,$$
and similarly for the sum of any finite number of statistically independent random variables.

21. Since $Y = [Y_1\ Y_2\ \dots\ Y_n]^T = \mathbf{R}_X^{-1/2}\mathbf{X}$ and $(\mathbf{R}_X^{-1/2})^T = \mathbf{R}_X^{-1/2}$, then
$$E\{\mathbf{Y}\mathbf{Y}^T\} = E\{\mathbf{R}_X^{-1/2}\mathbf{X}\mathbf{X}^T\mathbf{R}_X^{-1/2}\} = \mathbf{R}_X^{-1/2}E\{\mathbf{X}\mathbf{X}^T\}\mathbf{R}_X^{-1/2}$$
$$= \mathbf{R}_X^{-1/2}\mathbf{R}_X\mathbf{R}_X^{-1/2} = \mathbf{R}_X^{-1/2}(\mathbf{R}_X^{1/2}\mathbf{R}_X^{1/2})\mathbf{R}_X^{-1/2} = \mathbf{I}$$
and, therefore,
$$E\{Y_iY_j\} = \begin{cases} 1, & i = j, \\ 0, & i \ne j. \end{cases}$$

30. For (2.47), we have

$$E\{E\{X \mid Y\} \mid Z\} = \iint_{-\infty}^{\infty} x f_{X\mid Y}(x \mid y) \, dx \, f_{Y\mid Z}(y \mid z) \, dy.$$

But, since $Z = g(Y)$, then

$$f_{X\mid Y}(x \mid y) = f_{X\mid Y,Z}(x \mid y, z).$$

Also, we have

$$\int_{-\infty}^{\infty} f_{X\mid Y,Z}(x \mid y, z) f_{Y\mid Z}(y \mid z) \, dy = f_{X\mid Z}(x \mid z).$$

Therefore,

$$E\{E\{X \mid Y\} \mid Z\} = \int_{-\infty}^{\infty} x f_{X\mid Z}(x \mid z) \, dx = E\{X \mid Z\}.$$

For (2.48), since $Z = g(Y)$, then

$$f_{Z\mid Y}(z \mid y) = \delta(z - g(y))$$

and, therefore,

$$E\{E\{X \mid Z\} \mid Y\} = \int_{-\infty}^{\infty} E\{X \mid z\} f_{Z\mid Y}(z \mid y) \, dz$$

$$= E\{X \mid g(Y)\} = E\{X \mid Z\}.$$

32. Since $E\{X_i\} = P(A)$, $i = 1, 2, \ldots$, and $E\{[X_i - P(A)]^2\} = \sigma^2$, then we have from exercise 31 in Chapter 2

$$E\left\{\left|\frac{K_n}{n} - P(A)\right|^2\right\} = \frac{\sigma^2}{n}.$$

From the Bienaymé-Chebychev inequality, we have

$$\text{Prob}\left\{\left|\frac{K_n}{n} - P(A)\right| > \varepsilon\right\} \le \frac{1}{\varepsilon^2} E\left\{\left|\frac{K_n}{n} - P(A)\right|^2\right\} = \frac{1}{\varepsilon^2} \frac{\sigma^2}{n}.$$

Therefore,

$$\lim_{n \to \infty} \text{Prob}\left\{\left|\frac{K_n}{n} - P(A)\right| > \varepsilon\right\} \le \lim_{n \to \infty} \frac{1}{\varepsilon^2} \frac{\sigma^2}{n} = 0$$

or, equivalently,

$$\lim_{n \to \infty} \text{Prob}\left\{\left|\frac{K_n}{n} - P(A)\right| < \varepsilon\right\} = 1.$$

33. We have

$$X_n = \begin{cases} n, & \text{Prob} = \alpha/n^2, \\ 0, & \text{Prob} = 1 - \alpha/n^2 \end{cases}$$

and, therefore,
$$\lim_{n\to\infty} X_n = \begin{cases} \text{does not exist}, & \text{Prob} = 0 \\ 0, & \text{Prob} = 1. \end{cases}$$

Thus, X_n converges to 0 with probability 1. On the other hand,
$$\lim_{n\to\infty} E\{(X_n - 0)^2\} = \lim_{n\to\infty} n^2 P(s = n) = \alpha \neq 0.$$

Hence, X_n does not converge in mean square. ∎

Chapter 3

4. Following the hint we obtain from (3.11)
$$Y(t) = \int_{\infty}^{-\infty} h(u_2) X(t - u_2)(-du_2)$$
$$= \int_{-\infty}^{\infty} h(u_2) X(t - u_2) \, du_2$$

and
$$Y(t + \tau) = \int_{-\infty}^{\infty} h(u_1) X(t + \tau - u_1) \, du_1.$$

Substitution of $Y(t + \tau)$ and $Y(t)$ into (3.2) yields
$$\hat{R}_Y(\tau) = \lim_{T\to\infty} \frac{1}{T} \int_{-T/2}^{T/2} Y(t + \tau) Y(t) \, dt$$
$$= \lim_{T\to\infty} \frac{1}{T} \int_{-T/2}^{T/2} \int\int_{-\infty}^{\infty} h(u_1) h(u_2)$$
$$\times X(t + \tau - u_1) X(t - u_2) \, du_1 \, du_2 \, dt$$
$$= \int\int_{-\infty}^{\infty} h(u_1) h(u_2) \left[\lim_{T\to\infty} \frac{1}{T} \int_{-T/2}^{T/2} X(t - u_2) \right.$$
$$\left. \times X(t + \tau - u_1) \, dt \right] du_1 \, du_2.$$

Letting $t - u_2 = v$, the bracketed factor results in
$$\lim_{T\to\infty} \int_{-T/2-u_2}^{T/2-u_2} X(v + \tau + u_2 - u_1) X(v) \, dv = \hat{R}_X(\tau + u_2 - u_1).$$

Therefore,
$$\hat{R}_Y(\tau) = \int\int_{-\infty}^{\infty} h(u_1) h(u_2) \hat{R}_X(\tau + u_2 - u_1) \, du_1 \, du_2.$$

Letting $s = u_1 - u_2$ yields

$$\hat{R}_Y(\tau) = \int_{-\infty}^{\infty} \int_{-\infty}^{\infty} h(s + u_2)h(u_2) \, du_2 \, \hat{R}_X(\tau - s) \, ds$$

$$= \int_{-\infty}^{\infty} r_h(s) \hat{R}_X(\tau - s) \, ds$$

$$= \hat{R}_X(\tau) \otimes r_h(\tau),$$

where $r_h(\tau)$ is defined in (3.13).

5. Inverse Fourier transformation of (3.9) with X replaced by Y yields

$$\hat{R}_Y(\tau) = \int_{-\infty}^{\infty} \hat{S}_Y(f) e^{i2\pi f\tau} \, df.$$

Evaluation of this equation at $\tau = 0$ and use of (3.17) and (3.18) yields

$$\langle P \rangle = \lim_{T \to \infty} \frac{1}{T} \int_{-T/2}^{T/2} Y^2(t) \, dt = \hat{R}_Y(0) = \int_{-\infty}^{\infty} \hat{S}_Y(v) \, dv.$$

Use of (3.14) with f replaced by v then yields

$$\langle P \rangle = \int_{f-\Delta/2}^{f+\Delta/2} \hat{S}_X(v) \, dv + \int_{-f-\Delta/2}^{-f+\Delta/2} \hat{S}_X(v) \, dv.$$

Thus, with $\Delta \to 0$, we have

$$\langle P \rangle \to \Delta \hat{S}_X(f) + \Delta \hat{S}_X(-f).$$

But, since $\hat{R}_X(\tau)$ is real and even, then $\hat{S}_X(f)$ is real and even and, therefore,

$$\langle P \rangle \to 2\Delta \hat{S}_X(f),$$

which is the desired result (3.10).

6. Consider the time-variant finite segment

$$X_T(t + u) \triangleq \begin{cases} X(t + u), & |t| \leq T/2 \\ 0, & |t| > T/2. \end{cases}$$

The time-variant correlogram, which is a generalization of the correlogram (3.5), is defined by

$$R_X(u, \tau)_T \triangleq \frac{1}{T} \int_{-\infty}^{\infty} X_T(t + u + |\tau|) X_T(t + u) \, dt$$

$$= \frac{1}{T} \int_{-T/2}^{T/2 - |\tau|} X(t + u + |\tau|) x(t + u) \, dt.$$

Use of the generalization of (3.4),

$$\frac{1}{T}|\check{X}_T(u, f)|^2 = \int_{-\infty}^{\infty} R_X(u, \tau)_T e^{-i2\pi f\tau} d\tau,$$

yields the time-averaged periodogram,

$$\lim_{Z\to\infty} \frac{1}{Z}\int_{-Z/2}^{Z/2} \frac{1}{T}|\check{X}_T(u, f)|^2 du = \int_{-\infty}^{\infty} \left[\lim_{Z\to\infty} \frac{1}{Z}\int_{-Z/2}^{Z/2} R_X(u, \tau)_T du\right] e^{-i2\pi f\tau} d\tau$$

$$= \int_{-\infty}^{\infty} \frac{1}{T}\int_{-T/2}^{T/2-|\tau|} \lim_{Z\to\infty} \frac{1}{Z}\int_{-Z/2}^{Z/2} X(t + u + |\tau|)$$

$$\times X(t + u) \, du \, dt \, e^{-i2\pi f\tau} d\tau$$

$$= \int_{-\infty}^{\infty} \frac{1}{T}\int_{-T/2}^{T/2-|\tau|} \hat{R}_X(\tau) \, dt \, e^{-i2\pi f\tau} d\tau$$

$$= \int_{-\infty}^{\infty} \left(1 - \frac{|\tau|}{T}\right) \hat{R}_X(\tau) e^{-i2\pi f\tau} d\tau.$$

Finally, taking the limit as $T \to \infty$ yields

$$\hat{S}_X(f) \triangleq \lim_{T\to\infty} \lim_{Z\to\infty} \frac{1}{Z}\int_{-Z/2}^{Z/2} \frac{1}{T}|\check{X}_T(u, f)|^2 du$$

$$= \lim_{T\to\infty} \int_{-\infty}^{\infty} \left(1 - \frac{|\tau|}{T}\right) \hat{R}_X(\tau) e^{-i2\pi f\tau} d\tau$$

$$= \int_{-\infty}^{\infty} \hat{R}_X(\tau) e^{-i2\pi f\tau} d\tau,$$

which is the desired result (3.9). ∎

Chapter 4

1. *Bernoulli process:* Given

$$P\{X_n = 1\} = p, \qquad P\{X_n = 0\} = 1 - p,$$

we obtain

$$m_X(n) = E\{X_n\} = P\{X_n = 1\}1 + P\{X_n = 0\}0 = p$$

and

$$R_X(n_1, n_2) = E\{X_{n_1} X_{n_2}\}$$

$$= \begin{cases} 1^2 p + 0^2(1 - p) = p, & n_1 = n_2 \\ E\{X_{n_1}\}E\{X_{n_2}\} = p^2, & n_1 \neq n_2, \end{cases}$$

and also
$$K_X(n_1, n_2) = R_X(n_1, n_2) - m_X(n_1)m_X(n_2)$$
$$= \begin{cases} p(1-p), & n_1 = n_2 \\ 0, & n_1 \neq n_2. \end{cases}$$

Binomial counting process: From (4.14), we obtain
$$m_Y(n) = E\{Y_n\} = E\left\{\sum_{i=1}^{n} X_i\right\} = \sum_{i=1}^{n} E\{X_i\} = \sum_{i=1}^{n} p = np.$$

Use of this result yields
$$K_Y(n_1, n_2) = \sum_{i=1}^{n_1} \sum_{j=1}^{n_2} E\{X_i X_j\} - (n_1 p)(n_2 p)$$
$$= \sum_{i=1}^{n_1} \sum_{j=1}^{n_2} K_X(i, j)$$
$$= \sum_{i=j=1}^{n_1} K_X(i, i) + \sum_{i=1}^{n_1} \sum_{\substack{j=1 \\ j \neq i}}^{n_2} K_X(i, j), \quad n_1 \leq n_2$$
$$= n_1 p(1-p) + 0, \quad n_1 \leq n_2.$$

Similarly,
$$K_Y(n_1, n_2) = n_2 p(1-p), \quad n_2 \leq n_1.$$

Thus,
$$K_Y(n_1, n_2) = p(1-p)\min\{n_1, n_2\}.$$

Random-walk process: From (4.20), $Z_i = 2(X_i - 1/2)$ and, therefore,
$$W_n = \sum_{i=1}^{n} Z_i = 2Y_n - n.$$

Thus,
$$m_W(n) = E\{W_n\} = 2E\{Y_n\} - n = n(2p - 1)$$
and
$$K_W(n_1, n_2) = E\{[W_{n_1} - m_W(n_1)][W_{n_2} - m_W(n_2)]\}$$
$$= E\{[2Y_{n_1} - n_1 - n_1(2p - 1)]$$
$$\times [2Y_{n_2} - n_2 - n_2(2p - 1)]\}$$
$$= 4E\{(Y_{n_1} - n_1 p)(Y_{n_2} - n_2 p)\}$$
$$= 4K_Y(n_1, n_2)$$
$$= 4p(1-p)\min\{n_1, n_2\}.$$

8. Since $Y(t)$ and Φ are independent, then we have
$$E\{Z(t)\} = E\{Y(t)\}E\{\sin(\omega_o t + \Phi)\} = m_Y(t)E\{\sin(\omega_o t + \Phi)\}$$
and
$$\begin{aligned}R_Z(t_1, t_2) &= E\{Z(t_1)Z(t_2)\}\\ &= E\{Y(t_1)Y(t_2)\}E\{\sin(\omega_o t_1 + \Phi)\sin(\omega_o t_2 + \Phi)\}\\ &= R_Y(t_1, t_2)\frac{1}{2}E\{\cos(\omega_o[t_1 - t_2]) - \cos(\omega_o[t_1 + t_2] + 2\Phi)\}\\ &= \frac{1}{2}R_Y(t_1, t_2)[\cos(\omega_o[t_1 - t_2]) - E\{\cos(\omega_o[t_1 + t_2] + 2\Phi)\}]\end{aligned}$$

If Φ is uniformly distributed on the interval $[-\pi, \pi)$, then we have
$$E\{\sin(\omega_o t + \Phi)\} = \int_{-\infty}^{\infty} \sin(\omega_o t + \phi)f_\Phi(\phi)\,d\phi$$
$$= \frac{1}{2\pi}\int_{-\pi}^{\pi} \sin(\omega_o t + \phi)\,d\phi = 0$$
and
$$E\{\cos(\omega_o[t_1 - t_2] + 2\Phi)\} = \int_{-\infty}^{\infty} \cos(\omega_o[t_1 - t_2] + 2\phi)f_\Phi(\phi)\,d\phi$$
$$= \frac{1}{2\pi}\int_{-\pi}^{\pi} \cos(\omega_o[t_1 - t_2] + 2\phi)\,d\phi = 0.$$

Therefore,
$$m_Z(t) = m_Y(t)0 = 0$$
and
$$R_Z(t_1, t_2) = \frac{1}{2}R_Y(t_1, t_2)[\cos(\omega_o[t_1 - t_2]) - 0]$$
$$= \frac{1}{2}R_Y(t_1, t_2)\cos(\omega_o[t_1 - t_2]).$$

10. Since $V(nT)$ and $V([n + k]T)$ are independent for $k \neq 0$ and have zero mean value, then we have
$$K_X(t_1, t_2) = E\{[X(t_1) - E\{X(t_1)\}][X(t_2) - E\{X(t_2)\}]\}$$
$$= E\left\{\sum_{n=-\infty}^{\infty}[V(nT) - E\{V(nT)\}]h(t_1 - nT)\right.$$
$$\left.\times \sum_{m=-\infty}^{\infty}[V(mT) - E\{V(mT)\}]h(t_2 - mT)\right\}$$

$$= \sum_{n,m=-\infty}^{\infty} E\{V(nT)V(mT)\}h(t_1 - nT)h(t_2 - mT)$$

$$= \sum_{n=-\infty}^{\infty} \sigma_V^2 h(t_1 - nT)h(t_2 - nT).$$

Also, $E\{X(t)\} = 0$ and, therefore, $K_X(t_1, t_2) = R_X(t_1, t_2)$. From (2.45) and using the above result, we have $E\{Y(t)\} = 0$ and, therefore,

$$\begin{aligned}
K_Y(t_1, t_2) &= E\{Y(t_1)Y(t_2)\} \\
&= E\{E\{Y(t_1)Y(t_2) \mid \Theta\}\} \\
&= E\{R_X(t_1 - \Theta, t_2 - \Theta)\} \\
&= \int_{-\infty}^{\infty} R_X(t_1 - \theta, t_2 - \theta) f_\Theta(\theta) \, d\theta \\
&= \frac{1}{T} \sigma_V^2 \sum_{n=-\infty}^{\infty} \int_{-T/2}^{T/2} h(t_1 - nT - \theta) h(t_2 - nT - \theta) \, d\theta \\
&= \frac{1}{T} \sigma_V^2 \sum_{n=-\infty}^{\infty} \int_{t_2 - nT - T/2}^{t_2 - nT + T/2} h(t_1 - t_2 + \phi) h(\phi) \, d\phi \\
&= \frac{1}{T} \sigma_V^2 \int_{-\infty}^{\infty} h(t_1 - t_2 + \phi) h(\phi) \, d\phi \\
&= \frac{\sigma_V^2}{T} r_h(t_1 - t_2).
\end{aligned}$$

15. (a) We have

$$X(t) = \sum_{n=-\infty}^{\infty} p(t - nT - P_n),$$

where $p(t)$ is the zero-position pulse and $\{P_n\}$ are independent of each other and identically distributed.

(b) We have

$$\begin{aligned}
m_X(t) &= E\left\{\sum_{n=-\infty}^{\infty} p(t - nT - P_n)\right\} \\
&= \int_{-\infty}^{\infty} \sum_{n=-\infty}^{\infty} p(t - nT - u) f_P(u) \, du \\
&= \sum_{n=-\infty}^{\infty} \bar{p}(t - nT),
\end{aligned}$$

where
$$\bar{p}(t) \triangleq p(t) \otimes f_P(t).$$
Thus,
$$m_X(t+T) = \sum_{n=-\infty}^{\infty} \bar{p}(t+T-nT)$$
$$= \sum_{m=-\infty}^{\infty} \bar{p}(t-MT) \qquad (m = n-1)$$
$$= m_X(t)$$

and $m_X(t)$ is, therefore, periodic. Furthermore, if
$$f_P(p) = \begin{cases} \dfrac{1}{T-\Delta}, & 0 \le p < T - \Delta, \\ 0, & \text{otherwise,} \end{cases}$$
and
$$p(t) = \begin{cases} 1, & 0 \le t < \Delta, \\ 0, & \text{otherwise,} \end{cases}$$
then
$$\bar{p}(t) = \frac{1}{T-\Delta}\int_0^{T-\Delta} p(t-u)\,du = \frac{1}{T-\Delta}\int_{t-(T-\Delta)}^{t} p(u)\,du$$

$$= \begin{cases} \dfrac{1}{T-\Delta}\int_0^t du, & 0 \le t < \Delta, \\[4pt] \dfrac{1}{T-\Delta}\int_0^{\Delta} du, & \Delta \le t < T - \Delta, \\[4pt] \dfrac{1}{T-\Delta}\int_{t-(T-\Delta)}^{\Delta} du, & T-\Delta \le t < T \end{cases}$$

$$= \begin{cases} \dfrac{t}{T-\Delta}, & 0 \le t < \Delta, \\[4pt] \dfrac{\Delta}{T-\Delta}, & 0 \le t < T - \Delta, \\[4pt] \dfrac{T-t}{T-\Delta}, & T-\Delta \le t < T. \end{cases}$$

(c) We have
$$X(t) = \sum_{n=-\infty}^{\infty} p(t - nT - P_n - \Theta),$$

where

$$f_\Theta(\theta) = \begin{cases} 1/T, & -T/2 \le \theta < T/2, \\ 0, & \text{otherwise.} \end{cases}$$

Therefore,

$$\begin{aligned}
R_X(t+\tau, t) &= E\{X(t+\tau)X(t)\} \\
&= E\{E\{X(t+\tau)X(t) \mid \Theta\}\} \\
&= E\left\{E\left\{\sum_{n,m=-\infty}^{\infty} p(t+\tau-nT-P_n-\Theta)\right.\right. \\
&\qquad \times p(t-mT-P_m-\Theta) \Big| \Theta \bigg\}\bigg\} \\
&= E\left\{\sum_{n=-\infty}^{\infty} E\{p(t+\tau-nT-P_n-\Theta) \right. \\
&\qquad \times p(t-nT-P_n-\Theta) \mid \Theta\}\bigg\} \\
&\quad + E\left\{\sum_{r\neq 0}\sum_{n=-\infty}^{\infty} E\{p(t+\tau-nT-P_n-\Theta) \right. \\
&\qquad \times p(t-nT-rT-P_{n+r}-\Theta) \mid \Theta\}\bigg\}. \\
&\qquad (r = m-n)
\end{aligned}$$

We have

$$\begin{aligned}
&E\left\{\sum_{n=-\infty}^{\infty} E\{p(t+\tau-nT-P_n-\Theta)p(t-nT-P_n-\Theta) \mid \Theta\}\right\} \\
&= E\left\{\sum_{n=-\infty}^{\infty} \frac{1}{T-\Delta}\int_0^{T-\Delta} p(t+\tau-nT-u-\Theta) \right. \\
&\qquad\qquad \times p(t-nT-u-\Theta)\,du \Big| \Theta\bigg\} \\
&= \frac{1}{T-\Delta}\int_0^{T-\Delta}\sum_{n=-\infty}^{\infty}\frac{1}{T}\int_{-T/2}^{T/2} p(t+\tau-nT-u-\theta) \\
&\qquad\qquad \times p(t-nT-u-\theta)\,d\theta\,du \\
&= \frac{1}{T-\Delta}\int_0^{T-\Delta}\frac{1}{T}\int_{-\infty}^{\infty} p(v+\tau)p(v)\,dv\,du = \frac{1}{T}r_p(\tau),
\end{aligned}$$

and we have

$$E\left\{\sum_{r\neq 0}\sum_{n=-\infty}^{\infty} E\{p(t+\tau-nT-P_n-\Theta)\right.$$
$$\left.\times p(t-nT-rT-P_{n+r}-\Theta)\mid\Theta\}\right\}$$

$$= E\left\{\sum_{r\neq 0}\sum_{n=-\infty}^{\infty}\int_{-\infty}^{\infty} p(t+\tau-nT-u-\Theta)f_P(u)\,du\right.$$
$$\left.\times \int_{-\infty}^{\infty} p(t-nT-rT-v-\Theta)f_P(v)\,dv\right\}$$

$$= \sum_{r\neq 0}\iint_{-\infty}^{\infty} f_P(u)f_P(v)\left[\sum_{n=-\infty}^{\infty}\frac{1}{T}\int_{-T/2}^{T/2} p(t+\tau-nT-u-\theta)\right.$$
$$\left.\times p(t-nT-rT-v-\theta)\,d\theta\right]du\,dv$$

$$= \sum_{r\neq 0}\iint_{-\infty}^{\infty} f_P(u)f_P(v)\frac{1}{T}\int_{-\infty}^{\infty} p(s+\tau+v-u+rT)$$
$$\times p(s)\,ds\,du\,dv$$

$$= \frac{1}{T}\sum_{r\neq 0}\iint_{-\infty}^{\infty} f_P(u)f_P(v)r_p(\tau+rT+v-u)\,du\,dv$$

$$= \frac{1}{T}\sum_{r\neq 0}\int_{-\infty}^{\infty} r_p(\tau+rT+z)\int_{-\infty}^{\infty} f_P(u)f_P(z+u)\,du\,dz$$

$$= \frac{1}{T}\sum_{r\neq 0}\int_{-\infty}^{\infty} r_p(\tau+rT+z)r_f(z)\,dz$$

$$= \frac{1}{T}\sum_{r\neq 0}\int_{-\infty}^{\infty} r_p(\tau+rT-z)r_f(z)\,dz \quad \text{(since } r_f(\cdot) \text{ is even)}$$

$$= \frac{1}{T}\sum_{r\neq 0} r_p(\tau+rT)\otimes r_f(\tau+rT)$$

$$= \frac{1}{T}\sum_{n\neq 0} r_p(\tau-nT)\otimes r_f(\tau-nT).$$

Therefore, we have the result (4.43). ∎

Chapter 5

2. Consider the linear combination

$$Z = \sum_{i=1}^{n} a_i X(t_i) = \sum_{i=1}^{n} \sum_{m=-\infty}^{\infty} a_i V(mT) h(t_i - mT)$$

$$= \sum_{m=-\infty}^{\infty} V(mT) c_m,$$

where

$$c_m \triangleq \sum_{i=1}^{n} a_i h(t_i - mT),$$

which is a nonrandom constant for each m. Thus, Z is a linear combination of jointly Gaussian random variables $\{V(mT)\}$ and is, therefore, a Gaussian random variable. Thus, $X(t)$ is a Gaussian random process.

8. From exercise 12d in Chapter 1, we know that $X \mid Y$ is Gaussian with parameters α'' and β'' given by

$$\beta'' = \beta + \frac{\gamma(y - \beta')\alpha}{\alpha'}$$

$$\alpha'' = \alpha\sqrt{1 - \gamma^2}.$$

Let $Y = X(t_1)$ and $X = X(t_2)$, and

$$\beta = E\{X\} = E\{X(t_2)\} = m_X(t_2)$$

$$\alpha^2 = E\{(X - \beta)^2\} = E\{[X(t_2) - m_X(t_2)]^2\} = \sigma_X^2(t_2)$$

$$\beta' = E\{Y\} = E\{X(t_1)\} = m_X(t_1)$$

$$\alpha'^2 = E\{(Y - \beta')^2\} = E\{[X(t_1) - m_X(t_1)]^2\} = \sigma_X^2(t_1)$$

$$\gamma = \frac{E\{(X - \beta)(Y - \beta')\}}{\alpha\alpha'}$$

$$= \frac{E\{[X(t_2) - m_X(t_2)][X(t_1) - m_X(t_1)]\}}{\sigma_X(t_2)\sigma_X(t_1)}$$

$$= \frac{K_X(t_1, t_2)}{\sigma_X(t_1)\sigma_X(t_2)};$$

then

$$E\{X(t_2) \mid X(t_1)\} = E\{X \mid Y\}$$

$$= \beta'' = m_X(t_2) + \frac{K_X(t_1, t_2)}{\sigma_X^2(t_1)}[X(t_1) - m_X(t_1)].$$

15. (a) From (5.6) we obtain

$$\int_{-\infty}^{\infty} f_{X(t_3)|X(t_2)}(x_3 \mid x_2) f_{X(t_2)|X(t_1)}(x_2 \mid x_1) \, dx_2$$

$$= \int_{-\infty}^{\infty} f_{X(t_3)|X(t_2)X(t_1)}(x_3 \mid x_2, x_1) f_{X(t_2)|X(t_1)}(x_2 \mid x_1) \, dx_2.$$

But then by use of the definition of conditional probability density and $X(t_3) \mid X(t_2) X(t_1) = [X(t_3) \mid X(t_2)] \mid X(t_1)$, this latter expression becomes

$$\int_{-\infty}^{\infty} f_{X(t_3)X(t_2)|X(t_1)}(x_3, x_2 \mid x_1) \, dx_2 = f_{X(t_3)|X(t_1)}(x_3 \mid x_1).$$

This verifies (5.9). The discrete-distribution counterpart of (5.9) is given by

$$P_{X(t_3)|X(t_1)}(x_3 \mid x_1) = \sum_{x_2} P_{X(t_3)|X(t_2)}(x_3 \mid x_2) P_{X(t_2)|X(t_1)}(x_2 \mid x_1).$$

17. (a) We consider the conditional density

$$f = f_{X_n|X_{n-2}X_{n-3}\ldots X_1}(x_n \mid x_{n-2}, x_{n-3}, \ldots, x_1).$$

From the definition of conditional probability density, we have

$$f = \int_{-\infty}^{\infty} f_{X_n|X_{n-1}X_{n-2}\ldots X_1}(x_n \mid x_{n-1}, x_{n-2}, \ldots, x_1)$$

$$\times f_{X_{n-1}|X_{n-2}X_{n-3}\ldots X_1}(x_{n-1} \mid x_{n-2}, x_{n-3}, \ldots, x_1) \, dx_{n-1}.$$

If

$$f_{X_n|X_{n-1}X_{n-2}\ldots X_1} = f_{X_n|X_{n-1}} \tag{*}$$

for all n, then

$$f = \int_{-\infty}^{\infty} f_{X_n|X_{n-1}}(x_n \mid x_{n-1}) f_{X_{n-1}|X_{n-2}}(x_{n-1} \mid x_{n-2}) \, dx_{n-1},$$

which is independent of $x_{n-3}, x_{n-4}, \ldots, x_1$. Therefore,

$$f_{X_n|X_{n-2}X_{n-3}\ldots X_1} = f_{X_n|X_{n-2}}.$$

By the same method it can be shown that (*) implies

$$f_{X_n|X_{n-m}X_{n-m-1}\ldots X_1} = f_{X_n|X_{n-m}}$$

for all $m \geq 0$. Furthermore, it follows from this that

$$f_{X_n|X_{n-m_1}X_{n-m_2}\ldots} = f_{X_n|X_{n-m_1}}$$

for all $m_1 \leq m_2 \leq m_3 \ldots$, since if $X_n \mid X_{n-m}$ is independent of $X_{n-m_1-1}, X_{n-m_1-2}, \ldots, X_1$, then it is certainly independent of a subset of these random variables. This final result can be re-expressed as

$$f_{X_{n_m}|X_{n_{m-1}}\ldots X_{n_1}} = f_{X_{n_m}|X_{n_{m-1}}} \qquad (**)$$

for all $n_m \geq n_{m-1} \geq \ldots \geq n_1$. Thus, we have proved that (**) follows from (*). It can also be seen from this argument that if (*) is translation invariant, then so too is (**).

(b) If the process of interest has starting time t_0 and known starting value $x_0 \neq 0$, then without loss of generality we can work with the process obtained by subtracting the constant x_0 and shifting in time by t_0 so that $X(0) = 0$. Since the events $[X(t_i) = x_i$ and $X(t_j) = x_j]$ and $[X(t_i) = x_i$ and $X(t_j) - X(t_i) = x_j - x_i]$ are identical, then their probability densities are equal. It follows that

$$f_{X(t_n)\ldots X(t_1)}(x_n, \ldots, x_1)$$
$$= f_{X(t_n)-X(t_{n-1}),\ldots,X(t_2)-X(t_1),X(t_1)}(x_n - x_{n-1}, \ldots, x_2 - x_1, x_1).$$

Now, if $X(0) = 0$, then $X(t_1) = X(t_1) - X(0)$, and if $X(t)$ has independent increments, then the above equation reduces to

$$f_{X(t_n)\ldots X(t_1)}(x_n, \ldots, x_1)$$
$$= f_{X(t_n)-X(t_{n-1})}(x_n - x_{n-1}) \cdots f_{X(t_2)-X(t_1)}(x_2 - x_1) f_{X(t_1)}(x_1).$$

By the same argument,

$$f_{X(t_i)-X(t_j)}(x_i - x_j) = \frac{f_{X(t_i)-X(t_j)}(x_i - x_j) f_{X(t_j)}(x_j)}{f_{X(t_j)}(x_j)} = \frac{f_{X(t_i)X(t_j)}(x_i, x_j)}{f_{X(t_j)}(x_j)}.$$

Consequently,

$$f_{X(t_n)\ldots X(t_1)}(x_n, \ldots, x_1) = \frac{f_{X(t_n)X(t_{n-1})}(x_n, x_{n-1}) \cdots f_{X(t_2)X(t_1)}(x_2, x_1)}{f_{X(t_{n-1})}(x_{n-1}) \cdots f_{X(t_2)}(x_2)}.$$

(c) From the definition of conditional probability density,

$$f_{X(t_n)|X(t_{n-1})\ldots X(t_1)} = \frac{f_{X(t_n)\ldots X(t_1)}}{f_{X(t_{n-1})\ldots X(t_1)}}.$$

Using the result from (b) in both the numerator and denominator yields

$$f_{X(t_n)|X(t_{n-1})\ldots X(t_1)} = \frac{f_{X(t_n)X(t_{n-1})}}{f_{X(t_{n-1})}} = f_{X(t_n)|X(t_{n-1})}.$$

Therefore, $X(t)$ is a Markov process.

23. (i) From (5.27), we obtain

$$0 = E\{X(t_n) - X(t_{n-1}) \mid X(t_1), X(t_2), \ldots, X(t_{n-1})\}$$
$$= E\{X(t_n) \mid X(t_1), X(t_2), \ldots, X(t_{n-1})\}$$
$$\quad - E\{X(t_{n-1}) \mid X(t_1), X(t_2), \ldots, X(t_{n-1})\}$$
$$= E\{X(t_n) \mid X(t_1), X(t_2), \ldots, X(t_{n-1})\} - X(t_{n-1}).$$

Therefore,

$$E\{X(t_n) \mid X(t_1), X(t_2), \ldots, X(t_{n-1})\} = X(t_{n-1}).$$

(ii) Since an independent-increment process with known initial value is a Markov process, then

$$f_{X(t_n)\mid X(t_1)X(t_2)\ldots X(t_{n-1})} = f_{X(t_n)\mid X(t_{n-1})}$$

and (5.28), therefore, reduces to (5.30).

27. (a) We have

$$X_n = aX_{n-1} + bX_{n-2} + Z_n,$$

where Z_n is independent of $X_{n-1}, X_{n-2}, \ldots, X_1$. Thus,

$$f_{X_n\mid X_{n-1}X_{n-2}\ldots X_1}(x_n \mid x_{n-1}, x_{n-2}, \ldots, x_1)$$
$$= f_{Z_n\mid X_{n-1}X_{n-2}\ldots X_1}(x_n - ax_{n-1} - bx_{n-2})$$
$$= f_{Z_n\mid X_{n-1}X_{n-2}}(x_n - ax_{n-1} - bx_{n-2})$$
$$= f_{X_n\mid X_{n-1}X_{n-2}}(x_n).$$

Hence, X_n is a second-order Markov process.

(b) For $n \leq n_0$, we have

$$E\{X_n\} = aE\{X_{n-1}\} + bE\{X_{n-2}\} + E\{Z_n\}$$

or

$$m_X = am_X + bm_X + m_Z.$$

Thus,

$$m_Z = m_X(1 - a - b).$$

For $n = n_0 + 1$, we have

$$E\{X_{n_0+1}\} = aE\{X_{n_0}\} + bE\{X_{n_0-1}\} + E\{Z_{n_0+1}\}$$
$$= am_X + bm_X + m_Z = m_X,$$

where the last equality follows from the result for $n \leq n_0$. Thus,

$$m_X(n_0 + 1) = m_X.$$

Similarly,
$$m_X(n_0 + i) = m_X, \quad \text{for } i = 2, 3, 4, \ldots.$$
Furthermore, let $n = n_0 + 1$ and $k \le n_0$; then we have
$$X_{n_0+1} = aX_{n_0} + bX_{n_0-1} + Z_{n_0+1}$$
and
$$R_X(n_0 + 1, k) \triangleq E\{X_{n_0+1} X_k\}$$
$$= aE\{X_{n_0} X_k\} + bE\{X_{n_0-1} X_k\} + E\{Z_{n_0+1} X_k\}$$
$$= aR_X(n_0 - k) + bR_X(n_0 - 1 - k) + m_Z m_X$$
$$= aR_X(n_0 + 1 - k - 1) + bR_X(n_0 + 1 - k - 2)$$
$$+ m_Z m_X,$$
which is a function of only $n_0 + 1 - k$. Therefore,
$$R_X(n_0 + 1, k) = R_X(n_0 + 1 - k).$$
Similarly,
$$R_X(n_0 + 2, k) = R_X(n_0 + 2 - k) \quad \text{for } k \le n_0 + 1$$
and, likewise,
$$R_X(n_0 + i, k) = R_X(n_0 + i - k)$$
$$\text{for } k \le n_0 + i - 1, \quad \text{for } i = 3, 4, 5, \ldots.$$
Therefore, since $E\{X_n\} = m_X$ for $n > n_0$ and $E\{X_n X_m\}$ depends only on the difference $n - m$ for $n, m > n_0$, $\{X_n\}$ is a WSS process for $n > n_0$.

30. (a) Since $X(t)$ is WSS, then we have
 (i)
$$R_X(\tau) = E\{X(t + \tau)X(t)\} = E\{X(t)X(t + \tau)\}$$
$$= E\{X(t + \tau - \tau)X(t + \tau)\}$$
$$= E\{X(t' - \tau)X(t')\} = R_X(-\tau), \quad t' = t + \tau.$$
 (ii) Using the Cauchy-Schwarz inequality, we have
$$|R_X(\tau)| = |E\{X(t + \tau)X(t)\}|$$
$$\le E\{X^2(t + \tau)\}^{1/2} E\{X^2(t)\}^{1/2} = R_X(0).$$
 (iii) Using the Cauchy-Schwarz inequality, we have
$$|E\{X(t)[X(t + \tau + \varepsilon) - X(t + \tau)]\}|$$
$$\le E\{X^2(t)\}^{1/2} E\{[X(t + \tau + \varepsilon) - X(t + \tau)]^2\}^{1/2},$$

which is equivalent to
$$|R_X(\tau + \varepsilon) - R_X(\tau)| \le [2R_X(0)\{R_X(0) - R_X(\varepsilon)\}]^{1/2},$$
which reveals that
$$\lim_{\varepsilon \to 0} R_X(\tau + \varepsilon) = R_X(\tau) \quad \text{if} \quad \lim_{\varepsilon \to 0} R_X(\varepsilon) = R_X(0).$$
Hence, R_X is continuous at τ if it is continuous at 0.

(b) We have
$$\text{MSE} \triangleq E\{[X(t) - X(t - \tau_0)]^2\}$$
$$= E\{X^2(t) + X^2(t - \tau_0) - 2X(t)X(t - \tau_0)\}$$
$$= 2[R_X(0) - R_X(\tau_0)].$$
Thus, if $R_X(\tau_0) = R_X(0)$, then MSE $= 0$ for all t. ∎

Chapter 6

4. From (6.13), we have
$$R_W(t_1, t_2) = \alpha^2 \min\{t_1, t_2\} = \alpha^2[t_1 u(t_2 - t_1) + t_2 u(t_1 - t_2)].$$
Therefore,
$$R_Z(t_1, t_2) = \frac{\partial^2}{\partial t_1 \partial t_2} R_W(t_1, t_2)$$
$$= \frac{\partial}{\partial t_2} \alpha^2 [u(t_2 - t_1) - t_1 \delta(t_2 - t_1) + t_2 \delta(t_1 - t_2)]$$
$$= \frac{\partial}{\partial t_2} \alpha^2 [u(t_2 - t_1) + (t_2 - t_1)\delta(t_2 - t_1)]$$
$$= \frac{\partial}{\partial t_2} \alpha^2 u(t_2 - t_1)$$
$$= \alpha^2 \delta(t_2 - t_1) = \alpha^2 \delta(t_1 - t_2).$$

6. (a) Substitution of $w = r\Delta_w$ and $t = n\Delta_t$ into (6.52) yields
$$P_{W_\Delta} \simeq \frac{1}{\sqrt{2\pi t/\Delta_t}} \exp\left\{-\frac{1}{2} \frac{(w/\Delta_w)^2}{t/\Delta_t}\right\}.$$
Then, use of the hint and (6.5) yields
$$f_{W(t)}(w) = \lim_{\Delta_w \to 0} \frac{P_{W_\Delta}}{\Delta_w}$$
$$= \lim_{\Delta_w \to 0} \frac{1}{\Delta_w \sqrt{2\pi \alpha^2 t/\Delta_w^2}} \exp\left\{-\frac{1}{2} \frac{(w/\Delta_w)^2}{\alpha^2 t/\Delta_w^2}\right\}$$
$$= \frac{1}{\sqrt{2\pi \alpha^2 t}} \exp\left\{-\frac{1}{2} \frac{w^2}{\alpha^2 t}\right\}.$$

Since $\Delta_w = \alpha\sqrt{\Delta_t}$, then

$$\frac{w}{r} = \Delta_w = \alpha\sqrt{\Delta_t} = \alpha\left(\frac{t}{n}\right)^{1/2}$$

and, therefore, $r/\sqrt{n} = w/\alpha\sqrt{t}$. Consequently, we have the limit

$$\lim_{n\to\infty} \frac{r}{\sqrt{n}} = \lim_{n\to\infty} \frac{w}{\alpha\sqrt{t}} = \frac{w}{\alpha\sqrt{t}}$$

for fixed w and t. Thus, r is indeed on the order of $\sqrt{n}$ as $n \to \infty$.

(b) Let $Z_i = W(t_i) - W(t_{i-1})$, $i = 1, 2, \ldots, n$; then $\{Z_i\}$ is an independent identically distributed sequence with Gaussian probability density. Thus, $\{Z_i\}$ are jointly Gaussian random variables and, therefore, for arbitrary $\{a_i\}$,

$$Y = \sum_{i=1}^{n} a_i Z_i = \sum_{i=1}^{n} a_i[W(t_i) - W(t_{i-1})]$$

is a Gaussian random variable. But Y can also be expressed as

$$Y = \sum_{i=0}^{n} b_i W(t_i),$$

where $b_0 = a_1$, $b_n = a_n$, and $b_i = a_i - a_{i+1}$, $i = 1, \ldots, n-1$. If we let $t_0 = 0$, then $W(t_0) = 0$, and also, since $\{a_i\}_1^n$ are arbitrary, $\{b_i\}_1^n$ are arbitrary. Therefore, since Y is a Gaussian random variable, then $W(t)$ is a Gaussian random process.

7. (a) Let $t_1 < t_2$; then, from (6.12) and (6.13), we have

$$E\{W(t_1)\} = 0 \quad \text{Var}\{W(t_1)\} = \alpha^2 t_1$$

$$E\{W(t_2)\} = 0, \quad \text{Var}\{W(t_2)\} = \alpha^2 t_2$$

$$\gamma = \frac{K_W(t_1, t_2)}{\sqrt{\text{Var}\{W(t_1)\}\text{Var}\{W(t_2)\}}} = \frac{\alpha^2 t_1}{\alpha^2 \sqrt{t_1 t_2}} = \left(\frac{t_1}{t_2}\right)^{1/2},$$

from which it follows that $W(t)$ is nonstationary. However, use of the result of exercise 12d in Chapter 1 for jointly Gaussian random variables yields

$$f_{W(t_2)|W(t_1)}(w_2 | w_1) = \frac{1}{\sqrt{2\pi\alpha^2 t_2(1-\gamma^2)}}$$

$$\times \exp\left\{-\frac{1}{2}\frac{(w_2 - \gamma w_1\sqrt{\alpha^2 t_2}/\sqrt{\alpha^2 t_1})^2}{\alpha^2 t_2(1-\gamma^2)}\right\}$$

$$= \frac{1}{\sqrt{2\pi\alpha^2(t_2-t_1)}} \exp\left\{-\frac{1}{2}\frac{(w_2-w_1)^2}{\alpha^2(t_2-t_1)}\right\}$$

$$= f_{W(t_2+t)|W(t_1+t)}(w_2 | w_1),$$

from which it follows that $W(t)$ is a homogeneous process. It follows from exercise 17b in Chapter 5 that $W(t)$ is a Markov process.

(b) It is shown in (a) that $f_{W(t)|W(t_0)}(w \mid w_0)$ is given by (6.53) with replacement of t with $t - t_0$ and w with $w - w_0$ for $t_0 \geq 0$:

$$f_{W(t)|W(t_0)}(w \mid w_0) = \frac{1}{\sqrt{2\pi\alpha^2(t - t_0)}} \exp\left\{-\frac{1}{2}\frac{(w - w_0)^2}{\alpha^2(t - t_0)}\right\}.$$

To see that this function satisfies the diffusion equation, we observe that

$$\frac{\partial}{\partial t} f_{W(t)|W(t_0)}(w \mid w_0) = \frac{1}{\sqrt{2\pi\alpha^2(t - t_0)}} \left[\frac{-1}{2(t - t_0)} + \frac{(w - w_0)^2}{2\alpha^2(t - t_0)^2}\right]$$

$$\times \exp\left\{-\frac{1}{2}\frac{(w - w_0)^2}{\alpha^2(t - t_0)}\right\}$$

and

$$\frac{\partial^2}{\partial w^2} f_{W(t)|W(t_0)}(w \mid w_0) = \frac{1}{\sqrt{2\pi\alpha^2(t - t_0)}} \left[\frac{-1}{\alpha^2(t - t_0)} + \frac{(w - w_0)^2}{\alpha^4(t - t_0)^2}\right]$$

$$\times \exp\left\{-\frac{1}{2}\frac{(w - w_0)^2}{\alpha^2(t - t_0)}\right\}.$$

Therefore, we have

$$\frac{\partial}{\partial t} f_{W(t)|W(t_0)}(w \mid w_0) = \frac{\alpha^2}{2}\frac{\partial^2}{\partial w^2} f_{W(t)|W(t_0)}(w \mid w_0).$$

To show that the diffusion equation follows from the Fokker-Planck equation, we observe that (5.66e) and (6.66f) yield

$$m(w, t) = \frac{\partial}{\partial s} m_{W(s)|W(t)=w}\bigg|_{s=t^+} = \frac{\partial}{\partial s} w = 0$$

$$\sigma^2(w, t) = \frac{\partial}{\partial s} \sigma^2_{W(s)|W(t)=w}\bigg|_{s=t^+} = \frac{\partial}{\partial s} \alpha^2(s - t) = \alpha^2.$$

Therefore, the Fokker-Planck equation (5.66a) reduces to

$$\frac{\partial}{\partial t} f_{W(t)|W(t_0)}(w \mid w_0) - \frac{1}{2}\frac{\partial^2}{\partial w^2}[\alpha^2 f_{W(t)|W(t_0)}(w \mid w_0)] = 0,$$

which is the diffusion equation.

10. Let $t_1 < t_2$ and $N(t_1) = n_1 \leq N(t_2) = n_2$. Then

$$R_W(t_1, t_2) = E\{W(t_1)W(t_2)\}$$

$$= E\{E\{W(t_1)W(t_2) \mid N(t_1) \text{ and } N(t_2)\}\}$$

$$= E\left\{E\left\{\sum_{i=1}^{N(t_1)} Z_i \sum_{j=1}^{N(t_2)} Z_j \mid N(t_1) \text{ and } N(t_2)\right\}\right\}$$

$$= E\left\{E\left\{\sum_{i=1}^{N(t_1)} Z_i \sum_{j=1}^{N(t_1)} Z_j\right.\right.$$

$$\left.\left.+ \sum_{i=1}^{N(t_1)} Z_i \sum_{j=N(t_1)+1}^{N(t_2)} Z_j \mid N(t_1) \text{ and } N(t_2)\right\}\right\}$$

$$= E\left\{\sum_{i=1}^{N(t_1)} E\{Z_i^2\} + \sum_{\substack{i,j=1 \\ i\neq j}}^{N(t_1)} E\{Z_i Z_j\}\right.$$

$$\left.+ E\left\{\sum_{i=1}^{N(t_1)} Z_i\right\} E\left\{\sum_{j=N(t_1)+1}^{N(t_2)} Z_j\right\} \mid N(t_1) \text{ and } N(t_2)\right\}$$

$$= E\{(\sigma_Z^2 + m_Z^2)N(t_1) + m_Z^2 N(t_1)[N(t_1) - 1]$$

$$+ m_Z^2 N(t_1)[N(t_2) - N(t_1)]\}$$

$$= (\sigma_Z^2 + m_Z^2) E\{N(t_1)\} + m_Z^2 E\{N^2(t_1) - N(t_1)\}$$

$$+ m_Z^2 E\{N(t_1)\} E\{N(t_2) - N(t_1)\},$$

where the independence of Z_i and Z_j for $i \neq j$ and of $N(t_1)$ and $N(t_2) - N(t_1)$ has been used. Carrying out the expectation in the above equation yields (with the use of $E\{N^2(t_1)\} = \lambda t_1 + (\lambda t_1)^2$)

$$R_W(t_1, t_2) = (\sigma_Z^2 + m_Z^2)\lambda t_1 + m_Z^2(\lambda t_1)^2 + m_Z^2 \lambda t_1(\lambda t_2 - \lambda t_1)$$

$$= (\sigma_Z^2 + m_Z^2)\lambda t_1 + m_Z^2(\lambda t_1)(\lambda t_2).$$

Hence, use of (6.38) yields

$$K_W(t_1, t_2) = R_W(t_1, t_2) - m_W(t_1)m_W(t_2) = \lambda(\sigma_Z^2 + m_Z^2)t_1, \quad t_1 < t_2.$$

Similarly, it can be shown that

$$K_W(t_1, t_2) = \lambda(\sigma_Z^2 + m_Z^2)t_2, \quad t_2 < t_1.$$

Thus,

$$K_W(t_1, t_2) = \lambda(\sigma_Z^2 + m_Z^2)\min\{t_1, t_2\}.$$

Finally, to verify that $E\{N^2(t)\} = \lambda t + (\lambda t)^2$, we proceed as follows:

$$E\{N^2(t)\} = \sum_{n=1}^{\infty} n^2 P_t(n)$$

$$= \sum_{n=1}^{\infty} n^2 \frac{(\lambda t)^n e^{-\lambda t}}{n!}$$

$$= e^{-\lambda t} \sum_{n=1}^{\infty} \frac{[n(n-1) + n](\lambda t)^n}{n!}$$

$$= e^{-\lambda t} \left[(\lambda t)^2 \sum_{m=0}^{\infty} \frac{(\lambda t)^m}{m!} + \lambda t \sum_{q=0}^{\infty} \frac{(\lambda t)^q}{q!} \right]$$

$$= (\lambda t)^2 + \lambda t.$$

11. From the definition of conditional probability and (6.35) and (6.25), we obtain the conditional probability,

$$\text{Prob}\{N(t_m) = n_m \mid N(t_{m-1}) = n_{m-1} \ldots N(t_1) = n_1\}$$

$$= \frac{\text{Prob}\{N(t_1) = n_1, N(t_2) = n_2, \ldots, N(t_m) = n_m\}}{\text{Prob}\{N(t_1) = n_1, N(t_2) = n_2, \ldots, N(t_{m-1}) = n_{m-1}\}}$$

$$= P_{t_m - t_{m-1}}(n_m - n_{m-1})$$

$$= \frac{\text{Prob}\{N(t_m) = n_m, N(t_{m-1}) = n_{m-1}\}}{\text{Prob}\{N(t_{m-1}) = n_{m-1}\}}$$

$$= \text{Prob}\{N(t_m) = n_m \mid N(t_{m-1}) = n_{m-1}\}.$$

Therefore, by definition [cf.(5.11)], the Poisson process is a Markov process. Furthermore, the preceding shows that the transition probability is time-translation invariant; therefore, this is a homogeneous Markov process.

17. (a) Since $[0, t + \Delta_t) = [0, t) \cup [t, t + \Delta_t)$, then from (6.32) we obtain

$$\text{Prob}\{N(t + \Delta_t) - N(t) = 0, N(t) - N(0) = 0\}$$

$$= P_{\Delta_t}(0) P_t(0)$$

$$= \text{Prob}\{N(t + \Delta_t) - N(0) = 0\}$$

$$= \text{Prob}\{N(t + \Delta_t) = 0\}$$

$$= P_{t + \Delta_t}(0).$$

Substitution of (6.33) into this equation yields

$$P_{t + \Delta_t}(0) = P_t(0) P_{\Delta_t}(0) = (1 - \lambda \Delta_t) P_t(0), \quad \text{as } \Delta_t \to 0,$$

or, equivalently,

$$\frac{P_{t + \Delta_t}(0) - P_t(0)}{\Delta_t} = -\lambda P_t(0).$$

Hence,

$$\lim_{\Delta_t \to 0} \frac{P_{t + \Delta_t}(0) - P_t(0)}{\Delta_t} = \frac{d}{dt} P_t(0) = -\lambda P_t(0).$$

The solution to this differential equation is
$$P_t(0) = ce^{-\lambda t}.$$
But $P_0(0) = 1$; thus $c = 1$ and
$$P_t(0) = e^{-\lambda t}.$$

(b) Use of (6.33) in
$$P_{t+\Delta_t}(n) = \sum_{k=0}^{n} P_t(n-k) P_{\Delta_t}(k)$$
yields
$$P_{t+\Delta_t}(n) = P_t(n) P_{\Delta_t}(0) + P_t(n-1) P_{\Delta_t}(1) + \sum_{k=2}^{n} P_t(n-k) P_{\Delta_t}(k)$$
$$= (1 - \lambda \Delta_t) P_t(n) + \lambda \Delta_t P_t(n-1) + 0$$
as $\Delta_t \to 0$ [since $P_{\Delta_t}(k) \to 0$ for $k > 1$ from (6.33)]. Thus,
$$\frac{P_{t+\Delta_t}(n) - P_t(n)}{\Delta_t} = \lambda P_t(n-1) - \lambda P_t(n).$$
Hence,
$$\lim_{\Delta_t \to 0} \frac{P_{t+\Delta_t}(n) - P_t(n)}{\Delta_t} = \frac{d}{dt} P_t(n) = \lambda P_t(n-1) - \lambda P_t(n),$$
which is (6.61). To solve (6.61), we can integrate $dP_t(n)$ or we can multiply both sides of (6.61) by $e^{\lambda t}$ to obtain
$$e^{\lambda t} \frac{d}{dt} P_t(n) + \lambda e^{\lambda t} P_t(n) = \lambda e^{\lambda t} P_t(n-1),$$
which leads to
$$\frac{d}{dt} e^{\lambda t} P_t(n) = \lambda e^{\lambda t} P_t(n-1).$$
For $n = 1$, substituting (6.60) into this equation yields
$$\frac{d}{dt} e^{\lambda t} P_t(1) = \lambda$$
or, equivalently,
$$P_t(1) = (\lambda t + c) e^{-\lambda t}.$$
But $P_0(1) = 0$. Thus,
$$P_t(1) = \lambda t e^{-\lambda t}.$$

Similarly, for $n = 2$, we have

$$\frac{d}{dt} e^{\lambda t} P_t(2) = \lambda e^{\lambda t} P_t(1) = \lambda^2 t$$

or, equivalently,

$$P_t(2) = \left(\frac{\lambda^2 t^2}{2} + c\right) e^{-\lambda t} = \frac{(\lambda t)^2}{2} e^{-\lambda t}.$$

Therefore, by induction, we obtain

$$\frac{d}{dt} e^{\lambda t} P_t(n) = \frac{\lambda^n t^{n-1}}{(n-1)!}$$

or, equivalently,

$$P_t(n) = \left(\frac{\lambda^n t^n}{n!} + c\right) e^{-\lambda t} = \frac{(\lambda t)^n}{n!} e^{-\lambda t}. \quad \blacksquare$$

Chapter 7

4. Since $X(t)$ is a WSS integrable process, then we have

$$E\{X(t)\} = m_X$$
$$E\{(X(t_1) - m_X)(X(t_2) - m_X)\} = K_X(t_1 - t_2).$$

The integrated process

$$Y(t) = \int_0^t X(u)\, du, \quad t > 0$$

has mean and variance given by

$$m_Y(t) = E\{Y(t)\} = \int_0^t E\{X(u)\}\, du$$
$$= \int_0^t m_X\, du = m_X t, \quad t > 0$$

and

$$\sigma_Y^2(t) = E\{[Y(t) - m_Y(t)]^2\}$$
$$= \int_0^t \int_0^t E\{[X(t_1) - m_X][X(t_2) - m_X]\}\, dt_1\, dt_2$$
$$= \int_0^t \int_0^t K_X(t_1 - t_2)\, dt_1\, dt_2.$$

Making the change of variables $\tau = t_1 - t_2$ and using the window function

$$h(u) = \begin{cases} 1, & 0 \leq u < t, \\ 0, & \text{otherwise}, \end{cases}$$

$\sigma_Y^2(t)$ can be further expressed as

$$\sigma_Y^2(t) = \int\int_{-\infty}^{\infty} h(t_1)h(t_2)K_X(t_1 - t_2)\, dt_1\, dt_2$$

$$= \int_{-\infty}^{\infty} K_X(\tau) \int_{-\infty}^{\infty} h(t_2 + \tau)h(t_2)\, dt_2\, d\tau$$

$$= \int_{-\infty}^{\infty} K_X(\tau)[h(\tau) \otimes h(-\tau)]\, d\tau$$

$$= t \int_{-t}^{t} \left(1 - \frac{|\tau|}{t}\right) K_X(\tau)\, d\tau, \qquad t > 0,$$

since

$$h(\tau) \otimes h(-\tau) = t - |\tau|, \qquad |\tau| \leq t.$$

15. (a) The autocorrelation of $X^{(-1)}(t)$ is given by

$$R_{X^{(-1)}}(\tau) = E\{X^{(-1)}(t + \tau)X^{(-1)}(t)\}$$

$$= \int_{-\infty}^{t+\tau}\int_{-\infty}^{t} E\{X(u)X(v)\}\, dv\, du$$

$$= \int_{-\infty}^{t+\tau}\int_{-\infty}^{t} R_X(u - v)\, dv\, du.$$

Making the changes of variables $\sigma = u - t$ and then $\mu = \sigma + t - v$ yields

$$R_{X^{(-1)}}(\tau) = \int_{-\infty}^{\tau}\int_{-\infty}^{t} R_X(\sigma + t - v)\, dv\, d\sigma = \int_{-\infty}^{\tau}\int_{\sigma}^{\infty} R_X(\mu)\, d\mu\, d\sigma$$

$$= \int_{-\infty}^{\tau}\int_{\sigma}^{\infty} R_X(\mu)\, d\mu\, d\sigma.$$

(b) For $R_X(\tau) = \delta(\tau)$, we have

$$R_{X^{(-1)}}(\tau) = \int_{-\infty}^{\tau} u(-\sigma)\, d\sigma = \infty.$$

For $R_X(\tau) = e^{-|\tau|}$, we have

$$R_{X^{(-1)}}(\tau) = \int_{-\infty}^{\tau} [2u(-\sigma) + e^{-\sigma}u(\sigma) - e^{\sigma}u(-\sigma)]\, d\sigma = \infty.$$

For $R_X(\tau) = 2\delta(\tau) - e^{-|\tau|}$, we have

$$R_{X^{(-1)}}(\tau) = \int_{-\infty}^{\tau} [e^{-\sigma}u(\sigma) - e^{\sigma}u(-\sigma)]\, d\sigma < \infty.$$

Chapter 8

2. From (8.19), we have

$$\mathrm{Var}\{\hat{m}_X^N\} = \frac{1}{N} \sum_{i=1}^{N} \sum_{j=1}^{N} K_X([i-j]\Delta). \quad (*)$$

Use of the window function $h(k)$ defined by

$$h(k) = \begin{cases} 1, & 1 \le k \le N, \\ 0, & \text{otherwise,} \end{cases}$$

and the change of variables $i - j = k$ in (*) results in

$$\mathrm{Var}\{\hat{m}_X^N\} = \frac{1}{N} \sum_{i,j=-\infty}^{\infty} h(i)h(j) K_X([i-j]\Delta)$$

$$= \frac{1}{N} \sum_{j,k=-\infty}^{\infty} h(k+j)h(j) K_X(k\Delta)$$

$$= \frac{1}{N} \sum_{k=-\infty}^{\infty} K_X(k\Delta) \sum_{j=-\infty}^{\infty} h(k+j)h(j).$$

But

$$\sum_{j=-\infty}^{\infty} h(k+j)h(j) = \begin{cases} N - |k|, & -N \le k \le N, \\ 0, & \text{otherwise.} \end{cases}$$

Hence, (*) simplies to (using $N = T/\Delta$)

$$\mathrm{Var}\{\hat{m}_X^N\} = \mathrm{Var}\{\hat{m}_X^{T/\Delta}\}$$

$$= \frac{1}{N} \sum_{k=-N}^{N} (N - |k|) K_X(k\Delta)$$

$$= \frac{1}{N} K_X(0) + \frac{2}{N} \sum_{k=1}^{N} \left(1 - \frac{k}{N}\right) K_X(k\Delta)$$

$$= \frac{\sigma_X^2}{N} + \frac{2}{T} \sum_{k=1}^{N} \left(1 - \frac{k\Delta}{T}\right) K_X(k\Delta)\Delta.$$

14. Since $X(t)$ and $X(t+\tau)$ become statistically independent as $\tau \to \infty$, then

$$R_X(\tau) = E\{X(t+\tau)X(t)\} \to E\{X(t+\tau)\}E\{X(t)\} = m_X^2.$$

Therefore, $K_X(\tau) \to 0$ and, as a result,

$$\lim_{T \to \infty} \frac{1}{T} \int_0^T K_X(\tau) \, d\tau = 0.$$

Consequently, (8.42b) is satisfied. Thus, $X(t)$ exhibits mean-square ergodicity of the mean. From (8.49) and the statistical independence of $X(t_1 + u)X(t_2 + u)$ and $X(t_1)X(t_2)$ as $u \to \infty$, we see that

$$\lim_{u \to \infty} K_{Y_\tau}(u) = \lim_{u \to \infty} E\{X(t + \tau + u)X(t + u)X(t + \tau)X(t)\} - R_X^2(\tau)$$

$$= \lim_{u \to \infty} E\{X(t + \tau + u)X(t + u)\} E\{X(t + \tau)X(t)\} - R_X^2(\tau)$$

$$= R_X(\tau)R_X(\tau) - R_X^2(\tau)$$

$$= 0.$$

Therefore, (8.48b) is satisfied and $X(t)$ exhibits mean-square ergodicity of the autocorrelation.

20. Let $W(t) \triangleq u[x - X(t)]$. Then

$$\hat{m}_W(T) = \frac{1}{T} \int_0^T u[x - X(t)] \, dt = \hat{F}_X(x)_T$$

is the finite-time fraction-of-time distribution, and

$$m_W = E\{u[x - X(t)]\} = F_X(x)$$

is the probabilistic distribution. It follows that

$$\lim_{T \to \infty} E\{|\hat{F}_X(x)_T - F_X(x)|^2\} = 0$$

if and only if

$$\lim_{T \to \infty} E\{[\hat{m}_W(T) - m_W]^2\} = 0,$$

which holds if and only if (cf. (8.42))

$$\lim_{T \to \infty} \frac{1}{T} \int_0^T K_W(\tau) \, d\tau = 0,$$

where

$$K_W(\tau) = E\{W(t + \tau)W(t)\} - [E\{W(t)\}]^2$$

$$= E\{u[x - X(t + \tau)]u[x - X(t)]\} - F_X^2(x)$$

$$= F_{X(t+\tau)X(t)}(x, x) - F_X^2(x).$$

21. The necessary and sufficient condition for mean-square ergodicity of the distributuion is, from exercise 20,

$$\lim_{T\to\infty} \frac{1}{T}\int_0^T K_W(\tau)\,d\tau = 0,$$

where

$$K_W(\tau) \triangleq F_{X(t+\tau)X(t)}(x, x) - F_X^2(x).$$

Thus, it is required that

$$\lim_{T\to\infty} \frac{1}{T}\int_0^T F_{X(t+\tau)X(t)}(x, x)\,d\tau = F_X^2(x). \quad (*)$$

Since we are given that $X(t + \tau)$ and $X(t)$ become statistically independent as $\tau \to \infty$, then

$$\lim_{\tau\to\infty} F_{X(t+\tau)X(t)}(x, x) = F_X^2(x)$$

and, as a result, (*) is satisfied.

23. (a) From (8.97), we have

$$\hat{F}_X(y) \triangleq \lim_{T\to\infty} \frac{1}{T}\int_{-T/2}^{T/2} u[y - X(t)]\,dt.$$

Use of (8.100) and an interchange of the order of the time-averaging and differentiation operations yields

$$\hat{f}_X(y) = \lim_{T\to\infty} \frac{1}{T}\int_{-T/2}^{T/2} \delta[y - X(t)]\,dt, \quad (*)$$

since

$$\frac{du(y)}{dy} = \delta(y).$$

Substitution of (*) into (8.99) and interchange of the order of the time-averaging and integration operations yields

$$\hat{m}_X = \lim_{T\to\infty} \frac{1}{T}\int_{-T/2}^{T/2} \int_{-\infty}^{\infty} y\delta[y - X(t)]\,dy\,dt$$

$$= \lim_{T\to\infty} \frac{1}{T}\int_{-T/2}^{T/2} x(t) \int_{-\infty}^{\infty} \delta[y - X(t)]\,dy\,dt$$

$$= \lim_{T\to\infty} \frac{1}{T}\int_{-T/2}^{T/2} X(t)\,dt,$$

which is (8.95), as desired.

(b) From (8.101), we have

$$\hat{F}_{X(t_1)X(t_2)}(y_1, y_2)$$

$$\stackrel{\Delta}{=} \lim_{T\to\infty} \frac{1}{T} \int_{-T/2}^{T/2} u[y_1 - X(t_1 + t)]u[y_2 - X(t_2 + t)]\, dt.$$

Use of (8.103) and an interchange of the order of the time-averaging and differentiation operations yields

$$\hat{f}_{X(t_1)X(t_2)}(y_1, y_2)$$

$$= \lim_{T\to\infty} \frac{1}{T} \int_{-T/2}^{T/2} \frac{\partial}{\partial y_1} u[y_1 - X(t_1 + t)] \frac{\partial}{\partial y_2} u[y_2 - X(t_2 + t)]\, dt$$

$$= \lim_{T\to\infty} \frac{1}{T} \int_{-T/2}^{T/2} \delta[y_1 - X(t_1 + t)]\delta[y_2 - X(t_2 + t)]\, dt. \quad (**)$$

Substitution of (**) into (8.102) and an interchange of the order of time-averaging and integration operations yields

$$\hat{R}_X(t_1 - t_2) = \lim_{T\to\infty} \frac{1}{T} \int_{-T/2}^{T/2} \int_{-\infty}^{\infty} y_1 \delta[y_1 - X(t_1 + t)]\, dy_1$$

$$\times \int_{-\infty}^{\infty} y_2 \delta[y_2 - X(t_2 + t)]\, dy_2\, dt$$

$$= \lim_{T\to\infty} \frac{1}{T} \int_{-T/2}^{T/2} X(t_1 + t)X(t_2 + t)\, dt,$$

which is (8.96), as desired. ∎

Chapter 9

3. For a WSS process $X(t)$, we have $m_X(i) = m_X$ and $R_X(i, j) = R_X(i - j)$. Therefore, (9.23) yields

$$m_Y(i) = m_X(i) \otimes h(i) = \sum_{j=-\infty}^{\infty} m_X(i - j)h(j) = m_X \sum_{j=-\infty}^{\infty} h(j),$$

which is (9.25). Also, (9.24) yields

$$R_Y(i, j) = h(i) \otimes R_X(i, j) \otimes h(j)$$

$$= \sum_{l,k=-\infty}^{\infty} R_X(i - j + l - k)h(k)h(l).$$

Use of the change of variables $k - l = r$ in this equation results in

$$R_Y(i, j) = \sum_{r,l=-\infty}^{\infty} R_X(i - j - r)h(l + r)h(l)$$

$$= \sum_{r=-\infty}^{\infty} R_X(i - j - r)[h(r) \otimes h(-r)]$$

$$= [R_X(k) \otimes r_h(k)]_{k=i-j} = R_Y(i - j),$$

where $r_h(k)$ is given by (9.27).

6. (b) For a time-invariant system, we have $h(t, u) = h(t - u)$. Therefore, from (9.43) with $R_X(r, s) = R_X(r - s)$, we obtain

$$R_Y(t, u) = \int\int_{-\infty}^{\infty} h(t - r)h(u - s)R_X(r - s) \, dr \, ds.$$

Use of the change of variables $t - r = z + v$ and $u - s = z$ in this equation yields

$$R_Y(t, u) = \int\int_{-\infty}^{\infty} h(z + v)h(z)R_X(t - u - v) \, dz \, dv$$

$$= \int_{-\infty}^{\infty} R_X(t - u - v)[h(v) \otimes h(-v)] \, dv$$

$$= [R_X(\tau) \otimes r_h(\tau)]_{\tau=t-u} = R_Y(t - u),$$

where $r_h(\tau)$ is given by (9.46).

10. (a) The input-output relationship for a discrete-time filter is

$$Y(i) = \sum_{k=-\infty}^{\infty} X(i - k)h(k).$$

The empirical mean of $Y(i)$ is defined by

$$\hat{m}_Y \triangleq \lim_{N\to\infty} \frac{1}{2N + 1} \sum_{i=-N}^{N} Y(i)$$

$$= \sum_{k=-\infty}^{\infty} \lim_{N\to\infty} \frac{1}{2N + 1} \sum_{i=-N}^{N} X(i - k)h(k)$$

$$= \sum_{k=-\infty}^{\infty} \hat{m}_X h(k) = \hat{m}_X \sum_{k=-\infty}^{\infty} h(k),$$

which is (9.65). The empirical autocorrelation of $Y(i)$ is defined by

$$\hat{R}_Y(k) \triangleq \lim_{N \to \infty} \frac{1}{2N+1} \sum_{i=-N}^{N} Y(i+k)Y(i)$$

$$= \lim_{N \to \infty} \frac{1}{2N+1} \sum_{i=-N}^{N} \sum_{l,j=-\infty}^{\infty} X(i+k-l)X(i-j)h(l)h(j)$$

$$= \sum_{l,j=-\infty}^{\infty} \hat{R}_X(k+j-l)h(l)h(j).$$

Use of the change of variables $r = l - j$ in this equation yields

$$\hat{R}_Y(k) = \sum_{j,r=-\infty}^{\infty} \hat{R}_X(k-r)h(j+r)h(j)$$

$$= \sum_{r=-\infty}^{\infty} \hat{R}_X(k-r)[h(r) \otimes h(-r)] = \hat{R}_X(k) \otimes r_h(k),$$

which is (9.66).

12. Since H is statistically independent of X, then

$$R_Y(\tau) = E\{Y(t+\tau)Y(t)\}$$

$$= E\left\{\int\int_{-\infty}^{\infty} H(t+\tau-u)X(u)H(t-v)H(v)\,du\,dv\right\}$$

$$= \int\int_{-\infty}^{\infty} E\{H(t+\tau-u)H(t-v)\}R_X(u-v)\,du\,dv.$$

Letting $t - v = w$ and $t - u = w - z$ results in

$$R_Y(\tau) = \int\int_{-\infty}^{\infty} E\{H(\tau-z+w)H(w)\}R_X(z)\,dw\,dz$$

$$= \int_{-\infty}^{\infty} E\left\{\int_{-\infty}^{\infty} H(\tau-z+w)H(w)\,dw\right\}R_X(z)\,dz$$

$$= \int_{-\infty}^{\infty} E\{r_H(\tau-z)\}R_X(z)\,dz = R_X(\tau) \otimes E\{r_H(\tau)\}.$$

16. To obtain (9.90), we simply take the expected value of both sides of (9.85a) to obtain

$$m_X(i+1) = E\{X(i+1)\}$$

$$= A(i)E\{X(i)\} + b(i)E\{U(i)\}$$

$$= A(i)m_X(i) + b(i)m_U(i), \quad i \geq 0.$$

To obtain (9.91), we simply multiply (9.86), with i replaced by $i - 1$, by $U(j)$ and take the expected value of both sides to obtain

$$E\{X(i)U(j)\} = \Phi(i, j)E\{X(j)U(j)\}$$
$$+ \sum_{m=j}^{i-1} \Phi(i, m + 1)b(m)E\{U(m)U(j)\}, \quad i > j,$$

or, equivalently,

$$K_{XU}(i, j) = \Phi(i, j)K_{XU}(j, j)$$
$$+ \sum_{m=j}^{i-1} \Phi(i, m + 1)b(m)K_U(m, j), \quad i > j.$$

Now, use of the conditions (9.88) and (9.89) in this equation yields

$$K_{XU}(i, j) = \sigma^2 \Phi(i, j + 1)b(j), \quad i > j,$$

which is the desired result, (9.91).

By analogy with the method used to obtain (9.80) from (9.77), we can obtain (9.92) from (9.86), together with the assumptions (9.88) and (9.89), as follows:

$$K_X(i + 1, i + 1) = E\{X(i + 1)X^T(i + 1)\}$$

$$= E\left\{\left[\Phi(i + 1, j)X(j) + \sum_{m=j}^{i} \Phi(i + 1, m + 1)b(m)U(m)\right]\right.$$

$$\left.\times \left[\Phi(i + 1, j)X(j) + \sum_{n=j}^{i} \Phi(i + 1, n + 1)b(n)U(n)\right]^T\right\},$$

since the cross terms are zero by virtue of (9.88). Thus,

$$K_X(i + 1, i + 1)$$
$$= \Phi(i + 1, j)K_X(j, j)\Phi^T(i + 1, j)$$
$$+ \sum_{m=j}^{i} \Phi(i + 1, m + 1)b(m)K_{XU}^T(j, m)\Phi^T(i + 1, j)$$
$$+ \Phi(i + 1, j)\sum_{n=j}^{i} K_{XU}(j, n)b^T(n)\Phi^T(i + 1, n + 1)$$
$$+ \sum_{n=j}^{i}\sum_{m=j}^{i} \Phi(i + 1, m + 1)b(m)K_U(m, n)b^T(n)\Phi^T$$
$$\times (i + 1, n + 1)$$

$$= \Phi(i + 1, j)K_X(j, j)\Phi^T(i + 1, j)$$

$$+ \sigma^2 \sum_{n=j}^{i} \Phi(i + 1, n + 1)b(n)b^T(n)\Phi^T(i + 1, n + 1),$$

$$i \geq j \geq 0,$$

since the middle two terms are zero by virtue of (9.88) and since the last term simplifies with the use of (9.89). Letting $j = i$ and using (9.87a) now yields

$$K_X(i + 1, i + 1) = A(i)K_X(i, i)A^T(i) + \sigma^2 b(i)b^T(i), \qquad i \geq 0,$$

which is the desired result, (9.92).

20. Letting $t \geq u > v \geq 0$, multiplying (9.107)—with its mean subtracted—by $U(u) - m_U$, and taking expected values yields

$$K_{XU}(t, u) = E\{[X(t) - m_X][U(u) - m_U]\}$$

$$= \Phi(t, v)E\{[X(v) - m_X][U(u) - m_U]\}$$

$$+ \int_v^t \Phi(t, w)b(w)E\{[U(w) - m_U][U(u) - m_U]\}\, dw$$

$$= \Phi(t, v)K_{XU}(v, u) + \int_v^t \Phi(t, w)b(w)K_U(w, u)\, dw.$$

Use of the conditions (9.105) and (9.106) in this equation yields

$$K_{XU}(t, u) = \int_v^t \Phi(t, w)b(w)\delta(w - u)\, dw, \qquad 0 \leq v < u \leq t,$$

which is (9.109). ∎

Chapter 10

1. (a) Use of the change of variables $\tau = t_1 - t_2$ in (10.7) yields

$$E\{\tilde{X}(f_1)\tilde{X}^*(f_2)\} = \iint_{-\infty}^{\infty} e^{-i2\pi(f_1 t_1 - f_2 t_2)} R_X(t_1 - t_2)\, dt_1\, dt_2$$

$$= \iint_{-\infty}^{\infty} e^{-i2\pi(f_2 - f_1)t_2} e^{-i2\pi f_1 \tau} R_X(\tau)\, d\tau\, dt_2$$

$$= \int_{-\infty}^{\infty} R_X(\tau)e^{-i2\pi f_1 \tau}\, d\tau \int_{-\infty}^{\infty} e^{-i2\pi(f_2 - f_1)t_2}\, dt_2$$

$$= S_X(f_1)\delta(f_2 - f_1).$$

(b) Substitution of (10.11) and (10.6) into the left-side of (10.142) yields

$$\tilde{X}^{(-1)}(f_1) - \tilde{X}^{(-1)}(f_2) = \int_{-\infty}^{f_1} \tilde{X}(v) \, dv - \int_{-\infty}^{f_2} \tilde{X}(v) \, dv$$

$$= \int_{f_2}^{f_1} \tilde{X}(v) \, dv = \int_{f_2}^{f_1} \int_{-\infty}^{\infty} X(t) e^{-i2\pi vt} \, dt \, dv$$

$$= \int_{-\infty}^{\infty} X(t) \left[\frac{e^{-i2\pi f_2 t} - e^{-i2\pi f_1 t}}{i2\pi t} \right] dt.$$

(c) To simplify the notation, we use the change of variables $f_2 = f - \Delta/2$ and $f_1 = f + \Delta/2$. Then, using (10.142), we obtain

$$MS \triangleq E\{|\tilde{X}^{(-1)}(f_1) - \tilde{X}^{(-1)}(f_2)|^2\}$$

$$= \iint_{-\infty}^{\infty} E\{X(t)X(s)\} \frac{\sin(\pi\Delta t)}{\pi t} \frac{\sin(\pi\Delta s)}{\pi s} e^{-i2\pi f(t-s)} \, dt \, ds,$$

where we have used the identity

$$\frac{e^{i\pi\Delta t} - e^{-i\pi\Delta t}}{2i} = \sin(\pi\Delta t).$$

Thus,

$$MS = \int_{-\infty}^{\infty} \frac{\sin(\pi\Delta t)}{\pi t} \int_{-\infty}^{\infty} [R_X(t-s) e^{-i2\pi f(t-s)}] \frac{\sin(\pi\Delta s)}{\pi s} \, ds \, dt,$$

which is of the form

$$MS = \int_{-\infty}^{\infty} a(t)[b(t) \otimes a(t)]^* \, dt.$$

Using Parseval's relation and the convolution theorem, we obtain

$$MS = \int_{-\infty}^{\infty} A(v)[B(v)A(v)]^* \, dv,$$

where

$$A(v) = \begin{cases} 1, & |v| \le \Delta/2, \\ 0, & \text{otherwise}, \end{cases}$$

$$B(v) = S_X(v - f).$$

Therefore,

$$MS = \int_{-\Delta/2}^{\Delta/2} S_X(v - f) \, dv = \int_{-f-\Delta/2}^{-f+\Delta/2} S_X(\mu) \, d\mu$$

$$= \int_{f-\Delta/2}^{f+\Delta/2} S_X(\mu) \, d\mu \quad (S_X \text{ is even})$$

$$= \int_{f_2}^{f_1} S_X(\mu) \, d\mu,$$

which is the desired result.

(d) With the use of (10.12), (10.144) can be expressed (formally) as

$$\tilde{Y}^{(-1)}(f) = \int_{-\infty}^{f} H(v)\tilde{X}(v) \, dv.$$

Differentiation of both sides of this equation yields

$$\tilde{Y}(f) = \frac{d}{df}\tilde{Y}^{(-1)}(f) = \frac{d}{df}\int_{-\infty}^{f} H(v)\tilde{X}(v) \, dv = H(f)\tilde{X}(f),$$

which is (10.3).

2. (a) Substitution of

$$g(t) = \int_{-\infty}^{\infty} G(f)e^{i2\pi ft} \, df$$

into the left member of (10.145) and interchange of the order of integration yields

$$\int_{-\infty}^{\infty} g(t)h^*(t) \, dt = \int_{-\infty}^{\infty} G(f) \int_{-\infty}^{\infty} h^*(t)e^{i2\pi ft} \, dt \, df$$

$$= \int_{-\infty}^{\infty} G(f) \left[\int_{-\infty}^{\infty} h(t)e^{-i2\pi ft} \, dt \right]^* df$$

$$= \int_{-\infty}^{\infty} G(f)H^*(f) \, df,$$

which is the right member of (10.145).

6. Substitution of (10.34) into (10.38) yields

$$R_X(k) = \int_{-1/2}^{1/2} S_X(f)e^{-i2\pi fk} \, df = \sum_{l=-\infty}^{\infty} R_X(l) \int_{-1/2}^{1/2} e^{-i2\pi fl}e^{i2\pi fk} \, df$$

$$= \sum_{l=-\infty}^{\infty} R_X(l) \frac{\sin[\pi(k-l)]}{\pi(k-l)} = R_X(k).$$

Thus, (10.34) and (10.38) are indeed a Fourier transform pair.

9. Since $X(t)$ exhibits mean-square ergodicity of the mean and $S_X(f)$ does not contain an impulse at $f = 0$, then (see (10.141)) we have $m_X =$

516 • **Solutions to Selected Exercises / Chapter 10**

0. The mean of the output of the filter with input $X(t)$ is also zero, since

$$m_Y = E\{Y(t)\} = \int_{-\infty}^{\infty} h(u) E\{X(t-u)\} \, du = 0.$$

The variance of the input is given by

$$\text{Var}\{X(t)\} = E\{X^2(t)\} = R_X(0) = \int_{-\infty}^{\infty} S_X(f) \, df = BS_o.$$

Since the input and output spectral densities are related by

$$S_Y(f) = |H(f)|^2 S_X(f),$$

and the variance at the output is given by

$$\text{Var}\{Y(t)\} = E\{Y^2(t)\} = R_Y(0) = \int_{-\infty}^{\infty} S_Y(f) \, df,$$

then

$$\text{Var}\{Y(t)\} = \int_{-\infty}^{\infty} |H(f)|^2 S_X(f) \, df = \int_{-b}^{b} (1 - |f|/B) S_o \, df$$

$$= \begin{cases} bS_o[2 - b/B], & |b| \le B \\ BS_o, & |b| > B. \end{cases}$$

10. Let the input be denoted by $X(i)$. We have

$$E\{X(i)\} = 0 \quad \text{and} \quad R_X(i-j) = E\{X(i)X(j)\} = \begin{cases} \sigma^2, & i = j, \\ 0, & i \ne j. \end{cases}$$

(a) Since the input spectral density is

$$S_X(f) = \sum_{k=-\infty}^{\infty} R_X(k) e^{-i2\pi fk} = \sigma^2$$

and the transfer function is

$$H(f) = \sum_{k=-\infty}^{\infty} h(k) e^{-i2\pi fk} = \sum_{k=0}^{K} e^{-i2\pi fk}$$

$$= \frac{1 - e^{-i2\pi f(K+1)}}{1 - e^{-i2\pi f}} = \frac{\sin[\pi f(K+1)]}{\sin(\pi f)} e^{-i\pi fK},$$

then from (10.33) the output spectral density is given by

$$S_Y(f) = S_X(f)|H(f)|^2 = \sigma^2 \left[\frac{\sin[\pi f(K+1)]}{\sin(\pi f)}\right]^2.$$

The output variance is given by

$$\text{Var}\{Y(i)\} = E\{Y^2(i)\}$$

$$= \sum_{k=-\infty}^{\infty} \sum_{l=-\infty}^{\infty} h(i-k)h(i-l)E\{X(k)X(l)\}$$

$$= \sum_{k=-\infty}^{\infty} \sum_{l=-\infty}^{\infty} h(i-k)h(i-l)R_X(k-l)$$

$$= \sigma^2 \sum_{k=-\infty}^{\infty} h^2(i-k) = \sigma^2 \sum_{k=1}^{K} h^2(k) = (K+1)\sigma^2.$$

Note: This approach is simpler than that based on

$$\text{Var}\{Y(i)\} = \int_{-1/2}^{1/2} S_Y(f)\, df.$$

(b) Since the input and output are related by

$$Y(i) = \sum_{k=-\infty}^{\infty} h(i-k)X(k),$$

then the cross-correlation of the input and output is

$$R_{YX}(i-j) = E\{Y(i)X(j)\} = \sum_{k=-\infty}^{\infty} h(i-k)E\{X(k)X(j)\}$$

$$= \sum_{k=-\infty}^{\infty} h(i-k)R_X(k-j) = \sum_{k=i-K}^{i} R_X(k-j)$$

$$= \sum_{l=i-j-K}^{i-j} \sigma^2 \delta(l) = \begin{cases} \sigma^2, & 0 \le i-j \le K, \\ 0, & \text{otherwise}. \end{cases}$$

The cross-spectral density is given by

$$S_{YX}(f) = \sum_{k=-\infty}^{\infty} R_{YX}(k)e^{-i2\pi fk} = \sum_{k=0}^{K} \sigma^2 e^{-i2\pi fk}$$

$$= \sigma^2 \frac{\sin[\pi f(K+1)]}{\sin(\pi f)} e^{-i\pi fK}. \qquad\blacksquare$$

Chapter 12

1. To verify the Fourier-coefficient formula (12.6), we substitute (12.4) into (12.6) (with α replaced by β) to obtain

$$R_X^\beta(\tau) = \lim_{Z\to\infty} \frac{1}{Z} \int_{-Z/2}^{Z/2} R_X\left(t + \frac{\tau}{2}, t - \frac{\tau}{2}\right) e^{-i2\pi\beta t}\, dt$$

$$= \sum_\alpha R_X^\alpha(\tau) \lim_{Z\to\infty} \frac{1}{Z} \int_{-Z/2}^{Z/2} e^{i2\pi\alpha t} e^{-i2\pi\beta t}\, dt$$

$$= \sum_\alpha R_X^\alpha(\tau) \lim_{Z\to\infty} \frac{\sin[\pi(\alpha-\beta)Z]}{\pi(\alpha-\beta)Z}$$

$$= \begin{cases} R_X^\beta(\tau), & \alpha = \beta, \\ 0, & \alpha \neq \beta. \end{cases}$$

Hence (12.6) is indeed the correct formula for the Fourier-coefficient.

5. (a) To obtain (12.41), we substitute (12.4) into (12.39) to obtain

$$\langle R_X \rangle_T(t,u) = \lim_{N\to\infty} \frac{1}{2N+1} \sum_{n=-N}^{N} R_X(t+nT, u+nT)$$

$$= \lim_{N\to\infty} \frac{1}{2N+1} \sum_{n=-N}^{N} \sum_\alpha R_X^\alpha(t-u) e^{i\pi\alpha(t+nT+u+nT)}$$

$$= \sum_\alpha R_X^\alpha(t-u) e^{i\pi\alpha(t+u)} \lim_{N\to\infty} \frac{1}{2N+1} \sum_{n=-N}^{N} e^{i2\pi\alpha nT}.$$

Let us consider the quantity

$$J(\alpha T) \triangleq \lim_{N\to\infty} \frac{1}{2N+1} \sum_{n=-N}^{N} e^{i2\pi\alpha nT}$$

$$= \lim_{N\to\infty} \frac{1}{2N+1} \left[\frac{1 - e^{i2\pi\alpha(N+1)T}}{1 - e^{i2\pi\alpha T}} + \frac{1 - e^{-i2\pi\alpha(N+1)T}}{1 - e^{-i2\pi\alpha T}} - 1 \right]$$

$$= \lim_{N\to\infty} \frac{1}{2N+1} \left[2\,\mathrm{Re}\left\{ \frac{1 - e^{i2\pi\alpha(N+1)T}}{1 - e^{i2\pi\alpha T}} \right\} - 1 \right]$$

$$= \lim_{N\to\infty} \frac{1}{2N+1} \left[2\cos(\pi\alpha NT) \frac{\sin[\pi\alpha(N+1)T]}{\sin(\pi\alpha T)} - 1 \right].$$

It is easily seen that for $\alpha T \neq$ integer, $J(\alpha T) = 0$. For $\alpha T =$ integer, we have

$$J(\alpha T) = \lim_{N\to\infty} \frac{1}{2N+1} \sum_{n=-N}^{N} 1 = 1.$$

Therefore,
$$\langle R_X \rangle_T(t, u) = \sum_m R_X^{m/T}(t - u) e^{i\pi m(t+u)/T},$$

where m represents all the integer values of αT.

(b) Substitution of (12.41) into (12.43), with T replaced by T_j, yields
$$R_X(t, u) = \langle R_X \rangle(t - u)$$
$$+ \sum_j \left[\sum_{m=-\infty}^{\infty} R_X^{m/T_j}(t - u) e^{i\pi m(t+u)/T_j} - \langle R_X \rangle(t - u) \right].$$

The fact that (12.6) with $\alpha = 0$ is equivalent to (12.37), that is,
$$R_X^0(t - u) = \lim_{Z \to \infty} \frac{1}{Z} \int_{-Z/2}^{Z/2} R_X(v + t, v + u) \, dv = \langle R_X \rangle(t - u),$$

can be used to further reduce (12.43) as follows:
$$R_X(t, u) = R_X^0(t - u) + \sum_j \sum_{\substack{m=-\infty \\ m \neq 0}}^{\infty} R_X^{m/T_j}(t - u) e^{i\pi m(t+u)/T_j}.$$

But this equation is equivalent to (12.4), which can be expressed as
$$R_X(t, u) = \sum_\alpha R_X^\alpha(t - u) e^{i\pi \alpha(t+u)}$$
$$= R_X^0(t - u) + \sum_{\alpha \neq 0} R_X^\alpha(t - u) e^{i\pi \alpha(t+u)},$$

where α represents all integer multiples of all the fundamental frequencies $\{1/T_j\}$.

9. (a) The synchronized averaging identity (12.42) can be expressed more generally as
$$\lim_{N \to \infty} \frac{1}{2N + 1} \sum_{n=-N}^{N} f(nT) = \sum_{m=-\infty}^{\infty} \lim_{Z \to \infty} \frac{1}{Z} \int_{-Z/2}^{Z/2} f(t) e^{-i2\pi m t/T} \, dt$$

for any function $f(t)$ for which these averages exist. By using the identity $\tilde{R}_X(nT + kT, nT) = R_X(nT + kT, nT)$ in (12.94) and then considering
$$f(nT) = R_X(nT + kT, nT) e^{-i2\pi\alpha(nT + kT/2)},$$

we obtain
$$\tilde{R}_X^\alpha(kT) = \sum_{m=-\infty}^{\infty} \lim_{Z \to \infty} \frac{1}{Z} \int_{-Z/2}^{Z/2} R_X(t + kT, t) e^{-i2\pi\alpha(t + kT/2)} e^{-i2\pi m t/T} \, dt$$
$$= \sum_{m=-\infty}^{\infty} \lim_{Z \to \infty} \frac{1}{Z} \int_{-Z/2}^{Z/2} R_X(t + kT, t) e^{-i2\pi(\alpha + m/T)t} \, dt \, e^{-i\pi\alpha kT}.$$

Letting $s = t + kT/2$, this equation becomes

$$\tilde{R}_X^\alpha(kT) = \sum_{m=-\infty}^{\infty} \lim_{Z \to \infty} \frac{1}{Z} \int_{-Z/2+kT/2}^{Z/2+kT/2} R_X\left(s + \frac{kT}{2}, s - \frac{kT}{2}\right)$$
$$\times e^{-i2\pi(\alpha+m/T)s} \, ds \, e^{i\pi mk}.$$

$$= \sum_{m=-\infty}^{\infty} R_X^{\alpha+m/T}(kT) e^{i\pi mk},$$

which is the desired result, (12.96).
(b) Substitution of (12.96) into the definition (12.95) yields

$$\tilde{S}_X^\alpha(f) = \sum_{k=-\infty}^{\infty} \tilde{R}_X^\alpha(kT) e^{-i2\pi kTf}$$

$$= \sum_{m=-\infty}^{\infty} \sum_{k=-\infty}^{\infty} R_X^{\alpha+m/T}(kT) e^{i\pi mk} e^{-i2\pi kTf}$$

$$= \int_{-\infty}^{\infty} \sum_{m=-\infty}^{\infty} \sum_{k=-\infty}^{\infty} S_X^{\alpha+m/T}(v) e^{i2\pi vkT} e^{i\pi mk} e^{-i2\pi kTf} \, dv.$$

Use of the identity

$$\sum_{k=-\infty}^{\infty} e^{i2\pi \mu kT} \equiv \frac{1}{T} \sum_{n=-\infty}^{\infty} \delta\left(\mu + \frac{n}{T}\right)$$

yields

$$\tilde{S}_X^\alpha(f) = \int_{-\infty}^{\infty} \frac{1}{T} \sum_{m=-\infty}^{\infty} \sum_{n=-\infty}^{\infty} S_X^{\alpha+m/T}(v) \delta\left(v - f + \frac{m}{2T} + \frac{n}{T}\right) dv$$

$$= \frac{1}{T} \sum_{n,m=-\infty}^{\infty} S_X^{\alpha+m/T}\left(f - \frac{m}{2T} - \frac{n}{T}\right),$$

which is the desired result, (12.97). From (12.97) we observe that

$$\tilde{S}_X^\alpha\left(f + \frac{1}{T}\right) = \frac{1}{T} \sum_{n,m=-\infty}^{\infty} S_X^{\alpha+m/T}\left(f - \frac{m}{2T} - \frac{n-1}{T}\right)$$

$$= \frac{1}{T} \sum_{n',m=-\infty}^{\infty} S_X^{\alpha+m/T}\left(f - \frac{m}{2T} - \frac{n'}{T}\right) = \tilde{S}_X^\alpha(f),$$

$$\tilde{S}_X^{\alpha+2/T}(f) = \frac{1}{T} \sum_{n,m=-\infty}^{\infty} S_X^{\alpha+2/T+m/T}\left(f - \frac{m}{2T} - \frac{n}{T}\right)$$

$$= \frac{1}{T} \sum_{n,m'=-\infty}^{\infty} S_X^{\alpha+m'/T}\left(f - \frac{m'-2}{2T} - \frac{n}{T}\right)$$

$$= \frac{1}{T} \sum_{n,m'=-\infty}^{\infty} S_X^{\alpha+m'/T}\left(f - \frac{m'}{2T} - \frac{n+1}{T}\right)$$

$$= \frac{1}{T} \sum_{n',m'=-\infty}^{\infty} S_X^{\alpha+m'/T}\left(f - \frac{m'}{2T} - \frac{n'}{T}\right) = \tilde{S}_X^{\alpha}(f),$$

and

$$\tilde{S}_X^{\alpha+1/T}\left(f - \frac{1}{2T}\right) = \frac{1}{T} \sum_{n,m=-\infty}^{\infty} S_X^{\alpha+1/T+m/T}\left(f - \frac{1}{2T} - \frac{m}{2T} - \frac{n}{T}\right)$$

$$= \frac{1}{T} \sum_{n,m'=-\infty}^{\infty} S_X^{\alpha+m'/T}\left(f - \frac{m'}{2T} - \frac{n+1}{T}\right)$$

$$= \frac{1}{T} \sum_{n',m'=-\infty}^{\infty} S_X^{\alpha+m'/T}\left(f - \frac{m'}{2T} - \frac{n'}{T}\right) = \tilde{S}_X^{\alpha}(f).$$

10. (a) We first express (12.103) in the form of (12.44):

$$X(t) = \int_{-\infty}^{\infty} A(u)\cos(2\pi f_o u + \phi_o)\delta(t - u)\, du$$

$$= \int_{-\infty}^{\infty} A(u)h(t, u)\, du,$$

where

$$h(t, u) = \cos(2\pi f_o u + \phi_o)\delta(t - u).$$

Then, from (12.45) and (12.46), we obtain

$$h(t + \tau, t) = \cos(2\pi f_o t + \phi_o)\delta(\tau)$$

$$= \frac{1}{2}[e^{i(2\pi f_o t + \phi_o)} + e^{-i(2\pi f_o t + \phi_o)}]\delta(\tau)$$

and

$$g_n(\tau) = \begin{cases} \dfrac{1}{2} e^{i\phi_o}\delta(\tau), & n = 1, \\ \dfrac{1}{2} e^{-i\phi_o}\delta(\tau), & n = -1. \end{cases}$$

From (12.52), we obtain

$$r_{nm}^{\alpha}(\tau) = \int_{-\infty}^{\infty} g_n(t + \tau/2)g_m^*(t - \tau/2)e^{-i2\pi\alpha t}\, dt$$

$$= \frac{1}{4} e^{i(n-m)\phi_o}\delta(\tau), \qquad n, m = \pm 1.$$

Finally, substitution of this result into (12.50) with $T = T_o = 1/f_o$ yields the cyclic autocorrelation of $X(t)$:

$$R_X^\alpha(\tau) = \sum_{n,m=\pm 1} 2[R_A^{\alpha-(n-m)/T_o}(\tau)e^{-i\pi(n+m)\tau/T_o}] \otimes \left[\frac{1}{4}e^{i(n-m)\phi_o}\delta(-\tau)\right]$$

$$= \frac{1}{4} \sum_{n,m=\pm 1} R_A^{\alpha-(n-m)/T_o}(\tau)e^{-i\pi(n+m)\tau/T_o}e^{i(n-m)\phi_o}$$

$$= \frac{1}{4}[R_A^\alpha(\tau)e^{-i2\pi f_o\tau} + R_A^\alpha(\tau)e^{i2\pi f_o\tau} + R_A^{\alpha-2f_o}(\tau)e^{i2\phi_o}$$

$$+ R_A^{\alpha+2f_o}(\tau)e^{-i2\phi_o}]$$

$$= \frac{1}{2}R_A^\alpha(\tau)\cos(2\pi f_o\tau) + \frac{1}{4}R_A^{\alpha-2f_o}(\tau)e^{i2\phi_o} + \frac{1}{4}R_A^{\alpha+2f_o}(\tau)e^{-i2\phi_o}.$$

(b) Fourier transformation of the cyclic autocorrelation (12.104) yields the cyclic spectrum

$$S_X^\alpha(f) = \int_{-\infty}^{\infty} R_X^\alpha(\tau)e^{-i2\pi f\tau}\,d\tau$$

$$= \frac{1}{4}[S_A^\alpha(f+f_o) + S_A^\alpha(f-f_o) + S_A^{\alpha+2f_o}(f)e^{-i2\phi_o}$$

$$+ S_A^{\alpha-2f_o}(f)e^{i2\phi_o}],$$

where

$$S_A^\alpha(f) = \int_{-\infty}^{\infty} R_A^\alpha(\tau)e^{-i2\pi f\tau}\,d\tau. \qquad \blacksquare$$

References

M.H. Ackroyd, "Stationary and cyclostationary finite buffer behavior computation via Levinson's Method," *AT&T Bell Laboratories Tech. J.*, Vol. 63, pp. 2159–2170 (1984).

B.G. Agee, "The property restoral approach to blind adaptive signal extraction," Ph.D. dissertation, Dept. Electrical Engineering and Computer Science, Univ. of California, Davis, CA 95616 (1989).

B.G. Agee, S.V. Schell, and W.A. Gardner, "Self-coherence restoral: A new approach to blind adaptation of antenna arrays," *Proc. Twenty-First Asilomar Conf. on Sig., Sys., and Comp.*, Pacific Grove, Nov. 2–4, 1987, pp. 589–593 (1988a).

B.G. Agee, S.V. Schell, and W.A. Gardner, "The SCORE approach to blind adaptive signal extraction: An application of the theory of spectral correlation," *Proc. Fourth IEEE ASSP Workshop on Spectrum Estimation and Modeling*, Minneapolis, Aug. 3–5, 1988, pp. 277–282 (1988b).

P.A. Albuquerque, O. Shimbo, and L.N. Ngugen, "Modulation transfer noise effects from a continuous digital carrier to FDM/FM carriers in memoryless nonlinear devices," *IEEE Trans. Commun.*, Vol. COM-32, pp. 337–353 (1984).

B.D.O. Anderson and J.B. Moore, *Optimal Filtering*, Englewood Cliffs, N.J.: Prentice-Hall, 1979.

L. Arnold, *Stochastic Differential Equations: Theory and Applications*, New York, John Wiley & Sons, 1974.

W.R. Bennett, "Statistics of regenerative digital transmission," *Bell System Tech. J.*, Vol. 37, pp. 1501–1542 (1958).

N.M. Blachman and G.A. McAlpine, "The spectrum of high-index FM: Woodward's theorem revisited," *IEEE Trans. Commun. Technol.*, Vol. COM-17, pp. 201–208 (1969).

A. Blanc-Lapierre and R. Fortet, *Theory of Random Functions*, Vol. I, New York: Gordon and Breach, 1965.

H.W. Bode and C.E. Shannon, "A simplified derivation of linear least square smoothing and prediction theory," *Proc. IRE,* Vol. 38, pp. 417–425 (1950).

R.A. Boyles and W.A. Gardner, "Cycloergodic properties of discrete-parameter non-stationary stochastic processes," *IEEE Trans. Inform. Theory*, Vol. 29, pp. 105–114 (1983).

L. Breiman, *Probability*, Menlo Park, Calif.: Addison-Wesley, 1968.

P. Brémaud, *Point Processes and Queues*, New York: Springer-Verlag, 1981.

D.R. Brillinger, *Time Series*, New York: Holt, Rinehart, and Winston, 1975.

R.G. Brown, *Introduction to Random Signal Analysis and Kalman Filtering*, New York: John Wiley & Sons, 1983.

W.A. Brown, "On the theory of cyclostationary signals," Ph.D. dissertation, Dept. Electrical Engineering and Computer Science, Univ. of California, Davis, CA 95616 (1987).

W.A. Brown and H.H. Loomis, Jr., "Digital implementations of spectral correlation analyzers," *Proc. Fourth IEEE ASSP Workshop on Spectrum Estimation and Modeling*, Minneapolis, Aug. 3-5, 1988, pp. 264–270 (1988).

C. Campbell, A.J. Gibbs, and B.M. Smith, "The cyclostationary nature of crosstalk interference from digital signals in multipair cable—Part I: Fundamentals," *IEEE Trams. Commun.*, Vol. COM-31, pp. 629–637 (1983).

N. Campbell, "Discontinuous phenomena," *Proc. Cambridge Phil. Soc.*, Vol. 15, pp. 117–136 (1909).

C.K. Chen, "Spectral-correlation characterization of modulated signals with application to signal detection and source location," Ph.D. dissertation, Dept. Electrical Engineering and Computer Science, Univ. of California, Davis, CA 95616 (1989).

K.L. Chung, *Lectures from Markov Processes to Brownian Motion*, New York: Springer-Verlag, 1982.

W.B. Davenport and W.L. Root, *An Introduction to the Theory of Random Signals and Noise*, New York: McGraw-Hill, 1958.

J.L. Doob, *Stochastic Processes*, New York: John Wiley & Sons, 1953.

J. Durbin, "The fitting of time series models," *Rev. Inst. Internat. Statist.* Vol. 28, pp. 233–243 (1960).

B. Epstein, *Linear Functional Analysis*, Philadelphia: W.B. Saunders, 1970.

T.H.E. Ericson, "Modulation by means of linear periodic filtering," *IEEE Trans. Inform. Theory*, Vol. IT-27, pp. 322–327 (1981).

W. Feller, *An Introduction to Probability Theory and its Applications*, Volume I, 3rd Ed., New York: John Wiley & Sons, 1968.

W. Feller, *An Introduction to Probability Theory and its Applications*, Volume II, New York: John Wiley & Sons, 1966; 2nd Ed., 1971.

E.R. Ferrara, "Frequency-domain implementations of periodically time-varying filters," *IEEE Trans. Acoust., Speech, and Sig. Proc.*, Vol. ASSP-33, pp. 883–892 (1985).

W.H. Fleming and R.W. Rishel, *Deterministic and Stochastic Optimal Control*, New York: Springer-Verlag, 1975.

L.E. Franks, *Signal Theory*, Englewood Cliffs, N.J.: Prentice-Hall, 1969.

L.E. Franks, "Carrier and bit synchronization in data communication—a tutorial review, *IEEE Trans. Commun.*, Vol. COM-28, pp. 1107–1121 (1980).

B. Friedlander, "Lattice filters for adaptive processing," *Proc. IEEE*, Vol. 70, pp. 829–867 (1982).

W.A. Gardner, "A sampling theorem for nonstationary random processes," *IEEE Trans. Inform. Theory*, Vol. 18, pp. 808–809 (1972).

W.A. Gardner, "The structure of least-mean-square linear estimators for synchronous M-ary signals," *IEEE Trans. Inform. Theory*, Vol. IT-19, pp. 240–243 (1973).

W.A. Gardner, "An equivalent linear model for marked and filtered doubly stochastic Poisson processes with application to MMSE linear estimation for synchronous M-ary optical data signals," *IEEE Trans. Commun.* Vol. COM-24, pp. 917–921 (1976a).

W.A. Gardner, "Structurally constrained receivers for signal detection and estimation," *IEEE Trans. Commun.*, Vol. COM-24, pp. 578–592 (1976b).

W.A. Gardner, "Stationarizable random processes," *IEEE Trans. Inform. Theory*, Vol. IT-24, pp. 8–22 (1978).

W.A. Gardner, "A unifying view of second-order measures of quality for signal classification," *IEEE Trans. Commun.*, Vol. COM-28, pp. 807–816 (1980).

W.A. Gardner, "Structural characterization of locally optimum detectors in terms of locally optimum estimators and correlators," *IEEE Trans. Inform. Theory*, Vol. IT-28, pp. 924–932 (1982).

W.A. Gardner, "Optimization and adaptation of linear periodically time-variant digital systems," Signal and Image Processing Lab. Tech. Rept. No. SIPL-85-9, Dept. Electrical and Computer Engineering, Univ. of California, Davis, CA 95616 (1985).

W.A. Gardner, "Measurement of spectral correlation," *IEEE Trans. Acoust., Speech, and Sig. Proc.*, Vol. ASSP-34, pp. 1111–1123 (1986a).

W.A. Gardner, "The role of spectral correlation in design and performance analysis of synchronizers," *IEEE Trans. Commun.*, Vol. COM-34, pp. 1089–1095 (1986b).

W.A. Gardner, "The spectral correlation theory of cyclostationary time-series," *Signal Processing*, Vol. 11, pp. 13–36, p. 405 (1986c).

W.A. Gardner, *Statistical Spectral Analysis: A Nonprobabilistic Theory*, Englewood Cliffs, N.J.: Prentice-Hall, 1987a.

W.A. Gardner, "Common pitfalls in the application of stationary process theory to time-sampled and modulated signals," *IEEE Trans. Commun.*, Vol. COM-35, pp. 529–534 (1987b).

W.A. Gardner, "Signal interception: A unifying theoretical framework for feature detection," *IEEE Trans. Commun.*, Vol. COM-36, pp. 897–906 (1988).

W.A. Gardner, "Frequency-shift filtering theory for adaptive co-channel interference removal," *Proc. Twenty-third Asilomar Conference on Signals, Systems, and Computers*, Pacific Grove, Oct. 30–Nov. 1, 1989. (1989)

W.A. Gardner, "A method for elimination of the effects of correlated input/output measurement noise on system identification using cyclostationary excitation," *IEEE Trans. Automat. Control.*, Vol. AC-35, 1990.

W.A. Gardner and C.K. Chen, "Interference-tolerant time-difference-of-arrival estimation for modulated signals," *IEEE Trans. Acoust., Speech, and Sig. Proc.*, Vol. ASSP-36, pp. 1385–1395 (1988a).

W.A. Gardner and C.K. Chen, "Selective source location by exploitation of spectral coherence," *Proc. Fourth IEEE ASSP Workshop on Spectrum Estimation and Modeling*, Minneapolis, Aug. 3–5, 1988, pp. 271–276 (1988b).

W.A. Gardner and L.E. Franks, "Characterization of cyclostationary random signal processes," *IEEE Trans. Inform. Theory*, Vol. IT-21, pp. 4–14 (1975).

A. Gelb (ed.), *Applied Optimal Estimation*, Cambridge, Mass.: MIT Press, 1974.

I.I. Gihman and A.V. Skorohod, *The Theory of Stochastic Processes, I*, Translated from Russian by S. Kotz, New York: Springer-Verlag, 1974.

B.V. Gnedenko and A.N. Kolmogorov, *Limit Distributions for Sums of Independent Random Variables*, Reading, Mass.: Addison-Wesley, 1954.

F.K. Graef, "Joint optimization of transmitter and receiver for cyclostationary random signal processes," *Proc. of the NATO Advanced Study Institute on Nonlinear Stochastic Problems*, Algarve, Portugal, May 16–28, 1982, Dordrecht, Netherlands: Reidel, pp. 581–592, 1983.

R.M. Gray, *Probability, Random Processes, and Ergodic Properties*, New York: Springer-Verlag, 1987.

R.M. Gray and J.C. Kieffer, "Asymptotically mean stationary measures," *Ann. Prob.*, Vol. 8, pp. 962–973 (1980).

P. Hall and C.C. Heyde, *Martingale Limit Theory and its Applications*, New York: Academic Press, 1980.

E.J. Hannan, *Multiple Time Series*, New York: John Wiley & Sons, 1970.

K. Hasselman and T.P. Barnett, "Techniques of linear prediction for systems with periodic statistics," *J. Atmospheric Science*, Vol. 38, pp. 2275–2283 (1981).

E.M. Hofstetter, "Random Processes," Chapter 3 of *The Mathematics of Physics and Chemistry*, Vol. II, edited by H. Margenau and G.M. Murphy, Princeton: Van Nostrand, 1964.

L. Isserlis, "On a formula for the product-moment coefficient of any order of a normal frequency distribution in any number of variables," *Biometrika*, Vol. 12, pp. 134–139 (1918).

A.H. Jazwinski, *Stochastic Processes and Filtering Theory*, New York: Academic Press, 1970.

A.J. Jerri, "The shannon sampling theorem—its various extensions and applications: A tutorial review," *Proc. IEEE*, Vol. 65, pp. 1565–1596 (1977).

R.H. Jones and W.M. Brelsford, "Time series with periodic structure," *Biometrika*, Vol. 54, pp. 403–407 (1967).

T. Kailath, "An innovations approach to least squares estimation, Part I: Linear filtering in additive white noise," *IEEE Trans. Automat. Control*, Vol. A-13, pp. 646–654 (1968).

T. Kailath "A view of three decades of linear filtering theory," *IEEE Trans. Inform. Theory*, Vol. IT-20, pp. 146–181 (1974).

T. Kailath, *Linear Systems*, Englewood Cliffs, N.J.: Prentice-Hall, 1980.

T. Kailath, *Lectures on Wiener and Kalman Filtering*, New York: Springer-Verlag, 1981.

T. Kailath, B.C. Levy, L. Ljung, and M. Morf, "Fast time-invariant implementations of Gaussian signal detectors," *IEEE Trans. Inform. Theory*, Vol. IT-24, pp. 469–477 (1978).

R.E. Kalman, "A new approach to linear filtering and prediction problems," *Trans. ASME, J. Basic Engrg.*, Vol. 82, pp. 34–45 (1960).

R.E. Kalman and R.S. Bucy, "New results in linear filtering and prediction theory," *Trans. ASME, J. Basic Engrg. Ser. D*, Vol. 83, pp. 95–107 (1961).

M. Kaplan, "Single-server queue with cyclostationary arrivals and arithmetic service," *Operations Research*, Vol. 31, pp. 184–205 (1983).

S. Karlin and H.M. Taylor, *A First Course in Stochastic Processes*, New York: Academic Press, 1975.

S. Karlin and H.M. Taylor, *A Second Course in Stochastic Processes*, New York: Academic Press, 1981.

A.Y. Khinchin, "Korrelationstheorie der stationären stochastischen Prozesse," *Math. Ann.*, Vol. 109, pp. 604–615 (1934).

L. Kleinrock, *Queueing Systems, Vol I: Theory*, New York: Wiley-Interscience, 1975.

F.B. Knight, *Essentials of Brownian Motion and Diffusion*, Providence, R.I.: American Mathematical Society, 1981.

A.N. Kolmogorov, "Interpolation and extrapolation of stationary random sequences," *Bull. Acad. Sci. USSR, Math. Ser.*, Vol. 5, pp. 3–14 (1941). (English translation published by W. Doyle and I. Selin, RAND Corp., Santa Monica, Calif., Memo. RM-3090-PR, April, 1962.)

L.H. Koopmans, *The Spectral Analysis of Time Series*, New York: Academic Press, 1974.

H.J. Larson and B.O. Shubert, *Probabilistic Models in Engineering Sciences*, Volume II, New York, John Wiley & Sons: 1979.

J.L. Larson and G. E. Uhlenbeck, *Threshold Signals*, Massachusetts Inst. of Tech., Radiation Lab. Series, Vol. 24, 1949. (Republished by Boston Technical Publishers, 1964.)

C.T. Leondes (ed.), *Theory and Applications of Kalman Filtering*, North Atlantic Treaty Organization, AGARD Report No. 139, 1970.

N. Levinson, "The Wiener RMS (root mean square) error criterion in filter design and prediction," *J. Math. and Phys.*, Vol. 25, pp. 261–278 (1947).

R.S. Lipster and A.N. Shiryayev, *Statistics of Random Processes, I: General Theory*, New York: Springer-Verlag, 1977.

L. Ljung, *System Identification: Theory for the User*, Englewood Cliffs, N.J.: Prentice-Hall, 1987.

M. Loève, *Probability Theory II*, 4th Edition, New York, Springer-Verlag, 1978.

R.E. Maurer and L.E. Franks, "Optimal linear processing of randomly distorted signals," *IEEE Trans. Circuit Theory*, Vol. CT-17, pp. 61–67 (1970).

P.S. Maybeck, *Stochastic Models, Estimation and Control*, Vol. 1, New York: Academic Press, 1979.

J.L. Melsa and A.P. Sage, *Introduction to Probability and Stochastic Processes*, Englewood Cliffs, N.J.: Prentice-Hall, 1973.

M.F. Mesiya, P.J. McLane, and L.L. Campbell, "Optimal receiver filters for BPSK transmission over a bandlimited nonlinear channel," *IEEE Trans. Commun.*, Vol. COM-26, pp. 12–22 (1978).

D. Middleton, *An Introduction to Statistical Communication Theory*, New York: McGraw-Hill, 1960.

D. Middleton, "Canonically optimum threshold detection," *IEEE Trans. Inform. Theory*, Vol. IT-12, pp. 230–243 (1966).

M. Moeneclaey, "Linear phase-locked loop theory for cyclostationary input disturbances," *IEEE Trans. Commun.*, Vol. COM-30, pp. 2253–2259 (1982).

R.A. Monzingo and T.W. Miller, *Introduction to Adaptive Arrays*, New York: John Wiley & Sons, 1980.

W.W. Mumford and E.H. Scheibe, *Noise Performance Factors in Communication Systems*, Dedham, Mass.: Horizon-House, 1968.

H. Nyquist, "Thermal agitation of electric charge in conductors," *Phys. Rev.*, Vol. 32, pp. 110–113 (1928a).

H. Nyquist, "Certain topics in telegraph transmission theory," *Trans. AIEE*, Vol. 47, pp. 617–644 (1928b).

M.J. Ortiz and A. Ruiz de Elvira, "A cyclo-stationary model of sea surface temperature in the Pacific Ocean," *Tellus*, Vol. 37A, pp. 14–23 (1985).

A. Papoulis, *Probability, Random Variables, and Stochastic Processes*, 2nd Edition, New York: McGraw-Hill, 1984.

E. Parzen, *Stochastic Processes*, San Francisco: Holden-Day, 1962.

E. Parzen and M. Pagano, "An approach to modeling seasonally stationary time-series," *J. Econometrics*, Vol. 9, pp. 137–153 (1979).

A. Paulraj, R. Roy, and T. Kailath, "Estimation of signal parameters via rotational invariance techniques," *Conference Record, Nineteenth Asilomar Conference on Circuits, Systems, and Computers*, Pacific Grove, CA, Nov. 6–8, 1985, pp. 83–89 (1986).

L. Pelkowitz, "Frequency domain analysis of wraparound error in fast convolution algorithms," *IEEE Trans. Acoust., Speech, and Sig. Proc.*, Vol. ASSP-29, pp. 413–422 (1981).

M.B. Priestley, *Spectral Analysis and Time Series*, London: Academic Press, 1981.

R. Price, "A useful theorem for nonlinear devices having Gaussian inputs," *IRE Trans. Inform. Theory*, Vol. IT-4, pp. 69–72 (1958).

S.O. Rice, "Mathematical analysis of random noise, Parts I and II," *Bell Sys. Tech. J.*, Vol. 23, pp. 282–332 (1944).

S.O. Rice, "Mathematical analysis of random noise, Parts III and IV," *Bell Sys. Tech. J.*, Vol. 24, pp. 46–156 (1945).

S.O. Rice, "Statistical properties of a sine wave plus random noise," *Bell Sys. Tech. J.*, Vol. 27, pp. 109–157 (1948).

R.S. Roberts, "Digital Architectures for Cyclic Spectral Analysis," Ph.D. dissertation, Dept. Electrical Engineering and Computer Science, Univ. of California, Davis, CA 95616 (1989).

J. Rootenberg and S.A. Ghozati, "Generation of a class of non-stationary random processes," *Internat. J. System Sci.*, Vol. 9, pp. 935–947 (1978).

S.M. Ross, *Stochastic Processes*, New York: Wiley, 1983.

R. Roy, A. Paulraj, and T. Kailath, "Direction of arrival estimation by subspace rotation methods—ESPRIT," *Proc. Internat. Conf. on Acoust., Speech, and Sig. Proc.*, Tokyo, Japan, pp. 2495–2498 (1986).

S.V. Schell, R.A. Calabretta, W.A. Gardner, and B.G. Agee, "Cyclic

MUSIC algorithms for signal-selective direction estimation," *Proc. Internat. Conf. on Acoust., Speech, and Sig. Proc.*, Glasgow, May 23–26, 1989, pp. 2278–2281 (1989).

R.O. Schmidt, "Multiple emitter location and signal parameter estimation," *Proc. RADC Spectrum Estimation Workshop*, Griffiths AFB, Rome, NY, Oct. 1979, pp. 243–258 (1979). Reprinted in *IEEE Trans. Antennas Propagat.*, Vol. AP-34, pp. 276–280 (1986).

Z. Schuss, *Theory and Application of Stochastic Differential Equations*, New York: Wiley, 1980.

M. Schwartz and L. Shaw, *Signal Processing*, New York: McGraw-Hill, 1975.

C.E. Shannon and W. Weaver, *The Mathematical Theory of Communication*, Urbana: The University of Illinois Press, 1962.

D.L. Snyder, *Random Point Processes*, New York: John Wiley & Sons, 1975.

M.D. Srinath and P.K. Rajasekaran, *An Introduction to Statistical Signal Processing with Applications*, New York: John Wiley & Sons, 1979.

T. Strom and S. Signell, "Analysis of periodically switched linear circuits," *IEEE Trans. Circuits and Systems*, Vol. CAS-24, pp. 531–541 (1977).

S.A. Tretter, *Introduction to Discrete-Time Signal Procesing*, New York: John Wiley & Sons, 1976.

G.E. Uhlenbeck and L.S. Ornstein, "On the theory of the Brownian motion," *Phys. Rev. Ser. 2*, Vol. 36, pp. 823–841 (1930).

H.L. Van Trees, *Detection, Estimation, and Modulation Theory, Part I*, New York: John Wiley & Sons, 1968.

H.L. Van Trees, *Detection, Estimation, and Modulation Theory, Part III*, New York: John Wiley & Sons, 1971.

A.V. Vecchia, "Periodic autoregressive-moving average (parma) modeling with applications to water resources," *Water Resources Bull.*, Vol. 21, pp. 721–730 (1985).

G. Walker, "On periodicity in series of related terms," *Proc. Roy. Soc. London Ser. A*, Vol. 131, pp. 518–532 (1931).

N. Wiener, "Generalized harmonic analysis," *Acta Mathematica*, Vol. 55, pp. 117–258 (1930) (Reprinted in *Selected Papers of Norbert Wiener*, Cambridge: MIT Press, 1964).

N. Wiener, *Extrapolation, Interpolation, and Smoothing of Stationary Time Series, with Engineering Applications*, New York: Technology Press and John Wiley & Sons, 1949. (Originally issued in February, 1942, as a classified Natl. Defense Res. Counsel Rept.)

N. Wiener and E. Hopf, "On a class of singular integral equations," *Proc. Prussian Acad. Math.-Phys. Ser.*, p. 696 (1931).

N. Wiener and R.E.A.C. Paley, *Fourier Transforms in the Complex Domain*, Amer. Math. Soc. Colloquium Publication, Vol. 19, 1934.

R.A. Wiggins and E.A. Robinson, "Recursive solution to the multi-channel filtering problem," *J. Geophys. Res.*, Vol. 70, pp. 1885–1891 (1965).

H.O.A. Wold, "On prediction in stationary time-series," *Ann. Math. Statist.*, Vol. 19, pp. 558–567 (1948).

P.M. Woodward, *The Spectrum of Random Frequency Modulation*, Great Malvern, Worcs., England: Telecommunications Research Establishment, Memo No. 666, December, 1952.

M.C. Yovits and J.L. Jackson, "Linear filter optimization with game theory considerations," *IRE Natl. Conv. Record*, Part 4, pp. 193–199 (1955).

G.U. Yule, "On a method of investigating periodicities in disturbed series, with special reference to Wolfer's sunspot numbers," *Phil. Trans. Roy. Soc. London, Ser. A*, Vol. 226, pp. 267–298 (1927).

L.A. Zadeh and J.R. Ragazzini, "An extension of Wiener's theory of prediction," *J. Appl. Phys.*, Vol. 21, pp. 645–655 (1950).

Author Index

Ackroyd, M. H., 371n.
Agee, B. G., 373, 374, 377, 389, 400
Albuquerque, P. A., 371n.
Anderson, B. D. O., 434, 450, 455, 458, 459
Arnold, L., 102n., 150n.

Barnett, T. P., 371n.
Bennett, W. R., 371n.
Blachman, N. M., 277
Blanc–Lapierre, A., 97n.
Bode, H. W., 444
Boyles, R. A., 179, 332n., 368
Breiman, L., 97n., 102n., 104n.
Brelsford, W. M., 371n.
Brémaud, P., 104n., 105n.
Brillinger, D. R., 180
Brown, R. G., 459
Brown, W. A., 372, 374, 375, 377, 400
Bucy, R. S., 451

Calabretta, R. A., 374, 377
Campbell, C., 371n.
Campbell, L. L., 371n.
Campbell, N., 257
Chen, C. K., 373, 381, 383, 385, 385n, 392, 396
Chung, K. L., 97n.
Cramér, H., 55

Davenport, W. B., 136, 241, 273

Doob, J. L., 104n., 140, 220n., 244, 305, 306, 322, 465
Durbin, J., 302

Epstein, B., 426
Ericson, T. H. E., 371n., 401, 462

Feller, W., 55, 97n., 102n., 105n., 139, 147
Ferrara, E. R., 371n., 400
Fleming, W. H., 102n., 150n.
Fortet, R., 97n.
Franks, L. E., 110, 354, 364, 365, 371n., 400, 442, 462
Friedlander, B., 307

Gardner, W. A., 45n., 61n., 110, 179, 231, 232n., 266, 286, 297n., 300, 324, 332n., 352, 355, 357, 364, 365, 368–375, 377, 381, 383, 385, 385n, 389, 394–396, 398n., 400, 402, 404, 434, 462, 463, 466
Gelb, A., 459
Ghozati, S. A., 462
Gibbs, A. J., 371n.,
Gihman, I. I., 150n.
Gnedenko, B. V., 55, 241
Graef, F. K., 371n., 401, 462

Gray, R. M., 112, 163, 332n.

Hall, P., 104n.
Hannan, E. J., 171n.
Heyde, C. C., 104n.
Hofstetter, E. M., 180
Hopf, E., 443

Isserlis, L., 114–115

Jackson, J. L., 450
Jazwinski, A. H., 150n.
Jerri, A. J., 266
Jones, R. H., 371n.

Kailath, T., 203, 208, 300, 380, 434, 444, 451, 455, 458–460
Kalman, R. E., 451
Kaplan, M., 371n.
Karlin, S., 97n., 102n., 104n., 105n.
Khinchin, A. Y., 229
Kieffer, J. C., 332n.
Kleinrock, L., 105n.
Knight, F. B., 102n.
Kolmogorov, A. N., 55, 241, 444, 451
Koopmans, L. H., 173

Larson, H. J., 97n., 102n., 105n., 305

Author Index

Lawson, J. L., 236n.
Leondes, C. T., 459
Levinson, N., 302
Levy, B. C., 300
Lipster, R. S., 104n., 150n.
Ljung, L., 300, 459
Loève, M., 55, 150, 152, 155
Loomis, H. H., 372, 375

McAlpine, G. A., 277
McLane, P. J., 371n.
Maurer, R. E., 442
Maybeck, P. S., 459
Mesiya, M. F., 371n.
Middleton, D., 234, 236n., 279, 299
Miller, T. W., 386
Moeneclaey, M., 371n.
Monzingo, R. A., 386
Moore, J. B., 434, 450, 455, 458, 459
Morf, M., 300
Mumford, W. W., 290

Ngugen, L. N., 371n.
Nyquist, H., 234n., 236, 264, 410

Ornstein, L. S., 236n.
Ortiz, M. J., 371n.

Pagano, M., 371n.

Paley, R. E. A. C., 445
Papoulis, A., 341, 353
Parzen, E., 55, 97n., 371n.
Paulraj, A., 380
Pelkowitz, L., 371n.
Price, R., 116
Priestly, M. B., 244

Ragazzini, J. R., 444
Rajasekaran, P. K., 459
Rice, S. O., 271
Rishel, R. W., 102n., 150n.
Roberts, R. S., 372
Robinson, E. A., 302
Root, W. L., 136, 241, 273
Rootenberg, J., 462
Ross, S. M., 97n., 105n.
Roy, R., 380
Ruiz de Elvira, A., 371n.

Scheibe, E. H., 290
Schell, S. V., 373, 374, 377, 389
Schmidt, R. O., 376, 377
Schuss, Z., 102n.
Schwartz, M., 459
Shannon, C. E., 306, 444
Shaw, L., 459
Shimbo, O., 371n.
Shiryayev, A. N., 104n., 150n.
Shubert, B. O., 97n., 102n., 105n., 150n., 305

Signell, S., 371n.
Skorohod, A. V., 150n.
Smith, B. M., 371n.
Snyder, D. L., 105n., 136, 241
Srinath, M. D., 459
Strom, T., 371n.

Taylor, H. M., 97n., 102n., 104n., 105n.
Tretter, S. A., 450, 459

Uhlenbeck, G. E., 236

Van Trees, H. L., 293, 299, 434, 471
Vecchia, A. V., 371n.

Walker, G., 302
Weaver, W., 306
Wiener, N., 67, 220, 229, 294, 435, 443, 445, 449
Wiggins, R. A., 302
Wold, H. O. A., 181
Woodward, P. M., 277

Yovits, M. C., 450
Yule, G. U., 302

Zadeh, L. A., 444

Subject Index

Admissible estimation function, 417, 428
Algorithms (*see* Direction-of-arrival estimation; Kalman filtering; Levinson-Durbin algorithm; Linear estimation; Linear prediction; Periodically time-variant filtering; Recursive algorithms; Signal detection; Signal extraction; Spatial filtering; System identification; Time-difference-of-arrival estimation; Wiener filtering)
Aliasing error:
  in cyclic spectral density, 344
  in Fourier transform, 264
  in power spectral density, 345
Almost periodic function, 324
Almost periodic transformation, 338
Amplitude modulation (AM), 89, 346-348
Analytic signal, 248
Angle in a vector space, 418
Antenna array, adaptive, 385-391
AR (*see* Autoregressive model)
Arcsine law, 117

ARMA (autoregressive moving average model), 205
Asymptotically mean cyclostationary process, 332n.
Asymptotically mean stationary process, 112, 325, 332n.
Asymptotically stationary process, 109
Asynchronous telegraph signal, 140-141, 248
Autocoherence function, 327
Autocorrelation:
  cyclic, 325-326
  (*See also* Cyclic autocorrelation)
  cyclostationary components of, 332-334
  empirical, 63, 171, 198
  empirical cyclic, 368
  ergodicity of, 170-173
  evolution of, for dynamical system, 204, 210
  extrapolation of, 302
  finite, 89, 193, 196, 198
  finite time, 65
  input-output relations for filters, 193, 195
  instantaneous probabilistic, 177
  measurement of, 185-186
  probabilistic, 83
  properties of, 104
  stationary components of, 332-333

Autocorrelation (*Cont.*):
  time-variant probabilistic, 177
Autocovariance, 83
  evolution of, for dynamical system, 204, 210
Autoregressive model, 205, 300-309
  autocorrelation extrapolation, 302
  innovations representation, 304-305
  lattice filter, 307-309
  Levinson-Durbin algorithm, 302
  linear prediction, 303-304
  maximum-entropy model, 306-307
  spectral factorization, 305
  Wold-Cramér decomposition, 304-306
  Yule-Walker equations, 301-302
Autoregressive moving average model, 205
Available power spectral density, 289
  gain, 289
Axioms of probability, 8

Bandpass processes, 266-273
Bandwidth:
  absolute, 243-244
  mean-squared incremental fluctuation, 258

535

Bandwith (*Cont.*):
  one-sided, 243-244
  rectangular, 243-244
  rms, 243-244
  two-sided, 243-244
Bayes' law, 10
Bernoulli process, 83-84
Bienaymé-Chebyshev
  inequality, 53
Bilateral Laplace transform,
  441
Binary channel, 23
Binomial coefficient, 85
Binomial process, 84-85
Birth-death process, 100
Boltzmann's constant, 235
BPSK (binary PSK)
  modulation, 360-363,
  408
Brownian motion, 124-125

Calculus, stochastic, 148-157
Campbell's theorem, 257
Canonical factorization, 447
Cauchy probability density,
  49
Cauchy-Schwarz inequality,
  50, 123, 292, 420
Causal-anticausal decomposition, 440-441
Causal-anticausal factorization, 444-447
Causal estimate, 434
Causal filter, 192, 442, 444
Causally invertible filter, 445
Central limit theorem, 47,
  241
Chain, Markov, 99
Channel, binary, 23
Chapman-Kolmogorov
  equation, 98-99
Characteristic function,
  33-34, 40, 351, 371
  time-averaged, 351
Chebyshev inequality, 53
Clipped process, 76-77,
  116-117
Coherence, 228-229, 327
  completely coherent
    process, 327
  completely incoherent
    process, 327

Coherence, (*Cont.*):
  measurement of, 466-467
  (*See also* Autocoherence
    function; Coherence
    function; Cross-
    coherence function)
Coherence function, 228,
  327, 435-436
Colored process, 444
Communications, xiv, 59-61
Complex envelope, 270
Complex process, 121-122
Conditional mean, 40-42,
  428, 465-466
  linear, 465
  as minimum-mean-
    squared-error
    estimator, 428
  as orthogonal projection,
    428
  random, 41-42, 428
  wide-sense, 465-466
Conditional probability,
  9-10
Convergence, stochastic,
  43-45
  almost surely, 43
  in distribution, 44
  in mean square, 44
  in probability, 44
  with probability one, 43
Convolution:
  continuous, 196
  discrete, 192
Convolution theorem, 49
Correlation, 35-36
Correlation coefficient, 36
  as measure of linear
    relationship, 429-430
  of spectral components,
    228, 327, 329-331
Correlation matrix, 39
Correlation time, 167
Correlator, 293
Correlogram, 65, 298
  cyclic, 331, 391
Covariance, 35-36
Covariance matrix, 36
Cramér-Wold decomposition, 304-306
Cross-coherence function,
  327, 347, 349-351, 355,
  365

Cross-correlation function,
  83
  generalized, 381-382
  of input and output of a
    filter, 193, 196
Cross-covariance function,
  83
Cross-spectral density, 225
CYCCOR methods for
  direction-of-arrival
  estimation, 375-381
Cycle detection:
  multi-, 391-392
  single, 393-397
Cycle frequency, 325
Cycle spectrum, 325
Cyclic autocorrelation, 325
  characterization:
    as cross-correlation,
      326
    of optimum detector
      for cyclostationary
      signals, 391-392
    of optimum estimator
      for cyclostationary
      signals, 399
    in terms of HSR
      cross-correlation
      matrix, 364
    in terms of in-phase
      and quadrature
      AM, 409
  for discrete time process,
    343
  empirical, 368
  input-output relations for
    LPTV transforma-
    tions, 336-338
  of modulated signals,
    346-357
  for Rice's representation,
    340-342
  self-determinant charac-
    teristic under LPTV
    transformation, 336
  of time-sampled signals,
    343-345
Cyclic correlogram, 330,
  391
Cyclic cross-correlation, 336
  use for direction-
    of-arrival estimation,
    375-381

Subject Index ■ 537

Cyclic cross-correlation (*Cont.*):
use for time-difference-of-arrival estimation, 381-385
Cyclic finite cross-correlation, 336
Cyclic mean, 367
empirical, 367
Cyclic periodogram, 331, 392
Cyclic-periodogram-cyclic correlogram relation, 405
Cyclic spectral density, 326-327
characterization:
as cross-spectral density, 326
of optimally identified periodic system, 401-402
of optimum detector for cyclostationary signals, 392
of optimum estimator for cyclostationary signals, 399
in terms of HSR cross-spectral density matrix, 364
for discrete-time processes, 344
empirical, 368
input-output relations for LPTV transformations, 336-338
interpretation as spectral correlation function, 329-331
of modulated signals, 346-363
for Rice's representation, 340-343
self-determinant characteristic under LPTV transformation, 336
support of, 331-332
of time-sampled signals, 344-345
Cyclic spectrum (*see* Cyclic spectral density)
Cyclic Wiener-Khinchin relation, 331, 368

Cycloergodic property, 179, 368-371
Cycloergodicity, 179, 367-371
of the autocorrelation, 368
of the mean, 368
of the time-variant (instantaneous) spectral density, 368
Cyclostationarity:
almost, in the wide sense, 324
applications of, 323, 371-402, 404
asymptotically mean, 332n.
exhibition of, 324, 332n.
in the wide sense, 324
Cyclostationary process, 110-112, 123, 178-179, 232-233, 260, 323-415, 462
detection of, 391-397
direction-of-arrival estimation for, 375-381
spatial filtering of, 385-391
stationary representations of, 111-112, 363-367
time-difference-of-arrival for, 381-385
waveform estimation of, 397-401
(*See also* Cyclostationarity)

Deflection, 297, 391-392
DeMoivre-Laplace theorem, 139
DeMorgan's laws, 6
Density:
conditional, 19, 22, 98, 465-466
fraction-of-time, 67-68, 180
cyclostationary, 369
Gaussian, 15, 19, 39-40
joint probability, 18-19
probability, 13-14
Rayleigh, 27, 184, 273
transition, 98
uniform, 27

Desiré André's reflection principle, 140
Detection (*see* Signal detection)
Deterministic process, 174n., 304
Deterministic theory of random processes:
cyclostationary processes, 369
stationary processes, 61
Difference, nth, 244
Difference equation model, 200, 205, 454
Differentiator circuit, 253-254
Diffusion, 102n., 120n., 139, 352
Dirac delta, 16
Direction-of-arrival estimation, 374-381
Dispersion (*see* Variance)
Distance:
in a linear vector space, 418
rms, 156
Distribution:
conditional, 19
estimation of, 465
fraction-of-time, 67-68, 180
cyclostationary, 369-370
joint probability, 18-19
probability, 12-13
DOA (direction-of-arrival estimation), 374-381
Doob decomposition, 305
Duality, 61, 179-181, 198, 199, 369-370, 469
examples, xii-xiii
Dynamical systems, 100-101
continuous-time, 206-211
first-order, 206-208
nth order, 208-211
discrete-time, 200-205
first-order, 200-202
nth order, 203-205

Eigenvalue, 218, 376, 379-380, 388
Eigenvector, 218, 376, 388

Subject Index

Empirical autocorrelation function, 63
Empirical spectral density, 229-230
Ensemble, 29, 60, 74
Entropy:
  maximum, model, 306-307
  relative, 321
Entropy rate, relative, 306
Envelope:
  complex, 271
  magnitude, 339
Equalizer, 437
Ergodic property, 165
  cyclo-, 179, 369-371
  of Poisson process, 187
Ergodicity, 163-188
  of the autocorrelation, 170-173
    necessary and sufficient condition, 172-173
    for a nonstationary process, 175, 187
  cyclo-, 179, 367-371
  of the distribution, 188
  of higher-order moments, 173-174
  of the mean, 168-170
    necessary and sufficient condition, 170
    for a nonstationary process, 175
  of the spectral density, 230-231
    for a nonstationary process, 232
  time averages, 166-168
    effective number of uncorrelated samples, 168
Error function, 15
ESPRIT, 380
Estimate:
  of cyclic moments, 367-368
  of distribution, 188
  least squares, 423
  minimum-mean-squared-error, 428

Estimate, minimum-mean-squared-error (*Cont.*):
  extrapolated, interpolated, predicted, smoothed, 434
  of moments, 166-173
  sample, 37-38
  of signal parameters (*see* Parameter)
  of signal waveforms (*see* Signal extraction)
  of spectral density, 282-286
  of state of system (*see* Kalman filtering)
Estimation (*see* Estimate)
Estimator, 428
Event, 7
Event indicator, 4
Expectation, 29-42
  conditional, 40-41
    factorization property of, 42
  (*See also* Conditional mean)
  defining formulas, 30-32
  fundamental theorem, 32
  linearity property of, 32
  random, 41
Expected value, 30-32
Experiment, 7
Extraction (*see* Signal extraction)
Extrapolation:
  autocorrelation, 302
  minimum-mean-squared-error, 300

Factorization, spectral:
  continuous-time, 444-447
  discrete-time, 305
  Paley-Wiener condition, 305, 445
Failure rate, 145-146
Field, 8
Filter, 66
  frequency shift, 400-401
  lattice, 307-309
  mean squared response of, 238-239

Filter (*Cont.*):
  minimum-phase, 305, 322
  periodically time-variant, 334-338
  random, 199-200
Filtering, 191-200
  causal, 192
  continuous-time, 195-198
  discrete-time, 191-195
  finite-memory, 192
  input-output relations for, 193, 195, 198-200, 219, 220, 222, 225
  optimum (*see* Optimum filtering)
  periodically time-variant, 334-346
  spatial, 385-391
FM (*see* Frequency modulation)
Fokker-Planck equation, 120-121
Fourier series transform, 225
Fourier transform, 63, 219-220
  generalized (integrated), 220, 222, 247
Fraction-of-time density:
  cyclostationary, 369
  stationary, 67-68, 180
Fraction-of-time distribution:
  cyclostationary, 369-370
  stationary, 67-68, 180
Frequency modulation (FM), 70-71, 251, 273-282, 351-353
  cyclostationarity of, 277-278, 351-353
  de-emphasis for, 281-282
  instantaneous frequency of, 273
  modulation index of, 275
  narrowband, 276
  noise immunity of, 278-282
  noise-quieting effect of, 281
  relation to phase modulation, 352-353
  SNR gain for, 280
  spectrum of, 273-278

Subject Index ■ 539

Frequency modulation
(FM), (Cont.):
threshold effect of, 281
wideband, 276-277
Woodward's theorem for, 277
FSK (frequency-shift keying)
modulation, 357-359
FST (Fourier series
transform), 225

Gauss-Markov process, 102
Gaussian characteristic
function, 40
Gaussian density:
bivariate, 19-20, 36-37
multivariate, 39-40
univariate, 15
Gaussian process:
cyclostationary, 371
definition of, 96
Isserlis's formula for, 114-115
Price's theorem for, 116-117
stationary, 107
Wiener process, 138-139
Gaussian random variable
(see Gaussian density)
Generalized cross-correlation, 381-382
Generalized function, 154n., 221
Generalized harmonic
analysis, 67
Geometric progression, 194
Geometry of inner-product
spaces, 418-427

Harmonic analysis:
generalized, 67-68
probabilistic, 68, 220-222, 247
Harmonic series representation, 363-367
Hilbert space:
of functions of a real
variable, 426
of random variables, 426-427
Hilbert transform, 255, 270, 339, 342

HSR (harmonic series
representation), 363-367

Identification (see
Modulation, identification of; System
identification)
Impulse fence, 39
Impulse function, 16
Impulse-response function, 66, 195
Independence (see entries
beginning with
Independent)
Independent events, 10-11
Independent-increment
process, 86, 101-102, 119, 139, 241
Independent random
variables, 19
Inner product:
of functions of a real
variable, 426
of l-tuples of real
numbers, 418, 422
of random variables, 427
Innovations representation, 304-305, 447, 451, 455, 463
In-phase and quadrature
components, 271
Input-output relations, 189-225, 334-338
of filters:
for autocorrelation,
autocovariance,
and mean, 193,
195-196, 198-200,
204, 210
in the frequency
domain, 220, 247
for spectral density, 222-225
in the time domain, 219
of periodically time-variant transformations, 334-338
for cross cyclic
autocorrelation and
cross cyclic
spectral density,
337-338

Input-output relations, of
periodically time variant
transformations (Cont.):
for cyclic autocorrelation and cyclic
spectral density,
336-337
Instantaneous frequency, 273
Interpolation:
minimum-mean-squared
error, 434
of time-sampled data,
69-70, 263-266
Isserlis's formula, 114-115

Jacobian, 21

Kalman filtering, 451-462
equivalence with Wiener
filtering, 458, 461
extended, 458
generalization of, 458
initial conditions for, 460, 472
periodically time-variant, 462
state filtering recursions, 457
state prediction recursions, 456
Karhunen-Loève transform, 217-218
Khinchin-Pollaczek
formula, 147
Khinchin relation, 223
Kronecker delta, 194

Laplace probability density, 28
Lattice filter, 307-309
Laws:
of cosines, 419
of large numbers, 45-48
strong, 46
weak, 46
Least squares estimation, 423-425
recursive, 424-425
recursive least squares
algorithm, 425

Lebesgue integral, 30
Level crossings, 140, 161
Levinson-Durbin algorithm, 302
Line mass, 39
Line spectra (see Spectral lines)
Linear combination, 418, 420
Linear dependence, 19
Linear estimation, 429-434
  infinite-dimensional, 431-432
  n-dimensional, 430-431
  one-dimensional, 429-430
  of waveforms, 432-434
  (See also Kalman filtering; Least squares estimation; Linear prediction; Periodically time-variant filtering; Wiener filtering)
Linear model, 229, 438, 467
Linear periodically time-variant transformation, 334-335, 462
  (See also Periodically time-variant filtering)
Linear prediction:
  continuous-time, 434, 470
  discrete-time, 73, 81, 303-306, 452-453
  innovations representation, 304-305
  Kalman, 451-453, 456
  lattice model, 307-309
  Levinson-Durbin algorithm, 302
  Wold-Cramér decomposition, 304-307
  Yule-Walker equations, 301-302
  (See also Autoregressive model)
Linear systems (see Systems, linear)
Linear transformations, 189-218
Linear vector space:
  of functions of a real variable, 426

Linear vector space (cont.):
  of l-tuples of real numbers, 418, 422
  of random variables, 427
  subspace, 420
Lorenzian spectrum, 248
LPTV (see Linear periodically time-variant transformation)

MA (moving-average) model, 205
Marked and filtered Poisson process, 134-137
Markov chain, 99
Markov process, 97-105
  chain, 99
  continuous state, 99
  Gaussian, 102
  homogeneous, 101, 139
  mth order, 103, 200
  (See also Birth-death process; Chapman-Kolmogorov equation; Dynamical systems; Fokker-Planck equation; Independent-increment process; Ornstein-Uhlenbeck process; Poisson process; Queue; Renewal process; Wiener process)
Markov property, 98-99
Martingale process, 103-104
Martingale property, 103-104
Matched filter, 71, 291-294
Maximum-entropy model, 306-307
Mean:
  conditional (see Conditional mean)
  cyclic, 367
  empirical, 38
    one- and two-sided, 170
  empirical cyclic, 367
  evolution of, for dynamical systems, 204, 207

Mean (Cont.):
  probabilistic, 38
  of random process, 82
  random sample, 46
  statistical, 38
Mean-square calculus, 150-157
  continuity, 150-152
    of WSS process, 151
  differentiability, 152-154
    broadened interpretation of, 153-154
    of WSS process, 153
  integrability, 154-156
    broadened interpretation of, 155-156
    of WSS process, 155
Mean-square regularity, 174-179
  of the autocorrelation, 174
  of the mean, 174
Mean squared fluctuation, 222-225
Mean squared incremental fluctuation, 258, 275
Mean squared response of linear system, 238-239
Median, 51
Minimum-mean-squared-error estimation, 427-434
  design equation, 433, 434, 443
  Kalman filtering, 451-462
  linear, 429-434
    infinite-dimensional, 431-432
    n-dimensional, 430-431
    one-dimensional, 429-430
  nonlinear, 427-428
  orthogonal projection, 428, 462
  performance formula, 433, 435, 443
  periodically time-variant filtering, 397-402
  of probability density, 465-466
  time-averaged, 469
  waveform, 432-434

Minimum-mean-squared-
  error estimation
  (*Cont.*):
  Wiener filtering, 294-296,
    434-450
Minimum-phase function,
  305, 322
Mixture process, 188
MMSE (*see* Minimum-
  mean-squared-error
  estimation)
Mode, 52
Modulation:
  amplitude (AM), 89,
    346-348
  characterization in terms
    of LPTV transforma-
    tion, 334-335, 346-363
  digital pulse and carrier,
    355-363
  double sideband, 350
  frequency (*see* Frequency
    modulation)
  frequency-shift keying
    (FSK), 357-359
  identification of, 346, 372
  minimum-shift keying
    (MSK), 408
  phase (PM), 351-353
  phase-shift keying (PSK),
    360-363, 408
    binary (BPSK),
      360-363, 408
  pulse-amplitude (PAM),
    260-263, 353-355
  pulse-position (PPM),
    92-93, 313-314, 357
  pulse-width (PWM),
    93-94, 357
  quadrature amplitude
    (QAM), 92, 338-343,
    348-351
  single sideband, 350
  staggered QPSK, 408
  vestigial sideband, 350
Moments, 34-40
  correlation, 35-36
  correlation coefficient, 36
  covariance, 35-36
  cyclic, 367-368
  estimation of, 166-179
  first, 34

Moments (*Cont.*):
  second centralized, 34
  standard deviation, 34
  variance, 34
Moving average (MA)
  model, 205
MSK (minimum-shift keying)
  modulation, 408
MUSIC, 376-379

Narrowband process (*see*
  Bandpass process)
Noise:
  atmospheric, 136, 287
  cosmic, 287
  man-made, 287
  radiation, 287
  shot, 135-136, 239-241,
    287
  thermal, 3-4, 234-238, 287
Noise immunity, 70-71,
  278-282, 372
Noise modeling, 286-291
  effective noise band-
    width, 289-290
  effective noise tempera-
    ture, 288-289
  noise figure, 289*n*.
Nonlinear estimation,
  427-428
Nonlinear transformations
  of random variables:
  clipper, 116-117
  Euclidean norm, 26
  expected value of, 32
  half-wave rectifier, 28
  inverse tangent, 26
  powers, 114-115
  to produce desired
    distribution, 27
  quantization, 28, 48
  saturating, 27, 117
Norm:
  of functions of a real
    variable, 426
  of l-tuples of real
    numbers, 418, 422
  of random variables, 427
Normal density (*see*
  Gaussian density)
Normal equations, 422-423

Normal process (*see*
  Gaussian process)
Normal random variable
  (*see* Gaussian density)
Normality, 423
Nyquist sampling rate, 264
Nyquist's theorem, 236-237
Nyquist's zero-intersymbol-
  interference criterion,
  409-410

Optimum filtering (*see*
  Kalman filtering;
  Periodically time-
  variant filtering;
  Wiener filtering)
Ornstein-Uhlenbeck process:
  finite-starting-time,
    216-217
  spectral density of, 248
  steady state, 102, 198,
    217
Orthogonal projection,
  420-423, 428, 462
Orthogonal random
  variables, 36
Orthogonality, 36, 419, 423,
  427
Orthogonality condition,
  462
  in Euclidean space, 420,
    422
  for least squares
    estimation, 423
  for linear MMSE
    estimation:
    infinite-dimensional,
      432
    n-dimensional, 430
    one-dimensional, 429
  for linear prediction, 303,
    452
  for noncausal Wiener
    filtering, 294
  for nonlinear MMSE
    estimation, 428
  for waveform estimation,
    432
Orthogonality principle (*see*
  Orthogonality condi-
  tion)

542 ■ Subject Index

Paley-Wiener condition, 305, 445
PAM (pulse-amplitude modulation), 260-263, 353-355
Parallelogram law, 419
Parameter:
  empirical, 38
  estimation:
    direction of arrival, 375
    time-difference of arrival, 381
  probabilistic, 38
  sample, 38
  statistical, 38
PARCOR coefficient, 309
Parseval's relation, 248
Partial correlation coefficient, 309
Periodic nonstationarity (*see* Cyclostationarity)
Periodic phenomena, 323
Periodic stationarity (*see* Cyclostationarity)
Periodically correlated process (*see* Cyclostationarity)
Periodically time-variant filtering, 374, 397-402
Periodically time-variant transformations, 334-338
Periodicity in random data, 323
Periodogram, 64, 72, 298
  cyclic, 331, 392
  frequency-smoothed, 286
  time-averaged, 286
  time-variant, 285
Periodogram-correlogram relation, 64-65
  cyclic, 329, 331
Phase modulation (PM), 351-353
Phase randomization, 111, 352, 370-371, 409-410
Photon detection, 136
Planck's constant, 235
PM (phase modulation), 351-353
Point process (*see* Poisson process)

Poisson counting process (*see* Poisson process)
Poisson distribution, 49
Poisson impulse sampling, 143, 312-313
Poisson point process (*see* Poisson process)
Poisson process, 105, 129-137, 140-146
  count probability, 143-144
  derivative of, 133-134
  doubly stochastic, 136
  ergodic properties of, 187-188
  Gaussian approximation, 241
  inhomogeneous, 137
  interarrival probability density, 144
  marked and filtered, 134-137
  mean rate, 140
  point probability density, 144
Poisson sum formula, 354
Poisson theorem, 130
Power spectral density (*see* Spectral density)
Power spectrum (*see* Spectral density of power)
PPM (pulse-position modulation), 91–93, 313–314, 357
Predictable process (*see* Singular process)
Prediction (*see* Linear prediction)
Price's theorem, 116-117
Probability, 8-9
  axiomatic properties, 8
  Bayes' law, 10
  conditional, 9-10
Probability density, 13-14
  fraction-of-time, 67-68, 180, 369
Probability distribution, 12-13
  fraction-of-time, 67-68, 180, 369-370
Probability function, 8
Probability mass function, 18

Probability space, 8
PSD (power spectral density) (*see* Spectral density, of power)
Pseudoinverse matrix, 464
PSK (phase-shift keying) modulation, 360-363, 408
Pulse modulation, 260-263, 311-314, 355-363
  amplitude, 260-263, 353-355
  position, 92-93, 313-314, 357
  width, 93-94, 357
PWM (pulse-width modulation), 93-94, 357
Pythagorean theorem, 419

QAM (quadrature amplitude modulation), 92, 338-343, 348-351
QPSK (quaternary PSK), 362-363
Quadrature amplitude modulation (QAM), 92, 338-343, 348-351
Quadrature and in-phase components, 271
Quantization, 28, 48
Queue, 105, 118
  Khinchin-Pollaczek formula, 147
  single-server, 146-147

Radar clutter, 137
Radar range estimation, 185
Radiometer, 298
  optimum, 298, 396-397
Random process(es), 73-74
  almost cyclostationary, 324
  asymptotically mean cyclostationary, 332n.
  asymptotically mean stationary, 112, 325, 332n.
  asymptotically stationary, 109
  asynchronous telegraph, 140-141, 248

Subject Index ■ 543

Random process(es) (*Cont.*):
  autoregressive, 205, 300-309
  bandlimited, 263-266
  bandpass, 243, 266-273
  Bernoulli, 83-84
  binomial, 84-85
  birth-death, 100
  clipped Gaussian, 116-118
  continuous-amplitude, 76
  continuous-time, 75
  cycloergodic, 367-371
  cyclostationary, 110-111, 323-415
  deterministic, 174*n*.
    (*See also* singular *below*)
  diffusion, 102*n*. 124*n*. 352
  discrete-amplitude, 76
  discrete-time, 75
  doubly stochastic, 136
  ergodic, 163-188
    that exhibits cyclo-
      stationarity, 324, 332
  Gaussian, 96
    (*See also* Gaussian process)
  independent-increment, 101-102
  low-pass, 243
  Markov, 97-105
    (*See also* Markov process)
  martingale, 104
  mixture, 173, 188
  Ornstein-Uhlenbeck, 102, 198, 216-217, 248
  persistent, 112
  Poisson, 105
    (*See also* Poisson process)
  random walk, 85-86, 125
  regular (complement of singular), 304-305
  regular (time-averageable), 174-179
  renewal, 118
  second-order theory of, xiii, 61, 89, 325
  singular, 304-305
  stationary, 105-112
  symbolic, 128, 154
  transient, 112
  Wiener, 102-103, 124-129

Random variables, 11-12
  continuous, 16
  discrete, 16-17
  mixed, 16-17
  nonlinear transformations of (*see* Nonlinear transformations of random variables)
  as a vector, 426-427
Random walk process, 85-86, 125
Rational spectral density, 238, 254, 445
Rayleigh density, 27, 146, 184, 273
Realization, 11
Recursive algorithms:
  for adaptive arrays, 389
  difference equation for evolution:
    of state covariance, 204
    of state mean, 204
  Kalman filter, 457
  Kalman predictor, 456
  for least squares estimation, 425
  for linear prediction order update, 302
Reflection principle, 140
Regression, 303
  (*See also* Least squares estimation)
Regular process, 174-179, 304-305
Relative frequency, 5, 46
Reliability, 145
Resonance, spectral density of circuit response, 256-257
Rice's representation, 267-273
  for cyclostationary processes, 338-343, 406-407
  for finite-energy functions, 267-271
  for Gaussian processes, 273
  spectral correlation properties of, 340-343, 406-407
  for stationary processes, 271-273, 214-215

Riemann integral, 149
RMS (root mean square) value, 148
Root mean square bandwidth, 243-244, 275
Root mean square deviation, 275
Root mean square distance, 156
Root mean square (RMS) value, 148

Sample function, 74
Sample path, 74
Sample point, 7
Sample space, 7
Sampled and held process, 87-89
  reconstruction, 218-219
Sampling, 260-266, 343-346
  Poisson, 312-313
Sampling theorems, 263-266
  for bandlimited finite-energy functions, 263-264
  for bandlimited stationary processes, 264-265
  for mean-square bandlimited stationary processes, 265-266, 311-313
  for nonstationary processes, 266
  for nonuniform sampling, 312-313
Scattering communication channels, 137
Scatterplot, 37-38
Schwarz inequality (*see* Cauchy-Schwarz inequality)
SCORE method of adaptive spatial filtering, 389-391
Second-order theory of random processes, xiii, 61, 89, 325
Sets, 5-7
  complement, 6
  DeMorgan's laws, 6
  field, 8
  intersection, 6
  null, 7

Sets (*Cont.*):
 proper subset, 5-6
 union, 6
Shot noise:
 Campbell's theorem, 257
 definition, 135-136
 model of, 239-241
 spectral density of, 240-241
Signal, definition, xiv
Signal classification, 346, 372
Signal detection:
 of cyclostationary random signals, 372, 391-397
 estimator-correlator structure, 300
 of known signals, 71-72, 80-81, 291-294
 SNR formulas, 319-320
 of stationary random signals, 71-72, 296-300
Signal estimation (*see* Signal extraction)
Signal extraction, 72, 81, 294-296, 397-401, 436-438, 450
 (*See also* Kalman filtering; Periodically time-variant filtering; Wiener filtering)
Signal prediction (*see* Linear prediction)
Signal processing, 59-61, 69-72
Signal-to-noise ratio, 60
 power spectral density, 437
Singular process, 304-305
Smoothing:
 minimum-mean-squared-error, 434, 458
 sliding window, 196-197
SNR (signal-to-noise-ratio), 60
Sonar range estimation, 185
Spatial filtering, 373, 385-391
SPECCOA method of time-difference-of-arrival estimation, 385

SPECCORP method of time-difference-of-arrival estimation, 385
SPECCORR method of time-difference-of-arrival estimation, 381-384
Spectral autocoherence function, 327
Spectral correlation function, 329-331
 (*See also* Cyclic spectral density)
Spectral density, 219-247
 cross, 225
 empirical, 229-230
 of energy, 226-227
 estimation of, 282-286
 expected, 225-226
 instantaneous probabilistic, 232, 327
 of mean squared fluctuation, 222-225
 measurement analysis, 282-286
 mean, 283
 reliability, 285
 resolution, 285
 signal-to-noise ratio, 284-286
 spectral density, 284
 variance, 284
 for nonstationary processes, 231-234
 of power, 65-66, 225-226, 229-230
 time-average, 229-230
 time-variant probabilistic, 232
Spectral estimation (*see* Spectral density, measurement analysis)
Spectral factorization:
 continuous-time, 444-447
 discrete-time, 305
Spectral frequency, 327
Spectral intensity, 235
Spectral lines, 244-245
 regeneration, 328, 393-397
 of singular component of a process, 305

Spectral representation, 220-222, 247, 363-365
Spectrum, 224
 finite-time, 64
 (*See also* Spectral density)
SQPSK (staggered QPSK), 408
Standard deviation, 34
State of a Markov process, 99
State variable models:
 continuous-time, 208-209
 discrete-time, 203
 for Kalman filtering, 454-455
Stationarity:
 asymptotic, 109
 asymptotically mean, 112, 174, 332n.
 of complex processes, 121-122
 cyclostationarity, 324
 of increments, 109, 139
 joint, of two processes, 106
 nth order, 106
 second-order, 106
 strict-sense, 106
 vector, 111
 wide-sense, 106
Stationary process (*see* Stationarity)
Stationary representations of cyclostationary processes, 111, 363-367
Statistic, 60
Statistical communication theory, xv, 59-61
Statistical sample, 11
Statistical signal processing, xv, 59-61
Stochastic calculus, 148-157
Stochastic convergence (*see* Covergence, stochastic)
Stochastic process, 11n., 61n.
 [*See also* Random process(es)]
Strict-sense stationarity (*see* Stationarity)

Subject Index ■ 545

Strict-sense white noise, 241
Subspace, 420
Synchronization, 371n.
Synchronized averaging identity, 333, 344
Synchronized time-averaging, 333
System(s):
  communication, xiv, 59
  dynamical, 100-101, 200-211
    (See also Dynamical systems)
  linear, 189-218
    continuous-time filters, 195-199
    discrete-time filters, 191-195
    dynamical, 200-211
    of equations, 189-191
    mean-squared response of, 238-239
    optimum (see Kalman filtering; Periodically time-variant filtering; Wiener filtering)
    random filter, 199-200
  receiving, 286-291
  signal processing, 60
  state-variable models of, 203, 208-209, 454-455
System function, 335
System identification:
  periodically time-variant, 338, 401-402
  time-invariant, 467
Szegö-Kolmogorov formula, 306

Tchebycheff inequality (see Bienaymé-Chebyshev inequality)
TDOA (time-difference-of-arrival) estimation, 372, 381-385
Telegraph signal, 140-141, 248
Thermal noise:
  in circuits, 236-239

Thermal noise (Cont.):
  definition of, 3-4
  model, 234-236
  Nyquist's theorem, 236
  spectral density of, 234-235
  Thevenin equivalent circuit, 236
Time-average autocorrelation, 62-66, 174-181
  (See also Duality)
Time-average mean, 68, 174, 179-180, 367-369
  (See also Duality)
Time-average probability density, 67-68, 180-181, 369-370
  (See also Duality; Fraction-of-time density)
Time-average spectral density, 65-67, 229-234
  (See also Duality)
Time-averaged autocorrelation, 174-179, 332-333
Time-averaged mean, 174
Time-averaged spectral density, 232-234
Time-difference-of-arrival estimation, 372, 381-385
Time-series representation of cyclostationary processes, 366
Toeplitz matrix, 182
Transfer function, 66, 220
Transformations (see Linear transformations; Nonlinear transformations of random variables)
Transition density, 98
Transition matrix:
  continuous-time state, 209
  discrete-time state, 203
Translation invariance, 106
Translation series representation of cyclostationary processes, 366

Triangle inequality, 50
TSR (translation series representation) of cyclostationary processes, 366

Union bound, 23
Unit-pulse response, 191

Variance, 34
Vector, 418, 422, 426

Weibull probability density, 146
WGN (white Gaussian noise), 234-236
White noise, 234-243
  discrete-time, 241-243
  Gaussian, 234-236
  models other than WGN and WPIN, 241
  nonstationary, 258
  Poisson impulse, 241
  strict-sense, 24, 242
  wide-sense, 241
Whitening filter:
  continuous-time, 444-447
  discrete-time, 305, 458
Wide-sense conditional mean, 465
Wide-sense cyclostationarity, 324
Wide-sense stationarity, 106
Wide-sense white noise, 241
Wiener filtering:
  causal, 442-450
  complementary, 471
  discrete-time, 450
  equivalence with Kalman filtering, 458, 461
  innovations representation, 447
  invertible whitening, 444-447
  spectral factorization, 444-447

Wiener filtering (*Cont.*):
noncausal, 434-442
  causal truncation, 439-440
  discrete-time, 450
  equivalent linear model, 438
  general formulas, 434-436
  joint equalizing and noise filtering, 436-438
Wiener-Hopf equation, 443, 447-450
Wiener-Khinchin relation: generalization for cyclostationary processes, 331, 368
  for stationary processes, 230

Wiener process, 101-102, 124-129
  autocovariance function for, 128
  derivative of, 128-129
  diffusion equation for, 139
  Gaussian property of, 138-139
  level crossings of, 140
  Markov property of, 139
  reflection principle of, 140
Wiener relation, 223
Wold-Cramér decomposition, 304-306
Wold's isomorphism: generalization for cyclostationary processes, 369

Wold's isomorphism (*Cont.*):
  for stationary processes, 181
Woodbury's identity, 425, 464
WPIN (white Poisson impulse noise), 241
WSS (wide-sense stationary), 106

Yovits-Jackson formulas, 450
Yule-Walker equations, 301-302

Zero-crossings (*see* Level crossings)

ABOUT THE AUTHOR

William A. Gardner, Ph.D., is Professor of Electrical Engineering and Computer Science at the University of California at Davis. President of Statistical Signal Processing, Incorporated, he is the author of two books and numerous research papers. Dr. Gardner is a member of the Institute of Electrical and Electronics Engineers, American Mathematical Society, Mathematical Association of America, European Association for Signal Processing, and American Association for the Advancement of Science. He is the recipient of the Best Paper of the Year Award from the European Association for Signal Processing in 1986, the Distinguished Engineering Alumnus Award from the University of Massachusetts in 1987, and the Stephen O. Rice Prize Paper Award in Communication Theory from the Institute of Electrical and Electronics Engineers in 1989.